Water Resources Development and Management

Indexed by Scopus

Each book of this multidisciplinary series covers a critical or emerging water issue. Authors and contributors are leading experts of international repute. The readers of the series will be professionals from different disciplines and development sectors from different parts of the world. They will include civil engineers, economists, geographers, geoscientists, sociologists, lawyers, environmental scientists and biologists. The books will be of direct interest to universities, research institutions, private and public sector institutions, international organisations and NGOs. In addition, all the books will be standard reference books for the water and the associated resource sectors.

Somnath Basu · Andrew R. Shaw ·
Mudumbai Venkatesh
Editors

Water Management in Petroleum Industries

Springer

Editors
Somnath Basu
Jewel Innovations in Energy
and Environment
Katy, TX, USA

Andrew R. Shaw
Global Practice and Technology
Sustainability and Wastewater
Houston, TX, USA

Mudumbai Venkatesh
Process, Energy and Environmental
Consultant
Andover, MA, USA

ISSN 1614-810X ISSN 2198-316X (electronic)
Water Resources Development and Management
ISBN 978-981-19-3158-1 ISBN 978-981-19-3159-8 (eBook)
https://doi.org/10.1007/978-981-19-3159-8

This Springer imprint is published by the registered company Springer Nature Singapore Pte Ltd.
The registered company address is: 152 Beach Road, #21-01/04 Gateway East, Singapore 189721, Singapore

Preface

Water is a precious resource which sustains life on earth and is critical for most activities of society. These include all types of economic activities - industrial production, power generation, transportation, agriculture, food processing, manufacturing, recreational activities, and more. For centuries, human civilization has been built up and flourished on the banks of large rivers and the shores of lakes, oceans, and other large water bodies to take advantage of this invaluable resource.

Water is at the center of the food-water-energy nexus and is critical to support agriculture for providing food security to the rapidly burgeoning population on earth, and energy security to fulfill the growing needs of modern society. Since the invention of the steam engine in 1777 that spurred the industrial revolution in England and throughout the whole world, water has been supporting phenomenal economic growth leading to rising living standards around the world. In its new role, water is currently at the forefront of the second industrial revolution as the feed material for the production of 'Green' hydrogen by electrolysis, to achieve the goal of carbon neutrality.

Water is present in abundance on earth, however, almost 97% of the water on earth (approximately 1,386 million cubic kilometers) is present in seas and oceans that is not readily usable due to its high salinity (Gleick, 1993), without expensive and energy-intensive treatment. This leaves a relatively small fraction as fresh water for use. A large part of this is also inaccessible due to geographically uneven distribution of rainfall and other climatic conditions, e.g. floods and droughts. Of the total water withdrawn for use globally, agriculture is the largest consumer, followed by industries (UN Water, 2020) and only about 10% for human use (Cosgrove and Rijsberman, 2000).

Energy production is a large consumer of fresh water (10%) among industrial users (IEA, 2016). This includes all facets of petroleum exploration (upstream), transportation (mid-stream) and refining (downstream) operations. The availability of water has been under severe strain due to population growth, domestic and industrial pollution, a depleting groundwater table, climate change, etc. As a result, it is important to conserve this valuable resource so that society can continue to rely on it for generations to come.

Water use is a function of both the *quality* of water available and the *quantity* of water needed. For example, water for irrigation can be of a lower quality than that used for human consumption. Depending on the type of industry, water quality may be critical (e.g. for food processing or boiler feed water) or less important than the quantity (e.g. water used for hydraulic fracturing in petroleum exploration). In regions where there is a shortage of water of the desired quality, the user must consider reducing water use or finding alternative water sources of a lower quality that will require treatment (which is a more expensive option). In many regions, water availability is diminishing due to more water being withdrawn from watersheds and aquifers than are naturally being recharged, resulting in significant water stress. Increased urbanization and industrial water use, coupled with the compounding impact of climate change are also contributing to more stress in many areas of the world.

Similar to the concept of carbon footprint, researchers are developing methods to quantify holistic water use using a "water footprint" concept. This idea was originally proposed in the early 2000s to account for the water used in the manufacture of different products and for food production; but this concept can also be applied to look at water used for all types of processing including oil exploration, refining, and petrochemical production. An important but contentious aspect of water footprint is accounting for water availability. For example, a cubic meter of water coming from an arid or water-stressed location needs to be considered differently than a cubic meter of water from a region with abundant water resources.

Water and the Petroleum Industry

Water plays a major role in the petroleum industry. It is essential from the exploration of crude to its processing to the production of the refined products that serve as the major source of energy that drives the global economy. Operation of petroleum industry is conducted in four business sectors in sequence:

- Upstream—which includes exploration and production
- Midstream—which includes crude oil storage, transmission, and transportation
- Downstream—which consists of oil refining, and
- Marketing—which includes product storage and distribution

In the upstream sector oil is extracted most commonly in one of four ways:

1. On-shore, where the oil is extracted from formations located on land that could be hundreds to thousands of feet deep.
2. Off-shore, where the oil is extracted from locations that could be tens of miles from shore and several thousands of feet deep. The oil is transported to shore either by undersea pipelines or by ship tankers.
3. Oil sands, primarily in Canada where the oil is extracted by heating up the sand with steam. The product from the oil sands is bitumen which is diluted and piped to refineries for further upgrades.

4. Hydraulic fracturing, commonly referred to simply as 'fracking'. In this method high pressure water is injected into underground rock formations to fracture rock formations and release the oil contained therein.

In the midstream sector, oil that has been extracted is stored to act as a transportation hub for the crude oil. The oil is transported out either by pipeline, ship tankers, or rail cars and sent to oil refineries for further processing.

The downstream sector consists of oil refineries where crude oil is separated and refined into various products such as gasoline, diesel, jet fuel, heating oil, and liquified petroleum gas (LPG). The products from the refinery are sent out by pipeline, ship tankers, rail cars, or tank trucks.

The marketing sector consists of distribution terminals that could be located in various urban areas. These terminals usually consist of storage tanks where different products such as gasoline and diesel are stored for further distribution. Additives are added to the products as required before shipment to retail stations. The products are usually shipped out by tank trucks. Distribution terminals are also located at airports to service the aviation market.

Water and energy are two of the most important resources for society to manage. While there is significant reference materials on industrial water and wastewater science and engineering in general, a consolidated guidance document specifically to the water management practices in all sectors of the petroleum industries is lacking. This gap has been traditionally filled by in-house training courses of operating and consulting firms practicing in this area, professional conferences, and journal papers. This book attempts to bring the knowledge from those forums and present it in a cohesive form. The profession has attracted persons from various backgrounds in its workforce and a comprehensive book dedicated to the subject of Water Management in Petroleum Industries is expected to help them supplement their knowledge gained through practice.

This book is written to provide guidance to practicing professionals in the petroleum industry such as design engineers, operating engineers, plant operators. maintenance personnel, and equipment engineers. It is also hoped that this book will prove to be a useful reference for educators and students who want to learn about the intersection of water and the petroleum industry. The book provides information on the contaminants of wastewater produced in the operation of petroleum industries and their sources. It also includes the treatment methods for producing water that is used in various operations of the petroleum industry. In addition, this book also provides information pertaining to the regulations that apply to water and wastewater treatment in all sectors of petroleum industries. It includes discussions on sustainable water management as recycle and reuse are being increasingly practiced by the industry. Some of the basic elements of water chemistry and microbiology, and petroleum chemistry, and its origins and history of exploration are provided in the form of appendices for professionals entering the field without prior formal training in such areas.

Katy, USA Somnath Basu
Andover, USA Mudumbai Venkatesh
Houston, USA Andrew R. Shaw

References

Cosgrove WJ, Rijsberman FR (2000) World water vision: making water everybody's business. Earthscan Publications Ltd., London

Gleick PH (1993) Water in crisis: a guide to world's fresh water resources. Oxford University Press, New York

IEA (2016) World energy outlook, OECD/IEA, Paris: www.oecd-library.org/energy/world-energy-outlook-2016_weo-2016-en

UNESCO (2020) Water and climate change report: the united nations water development report 2020, UNESCO Press, Paris

Acknowledgements

Water and petroleum are two very important resources on earth that support human life and the development of a modern society. A great amount of academic and field scale research has been conducted on each of these two subjects over the last century to process water and oil into their usable forms, amounting to a large body of technical literature. One iconic textbook on petroleum processing, *Petroleum Refinery Engineering,*, authored by W. L. Nelson of the University of Tulsa, was first published in 1949 by McGraw Hill. The first major textbook dealing with petroleum and petrochemical wastes *Aqueous Wastes from Petroleum and Petrochemical Plants* by Milton Roy Beychok was published by John Willey & Sons in 1967. Since then, many of the technical reports and study results on petroleum industry wastes were published but those were primarily limited to publications in trade and research journals until the book *Activated Sludge Treatment of Industrial Wastewater* by W. W. Eckenfelder and J. L. Musterman, published by CRC Press, appeared in 1998.

The first effort to develop a training program to prepare professionals in the science and engineering of petroleum industry water and wastewater treatment was undertaken by Water Environment Federation (WEF), Alexandria, VA in 2010 which resulted in two parallel sets of workshops—one in produced water treatment from upstream operations, and the other in refinery and petrochemical wastewater treatment from downstream operations. Since then, the two workshops have merged and successfully continue each year and are always sold out events to date. The editors of this book express their gratitude to the group of professionals who dedicated their time and effort to develop and deliver the WEF Workshops each year as volunteers that helped shape the emerging science and engineering at the interface of water and petroleum. Since then, other instructors came forward and delivered the course modules over the years that created a self-sustaining and successful workshop program till today. The editors also wish to express their thanks and appreciation to all the authors and co-authors of the various chapters of this book for their invaluable contribution. Without their significant and tireless contributions this book would not be possible.

In addition to fundamental principles, this book also incorporates results from the latest developments in technology as reported in various publications. Information from industrial practices has also been incorporated for the readers to gain knowledge from a wide spectrum of perspectives. Numerous examples and case studies from equipment manufacturers and system operators have been presented to facilitate an in-depth understanding of the subject. This includes inputs and review comments from industry leaders, consultants, and academics in this field.

Several organizations and individuals contributed to enhance the content of this book. These include Ian Fife and Dallas Torgersen of WesTech Engineering Co., Salt Lake City, UT and Wayne Bates of Hydranautics/Nitto. Mr. Bates also conducted a thorough review of the section on Reverse Osmosis and helped improve its quality. Other contributors include John Trombka of Forum Energy Technologies, Houston, TX and Siva Angappan of Headworks International, Houston, TX. Specifically, Siva shared his knowledge and experience in the Moving Bed Biological Reactor (MBBR) and Integrated Fixed Film Activated Sludge (IFAS) technologies. Jeff Peeters of Veolia Water Technologies & Solutions provided case studies for membrane bioreactor (MBR) technology. Ian Fife and Dallas Torgersen of WesTech Engineering Co., Salt Lake City, UT, and Michael Doyle and Robert Wenta of Evoqua (currently Xylem) of Waukesha, WI also contributed to several figures in this chapter.

The Editors would like to express their appreciation to Gerald Seidl of Headworks International, Houston, TX for providing the necessary logistical support to produce this large volume of work. Contributions of graphic designers Matthew Banzer of Banzer Films, Houston, TX; Zach Cantrell of Strategy Engineering and Consulting, Houston, TX and Subroto Bhoumik of SMS Plus Marcom Advisory, Calcutta, India are gratefully acknowledged. They created new figures and adapted to recreate others from various sources and documents.

This effort started back in July 2019 when the Senior Editor of Springer Nature, Dr. Loyola D'Silva, met the Corresponding Editor of this book to discuss the idea of a volume on industrial water and wastewater treatment during the conference on Advancement in Water and Wastewater Treatment and Reuse (AWWTR) in Singapore. Since then, the manuscript was prepared, edited many times and revised until the latest version was finalized in 2025. The editors sincerely thank Dr. D'Silva for his guidance in preparing the manuscript. Every step of the way the Springer team comprising of Prasanna Naryana Samy and Deepa Varadrajan (Straive) was continuously engaged with the editors to improve it and make it suitable for publication.

Finally, we extend a big 'Thank You' to the readers who have been waiting patiently for the book to hit the shelves. Please read it, enjoy it, and give us your feedback so that we can enhance it further for the next edition.

June 2025

Somnath Basu
Andrew R. Shaw
Mudumbai Venkatesh

Endorsements for *Water Management in Petroleum Industries*

"I am very pleased to see the publication of this important book on water management in the petroleum industry. Responsible industries are increasingly recognizing that effective water management is key to their long-term viability and embedded in delivery of their duties to meet the social license to operate. A "go to" reference on water management and reuse for the petroleum industry has been missing but that gap is met by this excellent book. The users will find it to be comprehensive in its treatment of the broad range of topics relevant to managing water in the petroleum industry, including water requirements, treatment methods, and used water treatment and reuse. This book achieves excellent balance presenting the fundamental scientific and engineering principles along with their practical application. Accomplishing this for the full range of relevant topics represents no small feat, but one which is accomplished by the authors in an excellent fashion. I commend this book to those already involved with water management in the petroleum industry, as well as those just becoming involved. There is always something more to learn, and those experienced in the petroleum industry will find the comprehensive coverage of the water management topic useful for extending their knowledge. Those just entering the petroleum industry will find this book useful because it is written in such an easy-to-understand fashion. It is a "must-have" resource for all interested in petroleum water management."

—Glen T. Daigger, *Ph.D., P.E., BCEE, National Academy of Engineering, CAE, Distinguished Fellow, IWA, Distinguished Member, ASCE, Fellow, WEF, Professor of Engineering Practice, University of Michigan, Department of Civil and Environmental Engineering, Ann Arbor, MI 48109-2125, July 24, 2024*

I am pleased to recommend Water Management in Petroleum Industries as a highly valuable resource for working professionals, students, and those looking to expand their knowledge and benchmark the impact of energy systems on the environment. The book is comprehensive in coverage of water management considerations for key sectors of oil and gas including traditional as well as unconventional production methods such as "fracking" and oil sands, while also considering midstream and petroleum refining, specific technologies for recycle and reuse, and water treatment. Stewardship and protection of water resources is a key challenge for energy systems, often overshadowed by mitigation of CO_2 and greenhouse gas emissions. The book describes water management for the current fossil-based energy economy, whose practices and benchmarks can help in management and optimization of future systems including expanded use of renewable energy. The book can serve as both a comprehensive reference on water management for the practicing engineer, while also providing grounding and context needed for those seeking a higher-level overview. It is great to see the full breadth of topics covered in a single, essential reference book.

—Joseph B. Powell, *Ph.D., U.S National Academy of Engineering, Executive Director and Shell Endowed Chair for Energy Transition, Professor, Department of Chemical and Biomolecular Engineering University of Houston*

"I had the pleasure of reviewing this new and excellent book on water treatment and reuse in the oil and gas industry. I have been involved in research on treatment of oil and gas produced water for more than 15 years, exploring the use of conventional and advanced water treatment processes to reclaim impaired produced water from the oil and gas industry for in-field reuse or reuse in other sectors. Over the years I have realized how important it is to continuously advance this field, and how difficult it is to acquire all the tools and knowledge necessary to understand how the oil and gas industry operates and what are the technological needs to solve water quality and treatment problems in the industry.

This new book is a great resource to help students, engineers and scientists, stakeholders, and regulators to deeply understand the needs of the industry, the possibilities, and the basic knowledge needed to approach and solve water treatment and reuse problems. The book contains the fundamental of science and engineering at the core of problem, relevant and meaningful examples and information to guide engineers in solving water problems in the oil and gas industry, but also how to solve water problems in other heavy industries that deal with complex waste streams (e.g., mining, metal, pulp and paper, etc.).

I will definitely recommend this book to my students and colleagues and hope that it will be widely distributed to both universities and industries in which water plays a central role in producing materials or energy."

—Dr. Tzahi Y. Cath, *Professor of Environmental Science & Engineering, Co-Director, The ConocoPhillips Center for a Sustainable WE^{2}ST, Director, Advanced Water Technology Center (AQWATEC)*

"It is a great accomplishment to have a book addressing Water Management in the Petroleum Industry. Water plays a pivotal role in the petroleum industry. Starting with the upstream products, continuing with midstream, and ending downstream with refineries and retail.

Regional and global demand for resources such as energy and water, continue to increase and are driven by the world's growing population. Energy and water resources are also intrinsically linked since water is required for energy production while energy is required for water treatment and supply. This intrinsic link between water and energy is commonly referred to as the Water-Energy Nexus.

Energy is a cornerstone of the modern industrial economy and provides an essential ingredient for almost all human activities. On the other hand, Water is paramount in the different stages of the Petroleum Industry. Within upstream energy, water is used for oil & natural gas exploration and production—especially in unconventional (fracking) natural gas developments. For downstream processing, water is necessary for cooling, steam generation, hydrogen production and processing crude oil into a final refined product.

Water scarcity is a global concern for the twenty-first century with a global water crisis threatening the security, stability, and environmental sustainability of all nations. Managing the competition between energy production, agriculture, other industries, and domestic applications for water resources is a growing challenge for various watersheds worldwide.

Water Management in the Petroleum Industry is essential for a sustainable Industry, so it's great to have book covering this important need. In the book you will find a wide range of information including water needs, different treatment technologies for water and wastewater, EH&S regulations and water reuse and recycle. It covers in-depth technical information together with practical applications. I recommend this publication, not only for newcomers to the Industry, but also for experienced professionals."

—Brian Arntsen, *B.Sc. MBM., Global Domain Leader—Veolia Water Technologies & Solutions, Water Environment Federation (WEF) Industrial Water Community (IWC) Past Chair, WEF Community Leadership Council (CLC) Director, Water Environment Association of Ontario (WEAO) Wastewater Committee Member*

Contents

Editors and Contributors

About the Editors

Somnath Basu, Ph.D., PE, MBA, BCEE is a technology and policy advisor in environment and energy. His practice areas are industrial water and wastewater treatment, nutrient control, renewable energy and water-energy nexus. He served the US Department of Energy (DOE), Washington, DC as a Science & Technology Policy Fellow of the American Association for the Advancement of Science (AAAS) till 2023. His prior affiliations were with Headworks International, Houston, TX where he was the Chief Technology Officer (CTO), the Water & Utility division of Shell Oil Company where he was a member of the Shell Global Solutions team, Siemens Water Technologies, and Massachusetts Water Resources Authority.

Somnath is a Fellow of the American Institute of Chemical Engineers (AIChE). He is a Fellow of the Water Environment Federation (WEF) and the current chair of the Editorial and Publications subcommittee of the Industrial Water Committee (IWC) of WEF. He is a member of the Board of Trustees of the American Academy of Environmental Engineers and Scientists (AAEES). He has over 40 years of experience in water, wastewater and chemical process industries and published and presented extensively in his areas of practice and research.

Andrew R. Shaw, Ph.D., PE, BCEE, ENV SP is a Global Practice & Technology Leader in Wastewater and Sustainability for Black & Veatch, based in their Houston office.

Andrew is the past chair of the Technical Practice Committee for the Water Environment Federation (WEF) and he has served on the board of trustees for the American Association for Environmental Engineers and Scientists (AAEES). He is a past board member for the Institute for Sustainable Infrastructure (ISI) Envision review board.

He has over 30 years of wastewater treatment experience from 4 continents including operations, design, and research.

Mudumbai Venkatesh, M.S. is an engineering consultant with more than 45 years of experience in the oil and chemical industry. This includes oil refineries, chemical and petrochemical plants both in the United States and around the world.

His experience includes plant design of oil refineries and petrochemical plants as well as extensive experience in the design of environmental systems including wastewater treatment plants at these facilities. He has managed multidisciplinary teams in the execution of projects in order to achieve the goals and objectives of projects. He is also experienced in equipment design, operations support and research and development in the oil and petrochemical industries. He has co-authored the preparation of plant design standards for a major US oil company and also authored a manual addressing Best Practices for water use in oil refineries for a major trade organization in the United Kingdom.

He has a Bachelor's and Master's degrees in Chemical Engineering and previously worked as the Vice President and Practice leader of Environmental Process Engineering for ENSR and AECOM (ENSR was acquired by AECOM) and as the Manager for Environmental Process Engineering at Badger Engineers Ltd prior to his affiliation with AECOM.

Contributors

Priyanka Ali Rice University, Houston, TX, USA

Shiladitya Basu Stantec, Houston, TX, USA

Somnath Basu Jewel Innovations in Energy and Environment, Katy, TX, USA

Fernando Almada Calvo Chevron Corporation, San Ramon, CA, USA

Tanmoy C. Chakravarty ZVI, Inc, Houston, TX, USA

James Cunningham Chevron Corporation, Houston, TX, USA

Saurabh P. Deshpande Black and Veatch, Toronto, ON, Canada

A. J. Gerbino AQSim, Glen Ridge, NJ, USA

Rajeev Goel Hatch, Ltd., Hamilton, ON, Canada

Richard T. Grubbs Chevron Corporation, Midland, TX, USA

Kristen Jenkins Kiewit Engineering Group Inc., Lenexa, KS, USA

Arvind D. Kannan Carollo Engineers, Dallas, TX, USA

Leslie Miller OLI Systems, Parsippany-Troy Hills, NJ, USA

Narayana Murthy Nadella Technip Energies USA, Houston, USA

Nicholas Piccolo McMaster University, Hamilton, ON, Canada

Anil Rai AECOM, Mumbai, India

Thomas E. Rehm GSOP Climate Collaboration, Inc., Houston, TX, USA

Robert S. Reimers Tulane University, New Orleans, LA, USA

Justin Scott Cove Environmental LLC, Stillwater, OK, USA

Andrew R. Shaw Black & Veatch, Houston, TX, USA

Spencer Snowling Hatch, Ltd., Hamilton, ON, Canada

Mudumbai Venkatesh Process and Energy Consultant, Andover, MA, USA

Yue Xu Veolia Water Solutions, New Orleans, LA, USA

Chapter 1
An Overview of Chemistry and Microbiology of Water

Somnath Basu, Yue Xu, Priyanka Ali, Robert S. Reimers, and Arvind D. Kannan

Abstract Water contains constituents that significantly influence its properties. Although water supports life on earth there are certain constituents that may render it unsuitable for beneficial use—both for human consumption, as well as for industrial uses. Proper characterization is necessary to devise appropriate treatment systems for the use of water in various applications. Wastewater, on the other hand, is traditionally viewed as a liability to a utility or industrial facility due to the health and environmental threat it poses to the receiving bodies of water. Proper treatment and disposal methods can minimize most of the risks. However, wastewater can also perform the role of an asset when the treated effluent is reused in various applications, particularly in the water-stressed regions of the world. To better understand the characteristics of water and wastewater associated with various facets of petroleum industry, it is essential to understand an extensive range of parameters, both biotic and abiotic, so that proper treatment schemes can be developed and implemented. A description of some of the important parameters, their effects, sources, and their detection methods form the basis of this chapter. It covers a wide range of subjects including water chemistry and microbiology, chemical thermodynamics, and chemical and biochemical reaction kinetics, each of which is a subject by itself. For an

S. Basu (✉)
Jewel Innovations in Energy and Environment, Katy, TX, USA
e-mail: som.basu@jewelenergyenvironment.com

Y. Xu
Veolia Water Solutions, New Orleans, LA, USA
e-mail: yue.chloe.xu@gmail.com

P. Ali
Rice University, Houston, TX, USA
e-mail: pa26@rice.edu

R. S. Reimers
Tulane University, New Orleans, LA, USA
e-mail: rreimers@tulane.edu

A. D. Kannan
Carollo Engineers, Dallas, TX, USA
e-mail: akannan@carollo.com

S. Basu et al. (eds.), *Water Management in Petroleum Industries*, Water Resources Development and Management, https://doi.org/10.1007/978-981-19-3159-8_1

in-depth understanding, interested readers are referred to the books listed at the end of this chapter as "Suggested Reading".

Keywords Aerobic microorganisms · Alkalinity and acidity of water · Anaerobic microorganisms · Anoxic microorganisms · Autotrophic organisms · BOD · COD · Dissolved Oxygen (DO) · Hardness · Heterotrophic organisms · Langelier Index · Metals and metalloids in water · Microbiology of water and wastewater · Nonmetals in water · Oxidation Reduction Potential (ORP) · Oxygen Uptake Rate (OUR) · pH · Plate count method · Silt Density Index (SDI) · Solubility product · Specific Oxygen Uptake Rate (SOUR) · TDS · TOC · Total Petroleum Hydrocarbons (TPH) · Turbidity · TSS · VOC · VSS · Water chemistry · Water QA/QC

1.1 Chemical Characteristics of Water

Water has many special properties that make it unique. The next section discusses the special properties of water, its benefits to the environment, and the influence of chemical characteristics on its uniqueness. A good understanding of the basic concepts of water chemistry is essential, which serves as the foundation of this chapter. This is covered in detail in Appendix A.

1.1.1 Molecular Structure of Water and Its Effects on Water Properties

The unique properties of water render it to be the key ingredient to support life in all forms. Molecular structure of water consists of an oxygen atom at the center and two hydrogen atoms symmetrically on the two sides forming an angle of 105 as shown in Fig. 1.1a. Due to this shape, water molecules are strongly dipolar. The polarity causes them to attract each other and form aggregates by hydrogen bonding as shown in Fig. 1.1b. This molecular structure contributes to some unique physical and chemical properties of water.

Some of the properties of water relevant to its role and importance in its wide range of beneficial uses are discussed below.

Water is an unusual liquid in that it expands upon freezing. Thus, ice is lighter than water. The density of water gradually increases upon cooling until it attains the highest density at 4 °C. Further cooling results in a gradual decrease in density. Therefore, lakes freeze at the top in winter, retaining water underneath, without which aquatic life would not exist.

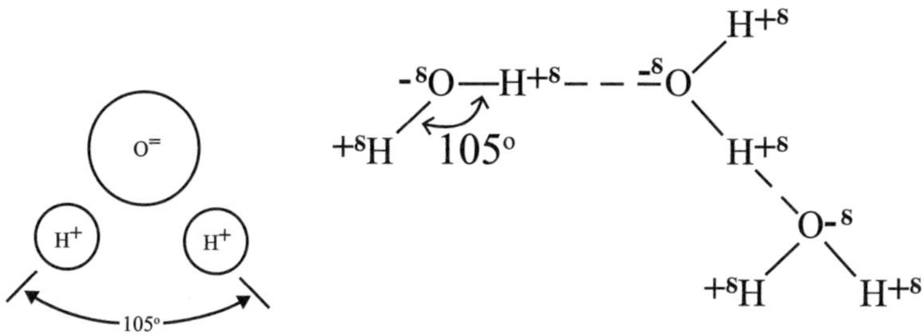

Fig. 1.1 **a** and **b** Molecular structure of water and polarity to form hydrogen bonding (Adapted from Snoeyink and Jenkins, 1991)

Due to the polarity of water molecules it interacts with many substances it comes in contact with. This makes water a universal solvent that becomes susceptible to contamination by other substances occurring in nature.

Two other important properties of water are its high specific heat, higher than many common liquids and most common solids, and high heat of vaporization, requiring very large amount of heat to vaporize and releasing large amount of energy when water vapor condenses. These two properties make water a useful medium for the transfer of energy and are utilized in industrial heating and cooling applications.

Following is a discussion on water chemistry and a few important parameters relevant to water and wastewater treatment.

1.1.2 Major Constituents of Water and Wastewater

Water is a universal solvent, and it dissolves a large number of elements in their molecular and ionic forms. These include metals, metalloids, and nonmetals. Due to its dipolar property, it acts as an electrolyte and dissolved polar substances exist as cations and anions in water. Their sources, effects on water properties and environmental impacts are discussed below.

1.1.2.1 Metals and Metalloids in Water

Metals are primarily present in water in ionic forms as parts of dissolved compounds and ions. Sodium, potassium, calcium, magnesium, iron and manganese are the major cations in both surface and groundwater resources. Sodium and potassium are monovalent cations. Sodium is present in high concentrations in the forms of sodium chloride and sulfate in seawater and water that comes in contact with petroleum crudes. Potassium is an important nutrient in living cells. It is added to chemical fertilizers along with nitrogen and phosphorus to stimulate plant growth in agricultural applications. Potassium is less common in natural sources of water and primarily occurs in agricultural run offs and effluents from fertilizer manufacturing industries.

Calcium and magnesium are divalent cations. They originate from the dissolution of minerals in nature that water comes in contact with. Both of these metals contribute to the hardness of water and at high concentrations they can form scale on the surfaces of equipment and pipes.

Manganese and iron occur in multivalent states in water. Manganese possesses seven oxidation states but only +2, +4 and +7 are of relevance to water chemistry. Similarly, the two important oxidation states of iron are +2 and +3. Primarily they are present as dissolved ions in groundwater in reduced states. When they come in contact with air, they are oxidized to higher oxidation states that are insoluble in water.

Iron is one of the most widely used materials of construction in both domestic and industrial applications because of its superior mechanical and structural properties. It is critical to the human body as an essential component of hemoglobin for transport of oxygen from lungs to the different parts of the body. It is also a micronutrient for cell growth. However, too much iron intake can cause adverse health effects. According to the US National Secondary Drinking Water Regulations the limit for iron is 0.3 mg/L.

Apart from the metals listed above a few others are also commonly present in water. These include copper, lead and zinc. Copper is extensively used in water pipes for both drinking water and industrial water applications. Due to their superior heat transfer property copper and its alloys find wide application in heat exchanger and condenser tubes in refining and chemical industries. It is a micronutrient for cell growth, but it is harmful at high levels. Some compounds are toxic when ingested or inhaled. The secondary MCL (maximum contaminant level) for copper set by EPA is 1 mg/L in drinking water. Above this concentration copper is toxic to aquatic plants and fish. Therefore, copper needs to be monitored carefully and controlled in industrial discharges to water bodies.

Lead has numerous uses as a material of construction for water pipes, an ingredient in paints and pigments, manufacture of battery, etc. However, it is very toxic by ingestion and is a cumulative poison. This has led to the discontinuation of lead from drinking water pipes. Also, the practice of addition of tetraethyl lead in gasoline as an octane enhancing measure has been abandoned with effect from January 1, 1996. EPA established the maximum concentration limit (MCL) of lead in water as 15 μg/L.

Zinc is an important metal for coating steel to protect it from corrosion. It is used to produce various alloys, e.g., galvanized steel, brass, bronze, batteries, and many other applications. Zinc is a micronutrient for cell growth. It also has many beneficial effects on human health and is an ingredient in several medicines. However, excessive intake of zinc can be harmful to health. Secondary Drinking Water Standard for zinc is 5 mg/L.

Several other metals, although not typically present in waters from natural resources, are associated with industrial effluents and contaminated groundwater plumes. These include chromium, nickel, cobalt, mercury, molybdenum and vanadium. Chromium is a heavy metal. In natural waters, chromium exists primarily in trivalent form as $Cr^{3+}$and hexavalent $CrO_4{}^{2-}$ and $Cr_2O_7{}^{2-}$ forms. Hexavalent chromium is an important metal in many industrial manufacturing applications, including steel making. Hexavalent chromium compounds were extensively used in petroleum industries as a corrosion inhibitor. These compounds are carcinogenic and are corrosive to tissue. Maximum permissible limit of chromium in water is 100 μg/L. If effluents from such industries contain hexavalent chromium above regulatory limit it needs to be removed by proper treatment prior to discharge.

Nickel is used for plating as a means for corrosion prevention. It is alloyed in combination with chromium for producing stainless steel. Worldwide its biggest use is in making coins. Also, it is an important ingredient for catalysts used for various chemical, petroleum refining and petrochemical processes. Nickel has no MCL enforced by EPA, but excessive ingestion poses adverse health effects. Maximum level for nickel in irrigation water is 200 μg/L.

Cobalt is present in effluents from paint and electroplating industries. It is typically not reported in water and wastewater streams in petroleum industries. It is an important micronutrient for cell growth. Cobalt dust is flammable and toxic by inhalation. There is no enforceable limit for cobalt or its compounds in water by EPA currently.

Mercury is used in thermometers, barometers, pharmaceuticals, pesticides, and fungicide, as well as a mold retardant in paper mills. Traditional manufacturing of caustic soda and chlorine is conducted electrochemically by mercury cells. Mercury is also present in crude petroleum from certain regions of the world, including crudes from Canada, as discussed in Chap. 3. Inorganic mercury can methylate in sediments and form dimethyl mercury. This form of mercury is very toxic and can concentrate in the aquatic food chain. Mercury can be lost readily from samples. It is necessary to preserve them by adding HNO_3 to reduce pH to <2. Due to its extreme toxic properties, mercury is highly regulated in water. The drinking water limit for mercury is 2 μg/L.

Molybdenum is a metal primarily used as a component for making various alloys, including stainless steel, to impart corrosion resistant properties. Molybdenum based catalysts are used in petroleum refining processes, such as desulfurization. It is an essential element as a micronutrient for cell growth. EPA imposes no regulatory limit on molybdenum in water. The recommended limit of molybdenum for long term use in reclaimed water for irrigation is 0.01 mg/L (Rowe and Abdel-Magid 1995).

Vanadium is used as an alloying ingredient of steel to make very high strength mechanical tools. Vanadium and its compounds are also used in catalysts for chemical and petroleum refining processes, most notably in the production of sulfuric acid. Injection of vanadium at concentration normally present in food is not considered harmful. EPA has no specific maximum contaminant level (MCL) established for vanadium.

A few metalloids are commonly encountered in water and wastewater streams that come in close contact with petroleum crudes and refined products. These include arsenic, boron, and selenium. Arsenic exists in water primarily in trivalent form as arsenious acid (H_3AsO_3), and in pentavalent form as H_3AsO_4. As discussed in Chap. 3, crudes from several regions of the world, including certain parts of the US, contain high levels of arsenic. It is harmful to health as exposure to arsenic can lead to skin cancer. According to EPA regulations the maximum permissible concentration limit of arsenic in water is 10 μg/L. If arsenic is present in the refinery effluent above the limit, it will need to be removed by proper treatment.

Boron in its most common form exists in natural waters as boric acid, H_3BO_3. Among the other industrial uses, primarily in glass and ceramic manufacturing processes, in petroleum industry boron is an important element for cross linking guar gum to create large polymeric chains in the exploration and production of petroleum from upstream unconventional resources as discussed in detail in Chap. 3. Flowback water from these operations contains boron. It is also associated with groundwater in several parts of the US.

Boron is an important micronutrient for plant cell growth but there is a narrow range within which boron is beneficial to plants. Excessive boron leads to leaf scorching and premature falling of leaves. For humans, ingestion of large amounts of Boron with drinking water can affect the central nervous system. Concentrations over the Long Term (2 mg/L) and Lifetime Health Advisory Levels (5 mg/L) pose adverse health effects on adults. Although boron is not regulated by the drinking water standards of EPA it is regulated in several states, typically in the range of 0.6 to 1 mg/L.

Selenium is a metalloid which is used in electronics, ceramics, and shampoos. It is considered a nonessential trace element for most plants, but is an essential trace nutrient for most animals, and selenium deficiency diseases are well known in veterinary medicine practice. Above trace levels selenium is toxic to animals and may be toxic to humans. It exists in water in two oxidation states—selenite (+4) and selenate (+6). Soluble selenium is leached from coal ash and fly ash at electric power plants that burn seleniferous coal. Selenium is present in petroleum crudes and wastewater streams from refineries. It is also present in wastewaters from mining operations. Many regions in the world, including the US, regulate selenium discharges in industrial effluents to very low levels (10 μg/L).

All of the above metals and metalloids present in water and wastewater are analyzed by Inductively Coupled Plasma—Mass Spectrometry (ICP-MS) following the procedure for SM 3125 A (APHA et al. 2023). Among these elements arsenic, copper, lead, iron, zinc, chromium, cobalt, molybdenum, mercury and vanadium fall under the category of heavy metals although arsenic is technically a metalloid, not a metal. The term 'heavy metals' is defined as metals with high atomic weight and a density at least 5 times greater than that of water.

1.1.2.2 Nonmetallic Constituents in Water

The majority of the non-metallic constituents in water include carbon, hydrogen, oxygen, nitrogen, sulfur, and phosphorus. They exist as parts of inorganic and organic molecules and ions. Carbon, hydrogen, and oxygen atoms make up most inorganic and organic constituents in water. Chloride, bromide, fluoride, nitrite, nitrate, and sulfate are the most common anions in natural waters and various industrial water and wastewater streams. These ions are analyzed by Ion Chromatographic method SM 4110 B (APHA et al. 2023) and EPA Method 300.1.

Molecules and ions containing nitrogen, phosphorus, potassium, and sulfur atoms play significant roles in water chemistry and microbiology. They are the nutrients for the growth of microbiological population in wastewater treatment and are building blocks of living cells. As listed in Table 1.2, sodium, and to a lesser extent, calcium, magnesium, chlorine, and iron play the role as micronutrients. In wastewater treatment, nutrient deficiency can lead to failure of metabolic activities, and the treatment process. The role of nutrients in the wastewater treatment is further discussed in Chap. 5. The significance of nonmetallic elements nitrogen, phosphorus and sulfur as nutrients is discussed in the following section.

Nitrogen is an essential nutrient for the growth of microorganisms as it is a component in the synthesis of protein. It supports the production of living tissues. Although sufficient nitrogen is essential to sustain biological wastewater treatment processes, however excess nitrogen, especially in receiving waters, causes nuisance aquatic growth of algal bloom and hyacinths leading to degradation of water body, also called eutrophication. Removal of nitrogen is a major concern and is discussed in Chap. 5.

The principal sources of nitrogen in raw water are nitrogenous compounds of plant and animal origin, from industrial and mining wastewater sources, and from atmospheric fixation by lightning and cosmic radiations. As an element, nitrogen exists in nature in nine oxidation states but three are relevant to aqueous environment. Forms of nitrogen that are major constituents of water are organic nitrogenous compounds, ammonia (-3), nitrite (+3), and nitrate (+5). Ammonia, at a high concentration, causes algal bloom which depletes the dissolved oxygen (DO) of the aquatic environment. In water, ammonia exists in equilibrium with ammonium ions according to the following relationship:

$$\begin{aligned} NH_4^+ &\leftrightarrow NH_3 + H^+,\ K_a = [NH_3] * \left[H^+\right]/\left[NH_4^+\right] \\ &= 10^{-9.25} \text{and } pK_a = 9.25 (\text{Metcalf \&Eddy/AECOM, 2014}) \end{aligned} \tag{1.1}$$

Note that $pK_a = -\log K_a$, where K_a = equilibrium constant.

Equilibrium distribution of ammonia and ammonium ions is presented as Fig. 1.2.

Below pH of 9.25 ammonia exists as ammonium ions, and above that pH it is present as ammonia dissolved in water. The relative amounts of ammonia and ammonium ions are determined as functions of pH and temperature based on the equilibrium relationship.

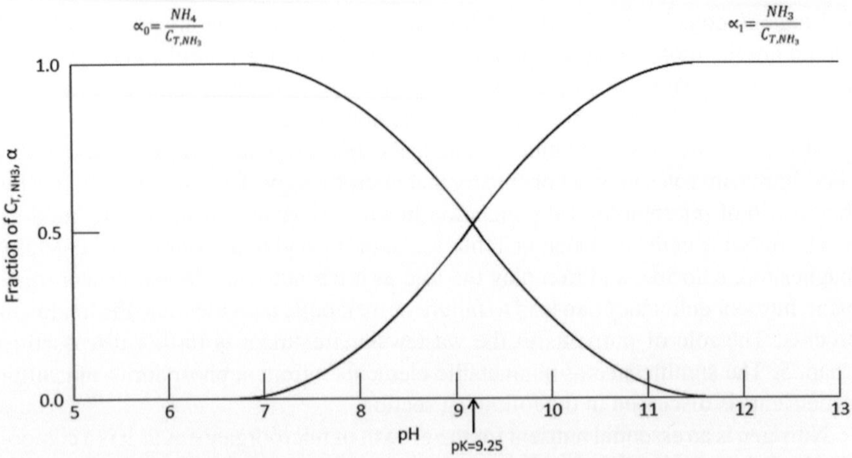

Fig. 1.2 Equilibrium distribution of ammonia and ammonium ions at 25 °C

Although nitrogenous compounds occur naturally in water sources, elevated levels usually come from human activity such as municipal and industrial wastes, runoff, or groundwater from fertilized agricultural lands, etc.

Domestic wastewater contains nitrogen in urea and proteinaceous organic matter such as amino acids and proteins, collectively called organic nitrogen. Organic nitrogen and ammonia together are designated as Total Kjeldahl Nitrogen (TKN) and is determined analytically by EPA Method 2687. This method consists of digesting the sample with a metal catalyzed acid to convert the organic nitrogen to ammonia, followed by colorimetric determination of ammonia.

Produced water from petroleum exploration and production (E&P) operations, and wastewater from petroleum refining also contains TKN consisting of ammonia-nitrogen and organic nitrogenous compounds. This is discussed in more detail in Chaps. 3 and 6. Organic nitrogen can be readily converted into ammonia through decomposition by microorganisms. In an aerobic environment, microorganisms can oxidize ammonia to nitrite and nitrate, through nitrification. Nitrite nitrogen is relatively unstable and is easily oxidized to nitrate form. Nitrate nitrogen is the most common oxidized form of nitrogen found in wastewater.

Total Nitrogen (TN) is comprised of all organic nitrogen, ammonia-nitrogen, nitrite-nitrogen and nitrate-nitrogen. Total inorganic nitrogen (TIN) is the combination of ammonia-nitrogen, nitrite-nitrogen, and nitrate-nitrogen. The measurement of ammonia-N in water can be made colorimetrically by EPA Method 350.1 and 350.2, or by ion selective electrode method SM 4500 NH_3-D (APHA et al. 2023). Methods SM 4500-NO_2^- and SM 4500-NO_3^- (APHA et al. 2023) are followed for nitrite and nitrate, respectively. The analytical methods for TKN are EPA Method 351.2 and 351.4. In water chemistry concentrations of all nitrogen species are expressed in terms of nitrogen as ammonia-N, nitrite-N and nitrate–N.

Phosphorus is also an essential nutrient for the growth of microorganisms. It supports cell growth. Similar to nitrogen, high concentration of phosphorus also aids in algal bloom, depleting the dissolved oxygen (DO) in receiving water bodies. Thus, phosphorus removal is of significant interest in domestic wastewater treatment. Discharges from petroleum refining, and exploration and production operations are generally deficient in phosphorus and such effluents typically do not contribute to nutrient load on account of phosphorus.

Some of the main sources of phosphorus in surface water are algae and other aquatic plants and organisms. Phosphorus is released from the decay of their body cells. It is classified as orthophosphates (OP), condensed phosphates (pyro-, meta-, and other polyphosphates), and organically bound phosphates. Orthophosphates occur in natural waters and wastewaters in pentavalent (oxidation state +5) state in three forms PO_4^{3-} (orthophosphate ions), $H_2PO_4^-$ (dihydrogen phosphate ions) and HPO_4^{2-} (monohydrogen phosphate ion) in equilibrium, as follows.

$$H_3PO_4 \leftrightarrow H_2PO_4^- + H^+, pK_1 = 2.1 \tag{1.2}$$

$$H_2PO_4^- \leftrightarrow HPO_4^{2-} + H^+, pK_2 = 7.1 \tag{1.3}$$

$$HPO_4^{2-} \leftrightarrow PO_4^{3-} + H^+, pK_3 = 12.1 \tag{1.4}$$

Note that pK_1, pK_2 and pK_3 are equal to—log K_1,—log K_2 and—log K_3, respectively where K_1, K_2 and K_3 are the equilibrium constants for reactions 1.2, 1.3 and 1.4, respectively. Equilibrium constant for a chemical reaction has been defined in the next section of this chapter.

The most encountered phosphate compounds of environmental significance are trisodium phosphate (Na_3PO_4), disodium phosphate (Na_2HPO_4), diammonium phosphate $(NH_4)_2HPO_4$) and monosodium phosphate (NaH_2PO_4). The combination of all forms of phosphorus—orthophosphates, polyphosphates and organically bound phosphorus compounds in the water environment is designated as total phosphorus (TP).

Determination of TP consists of converting all forms of phosphorus to orthophosphate (OP) by UV assisted persulfate oxidation, followed by colorimetric determination following method SM 4500-P (APHA et al. 2023).

Sulfur occurs naturally in water and wastewater. Groundwater and oilfield produced water at certain locations contain sulfate ions in contact with gypsum in the natural rock formation. Sulfate can be biologically reduced to sulfide under anaerobic conditions. The sulfide ions thus produced combine with hydrogen to form hydrogen sulfide (H_2S).

Hydrogen sulfide is often present in ground water and sediment formed by the biological reduction in the anaerobic environment. It is quite common in produced water from petroleum exploration and production (E&P) operations. Gaseous hydrogen sulfide is very toxic to human health. At high levels, it interferes with the olfactory system, giving a false sense of safe environment due to absence of H_2S.

Operationally, four categories of sulfide in water, wastewater, and sediments can be defined: (i) total sulfide, (ii) dissolved sulfide (iii) acid-volatile sulfide (iv) un-ionized hydrogen sulfide. Hydrogen sulfide-bisulfide-sulfide equilibrium has been discussed later in this chapter and depicted in Fig. 1.4. Sulfide ions in water can be analyzed by using sulfide ion selective electrodes. A detailed discussion on sulfide in produced water and refinery wastewater, its regulatory implications and treatment processes are presented in Chaps. 3 and 5.

Nonmetallic elements constitute the major anions—carbonate (CO_3^{2-}), bicarbonate (HCO_3^-), hydroxide (OH^-), sulfate (SO_4^{2-}), sulfide (S^{2-}), nitrate (NO_3^-), chloride (Cl^-) and fluoride (F^-) in water and wastewater. Fluoride and nitrate ions are present in certain sources of natural water. They have important public health implications relating to their presence in drinking water. Higher and lower than the optimum fluoride concentration of 1 mg/L in drinking water have been identified as the root cause of various dental diseases. Locations where groundwater is used as a drinking water resource, the presence of excessive nitrate ions can adversely affect the oxygen carrying capacity of blood (methemoglobinemia). The maximum contaminant level (MCL) for nitrate ions in set by the US EPA is 10 mg/L.

1.1.2.3 Solubility of Solids in Aqueous Phase and Solubility Product

The dissolution of a solid substance in a liquid may be considered as a chemical reaction where the solute exists in equilibrium between the solid and the dissolved phases in a solution. In the case of an aqueous solution, solubility defines the extent of dissolution of a solid (solute) in equilibrium with water (solvent).

Water is an electrolyte, and dissolved solids in water exist as ions. The dissolution of a solid 'A' into ions 'B' and 'C' may be represented by the reversible chemical reaction 1.5:

$$aA_{(s)} \leftrightarrow bB_{(aq)} + cC_{(aq)} \tag{1.5}$$

where 'A' is the reactant and 'B' and 'C' are the products.

Solubility product is the product of its dissolved ion concentrations raised to the power of their stoichiometric coefficients and is denoted as K_{sp}:

$$K_{sp} = [B]^b * [C]^c \tag{1.6}$$

[A], [B] and [C] are the concentrations of 'A', 'B' and 'C', and a, b and c are the respective stoichiometric coefficients in chemical reaction 1.5. For this reaction the chemical equilibrium of the solute 'A' between the solid and dissolved phases may be expressed as Eq. 1.7:

$$K_{eq} = [B]^b * [C]^c/[A]^a \tag{1.7}$$

Combining Eqs. 1.6 and 1.7 the equilibrium constant K_{eq} can be correlated with solubility product K_{sp} by Eq. 1.8:

$$K_{sp} = K_{eq} * [A]^a \tag{1.8}$$

Being a solid substance, the concentration of 'A' is a constant as it does not change in an aqueous solution in course of the reaction, and hence the term $[A]^a$ is also a constant. Thus, the solubility product K_{sp}, a product of two constant terms, is a constant. It is a function of temperature. Tables of solubility product values under standard conditions (25 °C) of inorganic solids dissolved in water are available in the literature (AWWA 2011).

1.1.2.4 Hydrogen Ion Concentration and pH

Hydrogen ion concentration is expressed in terms of pH. It is one the most important and frequently used parameters in water chemistry and microbiology as it indicates the intensity of acidic and alkaline condition of any aqueous solution. pH is expressed as the negative logarithm of hydrogen ion concentration.

$$pH = -\log_{10}[H^+] \text{ (Sorenson, 1909)} \tag{1.9}$$

The hydrogen ion concentration in water is related to the extent to which water molecules dissociate into hydrogen and hydroxyl ions.

$$H_2O \leftrightarrow H^+ + OH^- \tag{1.10}$$

$$\text{Thus } K_{eq} = \frac{[H^+] \cdot [OH^-]}{[H_2O]} \tag{1.11}$$

Since the concentration of water in dilute aqueous solutions is constant, the equilibrium constant K_{eq} is denoted by K_w. Its value is approximately equal to 1 X 10^{-14} at a temperature of 25 °C.

$$K_w = [H^+]*[OH^-] = 10^{-14} \tag{1.12}$$

For neutral water (non-acidic or non-alkaline)

$$\begin{aligned} &[H^+] = [OH^-], \text{ therefore } K_w = [H^+]^2 = [OH^-]^2 = 10^{-14}, \\ &\text{Or, } [H^+] = [OH^-] = 10^{-7} \end{aligned} \tag{1.13}$$

Hence, for neutral water, pH = 7.

For acids, $[H^+] > [OH^-]$, and for bases $[OH^-] > [H^+]$.

Therefore, $[H^+] > 10^{-7}$ for acids, and $[OH^-] > 10^{-7}$ for bases.

Consequently, pH for acidic solutions in water is < 7 and that for basic solutions in water is > 7.

For most biological life to survive, the pH in the aquatic environment needs to be in a narrow range of 6 to 9. For treated effluent discharged to the environment, the allowable pH usually falls in the range of 6.5 to 8.5. pH is an important parameter to judge the corrosivity of water to metals. Water with a pH below 6.5 can be corrosive. Alkalinity is another important parameter for water quality. This is further discussed later in this chapter.

Measurement of pH is done by electrometric method using a combination of a pH probe and meter following the procedure SM 4500-H^+ as described in **Standard Methods** by American Public Health Association (APHA et al. 2023).

1.1.2.5 Acids, Bases and Salts

Based on the above discussions, acids are chemicals that release hydrogen ions (H^+) when dissolved in water and bases are chemicals releasing hydroxide ions $\left(OH^-\right)$ in aqueous solution. In water and wastewater applications, the acids and bases are present as dilute solutions.

Strong acids, also called mineral acids, are hydrochloric (HCl), sulfuric (H_2SO_4) and nitric (HNO_3) acids. They almost completely ionize into hydrogen cations (protons), and chloride $\left(Cl^-\right)$, sulfuric $\left(SO_4^{2-}\right)$ and nitrate $\left(NO_3^-\right)$ anions in water. Weak acids only partially ionize in water. Examples of weak acids are carbonic acid (H_2CO_3), hydrogen sulfide (H_2S), phosphoric acid (H_3PO_4), acetic acid (CH_3COOH), etc.

Like the strong acids, strong bases also almost completely ionize. Sodium hydroxide (caustic soda, NaOH) is a strong base which ionizes into hydroxide anions $\left(OH^-\right)$ and sodium cations $\left(Na^+\right)$ in water. Ammonium hydroxide is a weak base that exists in water partially as ammonium ($NH_4{}^+$) cations and hydroxide ions $\left(OH^-\right)$ in equilibrium with unionized ammonia. Examples of equilibrium relationships between the unionized species and their constituent ions for several weak acids and bases are presented later in this chapter.

Salts are products of reactions between acids and bases. The cations of bases bond with the anions of acids to form salts. The other product of acid–base reaction is water formed by a combination of hydrogen ions released from acids and hydroxide ions from bases. Examples are as follows:

Salt of strong acid and strong base:

$$HCl + NaOH \rightarrow NaCl\,(salt) + H_2O \tag{1.14}$$

Salt of weak acid and strong base:

$$H_2CO_3 + NaOH \rightarrow NaHCO_3(baking\ soda) + H_2O \tag{1.15}$$

$$H_2CO_3 + 2\,NaOH \rightarrow Na_2CO_3(soda\ ash) + 2H_2O \tag{1.16}$$

Salt of strong acid and weak base:

$$HCl + NH_4OH \rightarrow NH_4Cl \text{ (ammonium chloride)} + H_2O \quad (1.17)$$

Reaction of weak acid and weak base:

$$H_3PO_4 + 2\,NH_3 \rightarrow (NH_4)_2HPO_4 \text{(diammonium phosphate)} + H_2O \quad (1.18)$$

Acid–base chemistry has been discussed in further detail in the light of Bronsted and Lowri theory and available in reference materials listed in the suggested readings section at the end of this chapter.

1.1.2.6 Vapor–Liquid Equilibrium in Aqueous Solutions

A Solubility of oxygen in water

Oxygen in the atmosphere dissolves in water and exists as dissolved oxygen (DO) under the conditions of vapor–liquid equilibrium. DO in water is necessary to sustain aquatic life and aerobic microorganisms for their respiration. It is an essential parameter for the design and operation of aerobic biological wastewater treatment plants as described later in this chapter and in Chap. 5 of this book. Solubility of oxygen in water is a function of pressure, temperature, and dissolved solids. Standard table of DO saturation as a function of these parameters is available in the literature (M&E/AECOM 2014). DO analysis can be conducted by a number of wet chemical methods but the most widely used is membrane electrode method following the procedure detailed in SM 4500-O (APHA et al. 2023).

B Carbon Dioxide-Carbonic Acid-Bicarbonate-Carbonate Equilibrium.

Carbon dioxide (CO_2) dissolves in water and exists in multiple forms. It is present in the atmosphere as a gas in equilibrium with water in the liquid phase.

Equilibrium distribution of carbon dioxide between air and water is expressed by Henry's Law as:

$$CO_{2aq} = K_H * P_{air} \quad (1.19)$$

where P_{air} = Partial Pressure of CO_2 in air = $10^{-3.4}$ atm

K_H = Henry's Constant for CO_2 in equilibrium between air and water at 25° C = $10^{-1.5}$M/atm.

Therefore, concentration of CO_2 in water (CO_{2aq}) under atmospheric conditions = $10^{-1.5} * 10^{-3.4}$

$$= 10^{-4.9}\text{atm} \quad (1.20)$$

CO_2 is an acid gas. When CO_2 is dissolved in water it reacts with water to form carbonic acid, which is a diprotic acid as it contains two hydrogen atoms. Carbonic acid, a weak acid, dissociates only partially in aqueous solution into bicarbonate and carbonate ions. The extent of dissociation is dictated by equilibrium.

Dissolved CO_{2aq} is almost entirely present as carbonic acid. Equilibrium relationship between CO_{2aq} and H_2CO_3 is expressed as:

$$CO_{2aq} + H_2O \leftrightarrow H_2CO_3(\text{carbonic acid}) \tag{1.21}$$

The equilibrium constant for conversion of CO_{2aq} to carbonic acid is

$$K = [H_2CO_3]/[CO_{2aq}] \sim 10^{-4}, \text{ and pK } = 4 \tag{1.22}$$

Subsequently, carbonic acid dissociates into bicarbonate (HCO_3^-) and carbonate (CO_3^{2-}) ions according to the following reactions:

$$H_2CO_3 \leftrightarrow H^+ + HCO_3^-(\text{bicarbonate ions}),\ K_1 \tag{1.23}$$

$$HCO_3^- \leftrightarrow H^+ + CO_3^{2-}(\text{carbonate ions}),\ K_2 \tag{1.24}$$

The equilibrium distributions of carbonic acid-bicarbonate, and bicarbonate-carbonate are depicted as Fig. 1.3. The equilibrium constants K_1 and K_2 are expressed as:

$$K_1 = [H^+]*[HCO_3^-]/[H_2CO_3], \text{ and} \tag{1.25}$$

$$K_2 = [H^+]*[CO_3^{2-}]/[HCO_3^-] \tag{1.26}$$

At 50% dissociation of H_2CO_3, the concentrations

$$[H_2CO_3] = [HCO_3^-] \tag{1.27}$$

At that point,

$$K_1 = [H^+], \text{ and pH } = pK_1 = 6.35 \tag{1.28}$$

Similarly, for bicarbonate-carbonate equilibrium, at 50% dissociation, the concentrations:

$$[HCO_3^-] = [CO_3^{2-}], \text{ and pH } = pK_2 = 10.33 \tag{1.29}$$

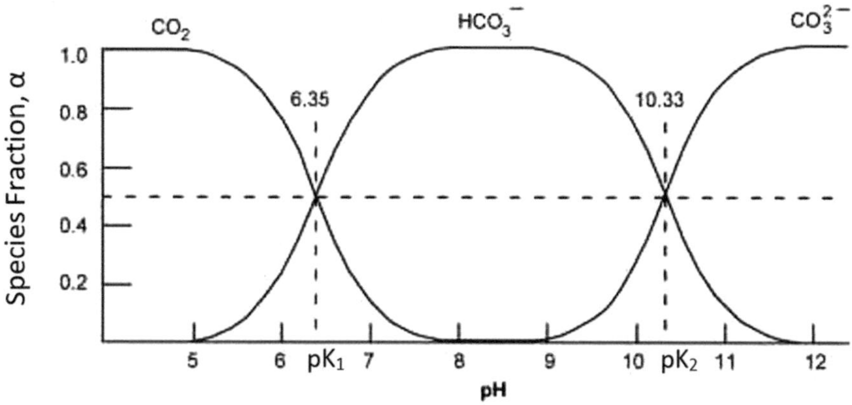

Fig. 1.3 Equilibrium distribution of carbonic acid-bicarbonate-carbonate versus pH at 25°C (Adapted from Snoeyink and Jenkins 1991)

Note that pK, pK_1 and pK_2 are—log K, –log K_1 and –log K_2, respectively. The significance of pK, pK_1 and pK_2 values is that at pH of 4 the dissolved CO_2 in water exists in equilibrium with carbonic acid at 50% molar ratio each; at pH of 6.35 carbonic acid exists in equilibrium with bicarbonate ions at 50% molar ratio each; and at pH of 10.33 bicarbonate and carbonate ions exist in equilibrium at 50% molar concentration each. At the typical pH range of 6.5 to 9 for water and wastewater bicarbonate ions dominate among the inorganic carbon species.

Carbonic acid-bicarbonate-carbonate equilibrium is a significant consideration in a large number of reactions in the natural environment, as well as in industrial applications. Several important water characteristics, e.g., alkalinity, hardness, corrosion and scaling potential are directly impacted by the relative concentrations of these constituents as dictated by the equilibrium.

C Hydrogen Sulfide-Bisulfide-Sulfide Equilibrium.

H_2S is also an acid gas. When dissolved in water it forms a diprotic acid, which is a weak acid. It exists in the form of hydrogen sulfide, bisulfide (HS^-) ions ad sulfide (S^{2-}) ions. The equilibrium constants are expressed as:

$$K_1 = [H^+]*[HS^-]/[H_2S] \text{ and} \tag{1.30}$$

$$K_2 = [H^+]*[S^{2-}]/HS^-] \tag{1.31}$$

The pK_1 and pK_2 values for H_2S-bisulfide-sulfide equilibrium are 7.1 and 13.9 respectively at 25 C, as depicted in Fig. 1.4.

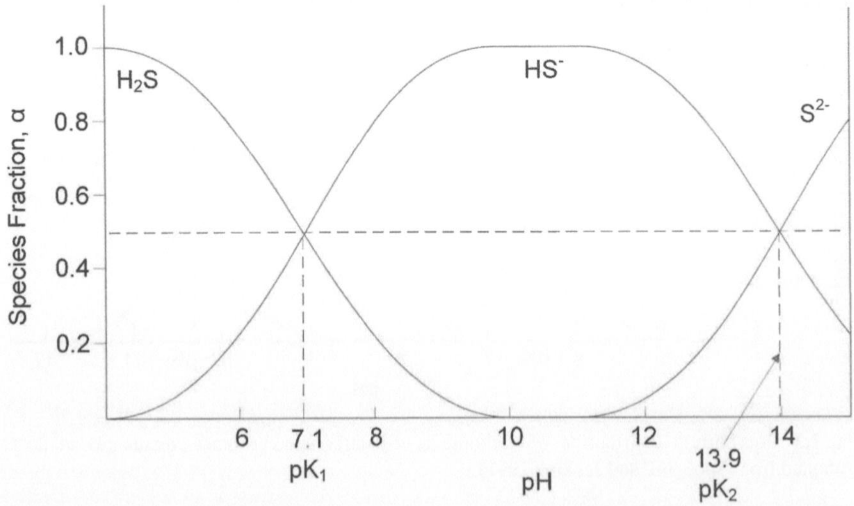

Fig. 1.4 Equilibrium distribution of hydrogen sulfide-bisulfide-sulfide versus pH at 25 C. (Adapted from Snoeyink and Jenkins 1991)

1.1.2.7 Alkalinity and Acidity

Alkalinity is a measure of acid neutralizing capacity of water. It results due to the presence of anions such as hydroxide $[OH^-]$, carbonate $[CO_3^{2-}]$ and bicarbonate $\left[HCO_3^-\right]$ in water. Calcium and magnesium bicarbonate are the most common compounds to produce alkalinity. Alkalinity is expressed as the sum of all the titratable bases, to a specified pH endpoint, typically to pH 4.5, using a standard acid of known strength. The measured value is expressed in terms of mg/L as $CaCO_3$.

Alkalinity is an important parameter of water and wastewater as it resists the change in pH occurring due to addition of acids and bases. From this viewpoint it plays the role of a buffer. A buffer is a solution that resists changes in pH when limited amounts of acid or base (diluted or undiluted) are added to it. Such a solution can be prepared by combining a weak acid, and its salt with a strong base (conjugated base), or, likewise, a weak base and its salt with a strong acid (conjugated acid). Buffer capacity (β) is defined as the amount of a strong acid or a strong base that must be added to 1 Liter of a buffered solution to cause a pH change of 1.0 unit. The buffer capacity is directly related to the slope of the titration curve. The buffer capacity is maximum at the point where the slope of the titration curve is the smallest. The titration is performed using phenolphthalein (P) and methyl orange (M) as indicators to obtain the end points of titration. Figure 1.5 represents the titration curve of a water sample using a standard H_2SO_4 solution. Typically, 0.25 or 0.5 N H_2SO_4 solutions are used for these titrations (Note: N represents for normality of the acid solution. For example equivalent weight of H_2SO_4 = molecular weight/2 i.e., 49. Therefore, concentration of 0.5 N $= 0.5 * 49\ g/L$, or 24.5 g/L). The inflection point at pH of 8.3 indicates the phenolphthalein end point, which accounts for the alkalinity due to hydroxide ions and one half of carbonate ions. The methyl orange end point at pH 4.5

marks the presence of carbonate, bicarbonate and hydroxide ions. Total carbonate alkalinity is the sum of carbonate and bicarbonate alkalinity. The total alkalinity of water is often equal to the carbonate alkalinity as the normal pH range of water and wastewater is well below the regime of hydroxide alkalinity (8.5 to 11) depending upon the temperature.

There can be five possible combinations of the three types of alkalinities:

- Case 1: Presence of hydroxide alkalinity alone
- Case 2: Presence of carbonate alkalinity alone
- Case 3: Presence of hydroxide and carbonate alkalinity
- Case 4: Presence of bicarbonate alkalinity alone
- Case 5: Presence of carbonate and bicarbonate alkalinity

It is assumed that hydroxide and bicarbonate alkalinities cannot coexist together as they combine to form carbonate according to the reaction:

$$OH^- + HCO_3^- \rightarrow CO_3^{2-} + H_2O \tag{1.32}$$

When hydroxide is the only form of alkalinity (Case 1) present in a water sample it can be calculated from the measure of pH according to the equation:

$$[OH^-] = K_w/10^{-pH} \tag{1.33}$$

For Case 2, when carbonate is the only form of alkalinity in water the pH is above 8.3. Titration to phenolphthalein end point marks the conversion of carbonate to bicarbonate ions by the reaction:

$$CO_3^{2-} + H^+ \rightarrow HCO_3^- \tag{1.34}$$

Continuing titration further to the methyl orange end point neutralizes the bicarbonate ions. Therefore, according to the reaction between bicarbonate and hydrogen ions, the consumption of acid to phenolphthalein end point (P) is half of that consumed for the total titration from the beginning of the titration to the methyl orange end point (M) and carbonate alkalinity $\left(CO_3^{2-}\right) = 2*$ Acid consumed from phenolphthalein end point (P) to methyl orange end point (M) based on the relationship represented by Eq. 1.35 (Fig. 1.5).

$$2\,HCO_3^- \rightarrow CO_3^{2-} + 2\,H^+ \tag{1.35}$$

For case 3, when hydroxide and carbonate alkalinity are present simultaneously, carbonate alkalinity can be determined from titration to P and M end points based on procedure followed for Case 2 and hydroxide alkalinity can be calculated from pH measurement following the procedure in Case 1.

In Case 4, when only bicarbonate alkalinity is present the pH of water is between 8.3 and 4.5. There is no acid consumption to reach phenolphthalein end point (P) and the bicarbonate alkalinity can be determined from the acid consumption from phenolphthalein end point (P) to methyl orange end point (M).

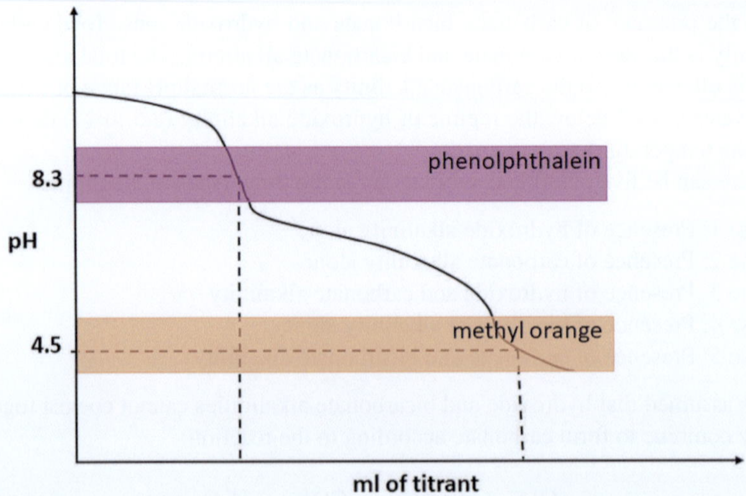

Fig. 1.5 Typical titration curve showing phenolphthalein (P) and methyl orange (M) Alkalinity

When both carbonate and bicarbonate alkalinities are present as in Case 5, the pH of the water is > 8.3

Carbonate alkalinity = 2∗ acid consumed to reach phenolphthalein end point (P).

Bicarbonate alkalinity = Total alkalinity as acid consumed to reach methyl orange end point (M)—carbonate alkalinity.

Details of the titration procedure for alkalinity determination is presented as method SM 2320 (APHA et al. 2023).

As noted earlier, alkalinity is expressed in terms of equivalent $CaCO_3$. Following is an example of conversion of alkalinity in mg/L to other units.

Example 1(a)

Consider a sample of water with Bicarbonate ions concentration = 30 mg/L.

In terms of Molarity, the concentration of Bicarbonate ions in the water sample = (30mg/L/1,000mg/g)/(61g/gmol of Bicarbonate) $\sim 5 \times 10^{-4}$M, or 50 mM.

In terms of equivalent $CaCO_3$, the Bicarbonate Alkalinity of the water sample = 30 mg/L*EW of $CaCO_3$/EW of Bicarbonate = $30 * 50/61$ i.e., ~ 24.6 mg/L as $CaCO_3$.

Example 1(b)

Consider another sample of water contains Carbonate ions concentration = 20 mg/L.

In terms of Molarity, the concentration of Carbonate ions in the water sample = (20mg/L/1,000mg/g)/(60g/gmol of Carbonate) $\sim 3.3 \times 10^{-4}$M, or 33.3 mM.

In terms of equivalent $CaCO_3$, the Carbonate Alkalinity of the water sample = 20 mg/L*EW of $CaCO_3$/EW of Carbonate = $20 * 50/30$ i.e., ~ 33.3 mg/L as $CaCO_3$.

Note:

Molecular Weight (MW) of Bicarbonate ions $\left(HCO_3^-\right) = 61$
Equivalent Weight (EW) of Bicarbonate ions $\left(HCO_3^-\right) = MW/1$, i.e. 61
Molecular Weight (MW) of Carbonate ions $(CO_3^{2-}) = 60$
Equivalent Weight (EW) of Carbonate ions $(CO_3^{2-}) = MW/2$, i.e. 30
Molecular Weight (MW) of Calcium Carbonate $(CaCO_3) = 100$
Equivalent Weight (EW) of Calcium Carbonate $(CaCO_3) = MW/2$, i.e. 50.

Alkalinity and acidity indicate the basic and acidic characteristics of water. Typically, natural pH of water and wastewater is in the range of 6 to 8.3, which is between the phenolphthalein and methyl orange end points of the titration curve in Fig. 1.5. A similar titration curve can be obtained by titrating a sample of acidic water of low pH with a standard sodium hydroxide (NaOH) solution. Acidity is the inverse of alkalinity which indicates the capacity of water to neutralize a strong base. Alkalinity and acidity of water are determined from the constituent molar concentrations by the following equations (Snoeyink and Jenkins 1991):

At pH 4.5:

$$\begin{aligned}\text{Total Alkalinity} &= \left[HCO_3^-\right] + 2*\left[CO_3^{2-}\right] + \left[OH^-\right] - \left[H^+\right] \\ &\quad + [\text{Other Bases in the Waste}]\end{aligned} \tag{1.36}$$

$$\begin{aligned}\text{Total Acidity} &= \left[H^+\right] - \left[HCO_3^-\right] - 2*\left[CO_3^{2-}\right] - \left[OH^-\right] \\ &\quad + [\text{Other Acids in the Waste}]\end{aligned} \tag{1.37}$$

Note that when conducting titration of an acidic water sample with a strong base from low pH to a pH of 4.5 the acidity is entirely due to the mineral acids. Therefore, the total acidity is also called mineral acidity.

At pH 8.3

$$\begin{aligned}\text{Total Alkalinity} &= \left[CO_3^{2-}\right] + \left[OH^-\right] - [H_2CO_3] - \left[H^+\right] \\ &\quad + [\text{Other Bases in the Waste}]\end{aligned} \tag{1.38}$$

$$\begin{aligned}\text{Total Acidity} &= \left[H^+\right] - [H_2CO_3] - \left[CO_3^{2-}\right] - \left[OH^-\right] \\ &\quad + [\text{Other Acids in the Waste}]\end{aligned} \tag{1.39}$$

At this pH the alkalinity is almost entirely due to carbonate and hence, the total alkalinity is approximately equal to carbonate alkalinity. On the other hand, for acidity determination, titration of a water sample from pH 4.5 to 8.3, almost the entire acidity is due to carbonic acid (H_2CO_3). Therefore, this acidity is also called carbon dioxide acidity.

The acidity and alkalinity determinations at pH 4.5 and 8.3 are considered here as the pH values of water and wastewater typically fall under this range. Also, it is important to note that the titrations described above are assumed as closed systems without the interference of CO_2 transfer between air and water, which affects the pH. The concepts above are presented in the context of inorganic carbon only, which is primarily responsible for acidity, alkalinity and buffering capacity of domestic water and wastewaters. The presence of sufficient alkalinity in wastewater is important to protect its pH from dropping during the certain wastewater treatment biochemical reactions, specifically nitrification of ammonia, which consumes alkalinity and causes potential dop in pH in absence of adequate buffering capacity provided by alkalinity. Various organic acids and bases also contribute to the acidity and alkalinity of water and wastewater streams of petroleum origin. These will require theoretical and experimental studies on a case-by-case basis at an advanced level.

1.1.2.8 Hardness

Hardness is a very important parameter in water chemistry. It is defined as the sum of the dissolved calcium and magnesium ion concentrations, both expressed in terms of calcium carbonate concentration, in milligrams per liter. Originally described as an undesirable property of water that consumes soap and form a difficult-to-remove scum, hardness has a much wider impact in both domestic and industrial application. Hard water forms scale in drinking water pipes in domestic water supply lines, and in cooling and boiler feed water lines in industrial facilities. Scaling reduces pipe sizes with time and increases pressure drop resulting in an increase in pumping energy. Calcium carbonate and magnesium hydroxide scales are difficult to remove and render the pipes unfit for use, ultimately requiring their replacement. The methods used to determine hardness are SM 2340B and SM 2340C (APHA et al. 2023).

As in the case of alkalinity, hardness is also expressed in terms of equivalent $CaCO_3$. The conversion of hardness from concentration units at equivalent concentration as $CaCO_3$ can be understood from the following example.

Example 2 Assume for a water sample, Ca^{2+} ion concentration = 15 mg/L and Mg^{2+} ion concentration = 10 mg/L.

Ca^{2+} Hardness = 15 mg/L * EW of $CaCO_3$/EW of Ca^{2+} i.e., 15* 50/20 = 37.5 mg/L as $CaCO_3$.

Mg^{2+} Hardness = 10 mg/L * EW of $CaCO_3$/EW of Mg^{2+} i.e., 10*50/12.2 ~ 41 mg/L as $CaCO_3$.

Total Hardness = 37.5 + 41 i.e., 78.5 mg/L as $CaCO_3$.

Hardness, and its expression in terms of equivalent $CaCO_3$ is very important in the design of water softening systems as discussed in Chap. 3.

1.1.2.9 Liquid–Liquid Equilibrium between Aqueous and Organic Phases: Octanol/Water Partition Coefficient

The octanol/water partition coefficient (K_{ow}) is defined as the ratio of the concentration of an organic compound in the octanol phase to its concentration in the aqueous phase, of a two-phase octanol/water equilibrium system and is expressed as.

K_{ow} = Concentration in octanol phase / Concentration in aqueous phase.

Chemicals with low K_{ow} values (e.g., less than 10) may be considered relatively hydrophilic, tend to have high water solubilities and small soil/sediment adsorption coefficients. Conversely, chemicals with high K_{ow} values (e.g., greater than 10^4) are very hydrophobic.

1.2 Solids in Water

Solids refer to matters suspended and dissolved in water and wastewater. Water containing high concentrations of solids are unsuitable for human consumption and also for industrial applications. Total solids content is one of the most important physical characteristics of water and wastewater, which is dependent on the amount of floating, settleable, colloidal and dissolved matters in water.

1.2.1 Total Solids (TS)

Total Solids (TS) is the term applied to the material residue left in a pre-weighed dish after evaporation of a measured volume of sample followed by its subsequent drying in an oven at a defined temperature (103 to 105°C). The weight of the solid is determined by subtracting the weight of the empty dish from the weight of the dish after drying. TS is expressed in mg/L by dividing the weight of the solids in milligrams (mg) by the volume of the sample in liters (L).

1.2.2 Volatile Solids (VS) and Fixed Solids (FS)

Volatile Solids (VS) are the portion of TS that can be volatilized and burned off at 550°C. **Fixed Solids (FS)** are the inert residue remaining (not volatilized) after ignition at 550 C. VS and FS are expressed in mg/L by dividing the weights of VS and TS (mg) by the volume of the sample (L) used for the test. The ratio of VS to TS is used to characterize water and wastewater for the fraction of organic matter in the solids contained in it. TS is the sum of VS and FS. These parameters are determined in the laboratory following the method SM 2540 B and E (APHA et al. 2023).

1.2.3 Total Suspended Solids (TSS), Volatile Suspended Solids (VSS), and Total Dissolved Solids (TDS)

Total solids (TS) include:

Total Suspended Solids (TSS), the portion of total solids retained by a filtration medium of 0.45 μm pore size, and.
Total Dissolved Solids (TDS), the portion that passes through the filter.

TSS measures the total concentration of suspended and undissolved solids in a sample, and it is a critical parameter in determining the operational behavior of a treatment system. For TSS determination the sample is conducted by a pre-weighed glass fiber filter (GFF) medium to avoid any potential physical or chemical interaction with the sample. After the filtration, the medium is dried in an oven at 103 to 105 C for over an hour. The weight of the suspended solids in the sample is obtained by subtracting the weight of the pre-weighed medium from the weight of the medium after drying. TSS is determined by dividing the weight of suspended solids (mg) by the volume of the sample (liter) and expressed in terms of mg/L.

TDS is the concentration of the solids in the filtrate after it is evaporated to dryness in a weighed dish and dried to constant weight at 180 ± 2 C. TDS is usually comprised of a combination of colloidal and dissolved solids in a sample. It is expressed in terms of mg/L by dividing the weight of solids (mg) left behind in the dish after evaporation by the volume of the sample (L). The formulation is shown below:

Total Solids (TS) = Total Suspended Solids (TSS) + Total Dissolved Solids (TDS)

In water and wastewater, the volatile fraction of solids is primarily the content of its organic matter. Thus, **Volatile Suspended Solids (VSS)** is essentially the quantity of organic matter in the TSS of a sample. Procedure for VSS measurement consists of igniting the filtration medium used for the TSS measurement at 550 C for 30 min. The weight loss on the ignition of the solids represents the amount of volatile suspended solids in the sample. VSS of the sample is determined by dividing this weight (mg) by the volume of the sample (liter) and expressed in terms of mg/L. The solids left behind after ignition is termed Fixed Solids (FS).

$$\begin{aligned}\text{Total Suspended Solids(TSS)} &= \text{Volatile Suspended Solids (VSS)} \\ &+ \text{Fixed Suspended Solids (FS)} \qquad (1.40)\end{aligned}$$

TS is comprised of both TSS and TDS, which are determined by following the methods SM 2540 D and C (APHA et al. 2023).

1.2.4 Turbidity

Turbidity is "an expression of the optical property that causes light to be scattered and absorbed rather than transmitted with no change in direction or flux level through the sample" (APHA et al. 2023). It is the reduction in clarity of water caused by the scattering of visible light by particles that are suspended in the sample. Thus, turbidity increases as the intensity of scattered light increases, where the scattering of light is dependent on the size, shape, refractive index, and composition of the suspended particle. Turbidity is measured following the method SM 2130 B (APHA et al. 2023) and expressed in Nephelometric Turbidity Units (NTU). In absence of a laboratory analysis for TSS sometimes turbidity is used as a surrogate of TSS.

1.2.5 Silt Density Index

To measure the resistance to flow offered by solids in water and wastewater, a variety of fouling indexes have been developed. Silt density index (SDI) is a static measurement of resistance to flow by the solids in water and wastewater. SDI denotes the fouling nature of water. It is an indicator of the susceptibility of reverse osmosis (RO) and nanofiltration (NF) membranes to fouling due to the presence of colloidal matter in clean water (see Chap. 2 for further discussion on SDI in the context of the flow of water through RO and NF membranes).

1.3 Electrical Properties of Water

1.3.1 Electrical Conductivity

Due to the hydrogen bonding of water molecules, it has dipolar characteristics. Water is an electrolyte as it contains dissolved substances in the form of cations and anions. Electrical conductivity (EC) is a measure of the ability of an aqueous solution to carry an electric current. Thus, it is also a measure of the concentration of dissolved constituents based on their ability to conduct electrical charge. Conductivity is expressed as microsiemens per centimeter (μS/cm) or micromhos per centimeter (μΩ/cm) in the water sample. Conductivity can also be used to estimate the total dissolved solid of a water sample based on the empirical relationship:

$$TDS = k * EC \tag{1.41}$$

where k is a proportionality constant which is often determined empirically or experimentally.

1.3.2 Oxidation and Reduction Reactions, and Oxidation–Reduction Potential (ORP)

Oxidation and reduction are chemical reactions involving transfer of electrons between reactants. In an oxidation–reduction (also called redox) reaction, an atom, molecule or ion that loses electron(s) is oxidized to a higher oxidation state. Simultaneously, the other reactant gains those electrons and is reduced to a lower oxidation state. Each of these reactions is called a 'Half Cell Reaction'. The positively and negatively charged ions created by the half-cell reactions combine together by ionic bonding under electrostatic force and form new compounds that are electrically neutral. A few examples are as follows:

$$2\,H_2 \rightarrow 4\,H^+ + 4\,e \text{ (Half Cell Reaction)} \tag{1.42}$$

$$O_2 + 4\,e \rightarrow 2\,O^{2-} \text{(Half Cell Reaction)} \tag{1.43}$$

Each hydrogen atom produces a hydrogen ion (proton) of single positive charge and an electron by a half-cell reaction and each oxygen atom produces an oxygen ion of double negative charge and two electrons. The positive hydrogen ions and negative oxygen ions combine together to form an electrically neutral water molecule:

$$2\,H_2 + O_2 \rightarrow 2\,H_2O \text{ (Overall Reaction)} \tag{1.44}$$

Hydrogen and chlorine atoms combine together by ionic bonding to form hydrogen chloride (HCl) molecules:

$$H_2 \rightarrow 2\,H^+ + 2\,e \text{ (Half Cell Reaction)} \tag{1.45}$$

$$Cl_2 + 2\,e \rightarrow 2\,Cl^- \text{(Half Cell Reaction)} \tag{1.46}$$

$$H_2 + Cl_2 \rightarrow 2\,HCl \text{ (Overall Reaction)} \tag{1.47}$$

In the above redox reactions, oxygen and chlorine gain electrons and reduce to a lower oxidation state or oxidation number—for oxygen it is from 0 to –2, and for chlorine from 0 to –1. These are termed oxidizing agents or oxidants. In each reaction hydrogen acts as the reducing agent of oxygen and chlorine respectively and is itself oxidized from 0 to + 1 oxidation state.

Oxidation reduction reactions are most common in chemical and biochemical processes in water and wastewater treatment. Many reactions are more complex than those presented above as they involve multiple oxidizing and reducing reactants. Potassium dichromate ($K_2Cr_2O_7$) is a strong oxidizing agent that is used in many industrial processes. Effluent from these industries contain dissolved dichromate ions with + 6 oxidation state of chromium. In this form chromium is harmful to the human body. Treatment of these wastewater streams consist of reducing chromium to oxidation state + 3, followed by removal and disposal of trivalent chromium which is less harmful. The reducing agent is ferrous ions (Fe^{2+}) which are themselves oxidized to ferric (Fe^{3+}) form. It is a multistep reaction as follows:

Step 1: Reduction of hexavalent chromium to trivalent form

$$Cr_2O_7^{2-} + 14\,H^+ + 6\,e \rightarrow 2\,Cr^{3+} + 7\,H_2O \text{ (Half Cell Reduction Reaction)} \tag{1.48}$$

$$6\,Fe^{2+} \rightarrow 6\,Fe^{3+} + 6\,e \text{ (Half Cell Oxidation Reaction)} \tag{1.49}$$

$$Cr_2O_7^{2-} + 14\,H^+ + 6\,Fe^{2+} \rightarrow 2\,Cr^{3+} + 6\,Fe^{3+} + 7\,H_2O \text{ (Overall Reaction)} \tag{1.50}$$

$$K_2Cr_2O_7 + 6\,FeSO_4 + 7\,H_2SO_4 \rightarrow K_2SO_4 + 3\,Fe_2(SO_4)_3 + Cr_2(SO_4)_3 + 7\,H_2O \text{ (Overall Reaction for Step 1)} \tag{1.51}$$

Step 2: Hydroxide precipitation of trivalent chromium

$$24\,NaOH \rightarrow 24\,Na^+ + 24\,OH^- \text{(Ionization Reaction)} \tag{1.52}$$

$$2Cr^{3+} + 6\,Fe^{3+} + 24\,OH^- \rightarrow 2\,Cr(OH)_3 + 6\,Fe(OH)_3 \text{(Hydroxide Precipitation)} \tag{1.53}$$

$$3\,Fe_2(SO_4)_3 + Cr_2(SO_4)_3 + 24\,NaOH \rightarrow 2\,Cr(OH)_3 + 6\,Fe(OH)_3 + 12\,Na_2SO_4 \text{(Overall Reaction for Step 2)} \tag{1.54}$$

This reaction is completed in multiple steps resulting in reduction of dissolved hexavalent chromium to trivalent state, followed by removal of trivalent chromium in the form of hydroxide. These steps are conducted by addition of sulfuric acid, ferrous sulfate in step 1, and sodium hydroxide in step 2.

ORP, the oxidation–reduction potential, is the potential difference measured in an aqueous solution between an inert indicator electrode and the standard hydrogen electrode. Water and wastewater treatment processes are based on oxidation and reduction reactions and ORP is a critical parameter. It is important in the design and operation of water and wastewater treatment plants. Applications of ORP is discussed in detail later in this chapter, and also in Chap. 2.

1.4 Water Stability

Two important considerations in petroleum industrial applications of water are corrosion and scaling. When in contact with solid surfaces water tends to dissolve metals or deposit certain minerals. When it dissolves a metal then the water is termed as corrosive, and on the contrary, if water tends to deposit mineral(s) then it is termed scaling.

A stable water is defined as a water that does not exhibit either scale forming or corrosive properties, although realistically, under stable conditions, a thin layer of calcium carbonate forms a passivation layer to prevent corrosion.

In natural water $CaCO_3$ solubility is used as the indicator of whether a water stream is corrosive or scaling. A number of indexes have been developed to quantify the scaling and corrosion potential of water streams. The commonly used indexes in describing water stability are listed below (Table 1.1).

The most commonly used index to define water stability is called the Langelier Saturation Index (LSI). Based on the degree of $CaCO_3$ saturation, this index provides a measure of the stability of water. If a water stream has a negative LSI value, it is under-saturated with respect to calcium carbonate and is potentially corrosive. Conversely, waters with positive LSI are oversaturated with $CaCO_3$ and thus a layer of calcium carbonate can form, resulting in scale. Water saturated with $CaCO_3$ is neither corrosive or scaling and has an LSI of zero. In this example of Fig. 1.6 where pHs is 7, if the actual pH of this water is above 7, it will be scaling and if it is below 7 it will be corrosive.

Table 1.1 Water stability index formulations

Langelier's saturation index for closed system	LSI = pH − pHs
Ryznar's saturation index for open system	RSI = 2pHs−pH
Puchorius saturation index for open system	PSI = 2pHs−pHeq

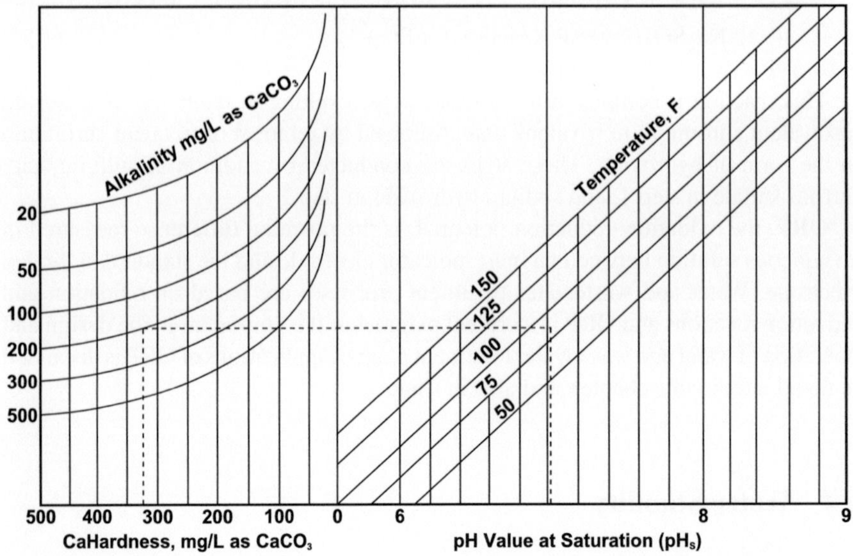

Fig. 1.6 pH_S values as functions of alkalinity, calcium hardness and temperature (Adapted from Kemmer 1988)

Ryznar Saturation Index (RSI) is another parameter to denote water stability with respect to corrosivity. The RSI value of water should be less than 10 for it to be considered stable and non-corrosive.

Puchorius Saturation Index (PSI) is calculated in a manner similar to the RSI. PSI uses an equilibrium pH rather than the actual system pH to account for the buffering effects.

where pH = measured water pH, and

$$pHs = \text{ pH at saturation with } CaCO_{3,}pH_{eq} = 1.465 \times \log_{10}[\text{Alkalinity}] + 4.54 \quad (1.55)$$

The parameter pH_s is a function of calcium hardness, alkalinity, and temperature of the water. It can be graphically determined based on the known values of the input parameters as shown in Fig. 1.6 (Kemmer 1988). In this figure pHs value has been graphically determined for a water sample with a calcium hardness with 325 mg/L as $CaCO_3$, and alkalinity of 200 mg/L as $CaCO_3$ at a temperature of 90 F to be 7. Subsequently, using the pH and pH_s values, the stability indexes for a sample of water can be determined.

All of the above indexes are based on the weak acid–base chemistry of the carbonate-bicarbonate system. However, other weak acids and bases (NH_3, H_2S, organic acids, borates, phosphates, silicates, etc.), which are negligible in natural water, are also sources of alkalinity in industrial water and wastewater. These weak acids and bases inevitably play a role in water stability. These indexes help in preliminary estimation of corrosion or scaling potential of a water stream but more accurate determination will require use of mathematical modeling techniques as detailed in Chap. 6.

1.5 Color

Reduction in the clarity of water caused by the absorption of visible light by substances, including organic and inorganic compounds, is an indication of **color** in water. The presence of organic substances including humic and fulvic acids, the presence of natural metallic ions such as iron and manganese, as well as turbidity in water contribute to color. Color can be classified into two kinds- apparent and true. Apparent color is when the measurement of color is done on an unfiltered sample and true color is when color is measured in filtered samples using 0.45 μm filters. Apparent color is therefore dependent on the turbidity of water and true color is the color of water from which turbidity has been removed when only the dissolved species are present. True color measurement is done by visual comparison method according to SM 2120 B and spectrophotometric method SM 2120 C (APHA et al. 2023). Both methods call for prefiltration to remove the interference by turbidity. The presence of color is reported in color units (c.u.) at the specific pH of the sample.

1.6 Odor

Odor, like taste, depends on contact of a stimulating substance with the appropriate human receptor cell. The adverse sensory response towards potentially toxic waters often provides the first warning of an environmental hazards. Odor in wastewater is usually caused by the gases produced during the anaerobic decomposition of organic wastes. The most characteristic odor of anaerobically treated wastewater is that of hydrogen sulfide produced by anaerobic microorganisms reducing sulfate to sulfide. Volatile organic compounds present in wastewater from petroleum industries also emit characteristic odors that can give rise to air pollution and human health risks.

1.7 Microbiology of Water and Wastewater

Microorganisms are very important to the environment as they have the ability to transform a great variety of inorganic and organic pollutants into harmless compounds, which can then be recycled back or disposed safely into the environment. Water and wastewater streams are ideal media for support and growth of microorganisms. Thus, microorganisms such as bacteria, algae, fungi, etc. are employed in wastewater treatment systems to alleviate the challenges posed by pollutants.

Microorganisms represent the simplest life forms consisting of single or multiple cells. Each cell is comprised of a nucleus in a cytoplasmic fluid protected by a cell wall. Metabolism and reproduction, the basic functions of a living cell are performed by microbial cells also. Metabolic activities include the assimilation of substrates (food) and respiration to produce energy to support their motility and reproduction. As opposed to the higher forms of life they reproduce asexually by cell division. Microorganisms are classified into different categories based on several different set of criteria as follows.

1.8 Classification of Microorganisms

Microorganisms are classified into various groups using the criteria as discussed below.

1.8.1 Phylogenic Classification

This is considered the most fundamental method of classification as it follows the evolution of the groups of microorganisms. Those present in wastewater fall under the evolutionary tree of life which was derived from comparative sequencing of 16S or 18S RNA (Madigan et al. 2010). Three major domains comprise all types of microorganisms. The Bacteria and Archaea domains consist of the prokaryotes which are single cell organisms. All prokaryotes are microorganisms (can only be seen under a microscope) that do not contain their chromosomes comprising the genetic code, inside a nucleus. The third major domain is Eukarya, consisting of single cellular as well as multicellular organisms, having their chromosome inside a nucleus. The characteristics of each domain are described in more detail below.

Bacteria

Bacteria are unicellular prokaryotic organisms ranging in size from about 0.1 to 10 μm containing a cell wall and cytoplasm. The deoxyribose nucleic acid (DNA) and ribose nucleic acid (RNA) are contained within the cytoplasm. They carry genetic information and support production of specific enzymes that conduct the metabolic and reproductive activities in wastewater treatment. A bacterial cell has about 80% water and 20% dry matter and among the dry matters, 90% is organic and 10% is inorganic. Typical composition of bacterial cells is shown in Table 1.2. The chemical composition of a bacterial cell is expressed by the simple empirical formula $C_5H_7O_2N$.

Numerous factors can affect the growth and death of bacteria during their life cycle. These include type of food or carbon source, abundance of food, nutrients, pH, temperature, presence or absence of oxygen, and toxic substances. Given the presence of optimal conditions, bacteria can grow in logarithmic proportions. A

Table 1.2 Typical composition of bacterial cells (Adapted from Metcalf & Eddy/AECOM 2014)

Constituents	Percent of dry weight (%)
Carbon	50
Oxygen	22
Nitrogen	12
Hydrogen	9
Phosphorus	2
Sulfur	1
Potassium	1
Sodium	1
Calcium	0.5
Magnesium	0.5
Chlorine	0.5
Iron	0.2
Other trace elements	0.3

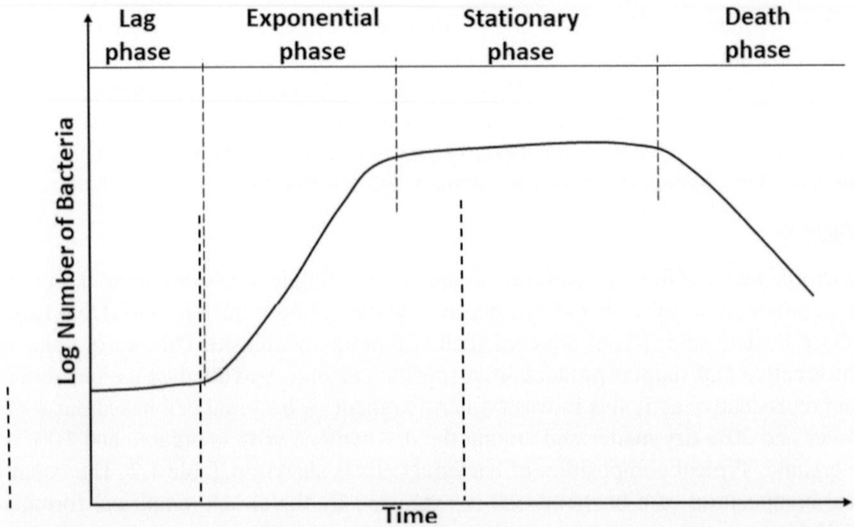

Fig. 1.7 Typical Bacterial Growth Curve Showing Different Phases of Growth and Decay in a Batch Test

batch experiment with a limited amount of food or substrate can produce a bacterial growth curve similar to the one illustrated in Fig. 1.7.

Archaea

Both archaea and bacteria are structurally similar. Archaea differ from bacteria based on their cell wall, cell material and RNA composition. They have a highly specialized function in wastewater treatment. Methanogenic archaea, such as methanobacterium, are dominant in anaerobic processes such as anaerobic digestion of sludge which produces methane gas from degradation of organic wastes.

Eukarya

Protozoa are unicellular, microscopic eukaryotes, ranging in size from a few to several hundred microns. The presence of aerobic protozoa in biological wastewater treatment indicates a well operating process.

Fungi are multicellular eukaryotes, which are non-photosynthetic, aerobic heterotrophs. They can degrade more recalcitrant organics like cellulose and therefore are more important in industrial waste treatment.

Algae can be unicellular or multicellular photosynthetic eukaryotes. Their size ranges from 5 μm to more than 100 μm, (Fig. 1.8) when they are visible as a green slime on water surfaces, e.g., *Pediastrum*. They produce oxygen in lagoons and ponds and can also remove soluble nitrogen and phosphorus by nutritional uptake.

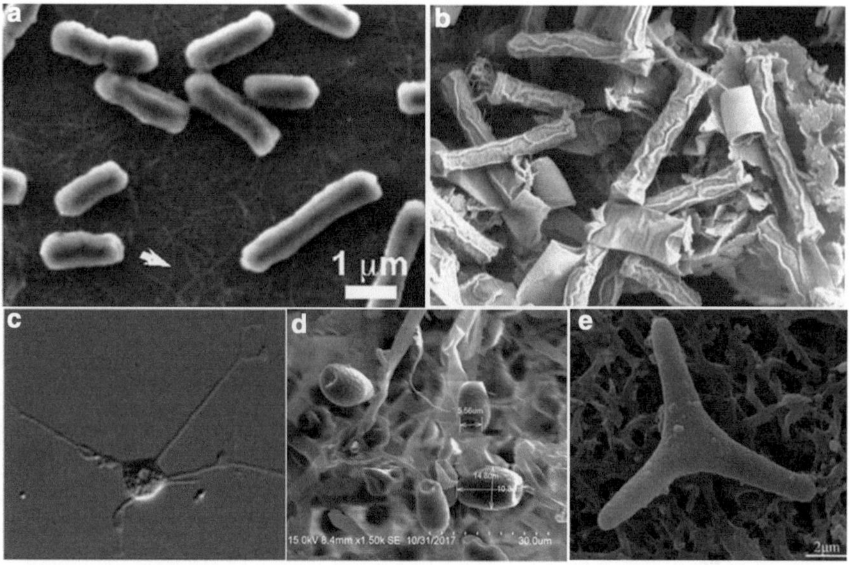

Fig. 1.8 Scanning electron microscopy images of A. Bacteria ***Salmonella Senftenberg*** (Golding et al. 2016) B. Methanotrophic archaea with individual cells size about 0.5 μm in diameter (Thiel. 2011) C. Protozoa ***Paracercomonas sp*** (Boenigk. 2011) D. ***Curvularia lunata*** fungii spores E. Algae ***N. closterium***

1.8.1.1 Classification of Microorganisms Based on Their Functions

Microorganisms play very important roles in maintaining human health, environment, and ecological balance. Bacteria are the most important players in the biological stabilization of wastewater. Treatment of wastewater is conducted by bacteria leading to biochemical oxidation, reduction and conversion to simple end products, as well as the production of new cells. The primary consumers of organic wastes are heterotrophic bacteria such as *Pseudomonas*. Aerobic nitrifying bacteria, such as *Nitrobacter* and *Nitrosomonas*, are responsible for nitrification whereas anoxic denitrifying bacteria such as *Pseudomonas, Bacillus* etc. can reduce nitrate to gaseous nitrogen through denitrification in an anoxic environment. For phosphorus removal, biological processes are configured for the growth of phosphorus-accumulating organisms (PAO) such as *Acinetobacter*, which can accumulate and store inorganic phosphorus.

While microorganisms contribute immensely to ensure the quality of life of the plant and animal kingdom, and also to the environment as described above, certain types are harmful to human health. One of the major concerns is the transmission of diseases by pathogenic microorganisms. Various types of pathogens that present a major health risk include enteric bacteria, viruses, parasitic protozoa, and helminths (Abdel-Raouf et al. 2012). Fecal matter is a major component of domestic sewage and the source of the majority of human pathogens in wastewater (Symonds and Breitbart

2014). Enteric pathogens enter the wastewater environment from human and animal fecal waste or fecal-contaminated water from other household uses. Industrial waste from food production, particularly from animal processing, can also be a source of pathogenic microorganisms. Water and wastewater from petroleum exploration and production operations, and refinery processing do not contain any pathogenic organisms but the sanitary wastes generated on-site from round-the-clock operation and maintenance of these facilities do contain pathogens that need proper disinfection prior to discharge.

Bacteria constitute the most diverse group of human pathogens in wastewater. Major human bacterial pathogens in wastewater include *Salmonella., Escherichia, Shigella, Yersinia, Klebsiella, Leptospira, Vibrio cholerae, Aeromonas hydrophila, Legionella pneumophila, Mycobacterium, and Pseudomonas* (Cai and Zhang 2013; Kristian Stevik et al. 2004; Maynard et al. 2005). These are hard to detect, isolate and identify in water environments, for which indicator organisms are commonly tested for target pathogens. Therefore, coliform bacteria have been accepted as an indicator for pathogenic organisms associated with feces. The absence of coliform bacteria is indicative of the water under consideration being free from disease-producing organisms. The coliform group consists of several genera of bacteria belonging to the family *Enterobacteriaceae*, and is defined as all facultative anaerobic, gram-negative, non-spore-forming, rod-shaped bacteria that ferment lactose with gas and acid formation within 48 hours at 35 °C.

E. coli (*Escherichia coli*) is a member of the indigenous fecal flora of warm-blooded animals. The presence of *E. coli* is considered a specific indicator of fecal contamination and the possible presence of enteric pathogens. Thus, it is used as a test organism for total coliform detection. Fecal coliform, also known as thermotolerant coliforms, ferment lactose to produce gas at 44.5 °C.

Certain types of *Eukarya* are also pathogenic. These include *cryptosporidium* and *Giardia* which are pathogenic protozoa that can pose problem due to their environmental persistence and their resistance to chlorine (*Cryptosporidium* in particular). There is another prokaryotic microorganism—the blue-green algae or *cyanobacteria* that can produce toxins harmful to aquatic life.

From the above discussions it can be concluded that microorganisms, even with their largely beneficial effects, include certain types that can cause adverse health effects. These need to be identified and the health risks be mitigated by implementing proper disinfection measures.

Some of the important groups of bacteria are classified in Table 1.3 based on their functions.

Table 1.3 Important groups of bacteria in water and wastewater (Adapted from Metcalf & Eddy/AECOM 2014)

Bacteria	Genus	Functions
Indicator bacteria	Escherichia, Enterobacter	Indicator of fecal pollution
Pathogenic bacteria	Salmonella	Salmonellosis causing
	Vibrio cholera	Cholera causing
Organics Oxidizing Bacteria	Wide range of heterotrophs	Oxidation of organic pollutants in wastewater
Nitrifying bacteria	Nitrosomonas	Oxidation of ammonia to nitrite
	Nitrobacter	Oxidation of nitrite to nitrate
Denitrifying bacteria	Pseudomonas, Bacillus	Reduction of nitrate to nitrogen gas
Phosphorus accumulating organism	Acinetobacter	Accumulation and storage of polyphosphate
Iron bacteria	Leptothrix, Crenothrix	Oxidation of ferrous iron to ferric iron
Sulfur oxidizing bacteria	Thiobacillus	Oxidation of sulfur and iron
Sulfur reducing bacteria	Thiobacillus, Pseudomonas	Reduction of sulfate to sulfide and hydrogen sulfide
Photosynthetic bacteria	Chromatium, Chlorobium	Reduction of sulfide to sulfur

1.8.1.2 Classification of Microorganisms Based on Metabolism and Energetics

For metabolic function and production of new cell materials, microorganisms require the supply of carbon and energy. Microorganisms in water and wastewater environments can thus be classified based on their source of cell carbon and source of energy for their growth, as depicted in Table 1.4.

Microorganisms can use either organic, or inorganic matter as their source of carbon for cell growth. Those which utilize organic compounds as their sole or primary carbon source are called heterotrophic organisms; whereas organisms that use inorganic species such as CO_2 or HCO_3^- as their carbon source are called autotrophic organisms.

Along with a source of carbon, microorganisms additionally require an energy source for cell synthesis, maintenance and motility. They can acquire the required energy through sunlight or oxidation of organic or inorganic compounds. Phototrophs are organisms that utilize light energy. Phototrophic organisms can be either heterotrophic (ones that reduce sulfur) or autotrophic (ones capable of photosynthesis). Chemotrophs obtain energy from oxidized compounds and can also be either heterotrophic or autotrophic. Chemoautotrophs use oxidation of inorganic compounds such as ammonia, nitrite, sulfide etc. for energy whereas chemoheterotrophs obtain energy from the oxidation of organic compounds.

Table 1.4 Broad metabolic classification of microorganisms in terms of carbon source, electron donor and acceptor (Adapted from Metcalf & Eddy/AECOM 2014)

Type of organism	Reaction name	Carbon source	Electron donor	Electron acceptor	End product
Chemoheterotrophic	Aerobic oxidation	Organic	Organic	O_2	CO_2, H_2O
	Denitrification	Organic	Organic	NO_2^-, NO_3^-	CO_2, H_2O, N_2
	Anaerobic fermentation	Organic	Organic	Organic	Volatile fatty acids (VFA)
	Anaerobic methanogenesis	Organic	Organic (VFA)	CO_2	Methane
	Sulfate reduction	Organic	Organic	SO_4^{2-}	H_2S, CO_2, H_2O
	Ferric reduction	Organic	Organic	Fe^{3+}	Fe^{2+}, CO_2, H_2O
Chemoautotrophic	Nitrification	CO_2	NH_3^-, NO_2^-	O_2	NO_2^-, NO_3^-
	Iron oxidation	CO_2	Fe^{2+}	O_2	Fe^{3+}, H_2O
	Sulfur oxidation	CO_2	H_2S, S^0	O_2	SO_4^{2-}, H_2O
Photoautotrophic	Algae	CO_2	H_2O	O_2	Organic

1.8.1.3 Classification of Microorganisms Based on Oxygen Requirement

As chemotrophs produce energy through oxidation-reduction reactions, electrons are transferred from a donor and accepted by an acceptor. Organisms that use molecular oxygen as the electron acceptor are aerobic and ones which need other electron acceptors are anaerobic. The term anoxic is used when organisms use combined oxygen (such as nitrite or nitrate) as an electron acceptor when molecular oxygen is not available. Facultative organisms are those which can switch their respiration mechanism between aerobic and anaerobic modes back and forth depending on the available environment. Additionally, organisms that use organic compounds as electron donors are called organotrophs, while those that use inorganic compounds are called lithotrophs.

In summary, based on the oxygen requirements and source of oxygen, the microorganisms are classified as aerobic, anaerobic, anoxic, and facultative. The biochemical reactions governing these types of processes are as follows:

Aerobic Heterotrophic:

$$\text{Organic Compounds (substrate)} + O_2 \rightarrow CO_2 + H_2O + \text{Growth of New Cells} + (-\Delta E) \quad (1.56)$$

For domestic wastewater (sewage) the empirical chemical formula for the organic substrate is $C_{10}H_{19}O_3N$. The empirical chemical formula for the new cells is $C_5H_7O_2N$. The term (-ΔE) represents the heat (energy) of reaction.

Refinery wastewaters contain a combination of various organic and inorganic chemical compounds derived from crudes and additives that are used for the exploration and processing of petroleum. The types of such compounds have been discussed in detail in Chap. 3. All of these organic substrates, regardless of their origin—domestic or industrial, are lumped together as biochemical oxygen demand (BOD) and chemical oxygen demand (COD). This topic is discussed in detail in a later section of this chapter.

Aerobic Autotrophic: The most common set of reactions mediated by aerobic autotrophs are nitrification of ammonia. It is a two-step biochemical oxidation process:

$$\text{(a) Ammonia} - \text{N (substrate)} + O_2\text{(by Ammonia Oxidizing Bacteria, AOB)} \rightarrow NO_2^-\text{(Nitrite)} \quad (1.57)$$

$$\text{(b) } NO_2^-\text{(Nitrite)} + O_2\text{(by Nitrite Oxidizing Bacteria, NOB)} \rightarrow NO_3^-\text{(Nitrate)} \quad (1.58)$$

Oxidation reactions are exothermic, and microorganisms utilize the energy evolved from the reactions to maintain their mobility and metabolic activities.

Anaerobic: These microorganisms do not require molecular oxygen for growth and survival; rather their growth and survival opportunities are inhibited in the presence of oxygen. Anaerobic oxidation is a four-step biochemical process in series.

(a) Hydrolysis of Complex Organic Compounds (substrate) → Simple Organic Molecules
(b) Fermentation by acid formers → Volatile Fatty Acids (VFA) (acetic, propionic, butyric and valeric) + H_2
(c) Acetogenesis → Formation of acetic acid and conversion of amino acids to ammonia
(d) Methanogenesis of acetic acid and methane synthesis from CO_2 and H_2 → CH_4 + CO_2 (Biogas) (1.59)

Anoxic Heterotrophic: These microorganisms do not require molecular oxygen and they use bound oxygen in NO_3^- and NO_2^- ions formed by autotrophic oxidation of ammonia. Nitrate and nitrite play the role of electron acceptors and organic substrates (BOD, COD) act as electron donors.

$$\text{Organic Compounds (substrate BOD or COD)} + NO_2^-/NO_3^{--} \rightarrow N_2 + CO_2 + H_2O + \text{Growth of New Cells} + (-\Delta E) \quad (1.59)$$

Biochemical reactions described above are associated with the respiration by the microorganisms. The heats of reaction (–ΔE) vary widely between the types of reaction, with heterotrophic oxidation producing the maximum amount of energy. Their respective order is:

Heterotrophic Oxidation > Heterotrophic Denitrification > Autotrophic Nitrification > Anaerobic Oxidation.

As the microbiological population in wastewater treatment processes derive their energy for their growth from the oxidation reactions the rates of growth also are in the decreasing order from heterotrophic oxidation to anaerobic reactions. The growth rates of anaerobic microorganisms, especially the methanogenic bacteria are orders of magnitude lowers than those carrying out heterotrophic oxidation reactions.

1.8.1.4 Role of Oxidation Reduction Potential (ORP) in Energetics of Biochemical Reactions

In biological treatment processes ORP gives an indication of whether a system is aerobic, anaerobic, or anoxic as ORP values increase with increasing dissolved oxygen concentration (DO) and decrease with declining DO concentration. ORP ranges for a few of the important biochemical reactions relevant to water and wastewater treatment processes are presented in Fig. 1.9.

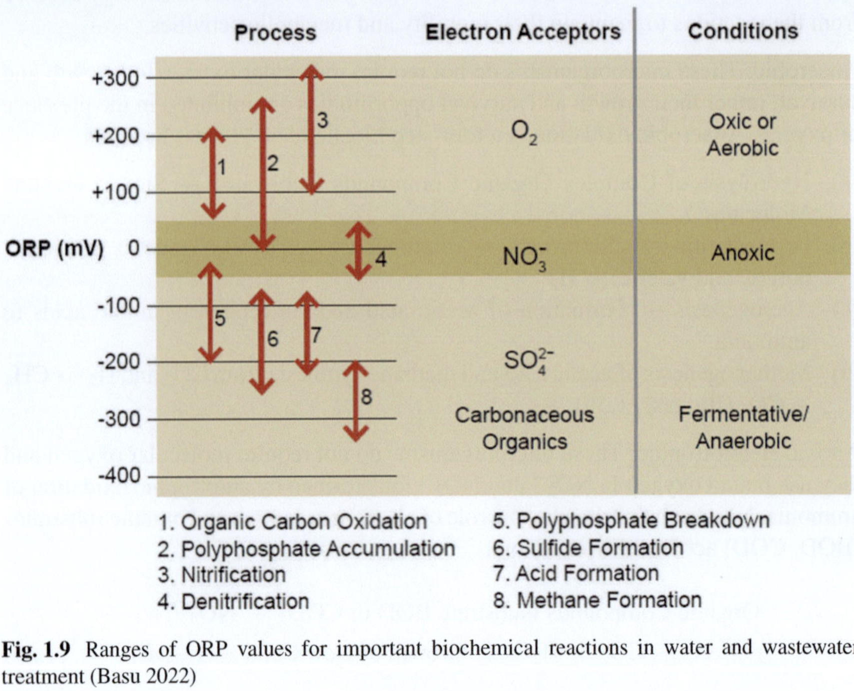

Fig. 1.9 Ranges of ORP values for important biochemical reactions in water and wastewater treatment (Basu 2022)

1.8.1.5 Classification Based on Active Temperature Ranges for Growth

Microorganisms (mostly bacteria) are further classified based on specific temperature ranges for their activity and growth. These are grouped as follows:

Psychrophilic- grows at temperatures between 12 and 18 °C.

Mesophilic- grows between 25 and 40 °C.

Thermophilic- grows between 50 and 75 °C.

1.9 Application of Biochemical and Chemical Methods for Characterization of Organic Constituents in Wastewater

Organic compounds in dissolved and suspended states form a large part of water and wastewater constituents. The complex matrix of organic matters derived from natural sources of water is described as Natural Organic Matter (NOM). NOM is present in both surface and groundwater that originates from the biological decomposition of organic matter in dead plant and animal bodies. Humic and fulvic acids are two large groups of NOM compounds present in surface water and groundwater. Presence of NOM can impact water quality parameters including color and odor and can also produce disinfection by-products (DBP) in treated wastewater by reacting with chlorine. In water quality analysis, NOM is most commonly indicated by total organic carbon (TOC) measurements.

The common analytical parameters for measuring the bulk concentration of organic matter in wastewater are biochemical oxygen demand (BOD), chemical oxygen demand (COD) and total organic carbon (TOC). Total oxygen demand measures oxygen demand (both organic and inorganic) by organic compounds, reduced nitrogenous compounds, reduced sulfur compounds and those by other inorganics.

1.9.1 Biochemical Oxygen Demand (BOD)

BOD measurement is a standardized laboratory procedure used to measure the molecular oxygen utilized during a specified incubation period for the biochemical degradation of an organic material. The 5-day BOD (BOD_5) is the most commonly used parameter to denote the biodegradable fraction of an organic waste in wastewater or other aquatic environments. Standard BOD test (BOD_5) is conducted at 20° C (Fig. 1.10).

Noncarbonaceous matter, such as ammonia, can also exert an oxygen demand as nitrifying microorganisms perform nitrification—the oxidation of ammonia to nitrite

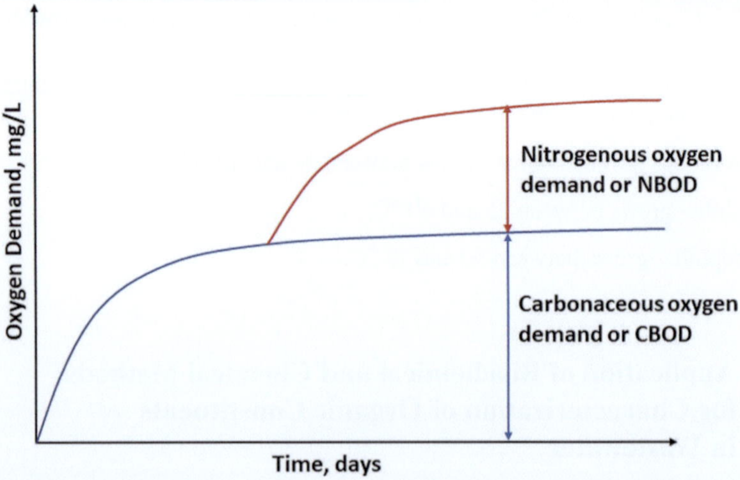

Fig. 1.10 Typical BOD curves denoting CBOD and NBOD

and subsequently to nitrate. Thus, the oxygen demand associated with the oxidation of ammonia is called the nitrogenous biochemical oxygen demand (NBOD). If nitrification is suppressed, by using a nitrification inhibitor, the resulting BOD is known as the carbonaceous biochemical oxygen demand (CBOD).

The amount of organic matter decays exponentially with time following the first order biochemical reaction kinetics, as shown in Fig. 1.11. Since L_o is the oxygen equivalent of the total amount of organic matter, the amount of oxygen used in the degradation of organic matter, or the BOD, can be determined from the L_t value, as expressed below.

$$BOD_t = L_o - L_t = L_o - L_o e^{-kt} = L_o\left(1 - e^{-kt}\right) \quad (1.60)$$

Oxidation of organic contents in wastewater should theoretically take an infinite amount of time, as it is a function of the amount of organic content remaining. However, it has been observed that within 20 days 95–99 percent of organics get oxidized, making this the 20-day or ultimate BOD (UBOD or L_o). The terms 't' and 'k' in the above equation denote time and first-order rate constant, respectively. The values of k range between 0.15 and 0.28 day^{-1} for untreated wastewater and between 0.06 and 0.10 day^{-1} for highly treated wastewater (Eckenfelder, et al. 2009).

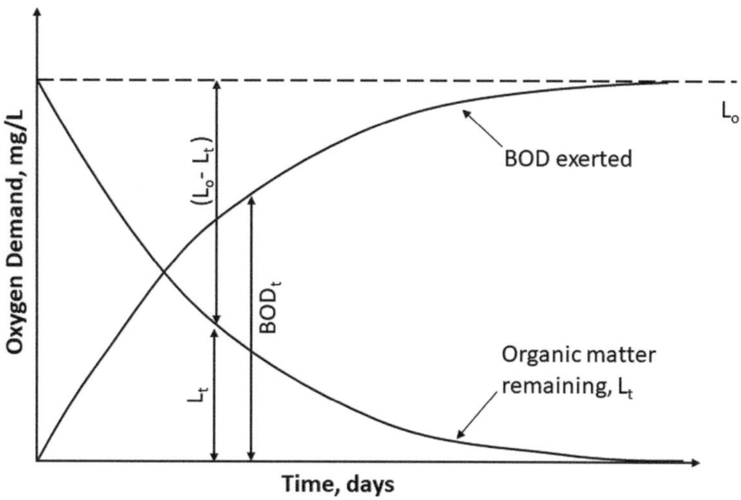

Fig. 1.11 Relationship between organic matter remaining with exertion of BOD

1.9.2 Chemical Oxygen Demand (COD), Theoretical Oxygen Demand (ThOD) and Total Oxygen Demand (TOD)

COD is a measure of the oxygen equivalent of the organic fraction in a water sample that can be oxidized by potassium dichromate in an acid solution. The benefit of the COD procedure is that it is not subject to as many interferences as BOD_5 and therefore is more time efficient. However, COD test results are also subject to certain uncertainties. For example, although COD measures both biodegradable and non-degradable portions of a sample, it may underestimate some readily biodegradable aromatics e.g., benzene, toluene, pyridine, etc., and aliphatic organics e.g., acetic acid, as these are not effectively oxidized by dichromate. Therefore, the presence of these types of organic compounds may lead to underestimation of COD. On the other hand, chloride ions, which are usually present in many water and wastewater streams in the petroleum industry, are oxidized by dichromate ions along with organic substances, and therefore, cause the COD value to be overestimated. This interference can be mitigated to some extent by using mercuric sulfate ($HgSO_4$). Mercury reacts with chloride ions and forms a very stable compound mercuric chloride which has poor solubility and therefore, does not dissolve under the reaction conditions, and thus removes the chloride ions from water that can lead to overestimation of COD. Also, other reduced inorganic species e.g., ferrous (Fe^{2+}), sulfide (S^{2-}) and sulfite (SO_3^{2-}) ions are oxidized by potassium dichromate in the COD test procedure and will account for the test results. It also must be noted that COD test conditions do not oxidize ammonia, and therefore COD results are not subject to interference by the presence of ammonia.

COD measurement can quantify both biodegradable and non-biodegradable organics.

As shown in Fig. 1.12, COD can be broadly classified into particulate and soluble forms. Soluble form can be further classified as—(a) readily biodegradable (rbsCOD) and (b) non-biodegradable soluble COD (nbsCOD). The soluble readily biodegradable COD (rbsCOD) consists of the small molecules such as volatile fatty acids, alcohols, amino acids, simple sugars etc. that are directly available for biodegradation by heterotrophic microorganisms. On the other hand, treated effluent may consist of non-biodegradable, yet soluble organics generated from the treatment process such as the soluble microbial product (SMP). SMP are defined as the pool of organic compounds that are released into the solution from substrate metabolism, usually with biomass growth and biomass decay. Typically, SMP accounts for 5% of the COD removed by biological treatment (Eckenfelder 2009) which represents a large part of the nonbiodegradable soluble COD (nbsCOD) in the treated effluent.

Particulate COD in wastewater arises from larger molecules and colloidal organic matters like carbohydrate and protein molecules that require structural breakdown before biodegradation. These can be subdivided into two fractions—slowly

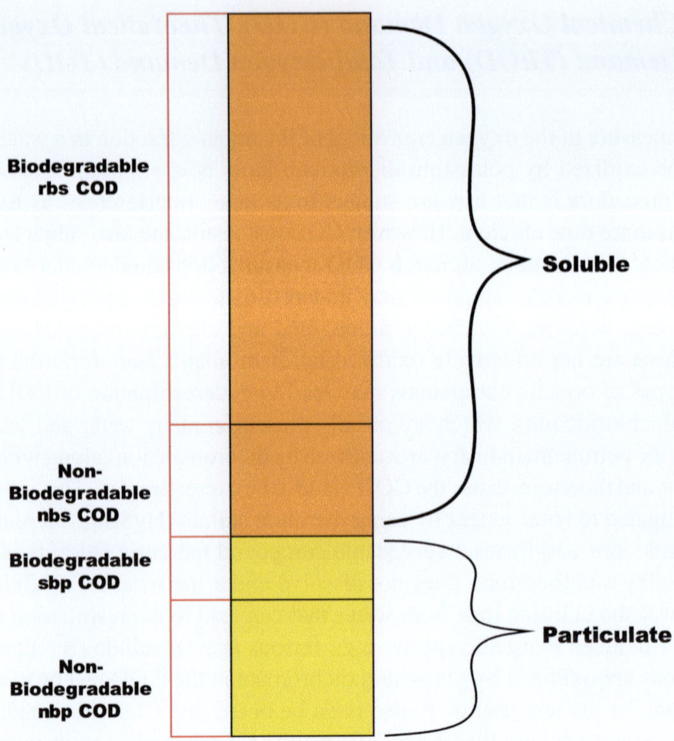

Fig. 1.12 COD partitioning of influent wastewater (Adapted from Marrs 2012)

biodegradable particulate COD (sbpCOD) and non-biodegradable particulate COD (nbpCOD).

The relative distribution of COD fractions changes during the course of biological treatment of wastewater. The degradable fractions rbsCOD and sbpCOD of the influent are consumed, and new cells are formed as biosolids. Figure 1.13 depicts the distribution of different fractions of COD in the treated effluent, which also takes into account the solids formed by the biodegradation of parts of influent COD. Another

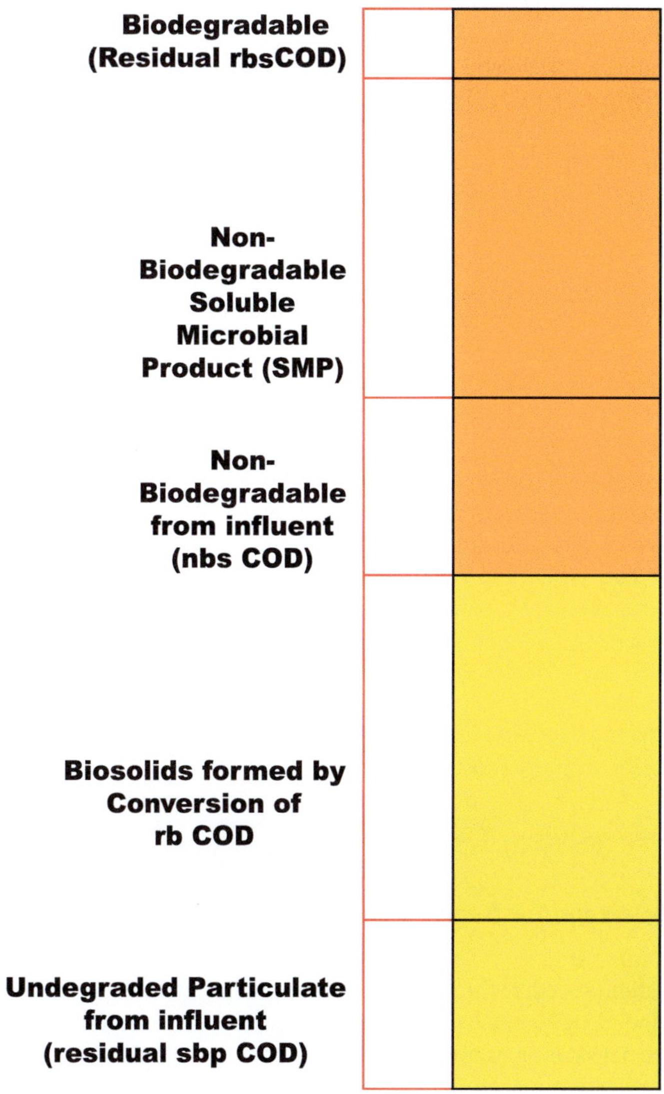

Fig. 1.13 Distribution of COD fractions in the effluent from biological treatment of wastewater

important empirical correlation is between biodegradable COD and the ultimate BOD of refinery wastewater. It is estimated as:

$$BOD_{ult} = 0.92 * \text{Biodegradable COD} \tag{1.61}$$

Theoretical oxygen demand (ThOD) is the theoretical estimation of oxygen demand based on the chemical formula of organic matter. It is the summation of the amount of O_2 required to completely convert all organic and inorganic compound to CO_2, H_2O, NH_3 and H_2SO_4 based on the stoichiometry of oxidation of individual compounds.

As mentioned earlier, apart from the organic compounds, various reduced inorganic species e.g., ferrous (Fe^{2+}), sulfide $\left(S^{2-}\right)$, sulfite $\left(SO_3^{2-}\right)$ ions also demand oxygen for their stabilization. Total Oxygen Demand (TOD) is the sum of the ThOD of the organic compounds and those exerted by the reduced inorganic species. Therefore, TOD is a larger number than ThOD.

1.9.3 Total and Volatile Organic Carbon (TOC and VOC)

Total Organic Carbon (TOC) is the total quantity of organic carbon in a sample. Analysis of TOC utilizes high temperature, catalysts and oxygen, or lower temperatures (<100°C) with ultraviolet irradiation, chemical oxidants, or combinations of these oxidants to convert organic carbon to carbon dioxide (CO_2). This method is quicker and has fewer interferences than BOD and COD measurements. It can also be adapted to acquire real-time data for water and wastewater treatment continuous monitoring purposes. TOC test results can also be used to estimate the BOD_5 or COD. An empirical relationship between TOC, BOD_5 and COD can be established for a given sample, although the relationship will be site-specific.

The parameter Volatile Organic Compounds (VOC) evaluates the types and concentrations of low-boiling-point organic materials present in a sample. USEPA SW 846 Method 3240 provides the extraction and analysis protocols. This method is applicable to quantify most of the organic compounds with boiling points below 200°C.

1.9.4 Interrelation Between Aggregate Organic Parameters

The calculation procedure for ThOD of a few common organic compounds and that of H_2S as an inorganic compound in wastewaters originating from petroleum industries is presented below as an example. TOC values of those organic compounds are also presented.

Example 3: ***Calculations for ThOD and TOC for some organic compounds commonly present in water and wastewater streams in petroleum industries***
n-Octane (C_8H_{18})

$C_8H_{18} + 12.5\ O_2 \rightarrow 8\ CO_2 + 9H_2O$.

MW of $C_8H_{18} = 114$.

MW of $O_2 = 32$.

ThOD = 12.5*32/114, i.e. 3.5 mg of O_2/mg of Octane.

TOC = 8*12/114, i.e. 0.84 mg of C/mg of Octane.
Benzene (C_6H_6)

$C_6H_6 + 7.5\ O_2 \rightarrow 6\ CO_2 + 3H_2O$.

ThOD = 3.1 mg of O_2/mg of Benzene.

TOC = 0.92 mg of C/mg of Benzene.
Monoethanol Amine (MEA, C_2H_7ON).

$C_2H_7ON + 2.5\ O_2 \rightarrow 2CO_2 + 2H_2O + NH_3$.

ThOD = 1.31 mg of O_2/mg of MEA.

TOC = 0.39 mg of C/mg of MEA.
Methyl Mercaptan (CH_4S)

$CH_4S + 3.5\ O_2 \rightarrow CO_2 + H_2O + H_2SO_4$.

ThOD = 2.33 mg of O_2/mg of Methyl Mercaptan.

TOC = 0.25 mg of C/mg of Methyl Mercaptan.

Hydrogen Sulfide (H_2S)

$H_2S + 2O_2 \rightarrow H_2SO_4$.

ThOD = 1.88 mg of O_2/mg of Hydrrogen Sulfide

Biomass in Wastewater Treatment (Empirical Formula $C_5H_7O_2N$)

$C_5H_7O_2N + 5\ O_2 + H^+ \rightarrow 5CO_2 + 2H_2O + NH_4^+$

ThOD = 1.42 mg of O_2/mg of Biomass.

TOC = 0.53 mg of O_2/mg of Biomass.

In the absence of analytical data for COD the Theoretical Oxygen Demand (ThOD) can serve as a good indicator for COD. Typically, for most of the organic contaminants present in refinery wastewaters COD is approximately 90 to 95% of ThOD. Note that the TOC of H_2S is zero as it is not an organic compound, and carbon is absent in its molecules. A summary of COD and BOD_5 with respect to ThOD of known refinery wastewater constituents is listed in Table 1.5.

Table 1.5 Typical BOD_5:COD ratios denoting degradability of refinery wastes (Marrs 2012)

BOD_5:COD	Types of wastewater and their biodegradability
1.0	Completely and readily biodegradable wastewater
0.35–0.55	Untreated refinery wastewater
0.05–0.15	Treated refinery effluent
0.00	Not biodegradable, refractory organics

A simple example of determination of influent soluble degradable COD from influent and effluent total COD is provided below.

Example 4: ***Calculations for fractions of petroleum refinery influent and effluent COD***

Influent and Effluent Analytical Data:

Influent total COD = 500 mg/L.

Influent soluble COD = 450 mg/L.

Effluent total COD = 150 mg/L.

Effluent soluble COD = 120 mg/L.

Effluent soluble BOD_5 = 5 mg/L.

What is the influent soluble biodegradable COD?

$BOD_t = BOD_{ult}\ (1–10^{-kt})$.

For k ~ 0.08 day^{-1} for treated wastewater (Eckenfelder et al. 1997) and t = 5 days.

$(1–10^{-kt}) = 0.60$.

$BOD_{ult} = BOD_5/(0.60)$.

Effluent Soluble BOD_5 = 5 mg/L.

Effluent Soluble BOD_{ult} = (5 mg/L)/0.60 = 8.30 mg/L.

For biodegradable fractions:

BOD_{ult} = 0.92*Soluble Biodegradable COD (Eckenfelder et al. 1997).

Therefore, effluent soluble biodegradable COD (rbsCOD) = (8.30 mg/L)/0.92 = 9.0 mg/L.

COD of Soluble Microbial Product (SMP) ~ 5% of COD removed biologically (Eckenfelder et al. 1997).

Influent soluble COD = 450 mg/L.

Effluent soluble COD = 120 mg/L.

Therefore, COD of effluent SMP = 0.05 (450–120) mg/L = 16.5 mg/L.

Influent non-biodegradable soluble COD (nbsCOD) =

Effluent soluble COD—COD of SMP—Effluent soluble biodegradable COD.

= 120—16.5—9.0 mg/L = 94.5 mg/L.

Influent non-biodegradable soluble COD = Effluent non-biodegradable soluble CD = 94.5 mg/L.

Influent biodegradable soluble COD (rbsCOD) = Influent soluble COD—Influent non-degradable soluble COD.

= 450—94.5 mg/L = 355.5 mg/L.

Influent rbsCOD is a very important parameter for the design and operation of a biological wastewater treatment plant. This analysis was possible when the influent and effluent total and soluble COD data and effluent soluble BOD_5 data were available.

In the absence of such information the best guidance for the design of a new biological treatment reactor is to use some standard ratios of the key parameters BOD_5, COD, ThOD and TOC. The typical BOD_5:COD ratio of refinery wastewater is ~ 0.5:1. Table 1.5 shows the BOD_5:COD ratio as a function of the type of refinery wastewater. The relationships between the type of wastewater and their BOD_5:TOC ratios are presented in Table 1.6.

Estimates of correlations between COD and ThOD, and BOD_5 and ThOD of a few representative organic compounds in refinery wastewaters are listed in Table 1.7. The numbers presented in these Tables are the best estimates and vary over wide ranges from site to site. These can be used as first estimates, but actual data should be established from comprehensive sampling and analytical programs.

As can be observed from Table 1.7 oxidative breakdown of organic compounds vary widely. This is dependent on the chemical structure of specific compounds. Those with resonance structures are less degradable than those with polar structures. The relative magnitudes of various types of oxygen demand by organic and nitrogenous compounds are depicted in Fig. 1.14.

Table 1.6 BOD_5:TOC Ratio for influent and effluent of typical domestic sewage and refinery/chemical wastewater (Reimers 2019)

BOD_5:TOC	Influent	Effluent
Domestic Sewage	1.62	0.47
Refinery/chemical wastewater	2.75	0.43

Table 1.7 Evaluation of COD and BOD_5 with respect to ThOD (Adapted from Eckenfelder 2009)

Chemical name	Chemical formula	ThOD mg/mg	COD/ThOD (%) (Estimated)	COD mg/mg	BOD_5/COD (%) (Estimated)	BOD_5 mg/mg
Aliphatics						
n-heptane	C_7H_{16}	3.52	95	3.34	80	2.68
n-octane	C_8H_{18}	3.51	95	3.33	80	2.66
Methyl Ethyl Ketone	C_4H_8O	2.44	90	2.20	82	1.80
Aromatics						
Benzene	C_6H_6	3.08	50	1.54	65	1.00
Toluene	C_7H_8	3.13	45	1.41	61	0.86
Benzaldehyde	C_6H_5CHO	2.42	80	1.94	82	1.59
Benzoic Acid	C_6H_5COOH	1.96	100	1.96	74	1.45
Organic nitrogenous compounds						
Monoethanol amine	C_2H_7NO	2.49	51	1.27	65	0.83

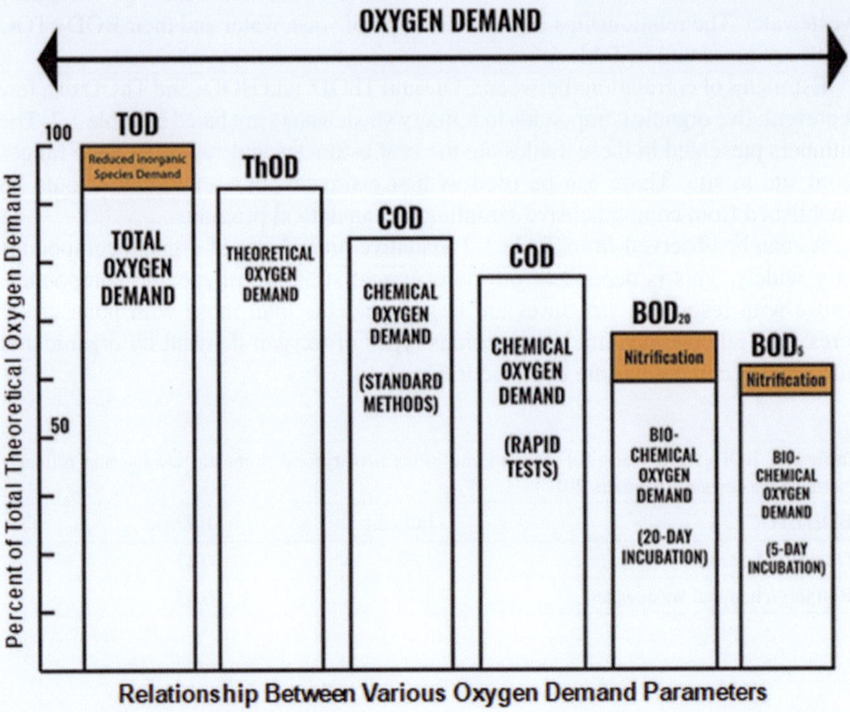

Fig. 1.14 Various types of oxygen demand by organic and nitrogenous constituents in wastewater (Adapted from Reimers 2019)

1.9.5 Oil and Grease

Oil and grease (O&G) are among the most widespread contaminants in water and wastewater streams originating from petroleum industries. Removal of O&G is essential for the effective treatment of water and wastewater streams to make them suitable for disposal and/or reuse. It requires a basic understanding of the chemistry of O&G to design and operate their treatment systems as a part of the overall water and wastewater treatment.

Oil and grease are not specific compounds, rather they represent groups of compounds with similar physical and chemical interactions with water. Compounds falling under the oil and grease category include free phase, emulsified and dissolved oil in water that can be separated from water by extraction by an organic solvent.

A prerequisite to O&G analysis is to separate the organic compounds from the aqueous phase, followed by their analysis. The accuracy of quantitative analytical results depends significantly upon the separation performance of the organics from water, which in turn depends upon their physical and chemical properties. An important property is the polarity of the molecules. The compounds that exist dispersed in water—both in the fee and emulsified phases, are non-polar. However, dissolved O&G consists of several groups of chemicals with varying polar characteristics. The aromatic compounds, except for the phenolics, are non-polar, whereas the aliphatic and naphthenic compounds, and the phenolic compounds in the aromatic group are all polar compounds.

Separation and concentration of organics from water samples is done by liquid–liquid extraction using n-hexane as the solvent following the Hexane Extraction Method (HEM) (EPA Method 1664 B). When the extract phase from HEM is treated with silica gel the polar compounds are removed by silica gel leaving behind the non-polar compounds in the extract phase. Thus, the silica gel—HEM (SGT-HEM) combination extracts organics excluding polar compounds. An alternative to the procedure of extraction by n-hexane is the use of solid-phase extraction (SPE) which uses solid adsorbent disks for extraction of organics. In summary, HEM or SPE can be used for the extraction of all types of organic compounds before their quantitative analysis for oil and grease. When the analyses of only non-polar compounds are desired combinations of SGT-HEM or SGT-SPE are used. Therefore, it is necessary to select the right method depending on the type of analysis desired for O&G from a water sample. Following this sample preparation step the extracted organic phase is quantified by a two-step procedure—(a) distilling out the solvent from the organic extract phase, (b) weighing the amount (W) of oil and grease left behind after all the solvent is distilled out, and (c) dividing the amount of separated oil and grease (W) by the volume of the sample (L), such that the concentration of oil and grease in the water sample = W/L mg/L.

1.9.6 Total Petroleum Hydrocarbon (TPH)

For waters in close contact with petroleum crudes and products Total Petroleum Hydrocarbon (TPH) is a very commonly used aggregate organic parameter to quantify the level of contamination. TPH is defined as the measurable amount of petroleum-based hydrocarbon in an environmental medium e.g., water, soil, sediment, etc. The term is used to describe several hundred petroleum compounds that come from crude oil, or products derived from it. TPH can be measured with EPA Method 418.1. Some chemicals that may be found in TPH are hexane, jet fuels, mineral oils, benzene, toluene, xylenes, naphthalene, fluorene, as well as other petroleum products and gasoline components. The TPH value can be a result from one of several analytical methods. The reliability of the TPH analytical results by method 418.1 depends largely on the effectiveness of the solvent in extracting the hydrocarbons from the water samples, and the absorption of infrared (IR) light by the organic extract phase. It must be noted that TPH analysis does not necessarily differentiate between hydrocarbons of petroleum versus non-petroleum origin. For example, humic acids in water samples, which are not hydrocarbons of petroleum origin, may also be detected by EPA Method 418.1. EPA Method 8440 is used for the measurement of total recoverable petroleum hydrocarbons (TRPH) extracted with supercritical carbon dioxide from sediment, soil and sludge samples using Method 3560.

In addition to the aliphatic and aromatic hydrocarbons, a few other groups of organic compounds containing phenols, phthalates and naphthenic acids are commonly present in petroleum refinery wastewater. **Phenols** are hydroxy derivatives of benzene and its condensed nuclei. The chloroform extraction method and direct photometric analytical method report phenol concentration in the form of C_6H_5OH since the relative amounts of various phenolic compounds in a given sample are unpredictable (Standard Methods 9065). Naphthenic acids and their impact on wastewater microbiology are discussed in Chap. 5.

1.10 Biodegradability of Wastewaters: Oxygen Uptake Rate (OUR), Specific Oxygen Uptake Rate (SOUR) and Die-Away Test Methods

Microorganisms consume oxygen as they break down available substrate or BOD in an aerobic aquatic system. Aerobic biological waste treatment process is based on the ability of the microorganisms to utilize dissolved oxygen in breaking down soluble organic substances. The rate at which oxygen is consumed by the microorganism is an indicator of the biological activity of the system and is called the **Oxygen Uptake Rate (OUR)**. OUR is used as a tool to evaluate the performance of process units, aeration equipment and biodegradability of the waste in treatment plants. Determination of OUR involves measuring the dissolved oxygen (DO) of the sample over a certain period and recording the depletion of DO in mg/L per hour, a method more commonly known as respirometry. Respirometric methods provide direct measurement of the oxygen consumed by microorganisms from an air or an oxygen-enriched environment in a closed vessel under conditions of constant temperature and agitation, by additionally monitoring the amount of CO_2 removed from the product gas. There are many commercially available respirometers to directly measure OUR, biodegradability, toxicity, biomethane potential etc. Challenge respirometer (Challenge AER-800 Research Respirometer) is one such example from Challenge Technology®, which facilitates automated and continuous recording of respiration rates for both aerobic and anaerobic laboratory scale cultures. There are other such respirometeric devices available commercially.

Specific Oxygen Uptake Rate (SOUR) is another method of measuring the respiration capability of microorganisms in wastewater environments. It describes the rate of oxygen consumed by microorganisms to biodegrade one gram of food and is expressed as the mg of oxygen utilized by one gram of VSS of sample per hour. Performing the SOUR instead of the OUR test eliminates the variation in uptake rate due to different amounts of mixed liquor and varying levels of volatile material in the mixed liquor. SOUR is a useful parameter when comparing different wastewater systems or if a single system's performance needs to be assessed over time. Some baseline performance metrics of the state of growth and/or decay of bacteria in wastewater can be correlated to SOUR values as listed below.

- SOUR > 20 mg O_2/g VSS/hr
  - Logarithmic growth, dispersed flock
  - Slow settling biomass
- SOUR 12–20 mg O_2/g VSS/hr
  - Declining growth, Flocks forming
  - Normal settling biomass
- SOUR < 12 mg O_2/g VSS/hr
  - Endogenous Respiration
  - Fast settling biomass

To assess the biodegradation of chemicals in water and wastewater, a common test is performed called the "**Die Away**" method (ASTM Method E1279—89, Reapproved 2001). Die-away biodegradation method is a simple test method that follows the disappearance of an added amount of the test substance (in terms of COD and TOC) in the water sample resulting from the activity of microorganisms in the sample (Fig. 1.15). The chemical-specific analytical technique used to follow the disappearance of the test substance are employed using relatively low initial concentrations, which is typically how most pollutants are present in the environment (at concentrations $\leq$ 1 mg/L). The use of low substrate concentration enhances the probability of observing first-order, or pseudo first-order kinetics as well as the half-life of the test compound under defined incubation conditions.

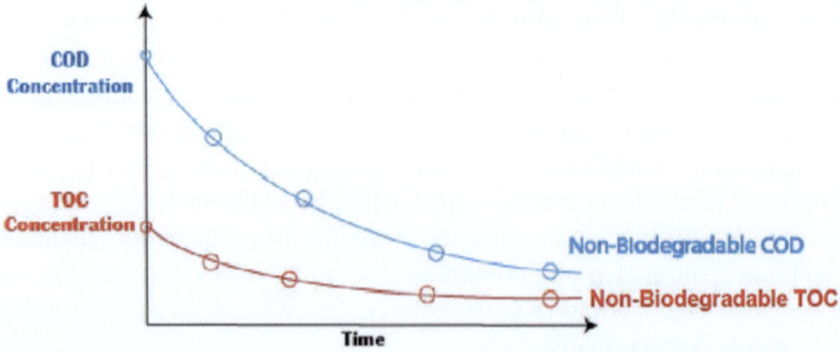

Fig. 1.15 Die-away test method indicating biodegradability of pollutants

1.11 Tools and Techniques for Microbial Community Analysis

The efficacy of biological treatment for refinery wastewater is closely tied to the metabolic activities of microorganisms, emphasizing the crucial need to understand their roles in process optimization. Despite the effectiveness of these treatments, unraveling the complexity of microbial communities, including their abundance and composition, remains challenging. Therefore, there is a growing interest in using a range of techniques to study these microbial communities. This section aims to briefly outline different methods, from traditional microscopy to modern molecular techniques used to analyze microbial communities in wastewater treatment.

1.11.1 Microscopic Techniques

Microscopic analysis is a traditional, yet essential method for identifying diverse microorganisms in wastewater. From simple light microscopes to advanced electron microscopes, these techniques enable quick identification of various taxa, including bacteria and archaea.

Microscopy allows for the detection of population changes, such as filamentous organisms and alterations in floc structures (Mesquita et al. 2013). Staining techniques, like India Ink, aid in identifying polysaccharides as indicators of biomass stress, facilitating necessary operational adjustments. Moreover, microscopic analysis helps identify key microbial groups involved in crucial wastewater treatment processes, such as nitrifying bacteria for nitrogen removal. Despite its advantages, microscopic techniques require skilled personnel and may have limited taxonomic resolution compared to molecular methods. Nonetheless, they remain vital for understanding microbial dynamics in wastewater treatment systems. Examples of microscopic methods to aid in the examination of microbiological population in water and wastewater environments are presented in Chapters 4 and 5.

1.11.2 Plate Count Method

The plate count method (PCM), also known as the standard plate count (SPC) or heterotrophic plate count (HPC), is a widely used microbiological technique for estimating the viable bacterial population in a sample. This technique requires diluting a sample of wastewater and then placing it on agar plates enriched with nutrients. The plates are incubated, usually at 35 C for a period of 24 to 48 hours. The colonies that appear on the plates are then counted. It is presumed that each colony originates from a single microorganism culture, allowing for an estimation of the viable cell count in the initial sample. The plate count method in its current form offers several approaches: pour plate (PP), spread plate (SP), and membrane filter (MF) (Reasoner 2004). Each technique has its specific applications and limitations. The PP method, involving the mixing of a water sample with tempered agar, is simple and cost-effective but may stress bacteria due to the heat of the agar. The SP method alleviates this issue by allowing bacteria to grow on the agar surface under less stressful conditions, typically yielding higher bacterial counts. The MF method, suitable for analyzing larger volumes of water, is invaluable for detecting low bacterial concentrations but requires more specialized equipment. Plate counts are used to assess the microbial population before and after treatment processes to gauge the efficacy of bacterial removal or reduction. This method helps in monitoring the performance of biological treatment units like activated sludge systems by quantifying the growth or reduction of microbial communities essential for treatment. The plate count method is limited by its inability to detect non-culturable organisms and its reliance on specific incubation conditions that may not support all bacterial types. This method is also time-consuming and labor-intensive, requiring careful manual counting and potentially leading to errors due to colony overlap or indistinct morphology.

1.11.3 ATP Assay Method

The ATP method measures adenosine triphosphate (ATP) concentration, an indicator of metabolic activity, thus signaling the presence of living microbial cells in wastewater. Utilizing bioluminescence, the enzyme luciferase catalyzes a reaction between ATP and luciferin, resulting in light emission, the intensity of which is directly proportional to ATP levels and measured quantitatively with a luminometer. ATP, essential for energy transfer in all living cells, serves as an indicator of cell viability. Its levels drop sharply upon cell death. Unlike DNA-based methods, the ATP assay determines the quantity of viable microorganisms in the sample. The assay is straightforward, requires only a minimal sample volume, and provides results quickly, often enabling on-site testing without the need for sample preparation. Nonetheless, the ATP assay has its drawbacks. It gives a general assessment of microbial quality but cannot differentiate between microorganism types without a prior isolation step. Moreover, ATP levels vary across different species and strains, and depend on environmental and metabolic conditions, which means the assay offers only an approximate measure of live cells in a sample (Linklater and Örmeci 2013).

1.11.4 Application of Principles of Molecular Biology in Identification of Microbial Communities

Molecular techniques are valuable tools for identifying and measuring non-culturable microorganisms. These are more common in domestic wastewater treatment but, given the potential for inhibition and very slow rates of biodegradation caused by certain chemicals in refinery wastes, they are increasingly finding their place in petroleum wastewater treatment applications also. A few such techniques are presented below based on their prior applications in refinery waste treatment as reported in the literature. For each only an outline of the working principle has been provided here. Interested readers are directed to the listed references for a deeper understanding. A few examples of their application in refinery wastewater treatment have been provided in Chap. 5.

1.11.4.1 Fluorescent In-Situ Hybridization (FISH)

FISH involves the use of fluorescently labeled nucleic acid probes to detect specific DNA or RNA sequences within intact microbial cells. By employing fluorescently labeled nucleic acid probes, FISH enables the detection of specific DNA or RNA sequences within intact microbial cells. The initial sample processing involves a step to enhance cell permeability. This is followed by the hybridization process of inserting the probes into the cells where they selectively bind to the complementary nucleic acid sequences in the target cells. Finally, the hybridized cells are visualized under a fluorescent microscope (Örmeci and Linden 2008). This technique offers insights into spatial and morphological information about microbial communities, complementing other molecular approaches. FISH has been widely used for identifying nitrifiers and denitrifiers in activated sludge systems (Nielsen et al. 2009).

1.11.4.2 Quantitative Polymerase Chain Reaction (qPCR)

Quantitative Polymerase Chain Reaction (qPCR) is a critical tool in analyzing microbial communities in wastewater, providing accurate quantification of specific DNA sequences. By targeting conserved regions like segments of the 16S rRNA gene, qPCR allows for both absolute and relative quantification of microbial taxa abundance. The qPCR process involves several key steps, starting with DNA extraction to isolate target DNA and primer design to ensure specificity. During qPCR, real-time DNA amplification occurs, with fluorescence signals monitored throughout cycles. The protocol includes controls to validate results and adjusts for variations in DNA extraction efficiency. Despite its advantages, qPCR may encounter limitations with primer specificity and DNA extraction efficiency, underscoring the need for careful optimization. Nonetheless, qPCR is essential for understanding microbial ecology in wastewater, guiding process optimization and environmental management. (Kim et al. 2013).

1.11.4.3 16S rRNA Gene Sequencing

16S rRNA gene sequencing plays a crucial role in microbial community analysis, providing valuable insights into microbial diversity, composition, and dynamics in various environmental settings, including refinery wastewater treatment systems. This technique involves sequencing a specific region of the 16S ribosomal RNA gene, which allows for the taxonomic classification of bacteria present in wastewater samples. Starting with DNA extraction and PCR amplification of the target region, followed by purification and sequencing on high-throughput platforms like MiSeq, HiSeq, or Ion Torrent, 16S rRNA gene sequencing enables researchers to explore the microbial ecology with depth and resolution. One of the key advantages of 16S rRNA gene sequencing is its ability to detect and quantify microbial diversity even among non-culturable microorganisms, which constitute a significant portion of microbial communities in refinery wastewaters. Additionally, this technique allows for the identification of microbial taxa that may be involved in important processes such as nutrient removal, organic substrate degradation, and pathogen removal. However, 16S rRNA gene sequencing encounters limitations such as PCR bias, sequencing errors, and challenges in capturing rare microbial taxa. Nevertheless, with meticulous experimental design, robust bioinformatics analysis, and careful interpretation of results, these limitations can be addressed, allowing for a comprehensive exploration of microbial communities in refinery wastewater. By unraveling microbial ecology, identifying key taxa, and elucidating microbial interactions, 16S rRNA gene sequencing contributes to optimizing wastewater treatment processes, enhancing environmental management practices, and fostering sustainable development in the petroleum industry. With the help of a case study in Chap. 5, this book has delved into the practical application of this technique for analyzing microbial communities in a refinery wastewater.

1.11.4.4 Shotgun Metagenomic Sequencing

Metagenomic sequencing unveils the genetic diversity within mixed environmental samples using short-read sequencing technologies. While 16S rRNA gene sequencing specifically targets a conserved region of the bacterial 16S ribosomal RNA gene to identify and classify bacterial taxa, metagenomic sequencing involves random sampling and sequencing of genetic material from environmental samples, including bacteria, viruses, and other microorganisms. This broader approach allows for the detection of a wider range of genetic information, including functional genes and non-bacterial organisms, providing a more comprehensive view of microbial diversity and functional potential in wastewater and other environmental samples (Garner et al. 2021).

None of these techniques is complete by itself. Rather they provide information complimentary to each other. For a complete and reliable characterization of the microbial population in a biological wastewater treatment reactor multiple methods may be necessary.

1.12 Quality Assurance and Quality Control (QA/QC) Procedures and Considerations

Commercial and in-house laboratories of water and wastewater treatment plants run QA/QC tests on each sample for precision and accuracy of results by conducting tests on blank samples and matrix spike on duplicate samples. Precision is the relationship among multiple measurements but does not necessarily elucidate accuracy, whereas accuracy is the relation between measured results and the real value of the parameter. Often the accuracy of results is limited by the detection limits of analytical instruments in the analyses of extremely low levels of contaminants of interest in the samples.

If QA/QC data are outside control limits, results must be rejected. Table 1.8 presents a list of standard test methods for a number of most commonly used parameters in water and wastewaters in petroleum industries.

Table 1.8 List of test methods and references

Parameter	Method
Alkalinity forms	Standard Methods: 2320
Ammonia Nitrogen	EPA Method 350.1, Hach (Note 4a)
Anions—Bromide, Chloride, Fluoride, Nitrite, Nitrate and Sulfate	EPA Method 300/300.1 and Standard Methods 4110
BOD	Standard Methods: 5210
COD	Standard Methods: 5220, ASTM 1252 B, Hach (Notes 1 and 4a)
Color	Standard Methods: 2120
Coliform Group	Standard Methods: 9222
Conductivity	Standard Methods: 2510
Die Away Test	ASTM Method E 1279—89, Reapproved 2001, Withdrawn 2013
Dissolved Oxygen (DO)	NERL 360.1, SM 4500 O
Hardness	Standard Methods: 2340
Metals and Metalloid	Standard Methods: 3125, EPA Method 6010 D and 6020 B
Nitrate-Nitrogen	SM 4500 NO_3^-
Nitrite-Nitrogen	SM 4500 NO_2^-
Odor	Standard Methods: 2150
Oil & Grease	EPA Metod 1664 B, Standard Methods 5520
ORP	Standard Methods: 2580
OUR	Standard Method 5210 D (Respirometry)
pH	Standard Methods 4500 H^+

(continued)

Table 1.8 (continued)

Parameter	Method
Phosphorus forms	EPA Method 365.1, Standard Methods 4500-P
Residual Chlorine in Water	SM 4500 Cl
Settling sludge volume	Standard Methods: 2710 C
SOUR	Standard Methods: 2710 B
Specific Gravity	ASTM Method D 1429_08_A and B
Sulfur forms	EPA NEMI Method 376.2, Hach Method (Note 4a)
TDS	Standard Methods: 2540 C
Temperature	Standard Methods: 2550
TOC	Standard Methods: 5310 B, EPA Method 415.1
Total Kjedahl Nitrogen (TKN)	EPA Methods 1687, 351.2 and 351.4
Total Petroleum Hydrocarbon (TPH)	EPA Method 418.1
Total Phenols	Standard Methods 5530 and 6420
Total Recoverable Petroleum Hydrocarbon (TRPH)	EPA Method SW 846_8440
Total Solids	Standard Methods: 2540 B
TSS	Standard Methods: 2540 D
Turbidity	EPA Method 180.1, Standard Methods: 2130
VSS	Standard Methods: 2540 E

Notes

1. COD analysis is very sensitive to chloride concentrations and may give erroneous results for high salinity waters. SM 5220 outlines the procedure to minimize the effect of chloride ion interference on COD analysis.
2. EPA methods— 40 CFR 136.3

EPA official website source: http://www.access.gpo.gov/nara/cfr/cfr-table-search.html

3. Standard Methods for the Examination of Water and Wastewater

The latest is the 24th edition—American Public Health Association (APHA), American Water Works Association (AWWA) and Water Environmental Federation (WEF) copyright 2023.

4. Vendor-Developed Methods

    (a) HACH Testing Method
    (b) Microtox Test instrument by Microbics Corporation.

Abbreviations and Glossary of Terms

16S rRNA Gene sequencing is identification of types of microorganisms by RNA sequencing.

ATP Standard Methods: 10200 I. This is a method of measuring living cells in a biomass population.

Alkalinity Standard Methods: 2320. This is a titration method that depends on the endpoint pH used. The results can be interfered by soaps, oily matter, suspended solids, or precipitates.

BOD Standard Methods: 5210. The biochemical oxygen demand determination is an empirical test for oxygen requirements of wastewaters, effluents, and polluted waters. The test has its widest application in measuring waste loadings to treatment plants and in evaluating the BOD-removal efficiency of such treatment systems. 5210 describes methods measuring oxygen consumed in: (i) a 5-d test period (BOD_5, by 5210B); (ii) after 60 to 90 d of incubation (Ultimate BOD, by 5210C); and (iii) continuous oxygen uptake by respirometric method, 5210D. BOD_5 and BOD have been used interchangeably in the book both mean 5 day BOD.

COD Standard Methods: 5220. Chemical oxygen demand is defined as the amount of a specified oxidant that reacts with the sample under controlled conditions. COD often is use as a measurement of pollutants in wastewater and natural waters. In a COD analysis, hazardous wastes of mercury, hexavalent chromium, sulfuric acid, silver, and acids are generated.

Color Standard Methods: 2120. Industrial wastewaters can contain lignins, tannins, dyes, and other organic and inorganic chemicals that cause color. Colored industrial wastewaters may require color removal before discharge into watercourses.

Coliform Group Standard Methods: 9222. The Membrane filter (MF) technique is reproducible, can be used to test relatively large sample volumes, and usually yields numerical results more rapidly than the multiple-tube fermentation procedure. The MF technique has limitations when testing waters that are of high turbidity or contain large numbers of non-coliform (background) bacteria. MF technique is not to be used to test wastewater containing high levels of toxic metals or toxic organic compounds (e.g., phenols) because such substances may be concentrated by the filter and inhibit coliform growth.

Conductivity Standard Methods: 2510. Conductivity instruments are used in pipelines, channels, flowing streams, and lakes and can be incorporated in multiple parameter monitoring stations using recorders. Most problems in obtaining good data with conductivity monitoring equipment are related to electrode fouling and to inadequate sample circulation. Conductivities greater than 10 000 to 50 000 μmho/cm or less than about 10 μmho/cm may be difficult to measure with usual measurement electronics and cell capacitance.

FISH Fluorescence In-Situ Hybridization is the use of fluorescently labeled nucleic acid probes to detect specific DNA or RNA sequences within intact microbial cells.

Hardness Standard Methods: 2340. Two methods are presented here: Method 2340B, hardness by calculation, is applicable to all waters and yields the higher accuracy. Method 2340C, the EDTA titration method, measures the calcium and magnesium ions and may be applied with appropriate modification to any kind of water.

Nitrogen forms Analyses by Hach Test Methods for Total Nitrogen, Ammonia, Nitrite, Nitrate and TKN. EPA Method 350.1 for ammonia-N and Method 351.2 for TKN and SM4500 Methods are used for Nitrite and Nitrate-N.

OUR Standard Method 5210 D for Oxygen Uptake Rate by Respirometry.

ORP Oxidation–Reduction Potential. Standard Methods: 2580. Electrometric measurements are made by potentiometric determination of electron activity (or intensity) with an inert indicator electrode and a suitable reference electrode. Standard solutions that are both chemically stable with known redox potentials are needed to check specific indicator electrodes for electrode response at temperature of measurement.

Odor Standard Methods: 2150. Odor, like taste, depends on contact of a stimulating substance with the appropriate human receptor cell. The stimuli are chemical in nature and the term “chemical senses” often is applied to odor and taste. Sensory tests are useful as a check on the quality of raw and finished water and for control of odor through the treatment process.

Oil and Grease Standard Methods: 5520 and EPA Method 1664 B. Certain constituents measured by the oil and grease analysis may influence wastewater treatment systems. A knowledge of the quantity of oil and grease present is helpful in proper design and operation of wastewater treatment systems and also may call attention to certain treatment difficulties.

qPCR Quantitative Polymerase chain reaction is a method to innumerate types of mircroorganisms based on DNA or RNA sequencing.

Specific Oxygen Uptake Rate (SOUR) Standard Methods: 2710 B. This test is applicable to the determination of Oxygen Uptake Rate of a biological sledge suspension in mg/L/hr per unit concentration of volatile suspended solids (VSS) in g/L. The unit of SOUR is mg/g/hr.

References

Abdel-Raouf N, Al-Homaidan AA, Ibraheem IBM (2012) Microalgae and wastewater treatment. Saudi J Biol Sci 19(3):257–275, ISSN 1319–562X

APHA, AWWA, WEF (2023) Standard Methods for the Examination of Water and Wastewater, 23rd Edition

Basu S (2022) On line monitoring and process control of SBR cycles for nutrient control from wastewater. Am J Water Sci Eng 8(4):79–88

Boenigk J (2011) Protozoa (Heterotroph, Eukaryotic). In: Reitner, J., Thiel, V. (eds) Encyclopedia of Geobiology. Encyclopedia of Earth Sciences Series. Springer, Dordrecht

Cai L., Zhang T (2013) Detecting human bacterial pathogens in wastewater treatment plants by a high-throughput shotgun sequencing technique. Environ Sci Technol 47:5433–5441

Eckenfelder, Ford, Englande (1997 and 2009) Industrial Water Quality, 4th Edition, McGraw Hill
Golding C, Lamboo L, Beniac D et al (2016) The scanning electron microscope in microbiology and diagnosis of infectious disease. Sci Rep 6:26516
Garner E, Davis BC, Milligan E, Blair MF, Keenum I, Maile-Moskowitz A, Pan J, Gnegy M, Liguori K, Gupta S, Prussin AJ, Marr LC, Heath LS, Vikesland PJ, Zhang L, Pruden A (2021) Next generation sequencing approaches to evaluate water and wastewater quality. Water Res 194:116907
Kemmer FN (ed) (1988) The Nalco water handbook, 2nd edn. McGraw Hill, NY
Kim J, Lim J, Lee C (2013) Quantitative real-time PCR approaches for microbial community studies in wastewater treatment systems: applications and considerations. Biotechnol Adv 31(8):1358–1373
Kristian ST, Aa K, Ausland G, Fredrik HJ (2004) Retention and removal of pathogenic bacteria in wastewater percolating through porous media: a review. Water Res 38:1355–1367
Linklater N, Örmeci B (2013) Evaluation of the adenosine triphosphate (ATP) bioluminescence assay for monitoring effluent quality and disinfection performance. Water Qual Res J 49(2):114–123
Madigan M, Martinko J, Stahl D, Clark D (2010) Brock biology of microorganisms, Thirteenth. Prentice Hall, New York
Maynard C, Berthiaume F, Lemarchand K, Harel J, Payment P, Bayardelle P, Brousseau R (2005) Waterborne pathogen detection by use of oligonucleotide-based microarrays. Appl Environ Microb 71:8548–8557
Marrs D (2012) Characteristics of refinery and chemical plant wastewater: testing and data analysis. Proceedings of Workshop 101 Back to Basics: Wastewater Treatment Fundamentals for Petroleum Refineries and Petrochemical Plants, WEFTEC 2012
Mesquita DP, Amaral AL, Ferreira EC (2013) Activated sludge characterization through microscopy: a review on quantitative image analysis and chemometric techniques. Anal Chim Acta 802:14–28
Metcalf & Eddy/AECOM (2014) Wastewater engineering: treatment and reuse, 5th edn. McGraw-Hill, NY
Microbics Corporation (1995) Microtox acute toxicity basic test procedure, Carlsbad, CA
Nielsen P, Daims H, Lemmer H (2009) FISH handbook for biological wastewater treatment, IWA Publishing
Örmeci B, Linden KG (2008) Development of a fluorescence in situ hybridization protocol for the identification of micro-organisms associated with wastewater particles and flocs. J Environ Sci Health Part A 43(13):1484–1488
Reasoner DJ (2004) Heterotrophic plate count methodology in the United States. Int J Food Microbiol 92(3):307–315
Reimers RS (2019) Chemistry and characteristics of refinery wastewater, workshop on wastewater management in petroleum refineries, 4C Health/Safety/Environmental Conference., February 7, 2019, Austin, Texas
Rowe DR, Abdel-Magid IM (1995) Handbook of Wastewater Reclamation and Reuse, CRC Press Inc., p 550
Symonds EM, Breitbart M (2014) Affordable enteric virus detection techniques are needed to support changing paradigms in water quality management. Clean—Soil, Air, Water 43:8–12
Sorenson S (1909) Uber die messing und die Bedeutung der Wasterstoff ionen Konzentration bei Enzymatischen Prozessen. Biochem Z 21:131
Thiel V (2011) Archaea. In: Reitner, J., Thiel, V. (eds) Encyclopedia of Geobiology. Encyclopedia of Earth Sciences Series. Springer, Dordrecht

Suggested Reading

American Water Works Association (2010) Water quality and treatment—a handbook of community water supplies, 6th edn. McGraw Hill

Anthony F Gaudy, Elizabeth T Gaudy (1988) Elements of bioenvironmental engineering, 1st Edition, Engineering Press Inc.

Levenspiel O (1998) Chemical reaction engineering, 3rd edn. Wiley

Sawyer CN, McCarty PL, Parkin GF (1994) Chemistry for environmental engineering, 5th edn. McGraw Hill

Snoeyink VL, Jenkins D (1991) Water chemistry. Wiley & Sons

Chapter 2
Principles of Water Treatment

Somnath Basu, Mudumbai Venkatesh, Shiladitya Basu, Anil Rai, and Saurabh P. Deshpande

Abstract Water is the most important resource on earth. It not only sustains life but also promotes the economic growth of a society by supporting industrial activities. No industry can operate without water, and petroleum industry is no exception. This chapter delves into the fundamental principles and crucial aspects of water treatment within the context of the petroleum industry. It discusses the various unit operations and processes used for water treatment and provides an overview of the typical design and operating guidelines. The purpose of this chapter is to serve as the foundation of the advanced levels of treatment processes in all sectors of the petroleum industry as described in the Preface section of this book.

Keywords Adsorption · Advanced Oxidation Processes (AOP) · Brine management · Chemical oxidation · Clarifiers · Coagulation and flocculation · Disinfection · Electrodialysis · Filtration · Flux · Granular Activated Carbon (GAC) · Ion exchange · Jar test · Lime-Soda process · Membrane fouling · Membrane separations · Multi effect distillation · Permeability · Permeate · Powdered Activated Carbon (PAC) · Reverse Osmosis (RO) · Seawater desalination · Sedimentation · Settling · Terminal settling velocity · Transmembrane pressure · Ultrafiltration (UF) · Water softening · Water treatment

S. Basu (✉)
Jewel Innovations in Energy and Environment, Katy, TX, USA
e-mail: som.basu@jewelenergyenvironment.com

M. Venkatesh
Process and Energy Consultant, Andover, MA, USA
e-mail: m.venkatesh1@yahoo.com

S. Basu
Stantec, Houston, TX, USA
e-mail: shiladitya.basu@stantec.com

A. Rai
AECOM, Mumbai, India
e-mail: anil.rai@aecom.com

S. P. Deshpande
Black and Veatch, Toronto, ON, Canada
e-mail: DeshpandeSP@bv.com

S. Basu et al. (eds.), *Water Management in Petroleum Industries*, Water Resources Development and Management, https://doi.org/10.1007/978-981-19-3159-8_2

2.1 Water Treatment

The purpose of water treatment is to make it suitable for use in sanitary and industrial applications. Water is obtained from two main types of resources:

- Groundwater, and
- Surface water
  - Rivers
  - Lakes
  - Sea

While groundwater is pumped to the surface from deep underground aquifers surface water is sourced from rivers, lakes, and sea. The generic term for water from all types of sources is 'Raw Water'.

Depending upon the source the raw water quality and characteristics vary widely. Typical contaminants found in various sources of raw water are listed in Table 2.1.

These contaminants need to be removed to various levels depending on the end use of the raw water. Water quality for drinking water needs to fulfill the US EPA Primary and Secondary Drinking Water Standards as dictated by the 1996 amendments of the Safe Drinking Water Act (SDWA). Similarly, the wastewater discharged from any source needs to comply with the provisions of the Clean Water Act (CWA) of the US EPA and any local regulatory requirements to protect the receiving bodies of water. For international operations similar regulatory requirements are mandated by authorities in the European Union (EU) and World Health Organization (WHO) that set the standards for drinking water and wastewater. The regulatory requirements are discussed in further details in Chap. 7. For industrial applications the treatment processes depend upon the quality requirements of the end user and are typically conducted at advanced levels beyond those for drinking water. In addition to sourcing raw water from various resources, as discussed above, there is a growing trend of reusing water to achieve sustainability goals in both municipal and industrial sectors. Examples of water reuse in petroleum industries have been presented in detail in Chap. 8. Figure 2.1 presents an overview of water treatment process for the production of water for sanitary and industrial uses.

2.1.1 *Unit Operations and Processes for Water Treatment*

Water treatment is a multi-step process as follows. The number of steps depends upon the desired goals for achieving the treated water quality.

Table 2.1 Contaminants in Raw Water (Qasim et al. 2000)

Constituent	Typical water source			
	Rainwater	Surface water	Ground water	Sea water
Physical				
TDS (mg/L)	25	150	250	35,000
Turbidity (NTU) (Note)	0	10	< 0.5	5
Color (units)	–	–	–	5
Biological				
Coliforms (MPN/100 mL)	0	2,000	100	–
Virus (pfu/100 mL)	0	10	1	0
Chemical				
BOD_5 (mg/L)	–	5	–	2
TOC (mg/L)	0.2	3	0.5	1
NH_3 (mg/L)	0.05	0.2	0.1	–
Total N (mg/L)	–	3	< 10	0.5
Total P (mg/L)	–	0.05	0.01	0.01
Hardness (mg/L as $CaCO_3$)	25	90	120	–
Alkalinity (mg/L as $CaCO_3$)	20	100	150	–
pH, su	7.0	7.5	7.5	7.9
Cation				
Ca^{2+} (mg/L)	6	20	50	400
Mg^{2+} (mg/L)	2	3	5	1,350
Na^{+} (mg/L)	5	20	5	10,500
K^{+} (mg/L)	–	2	2	350
Fe^{2+} (mg/L)	0.05	0.1	0.1	0.1
Anion				
HCO_3^- (mg/L)	18	90	120	150
Cl^- (mg/L)	5	25	25	20,000
SO_4^{2-} (mg/L)	4	20	10	2,800
SiO_2 (mg/L)	1	5	10	20
NO_3^- (mg/L)	0.1	0.5	< 10	–
F^- (mg/L)	–	0.2	0.1	1

Note NTU stands for Nephelometric Turbidity Unit. It is a measure of turbidity of water. It is determined by measuring the strength of scattered light at 90° to the source when a water sample is illuminated by a white beam of light. Refer to the definition of Turbidity in Sect. 1.2.4 of Chap. 1, and the EPA Method 180.1 and Standard Method 2130 B for details of the analytical methods

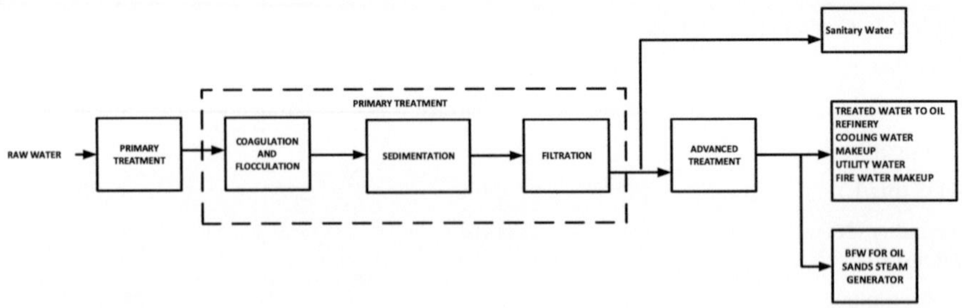

Fig. 2.1 Raw water treatment strategy

- Preliminary Treatment
- Primary Treatment
- Advanced Treatment
- Brine Management and Zero Liquid Discharge.

The unit operations and processes for raw water treatment are discussed in detail in the following sections.

2.1.1.1 Preliminary Treatment

When raw water is sourced from natural water bodies e.g., lakes, rivers, sea, etc., it typically contains debris, large objects, aquatic plants, and branches of trees that drop from adjacent land resulting from high winds and storms. The purpose of preliminary treatment is to remove these objects in order to protect the downstream treatment process equipment from fouling and mechanical damage. Screens are placed at an angle between 30 and 45° to the horizontal plane and typically have bar spacing between 12 and 36 mm. These are mechanically cleaned by rakes and water jets. As the solids drop to the bottom these are collected in bins and disposed in landfills or incinerated. The solids passing through the screens are sometimes ground by in-line grinders into small pieces so those can be removed by settling sedimentation step in the clarifier downstream by settling sedimentation.

In order to prevent excessive pressure drop water velocity through a bar screen must be limited to 0.6 m/s under design flow and 0.9 m/s under peak flow (Reynolds and Richards 1996). The number of screens to be deployed in parallel depends upon the total flow to be handled. Also, a standby screen is provided to account for shutdown of one for breakdown and/or maintenance.

Screens are also used for wastewater treatment applications and are discussed in further detail in Chap. 5.

2.1.1.2 Primary Water Treatment

Primary treatment of water constitutes physicochemical processes that remove suspended solids and a part of the dissolved solids to make the final product fulfill all the applicable regulatory requirements when used in sanitary applications, and useful as service water, fire water and cooling tower make-up water in case of industrial applications. Additionally, primary treatment serves as the basic step leading to the advanced treatment processes for production of high-quality water in specific industrial applications e.g., petroleum industries, pharmaceutical industries, semiconductor industries, power plants, etc.

Aeration

Aeration is an important primary treatment process. Many contaminants in water are converted to their oxidized forms by aeration. Some elements in the underground rock structure exist in their reduced forms in absence of air under the action of anaerobic microorganisms. Iron and manganese are two such elements that are present in reduced state as ferrous and manganous ions which are highly soluble in water. Groundwater at many locations contains iron, manganese, and sulfur in dissolved forms such as ferrous (Fe^{2+}), manganous (Mn^{2+}) and sulfide (S^{2-}) ions. These are oxidized to stable states such as ferric (Fe^{3+}), manganese (Mn^{4+}) and sulfate ($SO_4{}^{2-}$) by reacting with oxidants.

Aeration is the first unit operation done to groundwater as it is extracted from underground by bringing air in contact with water. It is very important to remove dissolved iron (ferrous) and manganese ions in groundwater by oxidation to particulate form. Without this step these ions oxidize and precipitate in presence of air in the downstream treatment processes to form solids and foul equipment and pipe surfaces. The oxidation reactions are represented by the following equations:

$$Fe^{2+} + {}^{1}\!/_{4}\,O_2 + 5/2\,H_2O \rightarrow Fe(OH)_3 \downarrow + 2\,H^+ \quad (2.1)$$

$$Mn^{2+} + {}^{1}\!/_{2}\,O_2 + H_2O \rightarrow MnO_2 \downarrow + 2\,H^+ \quad (2.2)$$

Hydrogen sulfide (H_2S), methane and volatile organic compounds (VOC) are among other reduced constituents in groundwater.

Aeration oxidizes hydrogen sulfide, which exists primarily as bisulfide (HS^-) ions under the prevailing pH conditions (as discussed in Chap. 1) according to the following reactions.

$$H_2S \rightarrow H^+ + HS^- \tag{2.3}$$

$$HS^- + 2\,O_2 \rightarrow SO_4^{2-} + H^+ \tag{2.4}$$

Hydrogen sulfide is odorous and volatile. Its presence in air poses human health and safety risk. Conversion to sulfate (SO_4^{2-}) makes it much safer to handle as sulfate salts are not a safety concern. Oxidation of hydrogen sulfide is a multistep process. Refer to the detailed discussion on hydrogen sulfide oxidation steps in Chap. 1.

Methane, however, is not easily oxidized by aeration and is stripped off and emitted to the atmosphere. Volatile organic compounds (VOC) impart foul odor and taste to groundwater. Aeration removes these by a combination of oxidation and stripping mechanisms. Typically, drinking water is sourced separately from refinery feed water and there is no cross mixing. However, in case drinking and sanitary water supply line for refinery personnel is tapped from the same source then removal of taste and odor by aeration becomes pertinent.

Various types of aerators used in the aeration process can be categorized as

- Cascade Aerators
- Spray Aerators
- Diffused Air Aerators
- Mechanical Aerators.

In cascade type of aerators, water flows down after spreading over horizontal tray surfaces (typically, 4 to 6 trays or baffles) in thin sheets and turbulence is created as water flows down through the series of trays.

In spray aerators, water is sprayed through nozzles in upward directions into the atmosphere and are broken into either a mist or droplets. This creates large air–water contact surface for oxidation and stripping. The major drawback with such aerators is losses due to evaporation and entrainment, and high-power consumption.

Diffused air aeration is conducted in a basin consisting of perforated pipes, porous tubes or plates which is used to inject fine compressed air into the tank. The rising air bubbles create turbulence and create large surface areas that help in transferring oxygen from air into water, thereby supporting the oxidation of dissolved ferrous, manganous and sulfide ions to respective oxidized forms.

Mechanical aerators generally employ motor driven impellers. Most common type of mechanical aerators are submerged paddle, propeller blades, turbine aerators etc.

Cascade and spray aerators have been used for decades in groundwater treatment to remove dissolved constituents. Diffused air and mechanical aerators are not typically used in water treatment applications but are most common in wastewater treatment and are discussed in Chap. 5.

Water Softening

Traditionally Lime Soda Softening is used to remove hardness in water. It is necessary to remove hardness and soften water to protect pipes and equipment from scaling. Hardness is comprised of dissolved divalent metal salts, primarily calcium and magnesium, in water. Total hardness (TH) is expressed as the sum of calcium and magnesium ions in terms of equivalent calcium carbonate concentration. Total hardness (TH) is also equal to carbonate hardness (CH) and non-carbonate hardness (NCH). Therefore,

$$\text{Total Hardness (TH)} = \text{Calcium Hardness (CaH)} + \text{Magnesium Hardness (MgH)} \tag{2.5}$$

$$= \text{Carbonate Hardness (CH)} + \text{Non-Carbonate Hardness (NCH)} \tag{2.6}$$

Carbonate hardness (CH) refers to hardness chemically equivalent to carbonate and bicarbonate alkalinities and non-carbonate hardness (NCH) is the part of total hardness (TH) excluding the carbonate hardness (CH).

There is a relationship between total hardness and total alkalinity. Both are expressed in terms of meq/L, or mg/L as $CaCO_3$. If the total hardness is less than total alkalinity it means, there is no non-carbonate hardness. If the total hardness is greater than the total alkalinity, the non-carbonate is equal to the difference between total hardness and total alkalinity and carbonate hardness is equal to the total alkalinity. Alkalinity is a measure of how much acid can be added to the water without resulting in a significant change in pH. Its definition is discussed in detail in Chap. 1.

Lime (calcium hydroxide—$Ca(OH)_2$) and Soda Ash (sodium carbonate—Na_2CO_3) are added to water to remove hardness by precipitation. Lime is used to treat carbonate hardness and both lime and soda ash are used to treat non-carbonate hardness. The dosing amounts are determined by the content of dissolved carbonate hardness (CH), non-carbonate hardness (NCH) and carbon dioxide in water. The chemical reactions are as follows:

$$CO_2 + Ca(OH)_2 \rightarrow CaCO_3 \downarrow + H_2O \tag{2.7}$$

$$Ca(HCO_3)_2 + Ca(OH)_2 \rightarrow 2CaCO_3 \downarrow + 2H_2O \tag{2.8}$$

$$Mg(HCO_3)_2 + Ca(OH)_2 \rightarrow CaCO_3 \downarrow + MgCO_3 + 2H_2O \tag{2.9}$$

$$MgCO_3 + Ca(OH)_2 \rightarrow CaCO_3 \downarrow + Mg(OH)_2 \downarrow \tag{2.10}$$

Although CO_2 does not contribute to hardness, it reacts with the calcium hydroxide, thus consuming lime in the process. Noncarbonate hardness is also removed by a combination of lime and soda ash according to the following reactions:

Table 2.2 Classification of hardness levels (Sawyer et al. 1991)

Range of hardness as mg/L of $CaCO_3$	Class
0–75	Soft
75–150	Moderately hard
150–300	Hard
>300	Very hard

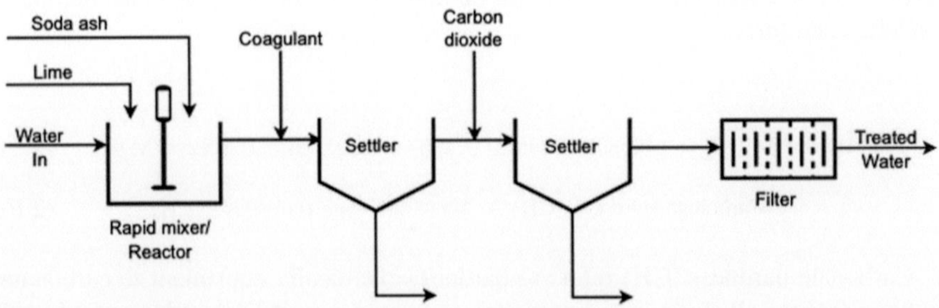

Fig. 2.2 Flow diagram for lime-soda softening with recarbonation for pH stabilization

$$MgSO_4 + Ca\,(OH)_2 \rightarrow Mg\,(OH)_2 \downarrow + CaSO_4 \tag{2.11}$$

$$CaSO_4 + Na_2CO_3 \rightarrow CaCO_3 \downarrow + Na_2SO_4 \tag{2.12}$$

To precipitate $CaCO_3$ the pH of water needs to be raised to about 9.5 and to precipitate Mg $(OH)_2$ a pH of about 10.8 is required. Therefore, an excess of lime is required to raise the pH for removal of magnesium hardness.

The amount of lime (meq/L) required = lime to remove dissolved carbon dioxide (meq/L) + lime to precipitate carbonate hardness (meq/L) + lime to remove magnesium hardness (meq/L) + Lime to raise the pH of water to 10.5.

The amount of soda ash (meq/L) required = soda ash to remove non-carbonate hardness (meq/L).

After softening, the pH of the water will be high and contain excess lime, magnesium carbonate, magnesium hydroxide the calcium carbonate that did not precipitate. Recarbonation (addition of CO_2) is used to stabilize the water and reduce the pH to ≤ 8.5.

With the use of the lime-soda ash process, residual hardness in the water can be reduced to 80 mg/L. The level of hardness of water is classified as shown in Table 2.2. Figure 2.2 illustrates a flow diagram for a lime soda softening system.

Feed water enters the Rapid Mixer/Reactor tank along with the lime and soda ash. The mixed effluent is sent to the settling tank to which coagulant is added to aid the settling. The hydraulic residence time in the settler is up to 3 h. The overflow from the settling tank is sent to a second settling tank and carbon dioxide is added

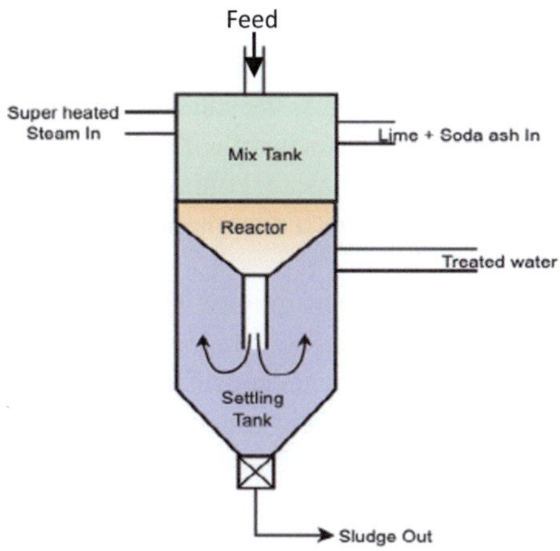

Fig. 2.3 Warm lime softener reactor

for recarbonization. The effluent from this settler is sent to a filter in which residual suspended solids are removed.

If the hardness needs to be lower than that can be achieved by lime-soda softening at ambient temperatures, the warm lime-soda softening process can be employed. Warm lime-soda softening is accomplished at temperatures up to 60 °C by adding superheated steam to the reactor. Further removal of hardness can be achieved by hot lime-soda softening operating at temperatures up to 150 °C. This process is typically installed at oil refineries and excess/exhaust steam from the plant is used in the process. In this process the residual hardness in the treated water can be reduced to 30 mg/L or lower. This can be achieved because the solubilities of the calcium and magnesium salts decrease at elevated temperatures. Figure 2.3 is a diagram of a warm lime softening reactor.

A residence time of about an hour is maintained in this equipment where chemical reactions, coagulation, flocculation and clarification are performed. Sludge generated is either centrifuged and disposed in landfill or collected in sludge ponds for evaporation of water and later deposition to a landfill. It can achieve silica content less than 50 mg/L as SiO_2 in the outlet water. The treated water is then filtered in media filters to remove any carried-over flocs before being sent to ion-exchange beds for removing the remaining hardness.

Coagulation and Flocculation

Raw water contains suspended solids that widely vary in size, concentration, chemical characteristics, and origin. These particles interfere with the beneficial use of water. The first step in water treatment is removal of suspended solids, and the associated turbidity. For groundwaters containing dissolved iron and manganese, the ferric hydroxide and magnesium oxide particles formed in the aeration step are removed along with the solids associated with raw water by settling in a sedimentation tank. If raw water contains high levels of hardness the solids formed by softening process are also co-precipitated with the solids associated with raw water in the same sedimentation tank. It is to be noted that not all water contains dissolved iron and manganese and/or hardness, and in such cases the sedimentation tank removes only the solids associated with the raw water.

Typically, solids in raw water settle by gravity. The principle of the different types of settling has been discussed in the sections below. However, very fine particles of size range between 0.45 μm and 1 μm are considered colloidal particles and do not settle by gravity. They remain dispersed in water as their surface charge repel them from each other and prevent them from flocking together to form larger size particles that can separate from water by settling.

Coagulation and flocculation are processes that aid in the removal of fine suspended solids and colloidal matter in raw water. These are chemical and physical processes which involve mixing coagulants and flocculant aids with water with the purpose to forming larger particles that are large enough to be removed by the downstream processes such as settling sedimentation and filtration.

Coagulation consists of the addition of coagulant chemical(s) to the raw water and mixing it thoroughly and uniformly in a Rapid Mixing Tank for 5–30 s. This results in destabilization of colloidal particles in the water. In this process the coagulant chemical neutralizes the stabilizing charges and enhances the destabilizing forces.

The most important parameter which provides evidence on the status of the coagulation process is the zeta potential (ZP). ZP is the electrical potential in the interfacial double layer (DL) at the location of the slipping plane relative to a point in the bulk fluid away from the interface. Figure 2.4 shows the schematic of Zeta potential with the ionic concentration and potential difference as a function of distance from the charged surface of the particles suspended in the dispersed medium of raw water.

Coagulant chemicals are inorganic or organic chemicals which hydrolyze upon dissolution in water. The charge of the ions interacts with the surface charge of colloids and causes effective particle destabilization and remove natural organic materials (NOM). Commonly used metal salts that are used as coagulants in water treatment are:

- Alum (aluminum sulfate), $Al_2(SO_4)_3.14H_2O$.
- Polyaluminum chloride, Al(OH)x(Cl)y.
- Ferric chloride, $FeCl_3$.
- Ferric sulfate, $Fe_2(SO4)_3$.

Cationic polymers have also been used as the primary coagulant or in combination with aluminum or iron coagulants (Delphos and Wesner 2005).

Coagulation by metal salts involves a set of chemical reactions. In simple terms it can be understood by following the example of alum, a commonly used coagulant:

Step 1; Hydrolysis of Alum

$$Al_2(SO_4)_3.xH_2O \rightarrow 2\ Al^{3+} + 3\ SO_4^{2-} + x\ H_2O$$
$$\text{(where x is the no. of moles of water of hydration)} \quad (2.13)$$

Fig. 2.4 Schematic of zeta potential with the ionic concentration and potential difference in solution as a function of distance from the charged surface

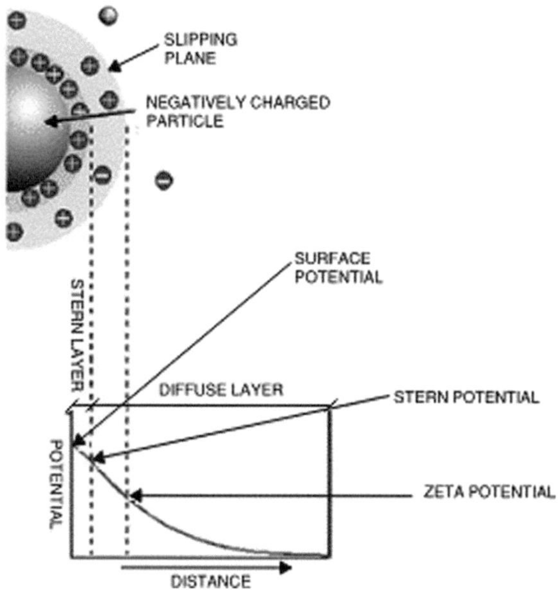

Step 2: Reaction of Aluminum Ions with Water

$$Al^{3+} + 3H_2O \rightarrow Al(OH)_3 + 3\,H^+ \tag{2.14}$$

Aluminum ions neutralize the surface charge of the colloidal particles and react with water molecules to form aluminum hydroxide, which has a very low solubility (10^{-6} mol/L, or 27 μg/L as Aluminum) in water at pH 6–7. Therefore, alum dosed as coagulant forms insoluble aluminum hydroxide particles and bridges with the destabilized colloidal matter in the flocculator. The chemical reactions produce acidity in the form of hydrogen ions. In absence of sufficient alkalinity (buffering capacity) it may lead to a depression of pH, and outside the pH range of 6–7 aluminum will resolubilize and lose its effectiveness as a coagulant.

Other metal salts e.g., ferric chloride ($FeCl_3$) also follow similar mechanism for coagulation. It must be noted that pH is a very important parameter. When using iron salts as a coagulant, pH should be maintained close to 7.5–8.5 which is the range for minimum solubility of iron. This is a very simplified representation of the chemical reactions involved in coagulation. In reality, it follows a multistep pathway involving several chemical reactions in series and parallel, and has been detailed in the literature (AWWA 2011).

The process of flocculation follows coagulation. Flocculation consists of gentle mixing which encourages collision of neutralized colloidal particles to form larger flocs that can be easily settled or filtered in the downstream processes. This process increases the particle size from submicroscopic (called microflocs) to visible flocs. Collisions of the microfloc particles cause them to bond and produce larger, visible

Table 2.3 Design criteria for flocculation Basin

		River water	Reservoir water
Minimum mixing time (mins)	Conventional treatment	20	30
	Direct filtration	15	15
Range of energy input, (G), sec^{-1}	Conventional treatment	10–50	10–175
	Direct filtration	20–75	20–100
Minimum number of stages	Conventional treatment	2–3	3
	Direct filtration	2	2

flocs called pinflocs. The floc size then continues to build by additional collisions and interactions, and macroflocs are formed. High molecular weight polymers termed as flocculant aids are sometimes added during this stage to help bond, bind, and reinforce the flocs. Once the flocs reach optimum size, they settle, and the raw water is then ready for further downstream process.

The flocculation process is governed by the rate of orthokinetic flocculation and is dependent on the velocity gradient (G) applied and the flocculation time. These parameters affect the rate and degree of particle aggregation and also the degree of breakup of these aggregates.

The general design criteria for flocculation basins are as listed in Table 2.3.

The parameter G is defined as Velocity Gradient and expressed as:

$$\mathrm{G} = \sqrt{\left(\frac{P}{\mu * V}\right)} \tag{2.15}$$

where,

G = Velocity Gradient, s^{-1}

P = Power for Mixing, W (Watts)

μ = Dynamic Viscosity of Water, $N.s/m^2$

V = Volume of Flocculator Tank, m^3

The symbol * denotes the operation of multiplication between numbers. This same convention has been followed throughout this document.

After completion of the flocculation process the raw water moves to the next step for removal of the solids by settling sedimentation.

The process of coagulation and flocculation promotes the formation and clustering of particles to enable the removal of such particles from raw water, however the type and dose of the required coagulant depends on several other factors. Jar test is a bench scale test which simulates the process of coagulation, flocculation and settling and helps select the type of coagulant and the minimum dose of coagulant required to achieve the required water quality.

Fig. 2.5 Jar test set with four (beakers) on a test bench (Source: https://www.waterboards.ca.gov/drinking_water/programs/districts/docs/mendocino/jar_testing_made_easy_aug7_2020.pdf)

In a jar test setup, a set of 4–6 jars/beakers with individual stirrers is used and a predetermined volume of the sample of raw water is added to each jar/beaker. A selected coagulant is added to each test jar and stirred. To simulate the action of coagulation, flocculation and settling in a single set up, the mixing is first conducted at a rapid rate for a given duration, followed by mixing at slow speed, and finally stopping the mixer and allowing solids to settle under a quiescent condition. Figure 2.5 demonstrates the development of flocs with various dosages of coagulant in a jar test procedure. As may be observed, the floc sizes vary from the left to the right jar with varying dosages of coagulant. The same set of tests are performed with several different types of coagulant chemicals. The minimum dose and the type of coagulant chemical that offers best removal of solids, as determined by testing TSS of the decant, is selected. Jar Test procedure is detailed in literature (ASTM, D 2035–19). The test results are utilized to select the type and dosing rate of coagulant for full scale application of settling and sedimentation of TSS in gravity clarifiers. Note, in this text the terms settling tanks, sedimentation basins and gravity clarifiers have been used interchangeably.

Sedimentation

Sedimentation is one of the oldest unit operations used in water treatment. It is a physical treatment process that uses gravity to separate suspended particles from water to reduce the solids load in downstream equipment. After coagulation and flocculation, water enters the gravity sedimentation basin, also referred to as clarifiers. Water moves very slowly through these basins which allows the flocs and particulate

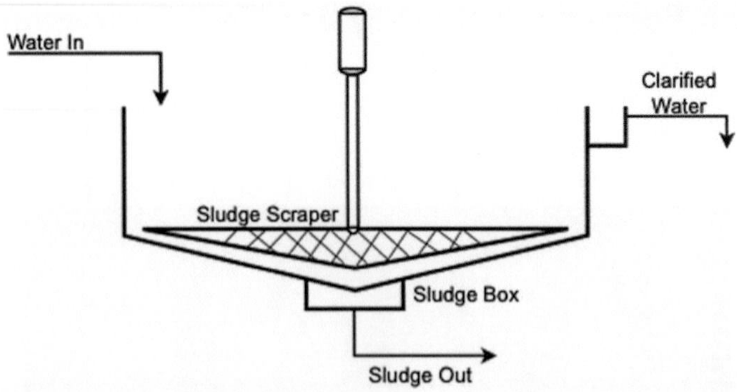

Fig. 2.6 Schematic view of a typical clarifier

matter in water to settle to the bottom of the basin. Settled solids are collected into a hopper with the help of a scraper mechanism and removed intermittently. The clarified water flows out from the top of the basin. Figure 2.6 shows a conceptual view of a typical clarifier.

Design of a sedimentation basin depends on the concentration, size, and behavior of the solids in suspension. There are four types of settling as described below.

Type 1 Settling—Discrete Settling

In discrete type of settling the particles settle as individual entities and there is no significant interaction with adjacent particles. Discrete settling refers to the sedimentation of particles in a suspension of low solids concentration.

Type 2 Settling—Flocculant Settling

In the flocculent type of settling the particles flocculate and by doing so the particles increase in size and settles at a faster rate. Flocculent settling refers to the settling of particles in a dilute suspension with solids concentration of less than 1,000 mg/L.

Type 3 Settling—Hindered Settling

When the particles are close enough so that inter-particle forces are adequate to hinder the settling of neighboring particles it results in hindered type of settling. These particles maintain their positions with respect to neighboring particles and the entire mass of particles settles as a unit or zone.

Type 4 Settling—Compression Settling

Compression type of settling is the settling type in which the solids concentration in water is so high that particles are in physical contact with each other and forming a structure with lower layers which are supporting the weight of upper layers of solids. Subsequently, further settling occurs by the compression of the whole structure of particles.

For raw water treatment in industrial applications Type 1 and Type 2 settling are most common. Type 3 and Type 4 are uncommon and are not considered for further discussion here. Preliminary sizing of clarifier is normally based on surface loading rate, which is expressed as:

$$S = Q/A \tag{2.16}$$

where,

S = Surface Loading Rate, $m^3/m^2.d$ or m/d

Q = Flow Rate into the Clarifier, m^3/d

A = Surface Area of Clarifier, m^2

Surface loading rate is equal to the settling velocity of a particle.

Particles settling in the clarifier encounter multiple forces—downward force of gravity, upward force of buoyancy and upward force of drag. They settle when the net downward force is positive i.e., the force of gravity is larger than the combined upward forces of buoyancy and drag. For solid spherical particles settling velocity is expressed as:

$$v = \frac{4g(\rho_s - \rho)d_p}{3C_d.\rho}^{0.5} \tag{2.17}$$

where,

v = Settling Velocity of Particles, m/s

g = Acceleration Due to Gravity, m/s^2

ρ_s = Density of Solid Particles, kg/m^3

ρ = Density of Liquid (Water in this Case), kg/m^3

d_p = Diameter of Settling Particle (Assuming Spherical Shape), m

C_d = Drag Coefficient

The parameter drag coefficient is a function of the flow regime of the settling particle, as defined by Reynolds Number, Re. In the laminar flow regime, where the Reynolds Number (Re) is in the range of 10^{-4} to 0.2, the drag coefficient is related to Reynolds Number (Re) as:

$$C_d = 24/Re \tag{2.18}$$

where,

$$Re = \frac{\rho * v * d_p}{\mu} \tag{2.19}$$

ρ = Density of Water, kg/m^3

v = Settling Velocity, m/s

d_p = Diameter of Settling Particle, m

μ = Dynamic Viscosity of Water, N.s/m^2, or Pa.s

The equation for settling velocity then simplifies to the following form:

$$v = \frac{g(\rho_s - \rho)d_p^2}{18\mu} \tag{2.20}$$

This is referred to as Stoke's Law for settling and the settling velocity is referred to as Terminal Settling Velocity. This is an important equation for the design of gravity clarifiers as settling of solids occurs under Type 1 settling conditions. All particles having settling velocities, equal to or greater than their terminal settling velocities will settle in the clarifier.

Various types of clarifiers are commercially used depending upon the type of water to be treated, site constraints, and preference of the treatment plant. These include the geometry of the basin—rectangular versus circular, surface loading rate, hydraulic retention time, weir loading rate and sludge collection and removal system. Surface loading rate, a very important parameter in sizing the clarifiers, is a function of clarifier geometry and feed configuration. In certain designs, as in the case of **Solids Contact Clarifiers**, the rapid mixing and flocculation functions are integrated into the clarifier tank. These are specifically applicable to circular clarifier tanks. In this design the influent and coagulant are introduced into an inner cylindrical concentric tank, also called a well, equipped with a vertically suspended and top mounted mixer. Here the influent mixes with the coagulant and previously settled solids within the well. The mixer promotes agglomeration of floc particles enmeshed with precipitated solids in the clarifier. This helps the aggregated solids settle rapidly and the clarifier to produce high effluent quality.

Upflow clarifiers are a special design of solids contact clarifiers where the influent is introduced from the bottom and rises through a thick blanket of settled sludge before entering the mixing zone in the well. The sludge blanket acts primarily to promote enmeshment of coagulated solids as well as a medium to capture the solids suspended in the raw influent. This type of clarifier is also termed Floc Blanket clarifiers in the literature. Table 2.4 presents typical values for key design parameters of clarifiers.

The treated effluent exits through weir plates along the perimeter of the clarifier tank and collects in launders leading to the effluent line. In the case of certain design of circular clarifiers, the weir plates may be placed along the diameter of the clarifier. Sludge scrapers transport settled solids to a collection hopper leading to the sludge discharge line. In the case of circular clarifiers, the hopper is located centrally at the bottom whereas in rectangular clarifiers it located toward the influent side of the basin. Figure 2.7 presents a schematic view of a circular solids contact clarifier with its important components.

Table 2.4 Sizing Guidelines for Clarifier Basins in Various Applications

Type of basin	Hydraulic retention time (Hrs.)	Surface loading rate m^3/m^2. d (gpd/ft^2)
Conventional rectangular basins		
Coagulationand flocculation	4–8	20–40 (50–1,000)
Softening	2–6	40–60 (1,000–1,500)
Solids contact units		
Coagulation and flocculation	2	40–60 (1,000–1,500)
Softening	1	60–100 (1,500–2,500)
Upflow basins		
Coagulation and flocculation	2	40–60 (1,000–1,500)
Softening	1	60–100 (1,500–2,500)

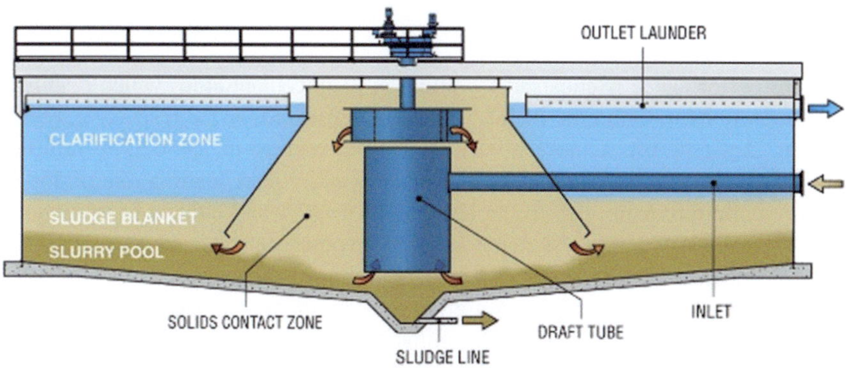

Fig. 2.7 Schematic view of solid contact clarifier (*Source* WesTech Engineering Co., Reprinted with Permission)

Clarifiers treating raw water by Type 1 and Type 2 settling are sized based on the results of settling column tests as detailed in literature (Metcalf and Eddy/AECOM 2014). Several variations in the design and operation of clarifiers are marketed by technology providers. These include ballasted clarifiers, which are also called high-rate clarifiers. Performance is significantly enhanced in these clarifiers by adding very fine particles externally in addition to the coagulant and flocculant chemicals that are normally added to traditional clarifiers. Two commonly available ballasted clarifier system are Actiflo marketed by Veolia Water Technologies, and Densa Deg marketed by Suez Water Technologies and Solutions. Actiflo uses fine micro-sand particles of diameter 45–100 μm as the external flocculant aid material. The sand settles in the clarifier with sludge which is recovered by cyclone separators and reused. The principle of operation of a ballasted clarifier is presented with the help of a schematic view as Fig. 2.8. Densa Deg clarifiers also operate on a similar principle where a fraction of the sludge from the clarifier bottom is recycled as the flocculation aide.

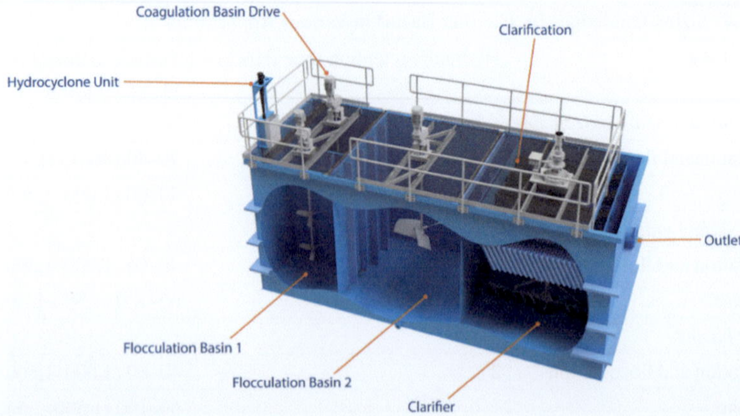

Fig. 2.8 Treatment scheme for ballasted clarification (*Source* WesTech Engineering Co., Reprinted with Permission)

Clarifier performance is enhanced by addition of flocculant aides which results in large reductions in hydraulic retention time. As reported by the respective system vendors, Actiflow can achieve a surface loading rate up to 2,450 L/m^2.min and Densa Deg has been reported to handle successfully a peak loading rate of 450 L/m^2.min while removing TSS to the level of 70 to 90% and reducing the footprint area (EPA, 2003). CoMag system marketed by Evoqua is a ballasted clarifier system that uses fine magnetite particles to enhance settling. As in the case of other ballasted clarifiers, ferric chloride or alum is used as the coagulant and magnetite is dosed as the ballasting material. Magnetite is a mineral containing oxides of iron and over 99% is recovered. The system needs a small amount of make-up for reuse in the system.

Inclined Plate Clarifiers are another type of high-rate clarifiers. While the surface loading rates are in the same range as those of the conventional clarifiers, the performance is enhanced by increasing the surface area multiple times with the help of inclined parallel plates. The influent is first mixed with coagulant in a rapid mix tank, and then it flows through a flocculator tank under low energy mixing condition, before being introduced into the clarifier consisting of parallel plates, typically inclined at 40 to 55°. As the flow is evenly distributed among the plates and it moves upward along the surface of the plates, the solids slide down by gravity and clarified water overflows into a collection trough and exits as treated effluent. Solids collect in a hopper at the bottom and are removed as sludge.

In summary, the types of clarifiers used in the water treatment industry can be classified as follows:

- Conventional clarifiers
  - Rectangular Geometry
  - Circular Geometry

- Special Design Clarifiers
  - Solids Contact Clarifiers
  - Upflow/Floc Blanket Clarifiers
  - Ballasted System Clarifiers

    Actiflo
    DensaDeg
    CoMag

- Inclined Plate Clarifiers

Filtration

Filtration is a process of separating fine suspended and colloidal solids from water by passing the water through a porous material or medium. Filtration (with or without a pretreatment process upstream) has been traditionally employed for the treatment of water to remove suspended solids, turbidity, hardness, iron and manganese (precipitated in aeration step), color, and microorganisms. After the flocs settle at the bottom of the sedimentation basin, the clarified water passes through filters for further polishing to achieve enhanced quality of treatment.

Typically, coagulation, flocculation and sedimentation are the upstream treatment processes of filtration. Functions of these upstream processes include removal of TSS associated with raw influent. Additionally, TSS is produced in the form of Fe $(OH)_3$ and MnO_2 by aeration when groundwater is the source of raw water. TSS is also produced as $CaCO_3$ and Mg $(OH)_2$ in the softening process when raw water contains significant amounts of hardness as Ca^{2+} and Mg^{2+} ions. A well-functioning clarifier effluent TSS ranges from 5 to ~20 mg/L. A well-functioning filter effluent TSS is expected to be between 1 and 2 mg/L. In terms of turbidity, clarifier effluent and filter effluent turbidities typically range from 2 to 100 NTU and 0.1 to 1.0 NTU, respectively. It is essential for the filtration systems to maintain the desired effluent TSS and turbidity levels to ensure that downstream process equipment are not fouled.

Filtration systems can be classified on the basis of several criteria—(a) conventional versus direct filters, (b) granular versus precoat filters, (c) rapid versus slow sand filters, (d) media types and beds of filters, and (e) gravity versus pressure filters. There are significant overlaps between the different types. For example, rapid sand pressure filters use granular media, which are most commonly used in industrial water treatment applications. A brief description of each classification is presented in the following sections.

(a) Conventional versus Direct Filtration

Conventional filtration systems follow coagulation, flocculation and sedimentation steps as described above. However, when treating surface waters of reasonably good quality many water treatment facilities eliminate the sedimentation step and proceed directly to the filtration step after coagulation, rapid mix, and flocculation. Therefore, even in the absence of a separate clarifier basin it is possible to achieve the desired effluent quality when the raw water is high quality as in the case of surface water at certain locations. American Water Works Association (AWWA) recommends a set of criteria (Committee Report, 1980) for direct filtration of surface water as follows:

Color	< 40 Color Units
Turbidity	< 5 NTU
Algae	< 2,000 asu/mL (asu = aerial standard unit of diatoms, where 1 asu = 400 μm^2)
Iron	< 0.3 mg/L
Manganese	< 0.05 mg/L

Direct filtration has an important advantage of reduced capital cost due to the elimination of the clarifier. Since settling step is avoided, coagulant dosage is also reduced compared to conventional filtration as the goal of coagulation prior to direct filtration is to create filterable pin flocs, as opposed to large settleable flocs required by high rate of coagulant dosing in case of clarifiers. This reduction in coagulant dosing also results in operating cost saving. Therefore, from both capital and operating cost perspectives, direct filtration is more economical than conventional filtration. However, there are certain disadvantages also. Due to the low retention time ahead of filtration any sudden fluctuations in water turbidity and TSS can rapidly affect the effluent quality. It is essential for the direct filtration systems to be equipped with on-line, real time monitoring instruments that prompt response by operators in the event of performance loss. Although more expensive than direct filtration, conventional filtration systems ensure reliable and robust performance. Conventional filters are most commonly used by oil refineries using surface water as the source of raw water for on-site water treatment.

(b) Granular versus Precoat Filters

Granular bed filters are comprised of filtration media contained in concrete or steel vessels. These can be of mono media or multimedia type. In multimedia granular filers media are arranged vertically in layers of anthracite, garnet and sand from top to the bottom of the vessel. Grains of media in a bed are within a specified size range which gradually decreases from the top bed to the bottom. This can be observed from Table 2.5.

Precoat filters use a thin layer of diatomaceous earth which serves as the filtration medium coated on a porous material, like textile or synthetic fabric, supported on a rigid steel frame. The filtration elements are configured in the form of circular disks mounted on horizontal shaft within a cylindrical steel casing. The influent water is introduced under pressure from one end and the filtered water exits from the other end. Suspended particles $\geq 2\ \mu m$ are arrested by the filter media. Precoating material is dosed with the influent from time to time to maintain the permeability of the media that helps increase cycle run length.

(c) Filtration Rate - Rapid versus Slow

Filters, based on the filtration rate can be classified as slow sand filters, rapid filters and high-rate filters. Filtration rates depend upon the applied hydraulic loading rates.

Table 2.5 Typical design parameters for media filters (Adapted from Reynolds and Richards 1955)

Media characteristics	Single media conventional filters	Dual media filters	Multimedia (Trimedia) filters
Anthracite Layer ($\rho = 1.60$)			
Effective grain size, mm	0.70–0.75	0.90–1.10	0.95–1.00
Uniformity coefficient	<1.75	1.60–1.80	1.55–1.75
Bed Depth, mm (in.)	610–760 (24–30)	460–610 (18–24)	420–530 (16.50–21)
Sand Layer ($\rho = 2.65$)			
Effective grain size, mm	0.35–0.70	0.45–0.55	0.45–0.55
Uniformity coefficient	< 1.70	1.50–1.70	1.50–1.65
Bed Depth, mm (in.)	610–760 (24–30)	150–205 (6–8)	150–230 (6–9)
Garnet Layer ($\rho = 4.2$)			
Effective grain size (mm)	N/A	N/A	0.20–0.35
Uniformity coefficient	N/A	N/A	1.6–2.0
Bed Depth, mm (in.)	N/A	N/A	75–115 (3–4.5)
Filtration Rate, L/m^2.s (gpm/ft^2)	1.36–3.40 (2–5)	2.04–5.44 (3–8)	2.72–6.80 (4–10)

Notes

1. ρ = Specific Gravity of Filtration Media
2. Uniformity Coefficient (UC) of media particles is used to indicate the variations of the grain size of particles in a mixture. Ideally, if all particles in a mixture have identical size the uniformity coefficient of the particles is 1. It is expressed as the ratio of the sieve size through which 60% by weight of the particles in a mixture pass, to the sieve size through which 10% pass. As the non- uniformity in size of a mixture of particles increases, the UC also increases
3. Effective size (ES) of the grains of media is determined from the sieve analysis of the media particles following the ASTM Standard Test Method C136-92 which is based on the plot of % of grains by weight (X – axis) passing through sieves versus the sieve opening size, mm (Y-axis). The sieve size that passes 10% of the grains by weight is the ES of the particles in that bed of media
4. A detailed description of the procedures to determine UC and ES of the grains in the bed of a filtration media is available from ASTM Test Method C136-92 and Water Quality and Treatment, A Handbook of Community Water Supplies (AWWA 2011)

Slow sand filters typically have a hydraulic loading rate of less than 10 m^3/m^2/d (0.17 gpm/ft^2). They may not require prior chemical coagulation but rate of production of treated water is low due to low hydraulic loading rates. Rapid gravity filters have hydraulic loading rate of around 120 m^3/m^2/d (2 gpm/ft^2), and the high-rate filters have hydraulic loading rate of 240 m^3/m^2/d (4 gpm/ft^2) (Kawamura 1975).

Results of several full-scale studies demonstrated that the treated effluent turbidity increases, and filtration run length decreases with increasing hydraulic loading rate. Also, pressure drop increases with increasing loading rate. However, the results of increased production rate and smaller footprint requirement outweigh the disadvantage posed by increase in effluent turbidity with high hydraulic loading rate which can be mitigated by using dual or multimedia filters in place of mono media filters.

(d) Media Types and Beds

Filters can further be classified as single media, dual media, or multi-media. In single media filters only one type of material (graded sand or anthracite) is used as filtration medium. The dual media filters use anthracite and sand as filter media. Multi-media are similar to dual media filters, however in these filters several types of materials in multiple layers are used as filtration media. Typically, three different materials e.g., anthracite, sand and garnet are used as filter media in multi-media filters. The media are arranged in beds from top to bottom with the coarsest grains in the top bed and decreasing grain sizes in the lower beds. The main advantage of using dual or multimedia over monomedia filtration is the reduction in pressure drop, and hence extending filter runs between regenerations. This is possible because, in monomedia filters only the medium surface actively removes suspended solids, and the lower depths of the media remain unused. But in multimedia filters suspended solids are trapped along the depth of the media in the direction of coarse to fine size grains. This allows better utilization of both the surface as well as the depth of filter media (AWWA 2011). Typical design parameters for various media are as presented in Table 2.5.

(e) Driving Force

Filters can also be classified as gravity versus pressure filters. The main difference between these two types of filters is the head required to force water through the media and the vessel type used. Gravity filters require a head of approximately 2–3 m and are typically housed in open concrete rectangular tanks or circular steel tanks. The sand or other media in the filter is usually less than a meter deep. These filters have a capacity of up to 50 m^3/hr. If more capacity is required multiple vessels in parallel are used.

Pressure filters are usually vertical closed steel tanks, with diameters up to 3 m. Both gravity and pressure filters operate under the same principle using the same types of media as shown in Table 2.5. However, pressure filters offer certain advantages over gravity filters:

- Since these filters operate under pressure, the effluent from the filter usually has adequate pressure to reach its destination; and
- Pressure filters are compact and easy to automate, requiring less operator attention.

In most cases, pressure filters are used in petroleum refinery water treatment applications. Capacity of pressure filters is up to 70 m^3/hr. Multiple filters are used if more capacity is required. Figures 2.9 and 2.10 are illustrations of a circular pressure filter and a rectangular gravity filter respectively with the views of their internal components.

Head Loss for Flow through Filtration Media

Head loss (pressure drop) for flow of water through a clean static filtration medium is estimated by the Kozeny equation, as given below:

Fig. 2.9 Circular pressure filter (*Source* WesTech Engineering Co., Reprinted with Permission)

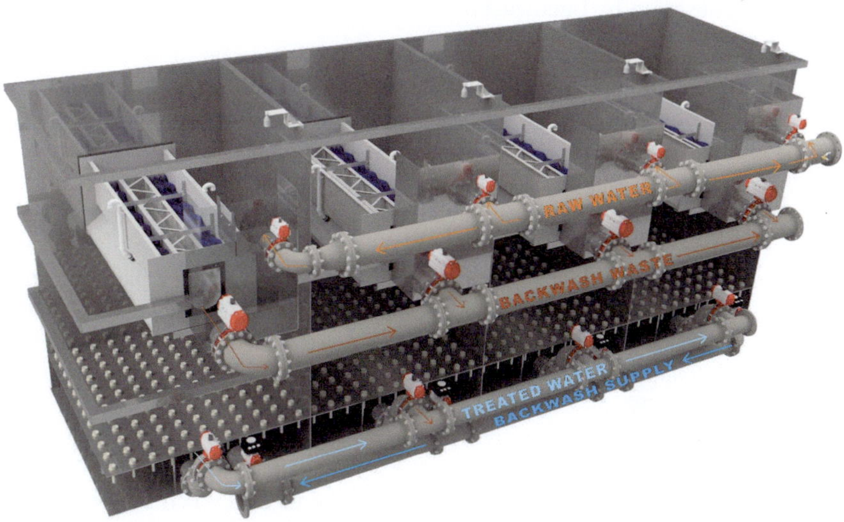

Fig. 2.10 View of a rectangular gravity filter with internal details (*Source* WesTech Engineering Co., Reproduced with Permission)

$$h/L = \frac{k * \mu}{\rho * g} * \frac{(1-\varepsilon)^2}{\varepsilon^3} * \left(\left(\frac{a}{v}\right)^2\right) * S \qquad (2.21)$$

where,

h/L = Head loss/unit bed depth (dimensionless)

k = Dimensionless Kozeny Constant ~ 5 (typical value)

μ/ρ = Kinematic Viscosity of Water at Operating Temperature, m^2/s

g = Acceleration due to Gravity = 9.8 m/s^2

ε = Porosity of medium bed (dimensionless)

a/v = Specific surface area/unit volume of grain making up the medium = 6/d for spherical grains

d = Grain diameter for spherical grains, m

S = Hydraulic loading rate, m/s

Kozeny equation applies to strictly laminar flow through the clean filtration medium which is truly the case for flow through a clean bed of medium with grain size between 0.5 mm and 1.0 mm and hydraulic loading rate of 5–12 m/hr. This equation is an expression of pressure drop for flow through a single medium. For multimedia filters beds the total pressure drop across the total depth of the filtration media is the sum of the pressure drop for each individual media bed.

Media Regeneration by Backwashing

During filtration the flocs of solids are intercepted by the grains of filter media and gradually accumulate in the media. This leads to increasing restriction to the flow, and gradual increase in pressure drop across the media bed. With time the media become fouled, and the effluent quality starts degrading due to breakthrough of solids. To sustain the quality of filtered effluent at the desired rate of filtration it is necessary to clean up the filters routinely by backwashing. There are various modes of backwashing but traditionally it is done by fluidizing the media beds using clean water at a high rate in a direction opposite to the normal flow. The backwash rate, frequency and duration are determined by the influent water quality and the design of the filtration system as recommended by the equipment suppliers based on their in-house test results. In general, the backwash operation is terminated when the waste wash water becomes reasonably clean with a turbidity ~ 10 NTU. In some filtration systems the backwashing is preceded by an air scouring step which dislodges the solids accumulated in the filtration media. After air scouring, backwash water is introduced which washes off the solids from the media. When backwashing is complete the media settle down as a clean bed, ready to start its new cycle of filtration operation. Treated effluent from filtration systems is collected in a tank for feeding to the next treatment steps. A part of this water is utilized as backwash water. The waste backwash water is thickened, and the sludge dewatered for disposal.

In most water filtration systems multiple units operate in parallel, with one or multiple unit(s) in service (depending upon flow) while others are in backwash and/or standby mode. When the operating unit(s) become sufficiently fouled, and pressure drop increases to a certain predetermined level, it is taken off-line for backwash and the standby bed is brought into service so that the flow through the filtration system as a whole is uninterrupted. The run time of a filter unit is chosen based on pressure drop or time, depending upon the type of water and operating experience. The number of parallel filtration units, run length of each filter, backwash rate, frequency and duration are parts of the overall design of the filtration system which depend on the characteristics of influent water, desired effluent quality, and desired production rate.

Other Types of Filtration Systems

A few other types of filters are also used in industry. Upflow filters are among the commonly used types of filters. As opposed to downflow filters described above, in the case of the upflow filters the influent flows up from the bottom of the filtration unit and the filtered effluent exits from the top. Several designs of upflow filters are commercially available. In one such design, the filtration system is a tall, vertical cylindrical tank. Influent is introduced at the bottom of the media through a distribution plate. As it rises through the bed of the media the suspended solids are trapped and removed, and filtered water flows out from the top. Solids removal in this case is by simple filtration mechanism in reverse direction of downflow filtration. During backwash, filtered water in combination with compressed air is supplied through a pipe which loosens the media and airlifts the trapped solids. The sand media remains in place due to its high density. In some designs a retainer screen is provided on the top to arrest media particles from escaping.

Disinfection

The treatment processes discussed above assist in removal of suspended solids along with the removal of microorganisms to a large extent. However, these processes may not be sufficient to ensure removal of microorganisms to the level necessary for beneficial use of treated water. The main purpose of disinfection of water is to remove disease causing organisms when a part of the treated raw water is used for drinking and sanitation in an industrial facility.

Boiling water prior to use has been practiced for many years as a means of protecting public health and hygiene. In large scale operations disinfection processes are as follows:

- Use of biocide chemicals, primarily chlorine, sodium hypochlorite, chlorine dioxide and ozone; and
- Application of UV Radiation.

In petroleum exploration and processing industries the purpose of disinfection are as follows:

- Kill iron and sulfur reducing bacteria in cooling water that can cause corrosion of heat exchanger and condenser tubes when groundwater is used as the source of raw water

- Perform as biocide to water injected for enhanced oil recovery, water flooding and hydrofracturing in oil and gas exploration and production operations
- Control microbiological fouling of ultrafiltration (UF) and reverse osmosis (RO) membranes.

Chemical biocides are predominantly used for disinfection. The chemicals used for the disinfection undergoes several transformations so that the disinfecting action is really exerted by the end products of reaction. As an example, when chlorine is used as the disinfectant it hydrolyzes in water according to the following reaction:

$$Cl_2 + H_2O \rightarrow HOCl + HCl \tag{2.22}$$

Hypochlorous acid (HOCl) dissociates in water in accordance with the following equilibrium reaction:

$$HOCl \leftrightarrow H^+ + OCl^- \tag{2.23}$$

$$K = \frac{[OCl^-] * [H^+]}{[HOCl]} \tag{2.24}$$

The relative distribution of HOCl and OCl^- is a function of pH and temperature with a pK value of ~ 7.6 at 25 °C (AWWA, 2011). The relative distribution of these two species is expressed as:

$$K = 10^{(7.5-pH)} \text{ at } 25^{\circ}C \text{ (AWWA, 2011)} \tag{2.25}$$

$$\text{and, } pK = -\log K \tag{2.26}$$

While both species have disinfection properties HOCl is a superior disinfectant in terms of its germicidal effect of killing bacteria, virus and other microorganisms that can cause adverse health effects.

Due to the health and safety risk of handling chlorine gas many facilities use sodium hypochlorite as the source of the disinfectants HOCl and OCl^-. It is commercially available as aqueous solution in various strengths. In water it dissociates as:

$$NaOCl \rightarrow Na^+ + OCl^- \tag{2.27}$$

$$OCl^- + H^+ \leftrightarrow HOCl \tag{2.28}$$

Factors affecting the effectiveness of disinfection are the type and concentration of disinfectant, contact time and characteristics of the water, as follows:

- Concentration:

Application of higher concentration of a disinfectant results in higher destruction of organisms. However, application beyond the optimum point, (as determined by post disinfection microbiological population count), results in waste of disinfectant.

- Characteristics of Water:

The presence of organic matter and nitrogenous compounds (TKN) in water consumes the active disinfection agent and reduces its availability. Also, the presence of suspended particles in water may shield the microorganisms and evade disinfection.

- Contact Time:

The higher the contact time available for disinfection, the higher will be the destruction of organisms. Typically, the contact period of no less than 30 min is considered for the design of contact tank for disinfection. If the sufficient contact time is not available, the dose of the disinfectant should be increased accordingly.

- Temperature of Water:

The effectiveness of the disinfection process is directly proportional to the temperature of the water. The higher the temperature of the water, the higher and faster will be the destruction of organisms.

Gaseous chlorine and sodium hypochlorite solutions carry out disinfection reaction by oxidative destruction of organic matters making up the microbiological species in water. Other oxidizing agents commonly used in water and wastewater treatment are chlorine dioxide, ozone and UV radiation.

Apart from the chemicals discussed above other chemicals also are used as biocides in specific applications. For example, in upstream water treatment applications glutaraldehyde (pentane dial, $C_5H_8O_2$) is used as a biocide in fracturing water (see Chap. 3 for details). This is a non-oxidizing type of biocide. For cooling water treatment too, some non-oxidizing biocides e.g., quaternary amines, glutaraldehyde, etc. are used in combination with oxidizing biocides, primarily chlorine and sodium hypochlorite.

2.1.1.3 Advanced Water Treatment

High quality water is used in certain industrial applications e.g., power generation, petroleum refining, semiconductor industries, food and pharmaceutical, etc. The role of advanced treatment is to remove dissolved and colloidal constituents from the product of primary treatment to fulfill the requirements for specific industrial feedwater quality. Unit operations and processes for advanced water treatment can be broadly classified as follows:

- Removal of Organic Constituents
  - Adsorption
  - Chemical Oxidation Processes
  - Membrane Separations
- Separation of Inorganic Constituents
  - Ion Exchange (IX)
  - Membrane Separations

These are not very strict classifications. For example, in water treatment, while adsorption is widely used in the removal of organic matters, it is also effective in removing metals in water present in ionic and colloidal forms. Membrane separations are applicable in the removal of both organic and inorganic constituents. Also, advanced processes are used for production of high purity water from effluents of both water and wastewater treatment systems.

Adsorption

Adsorption is a solid–fluid operation involving transfer of dissolved substances from a fluid to a solid phase. It is a surface phenomenon. The substance being transferred is the adsorbate which concentrates on the surface of the solid absorbent. This principle is utilized in water and wastewater treatment applications to remove dissolved contaminants from the liquid phase and concentrate those on to the surface of the adsorbent.

Activated carbon is the most widely used adsorbent in water and wastewater treatment applications. It is made by heating materials of carbonaceous origin e.g., wood, coconut shell, coal, etc., in an inert atmosphere at a temperature of 300–1000 °C. Under this condition moisture and some amount of light organic compounds leave as vapor creating a porous structure on the solid surface. The pores are numerous in the size range of 10–1000 Å providing a very large surface area of the order of 1000 m^2/gram of activated carbon, for the adsorbate substances to deposit on in multiple layers. This is the key to the effectiveness of activated carbon as an adsorbent.

Activated carbon is very effective in removing organic compounds from water due to its selective affinity for organic compounds. Although chemical bonding, called chemisorption, is quite prevalent in solid–fluid adsorption, but it is not relevant to water treatment applications. Adsorption of organic molecules in aqueous phase on to carbon surfaces results from weak physical bonding and van der Waals' forces following the mechanism of physical adsorption.

Adsorption of an adsorbate A on to a solid adsorbent surface is a reversible process, expressed as:

$$C_L \leftrightarrow C_S \tag{2.29}$$

where, C_L is the concentration of adsorbate A in Liquid (Water) Phase, and

C_S is the concentration of adsorbate A in Solid Adsorbent Phase.

Capacity of an adsorbent to remove a specific organic compound is determined by its equilibrium distribution between water and activated carbon phases, which is a function of temperature and pressure. Carbon adsorption of contaminants in water is favored at low temperature and high pressure. The equilibrium relationship is expressed by several equations:

- Langmuir Isotherm
- Brunauer, Emmet and Teller (BET) Isotherm
- Freundlich Isotherm

In water treatment applications, Freundlich isotherm is the most widely used equilibrium relationship expressing the distribution of a solute between the aqueous and activated carbon phases at a given temperature and pressure. Mathematically, it is expressed as:

$$\frac{x}{\mathrm{M}} = \mathrm{K} * \mathrm{C}^{(1/\mathrm{n})} \tag{2.30}$$

where,

x/M = Mass of the organic compounds adsorbed per unit mass of activated carbon

K = Equilibrium constant

C = Concentration of organic compound in water in equilibrium with activated carbon

n = Freundlich Constant

It is an empirical equation and for any full-scale application, it is essential to conduct a number of batch tests with an aqueous solution of the organic compound under consideration in contact with activated carbon in varying amounts in each batch. To ensure attainment of equilibrium it is necessary to run the tests on a shaker table for several days such that there is no further change in the aqueous phase concentration in each batch after the conclusion of the tests. At that point concentration of the compound in water is measured, and the amount adsorbed (x) is calculated from the difference in concentration and the volume of the solution, as follows:

$$\mathrm{x} = (\mathrm{C_o} - \mathrm{C}) * \mathrm{V} \tag{2.31}$$

where,

x = mass of solute (organic adsorbate) adsorbed, mg

Co = Initial concentration of the solute in aqueous solution, mg/L

C = Concentration of organic solute in aqueous solution after conclusion of test, mg/L

V = Volume of the aqueous solution in each test batch, L

The ratio x/M for each batch is then calculated by dividing x (mg) by the amount activated carbon M (g) in the test batch. In the next step x/M for each batch is plotted on the y-axis against C on the x-axis in log–log scale. This yields a straight line with the slope 1/n and intercept log K. The constants K and n for adsorption of the specific organic compound by activated carbon obtained from this graphical procedure is used in sizing full scale adsorption systems. A list of Freundlich Isotherm parameters (K and 1/n) for several commonly encountered organic contaminants in water are available in the literature (AWWA 2011). In water and wastewater applications COD is widely used as a means of expressing the concentration of bulk organic contaminants.

Carbon adsorption is conducted in two configurations - granular activated carbon (GAC) and powdered activated carbon (PAC) that differ on the basis of the size of the carbon particles, system configuration and the mode of application. GAC consists of carbon particles in the size range of 0.2–5 mm, and PAC particles are smaller than 0.5 mm size. Typically, GAC adsorption systems are packed vertical columns which treat contaminated water streams continuously in downflow mode. GAC column treats large volume of water until the surfaces of the activated carbon particles are saturated, the column loses its effectiveness, and contaminants break through the column untreated.

In addition to batch tests, it is always advisable to conduct pilot tests using GAC columns running with the actual water at site, or in a laboratory. These tests can give valuable information about GAC bed depth, surface and volumetric loading rates, empty bed contact time (EBCT) and estimated treatment capacity of a given volume of GAC prior to breakthrough. EBCT is defined as the volume of the empty bed that houses the GAC adsorbent divided by the flow rate of contaminated water through the GAC bed and is reported in the unit of time (minutes or hours).

In full scale installations, GAC columns are set up in series or parallel configuration. Based on pilot test results and operational experience on site the breakthrough time of each column is projected. On the basis of such a projection, in parallel configuration, the operating column is taken offline for regeneration and a standby column with regenerated GAC is brought into service. Alternatively, when the columns operate in lead-lag mode in series, the lead bed is taken offline, and the spent GAC is removed and replaced with either new or reactivated GAC.

Carbon adsorption is a reversible process. High temperature and low pressure promote the desorption of adsorbed species. This is the principle utilized for regeneration of spent carbon media. Typically, the spent GAC media are transported to the supplier site and thermally reactivated for reuse.

Carbon adsorption is used in both water and wastewater treatment. In case of water treatment, it is primarily used in powdered activated carbon (PAC) form to remove naturally occurring organic materials (NOM) at the beginning of a treatment train as they interfere with the downstream treatment processes. Another application in water treatment is removal of free chlorine or hypochlorite ions from water by surface reaction on GAC granules.

In wastewater treatment applications, GAC is primarily used as a post biological tertiary treatment step for removal of residual COD and TOC, which may be carried over due to the refractory nature of organic chemicals originating from industrial wastes.

PAC differs from GAC not only in particle size but also in the mode of application. GAC is packed in static columns through which contaminated water flows for treatment. In case of PAC, it is dosed into the water stream in a slurry form. Adsorption takes place under continuous mixing at one or multiple dosing locations including raw water intake station, rapid mix tank, flocculation tank, biological treatment reactor, and sedimentation tank. In biological treatment reactors the organics concentrate onto the powdered carbon surfaces that provide sites for microbiological growth and attachment leading to breakdown of those compounds. Good mixing and sufficient contact time are the key to good performance of PAC adsorption process. These operating parameters are determined by field pilot testing with the actual wastewater on site. PAC settles and is disposed with the sludge in the settling basin. Used PAC is wasted and cannot be regenerated. The main application of PAC is in the removal of NOM, taste and odor from water used in drinking and industrial water applications.

Further discussion on application of PAC adsorption in refinery wastewater treatment has been presented in Chap. 4.

Chemical Oxidation

Chemical oxidation is an important group of unit processes in water treatment. Many contaminants in water are broken down to harmless forms by oxidation reactions. These reactions are relevant to both water and wastewater treatment.

Chemical reactions involving exchange of electrons between the reactants are oxidation–reduction, or redox reactions. The oxidizing agent(s), or oxidant gains electron(s) and relegates to a lower oxidation state. In exchange, the reducing agent(s) loses electrons and elevates to a higher oxidation state. Most commonly the oxidation reactions of ferrous, manganese and sulfide ions are conducted by aeration where air is used as the source of oxygen.

Oxidants are classified in order of their strengths, expressed as Standard Electrode Potential (E^o), listed in Table 2.6. It may be observed from this table that the oxidation potential of hydroxyl radicals (OH^*) is far above those of classic oxidants such as oxygen, ozone, and hydrogen peroxide, and second only to fluorine. Oxidation reactions conducted by hydroxyl radicals are Advanced Oxidation Processes (AOP).

Hydroxyl radicals are highly energized chemical species that do not freely exist by themselves and are generated in-situ at the point of application. AOP based processes are deployed to treat some of the very hard to degrade chemicals present in certain natural waters and in industrial effluents. Water from both surface and ground water resources sometimes contains natural organic matters (NOM), primarily in the form

Table 2.6 Standard electrode potentials of various oxidants at 25 °C (Reconstructed from Basu 1996; Skaggs 2012)

Oxidant	Half cell reaction	Standard potential, E° (Volts)
Fluorine	$F + e \rightarrow F^-$	3.03
Hydroxyl radicals (OH^*)	$OH^* + H^+ + e \rightarrow H_2O$	2.80
Singlet oxygen	$^1O_2 + 4\,H^+ + 4e \rightarrow 2H_2O$	2.42
Ferrate	$FeO_4^{2-} + 8H^+ + 3e \rightarrow Fe^{3+} + 4H_2O$	2.20
Ozone	$O_3 + 2\,H^+ + 2e \rightarrow O_2 + H_2O$	2.07
Persulfate ($S_2O_8^{2-}$)	$S_2O_8^{2-} + 2e \rightarrow 2\,SO_4^{2-}$	2.01
Hydrogen peroxide	$H_2O_2 + 2H^+ + 2e \rightarrow 2H_2O$	1.78
Permanganate	$MnO_4^- + 4H^+ 3e \rightarrow MnO_2 + 2H_2O$	1.68
Chlorine dioxide	$ClO_2 + e \rightarrow ClO_2^-$	1.50
Hypochlorous acid	$HOCl + H^+ + 2e \rightarrow Cl^- + H_2O$	1.49
Chlorine	$Cl_2 + 2e \rightarrow Cl^-$	1.36
Molecular oxygen	$O2 + 4H + + 4e \rightarrow 2H_2O$	1.27
Bromine	$Br_2 + 2e \rightarrow Br^-$	1.09

of various humic and fulvic acids, which add odor, color, and taste that are difficult to remove. AOPs are one of the processes to break down these chemicals in water.

Several AOPs are used in various water and wastewater treatment applications. Most commonly these are UV-Peroxide Treatment, UV-Ozone Treatment, and combinations of UV-Peroxide and Ozone. In the case of UV-Peroxide treatment UV radiation induces in-situ generation of high energy hydroxyl radicals from hydrogen peroxide. The simplified reaction scheme is represented below:

$$H_2O_2 + h\nu\ (at\ \lambda = 254\ \text{nm}) \rightarrow 2\ OH^* \tag{2.32}$$

where,

h = Planck's Constant, $6.62607015 \times 10^{-34}$ Joule/Hertz

λ = Radiation Wavelength, nm

ν = Frequency, Hertz

The oxidizing reactions are promoted by the highly energized hydroxyl (OH^*) and perhydroxyl (HO_2^*) radicals, which are formed as intermediates in the catalytic decomposition of hydrogen peroxide. UV radiation also generates hydroxyl and perhydroxyl radicals from multistep decomposition of ozone, which is the principle behind UV-Ozone AOP. UV-Peroxide and UV-Ozone treatments have been applied extensively to treat hard-to-degrade organic compounds in water, wastewater and contaminated groundwater. For industrial wastewaters AOPs are sometimes used to breakdown certain hard-to-degrade chemicals to make them amenable to biological treatment.

Ion Exchange

Ion exchange (IX) process removes the cationic and anionic constituents in water that make up the dissolved solids (TDS). An IX system consists of a cation exchanger that removes the total cations sodium (Na^+), calcium (Ca^{2+}), magnesium (Mg^{2+}), etc. and an anion exchanger that removes anions e.g., chloride (Cl^-), sulfate ($SO_4{}^{2-}$), nitrate (NO_3), etc. from water. The cation and anion exchange reactions are conducted by ionic functional groups attached on the surfaces of beads of polystyrene resins that measure up to 1 mm in diameter. Multiple polystyrene chains are cross-linked together to form a rigid backbone that supports the mobile functional group on the surface. The cross linking provides the strength of the resin beads and prevents those from attrition. The mobile ions on the resin surfaces are exchanged with ions in the water. When the active ions in the functional groups are exhausted the spent resin is regenerated and reused.

The ion exchange process can be operated in a batch or continuous mode. In batch process the resin is stirred in water until the reaction is completed and equilibrium is established. Then the spent resin is settled, removed, and regenerated. In case of continuous mode of operation, water is introduced from the top of a vertical tank, also referred to as a column, which is packed with the specific type cationic or anionic resins. As it flows down the column it comes in intimate contact with the functional groups on the surfaces of the ion exchange resins which carry out the chemical reactions.

The media bed depth in a continuously operating ion exchanger column is typically about one meter. When the bed is exhausted, it is regenerated using salt (sodium chloride), acid, or caustic soda (sodium hydroxide) depending upon the type and function of the resin. The functional groups on the ion exchange (IX) resin surfaces dictate which ions are exchanged. There are four types of ion exchangers namely,

- Strong acid cation (SAC)
- Strong base anion (SBA)
- Weak acid cation (WAC)
- Weak base anion (WBA)

Sulfonate (-R-$SO_3{}^-$) is the functional group of Strong Acid Cation (SAC) exchange resins. SAC resin behaves in a manner similar to a strong acid and are highly ionized over the entire pH range. It exchanges hydrogen ions of its functional group with all types of cations present in feed water including metal ions that contribute to hardness.

Weak Acid Cationic (WAC) resins use carboxylate (-COO^-) ions as the functional group. These resins behave like weak organic acids that are weakly dissociated. They have certain limitations in their use as WAC resins exchange active ions in their functional group with hardness ions calcium and magnesium only, but not sodium. So, they are not useful for desalination applications. Moreover, the functional group is active in alkaline pH range only.

Cation exchange systems containing both SAC and WAC resins are quite effective in removing calcium and magnesium ions and can reduce the hardness to about 2 mg/L as $CaCO_3$.

Strong Base Anion (SBA) exchange resins use quaternary amine group in hydroxide form. This group of resins is highly ionized, having strong base functional groups such as hydroxide (OH^-) and is effective for ion exchange through the entire pH range of 0 to 13. SBA resins exchange all anions of both strong and weak acids. In addition to hydroxide (OH^-) these resins are also used in chloride (Cl^-) form for water deionization.

Weak base anion (WBA) resins have weak organic base as functional group. Its degree of ionization is dependent on pH. WBA is effective only in acidic pH range and the resin removes anions of strong acids only that are produced by cation exchanger upstream, but not those of the weak acids, e.g., carbonic acid and silicic acid. In ion exchange reactions the organic base incorporates (as opposed to exchange) acids to form salt at the active sites instead of exchanging the anions. WBA cannot conduct any ion exchange or incorporation reaction at neutral or basic pH. Application of WBA in water treatment is very limited, especially in refinery water treatment applications.

In industrial water treatment applications where high quality product is necessary for raising steam in high pressure boilers for power generation and for use as process steam the IX systems typically consist of a SAC cationic bed and an SBA anionic bed. The reactions that take place in the beds are as follows:

Cation Bed

$$2\,RH^+ + Ca^{++} \rightarrow R_2Ca^{++} + 2\,H^+ \tag{2.33}$$

$$2RH^+ + Mg^{++} \rightarrow R_2Mg^{++} + 2\,H^+ \tag{2.34}$$

$$RH^+ + Na^+ \rightarrow RNa^+ + H^+ \tag{2.35}$$

As hydrogen ions from the cation exchange resin are released into the water in exchange for cations the effluent from a cation exchanger contains a large hydrogen ion concentration. It is highly acidic and has low pH. If all the incoming cations are replaced by hydrogen ions the acidity of the cation exchanger effluent is called Theoretical Mineral Acidity (ThMA). However, the actual acidity of this water, also called Free Mineral Acidity (FMA) is always a little less than ThMA as the ion exchange is never complete and there is always some leakage of cations into the effluent stream.

Anions in the water exiting the cation exchanger are replaced by the hydroxide ions that are embedded in the anionic resin structure resulting in the following reactions.

Anion Bed

$$R^1OH^- + Cl^- \rightarrow R^1Cl^- + OH^- \tag{2.36}$$

$$2\,R^1OH^- + SO_4^{2-} \rightarrow R_2^1SO_4^{2-} + 2\,OH^- \tag{2.37}$$

$$2\,R^1OH^- + CO_3^{2-} \rightarrow R_2^1CO_3^{2-} + 2\,OH \tag{2.38}$$

In addition to the various constituents discussed above, water in certain regions also contain silica in both colloidal and dissolved forms. Colloidal silica originates from erosion of sandstone, and attrition of sand and silt in contact with source water by mechanical forces. These are removed in the pretreatment step by coagulation, flocculation, sedimentation, and filtration processes. However, a part of silica is present in dissolved form as silicate (SiO_3^{2-}) ions. These are weak anions and are arrested by the anion exchange resin according to the following reaction.

$$2\,R^1OH^- + SiO_3^{2-} \rightarrow R_2^1SiO_3 + 2\,OH^- \tag{2.39}$$

In the above reactions R denotes the SAC resin support structure and R^1 denotes the SBA resin support structure. The effluent from the anion exchanger is neutral as the free hydrogen ions released from the cation exchanger combine with the free hydroxide ions from the anion exchanger to form water according to the reaction:

$$H^+ + OH^- \rightarrow H_2O \tag{2.40}$$

After the resins in the bed are spent both the beds are regenerated to restore the cation and anion exchange functions using a dilute acid (such as sulfuric acid) and a dilute base (such as sodium hydroxide) solution respectively. The regeneration reactions are as follows:

Cation Bed

$$R_2Ca^{++} + 2H^+ \rightarrow 2RH^+ + Ca^{++} \tag{2.41}$$

$$R_2Mg^{++} + 2H^+ \rightarrow 2RH^+ + Mg^{++} \tag{2.42}$$

$$RNa^+ + H^+ \rightarrow RH^+ + Na^+ \tag{2.43}$$

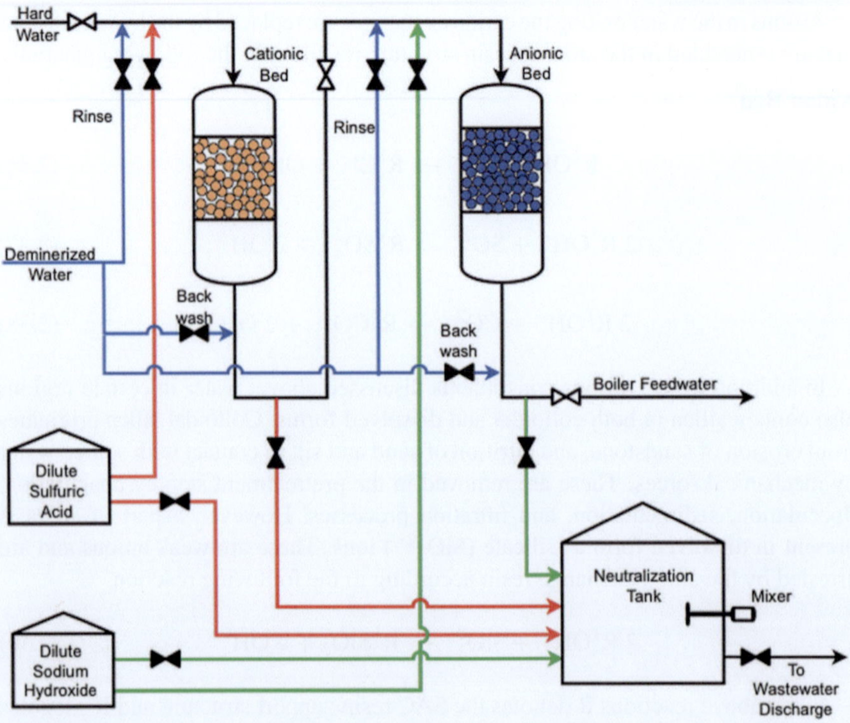

Fig. 2.11 Ion exchange treatment system

Anion Bed

$$R_2^1SO_4^{2-} + 2OH^- \rightarrow R_2^1OH^- + SO_4^2 \quad (2.44)$$

$$R^1Cl^- + OH^- \rightarrow R^1OH^- + Cl^- \quad (2.45)$$

$$R_2^1SiO_3^{2-} + 2\,OH^- \rightarrow R_2^1OH^- + SiO_3^{2-} \quad (2.46)$$

The spent regenerant solution from cation exchange resin consists of an acidic solution containing sulfate salts of calcium, magnesium, and sodium cations, and that from the anion exchange resin contains an alkaline solution of sodium chloride, sulfate, and other anions in high concentration. They are mixed together, and pH is further adjusted by adding necessary amount of acid and/or alkali to bring it to neutral range before disposal to the wastewater treatment plant. Figure 2.11 is a flow diagram depicting the principle of ion exchange process. A detailed discussion and process description follows in the Boiler Feed Water Treatment section in Chap. 5.

Membrane Separations

Membrane separation technologies offer the ability to design robust, compact, cost effective, and versatile treatment systems that meet a wide range of process requirements for removal of suspended and dissolved solids in water. There are many uses of water within both upstream production and downstream processing of petroleum where membranes are routinely applied as the most effective and optimal solution to meet these requirements.

Background and Operating Principle

Membrane separation emerged as a viable means of water purification in the 1960s with the development of high-performance synthetic membranes. Implementation of membranes for water treatment has progressed using advanced membranes made from new materials and employed in various configurations. The increasing scarcity in fresh water sources fueled a push towards alternative resources including waters of various levels of salinity from brackish water to ocean water as the source of both drinking and industrial water. In the 1970s the use of membranes for water desalination was being explored. For producing purified water from saline water, membranes became a viable alternative to evaporation-based technologies in the water treatment market (Sagle and Freeman 2004). Membrane separations processes use a semi-permeable membrane to filter fluids from dissolved or suspended solids (TDS, TSS) including, but not limited to water. For water treatment applications, membrane systems produce a high-quality product stream and a highly concentrated reject stream consisting of the undesired solids.

In comparison with conventional sand or media filtration, which relies on several mechanisms such as adsorption, settling, and straining to remove particles, micro and ultrafiltration (MF and UF) membranes provide an absolute mechanical barrier for particles that are larger than its pores to be rejected, guaranteeing a certain filtered water quality as depicted conceptually in Fig. 2.12.

Reverse Osmosis (RO) systems were the first type of membrane systems to be used to remove dissolved solids in desalination applications and first saw commercial application in the 1970s. Since that time there has been tremendous growth in the types of membrane processes and applications due to the growth in demand for high purity water for the general population and industry.

There are two major types of membrane which include products in the microfiltration (MF) and ultrafiltration (UF) range and, products in the nanofiltration (NF) and reverse osmosis (RO) range, for separation of suspended and dissolved solids in water respectively. MF and UF systems can either be configured by installing a membrane train downstream of a pressurized feed source or upstream of the suction of a pump. NF and RO systems can only be configured downstream of a pressurized feed.

These different types of membranes operate on two fundamental principles: (1) size exclusion, and (2) osmosis. Size exclusion means that particles that are larger than the pore size of the membrane will not pass through the pores. MF and UF operate utilizing the size exclusion principle.

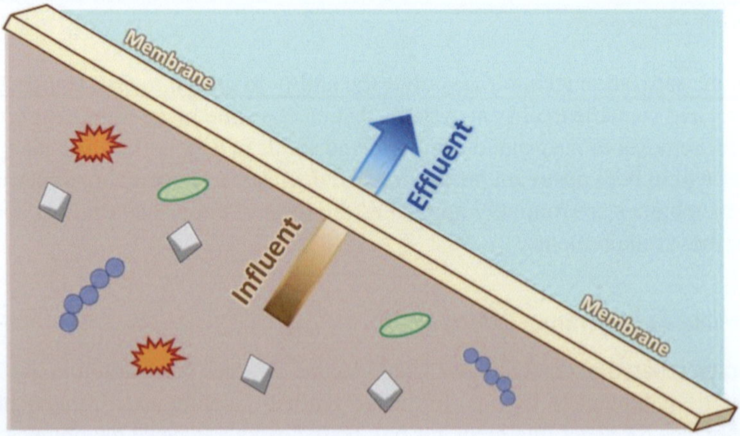

Fig. 2.12 Membrane filtration mechanism (Photo Courtesy of Nitto Global Membranes Division—Reprinted with Permission)

MF and UF processes are used extensively for the removal of suspended solids. MF membranes have the highest pore size (≥0.1 micron) whereas UF membranes have pore size smaller than those of MF membranes. These membranes are typically used in similar applications in water treatment (except that when used in process separations application UF membranes are often used to target specific size molecules) such as in the removal of bacteria and suspended solids that are < 500 micron which is the typical minimum required pretreatment. Nanofiltration (NF) membrane sizes are between those of MF/UF and RO membranes. They are typically used to remove dissolved organic molecules (e.g., color or TOC removal) and larger dissolved solids such as those containing divalent cations.

Figure 2.13 highlights the application of different membrane processes based on their pore sizes. Table 2.7 lists a summary of the characteristics and functions of different types of membranes.

Osmosis is defined as the natural passage of water through a semi permeable membrane, from a dilute solution to a concentrated solution until the chemical potential is equalized in the solutions. In reverse osmosis, an external pressure which is higher than the osmotic pressure is applied to the solution, which causes the water to flow against the natural direction through the membrane, thereby producing a very high-quality water. This process is widely used for demineralization of industrial water e.g., production of high-quality water for boiler feed make up, cooling tower make up, and other process waters, as well as for desalination of sea water. The working principle of reverse osmosis is discussed in detail later in this chapter.

It is important to note that all these processes require proper pretreatment. If RO is required in a process train, it is necessary to include pretreatment steps using membrane or non-membrane processes in the design to ensure proper feedwater quality which will allow for sustainable operation of the RO system. NF and RO membranes have been known to last 10–20 years with proper pretreatment which results in no fouling or scaling of the membrane. Poorly pretreated feedwater will lead

Table 2.7 Characteristics and functions of different types of membranes (Adapted from Membrane Systems for Wastewater Treatment, WEF 2006)

Type of membrane	Driving force	Separation principle	Molecular weight Cut-off size[1,3], MWCO Range (kDA)	Pore size range[2] (μm)	Operating pressure (bar)
MF	Pressure or vacuum	Size exclusion	>100	0.05–2.0	0.01–2.0
UF	Pressure	Size exclusion	2–100	0.002–0.1	2.0–5.5
NF	Pressure	Size Exclusion + Diffusion	0.3–1	0.001–0.008	4.8–15.0
RO	Pressure	Size Exclusion + Diffusion	0.1–0.2	0.0001–0.001	55–83

Notes

1. kDA = kilo Dalton, where Dalton is an alternate means to express molecular mass of a molecule, especially for large molecules. 1 Dalton is the mass of one twelfth of a Carbon 12 atom
2. Performance of MF and UF are better expressed in terms of their pore sizes in μm and for NF and RO in terms of their molecular cut-off size kDA
3. Molecular weight cut-off (MWCO) size is the molecular weight (in terms of kDA) of a compound, 90% of which are retained by the membrane system. Pore size range is the size distribution of the pores (openings) on the membrane surface

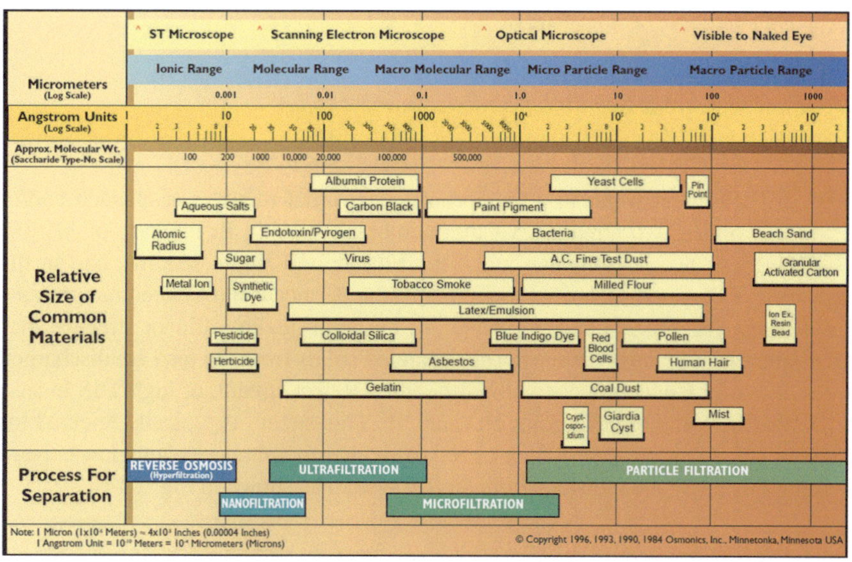

Fig. 2.13 Application of different separation processes based on size range

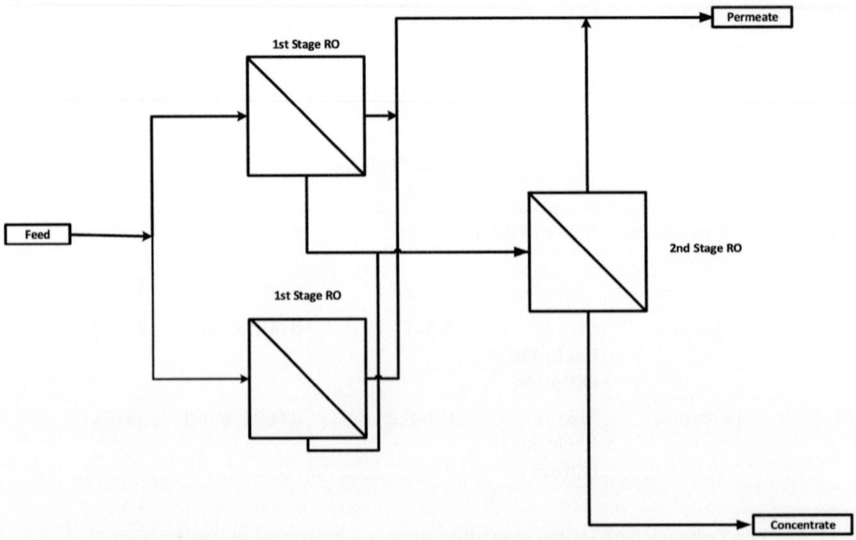

Fig. 2.14 RO system to increase product recovery in second stage from concentrates from first stages

to excessive downtime for cleaning, excessive need for chemical dosing to mitigate fouling, and the need for frequent membrane replacement. Operating principles and requirements for proper functioning of four types of membranes are discussed further in the later sections of this chapter.

Important Membrane Separation Process Terminologies

Streams

Feed is the influent water to a membrane system that removes suspended and/or dissolved solids. Two streams exit the membrane system. In the case of MF/UF systems, these two streams are called the Filtrate and Waste streams, and in the case of NF/RO systems the streams are called the Permeate and Concentrate streams (Fig. 2.14). The Filtrate and Permeate are the high-quality effluent streams after solids are removed from the feed. The separated solids from the feed are discharged in the Waste or Concentrate stream, also called Reject stream, of high TSS in case of UF and MF, and high TDS for RO and NF. This stream is typically rejected for disposal but, in some cases, it may be fed to another membrane system downstream to further increase the recovery from an RO system as shown in Fig. 2.14.

Recovery:

Recovery is the percentage of feed to a membrane system that is recovered as filtrate or permeate. It is expressed as:

Percent Recovery, $R = (Q_P/Q_F)*100$, and

$$Q_F = Q_P + Q_C \tag{2.47}$$

where, Q_F, Q_P and Q_C are the flow rates of the Feed, Permeate and Concentrate streams, respectively.

Flux

Flux is the flow rate of water passing through a membrane per unit surface area. In SI units it is expressed as Liters/hour/square meter (L/h/m^2), and in customary US units as gallons/day/square foot (g/d/ft^2 or gfd). Flux is a function of temperature. Since the viscosity of water decreases with increase of temperature, 20 °C is accepted as the standard temperature for specifying flux.

Transmembrane Pressure (TMP)

TMP is the pressure drop across the membrane and is the difference between the feed and the permeate or filtrate pressure and is expressed as:

$$TMP = P_F - P_P \tag{2.48}$$

where,
P_F and P_p are the pressures of the feed and permeate/filtrate streams respectively. It is the required driving force for flow across a membrane. TMP gradually increases with fouling and/or scaling of membrane.

Permeability

Permeability of a membrane system is one of the most critical parameters and is related to the ease of water passing through it. It is a function of the membrane material including the material properties such as porosity, hydrophilicity, and thickness, and of the environment in which it is operating such as, temperature, and fouling. During filtration by UF and MF membranes, foulants accumulate and TSS forms a layer on the surface of the membrane which adds resistance to the flow of water through it, reducing permeability, and reducing flux at a given pressure, requiring that the pressure be increased to maintain the same flux level. The initial permeability of a membrane is known as the clean water permeability. In course of its operation as the permeability declines from the clean water permeability solids deposit on the surface of the membrane. Various periodic cleanings are programmed into the membrane system design to restore the permeability and maintain sustainable operation.

Similarly, in case of RO and NF separations scale formation due to exceedance of the solubility limits of dissolved solids can reduce the permeability, resulting in loss of flux. Expressions for Rejection and Flux specific to RO membranes are further discussed in a later section on RO in this chapter.

Membrane Treatment System Configurations and Design Considerations

Main design variables in a membrane system are as follows:

- Membrane materials and their selection based upon water treatment application
- Membrane system set-up and modularity

- Water quality and source
- Treated water quality goals
- Flowrate
- Requirements for redundancy and on-stream time.

Membrane Materials

Common membrane materials, and their advantages and disadvantages are listed in Table 2.8. These are organic polymeric materials and are used stand alone or in combinations for the fabrication of various types of membranes.

Apart from the polymeric materials listed in Table 2.8, there are also ceramic membrane materials such as titanium dioxide (TiO_2), silicon carbide (SiC), and zirconium dioxide (ZrO_2). These have some advantages over polymeric membrane including high chemical resistance, thermal stability, and longevity. Also, ceramic membranes are much more permeable than polymeric membranes which translates into lower energy requirement. However, these are more expensive and typically have larger footprint and weight compared to polymeric membranes. At this point, ceramic membranes have niche applications and are not yet widely used. Manufacturers are making improvements to make the products more competitive so it can be a viable alternative to polymeric membranes from both technical and commercial viewpoints.

Modular Set-up of Membrane Systems and Module Types

Membrane elements are sheets of membranes that are arranged and installed in the form of modules. Depending upon membrane element and the treatment system configuration a membrane module can be selected. Common membrane modules are configured as:

- Flat sheet
- Tubular
- Spirally Wound
- Hollow Fiber

Table 2.8 Advantages and disadvantages of common membrane materials

Material	Advantages	Disadvantages
Polyamide	Tolerant to high temperatures >50 °C, good chemical stability and applicable to wide pH range	Chlorine can damage membrane structure
Polysulfone	Good thermal stability to 75°F. Tolerant to wide pH range, chlorine, aliphatic hydrocarbons, halogenated hydrocarbons, alcohols and acids. Easy fabrication	Poor resistance to aromatic hydrocarbons, ketones, ethers and acids. Low operating pressure limits compared to other types of membranes
Polyvinylidene fluoride (PVDF)	Good resistance to chemical solvents	Moderate chemical resistance. Applicable only to MF and UF pore sizes

These four configurations are shown in Fig. 2.15a, b, c and d.

Flat sheet membranes are the simplest in construction. They comprise of flat membrane sheets in parallel and are held in place by support plates in multiple cartridges in a plate and frame configuration. The spacers separate the membranes. Feed is introduced from an end of the module and the concentrate (retentate) stream flows out from the other end. Filtration occurs perpendicular to direction of flow as the filtrate (permeate) stream is collected in a common channel and leaves the module. Figure 2.15a depicts a schematic view of a flat plate membrane module. This type of module is not the optimal configuration and is therefore not commonly used.

A tubular module configuration consists of a number of tubes cast within a support tube as shown in Fig. 2.15b. Feed enters from one end of each tube and the concentrate (retentate) exits from the other end. The filtrate (permeate) stream flows out radially through the pores and collects in the outside tube. As with the flat sheet configuration, this arrangement is not optimal either and therefore not commonly used.

In spirally wound module, flat plate membranes separated by spacers are wrapped around a cylindrical perforated tube in spiral form as shown in Fig. 2.15c. Feed is introduced from one end and concentrate stream exits from the other end. Permeate collects and exits from the central perforated tube. Spirally wound modules are most commonly used in NF and RO applications.

The hollow fiber (HF) module type is similar to the configuration of tubular modules as shown in Fig. 2.15d. The main differences are in the tube diameter and,

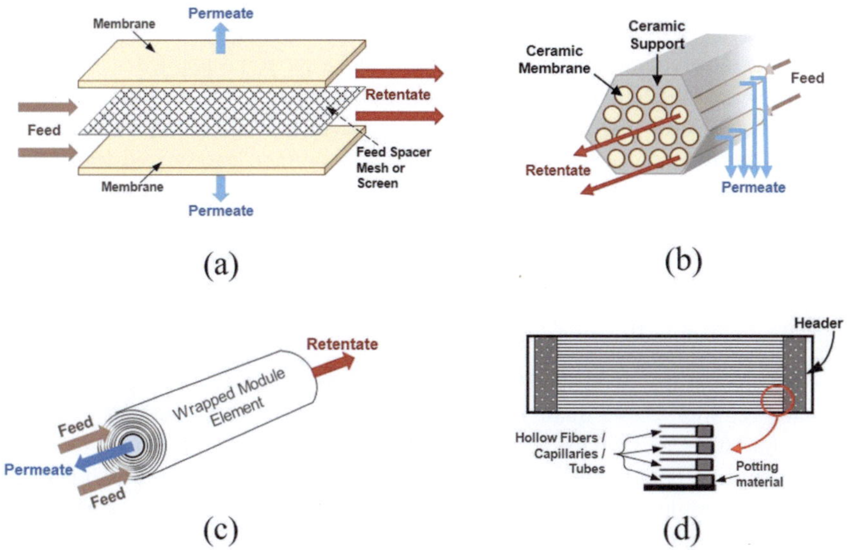

Fig. 2.15 **a**, **b**, **c** and **d**: Configurations for **a** Flat Plate, **b** Tubular, **c** Spirally Wound and **d** Hollow Fiber (HF) Membrane Modules (Photo Courtesy of Nitto Global Membranes Division—Reprinted with Permission)

in some cases, in the flow path. In case of tubular module, the tube diameter is on the order of 25 mm, and it is on the order of 1 to 2 mm in case of hollow fiber membranes (WEF 2006). For HF modules, typically thousands of small diameters hollow fibers are bundled together within a casing of the HF module. There are also different flow configurations of outside-to-in or inside-to-out within HF membranes depending upon the direction of the flow through the membranes. This dense packing of membrane fibers allows membranes to process larger flows than tubular or flat sheet membranes. This, combined with low transmembrane pressure (TMP), makes HF the most widely used module in water treatment applications by MF and UF.

Operating Principles and Operational Requirements of Membrane Separation Systems

A discussion on the operating principles and operational requirements of UF/MF and RO membranes are presented in the following sections since these are the two main types of membranes that are most widely used in the industrial applications including water treatment in petroleum industry.

Microfiltration and Ultrafiltration Systems

Typically, MF and UF membranes are made of polyvinylidene fluoride (PVDF) and polysulfone/polyethersulfone (PES) materials and can be installed as immersed or pressurized systems.

In immersed systems, a series of membrane modules are assembled onto membrane frames that are mounted and immersed inside the process tank. The membrane fibers or sheets are exposed and do not have any exterior shell. Feedwater is introduced into the tank for treatment and a suction is applied on the filtrate side of the membrane to draw filtered water through the membrane fibers.

In contrast, pressurized systems are arranged such that a group of membrane modules are mounted on a skid or a frame with common pressurized feed source and common piping headers for filtrate, drain, scouring air and cleaning solutions. The feed water is pumped and distributed across the membrane modules. The filtrate from each module connects to a common manifold and flows to the downstream processes or storage tanks under pressure through dedicated piping systems.

Feed Water Quality Requirements for MF and UF

The UF membrane performance depends on feed water quality, operating flux, cleaning regime, filtrate quality goal and membrane life. It is difficult to predict the performance of UF system precisely since a wide range of contaminants are present in water bodies which may also experience seasonal variation, but manufacturers have long term experience and are typically able to guide design efforts. In more challenging applications, such as in the oil and gas industry, pilot testing is often conducted on actual feed streams to verify membrane performance. Table 2.9 summarizes the typical feed water characteristics for optimum performance of UF.

Table 2.9 Recommended feed water characteristics for MF/UF systems

Parameter		Pressurized system	
		Recommended	Allowable limits
Turbidity	NTU	<50	300
Total organic carbon (TOC)	mg/L	<10	40
Chemical oxygen demand (COD)	mg/L	<20	60
Oil and greases	mg/L	0	5
Chlorine (continuous)	mg/L	<2 (Note 1)	200,000–2,5000 mg/L*hr (Note 2)
pH	su	2–11	1–14
Maximum temperature	°C	25	40

Notes

1 For continuous feed 2 mg/L is the tolerable limit and 500 is the maximum. 200 to 2000 mg/L is the limit during cleaning. These values vary depending upon the type, manufacturer and model including membranes from certain manufacturers that are not chlorine tolerant at all

2 This is the limit over lifetime. This means that the upper limit can be just one pulse load of 200,000–250,000 mg/L, or consistently an average of 200,000 to 250,000/(3* 365* 24), i.e., an average of 7 to 10 mg/L for 3 years, as an example, or anything over the number of years in its entire life time of that membrane that amounts to 200,000–250,000 mg/L. The range 200,000–250,000 represents the range of limits followed by various manufacturers

Design Considerations for MF/UF Systems:

Technology and system vendors will ultimately be responsible for sizing the system and providing all the specific parameters considering the system constraints. The information that is given to the vendor is important to ensure that the system is designed correctly and will operate as expected. The following considerations should be taken into account when initiating the design process.

- Key design parameters include knowledge of source of water, flowrate, temperature, TSS, turbidity, organics (TOC, BOD, COD), and metals concentrations (Fe, Mg, etc.). Ranges should be provided to consider variations in the design parameters. Peak conditions should also be considered in order for the system to remain in operation during these events (e.g., in case of upstream pretreatment failure or upset events).
- Provide site constrains such as available space, height limitations, or lack of availability of certain chemicals that may be typically used in cleaning processes.
- Provide additional system requirements such as redundancy, required uptime, turndown requirements, or the need for constant flow (typically systems are designed to ensure net daily average flow so if constant instantaneous flow is required, the system provider should know about it prior to design).

Advantages of MF and UF Membrane Technology

(a) TSS Filtration

MF and UF membrane systems are usually designed in modular skids, which are compact and light, produce stable water quality, and in many cases, consume less chemicals compared to mono and multimedia filters while producing better quality effluent than the traditional filtration systems.

(b) Oil Removal

Certain UF membranes can be applied for oil removal as a polishing step but only limited to influent emulsified oil concentration in the < 10 mg/L range. In these instances, membrane selection is critical because many membranes can foul irreversibly in treating oily water.

Reverse Osmosis (RO) Membrane Systems

RO membrane acts as a barrier to all dissolved salts and inorganic molecules, as well as organic molecules with a molecular weight greater than approximately 0.1 kDA (100 Daltons). The dissolved solids that do not pass through the membrane form the concentrate stream. Water molecules on the other hand, pass freely through the membrane creating a purified product, or permeate stream. Usually required recovery rate for RO systems is >75%. Rejection of dissolved salts is typically >95% and often >99% depending on the membrane selected. The term 'Rejection' refers to the percentage of dissolved solids in the feed that is rejected with the concentrate stream. It is expressed as:

$$\text{Rejection, \%} = [(\text{TDS}_{\text{Feed}} - \text{TDS}_{\text{Permeate}})/\text{TDS}_{\text{Feed}}] * 100 \qquad (2.49)$$

Rejection is a function of the type of membrane being used, feed pressure, feed TDS, and temperature.

When two aqueous solutions of different concentrations are separated by a semipermeable membrane the natural tendency is for the water to pass through the membrane from low concentration to high concentration side until the two concentrations equalize. The driving force for such a movement of the water molecules is the osmotic pressure (P_o). However, the direction of movement of the water can be reversed by applying a pressure (P) higher than the osmotic pressure in opposite direction. This causes water to pass through the membrane from high to the low concentration, thus creating a concentrate stream of small volume and a dilute permeate stream of larger volume. This is the principle of recovering high quality water from a high TDS feed e.g., sea water, or brackish water, by application of RO. Figure 2.16 depicts the principle of reverse osmosis process to produce high purity water. It should be noted that there are non-permeable membranes also available commercially for other applications but those are not useful for as materials for building RO systems for water purification. Saran wrap, made of polyethylene, is

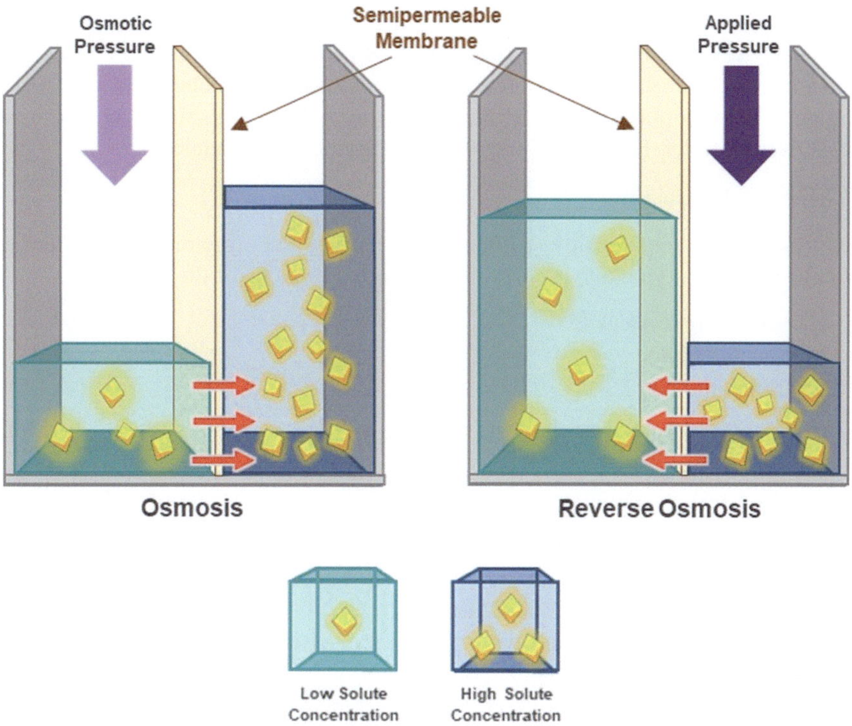

Fig. 2.16 A conceptual diagram of osmosis and reverse osmosis processes (Photo Courtesy of Nitto Global Membranes Division—Reprinted with Permission)

such an example which is used as a waterproofing material protecting household and industrial goods from damage or degradation by coming in contact with water.

The rate of transport of water through the membrane surface is expressed as the flux rate:

$$J = k * (P - P_o) \tag{2.50}$$

J = Flux Rate in L/m^2.hr.

P = Applied Pressure, mPa
P_o = Osmotic Pressure, mPa

k = Proportionality Constant

The operating principle of NF is not discussed separately since the pore size range of NF is between those of UF and RO and their functionality is intermediate between those two types. The energy consumption for nanofiltration (~0.3 to 1.0 kWh/m^3) is larger than that for UF and MF but smaller than that for RO. NF membranes are ion and charge selective based on the ionic size as they remove > 90% of multivalent ions (0.23 nm size) but they are not effective to remove smaller monovalent ions (0.12 nm size). RO is non-selective in terms of ionic size and very effectively remove most of TDS from water. Figure 2.17 shows the picture of a RO module.

RO Feed Water Quality

RO systems are very sensitive to feed water quality and a pretreatment system is usually required. Typically, pre-treatment is done using one of the following:

- Ultra-filtration system
- Multimedia filters
- Cartridge filter, etc.

Pre-filtration requirements for RO/NF systems are as follows:

- Turbidity <0.4 NTU ideally (<1 NTU max) although 0.2 to 0.4 NTU is preferrable to ensure an extended membrane life.
- SDI < 5 although SDI < 3 is preferrable for extended membrane life. Typically, manufacturer's warranties for RO membranes state that SDI should be < 4 at 100% of the time, and < 3 at 95% of the time.
- Suggested guidelines for conventional polyamide base NF and RO for continuous service operation is < 0.04–0.05 ppm of chlorine in the absence of iron and other transition metals. Composite polyamide membranes can tolerate about 1000 ppm-hrs before significant membrane degradation is exhibited by a quick decline in rejection capability. This means that the upper limit can be just one pulse load of 1,000 mg/L, or consistently an average of 1000/(3*365*24), i.e., ~ 0.04 mg/L for 3 years, or anything in between that amounts to 1,000 mg/L over a period of 3 years. This is the standard followed by Nitto Global Membranes Division (Bates 2024), which is typical in the industry. Similar in-house standards are established and followed by other RO membrane manufacturers also. Because of the sensitivity of NF and RO membranes to free chlorine it is strongly recommended to dechlorinate feed water by dosing sodium bisulfite (SBS) prior to its contact with the membrane surface. SBS acts as a reducing agent of the chlorine, which is a strong oxidant, according to the reaction:

$$NaOCl + Na_2SO_3 \rightarrow Na_2SO_4 + NaCl \tag{2.51}$$

Design Considerations for RO Membrane Separation Systems

As in the case of MF and UF systems, technology vendors will ultimately be responsible for sizing and configuration of the NF and RO systems also and it is critical to provide the vendor accurate information to ensure that the system is designed correctly and will operate as expected. The following considerations should be taken into account when initiating the design of a RO system.

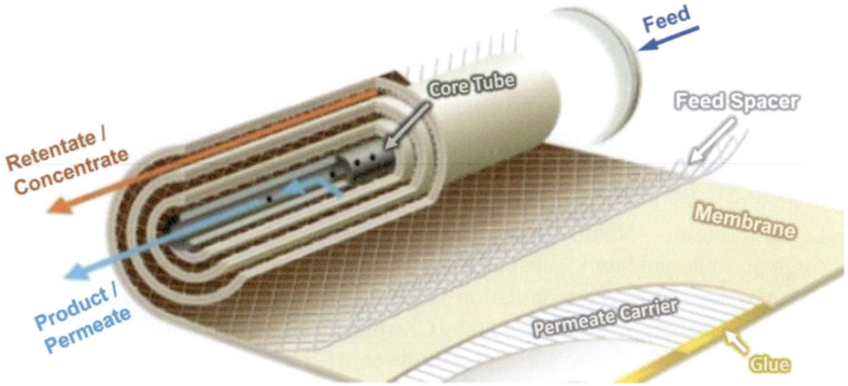

Fig. 2.17 Functioning of RO module (Photo Courtesy of Nitto Global Membranes Division—Reprinted with Permission)

- Establish key inputs parameters including source water characteristics, pretreatment, feed flowrate, desired product flowrate and quality, temperature range, guaranteed turbidity and SDI, and organic concentrations (TOC, BOD, COD). A full ionic analysis of feed water is required for NF and RO systems. Ranges should be provided to consider variations in the design parameters.
- Select system configuration (pressurized or submerged)
- Select membrane type, membrane area and operating window
- Select design flux rate, feed water quality normalized with respect to operating temperatures
- Provide site constrains such as available space.
- Provide additional system requirements such as redundancy, required uptime, and turndown requirements.

For all types of membrane treatment systems—MF/UF and NF/RO, the first step consists of the process design using the vendor's in-house or a commercially available process simulator. This is followed by the detailed design and system engineering based on the specific site considerations.

Membrane Fouling

All membrane separation processes have a common risk of fouling which results in gradual decrease in flux. This impacts the system performance leading to gradual decrease in permeability of the membranes and resulting in a decrease in the production rate. In most cases fouling can be controlled with pre-treatment; chemical conditioning; and optimization of design parameters such as flux and cleaning program.

There are four types of fouling and inorganic scaling commonly encountered:

- Particulate fouling—This is caused by suspended solids and colloidal materials which can be reduced by primary solid separations processes such as coagulation, clarification, and media filtration. Particulate built-up on membrane surfaces can be washed off using backwashing and air scouring.

- Biological fouling—This is due to the biological growth on membrane surfaces, which can be prevented by pre-oxidation, application of biocides or disinfectant chemicals injected into the feed water stream. Membrane cleaning methodology involves using oxidants such as NaOCI or other trade-marked biocides in addition to high pH cleaning.
- Organic fouling—Natural organic matter (NOM) originating from source water (organic acids and humic substances) adsorb on membrane surface. It is generally removed upstream via coagulation using aluminum or iron-based coagulants. Effective cleaning can be achieved by Chemically Enhanced Backwash (CEB) or Clean In Place (CIP) with alkaline solution at pH $\geq$ 12. Typically, dilute solutions of caustic soda (sodium hydroxide) or other proprietary solutions are used as alkali for membrane cleaning.
- Inorganic Scaling—caused by precipitation of inorganic matters on membrane surface, usually due to high iron and manganese (Fe, Mn) and hardness (Ca, Mg) content in source water The scaling potential for RO membranes is mitigated by suitable pretreatment steps, proper design of the recovery, limiting the reject stream concentration, and addition of scale inhibitors. Common cleaning methods include CEB or CIP procedures with low pH solution. Typically, low pH cleaning agents e.g., citric or mineral acids, in combination with proprietary surfactant solutions as recommended by the system provider, are used for removing inorganic scales.

As discussed above, fouling and scaling are operating challenges to all the membrane processes. But systems have been developed, along with the necessary instrumentation and controls, to produce high quality filtrate/permeate without affecting membrane life. Membrane system vendors provide the routine cleaning procedure and recommend the types of cleaning chemicals. Membranes are routinely regenerated using CIP and CEB procedures to maintain adequate performance. Typical membrane life ranges between 3 and 10 years depending on the type of membrane and its application before replacement is required since routine cleaning procedures cannot regenerate the membrane system enough to return its performance to acceptable levels. It is also essential to follow the cleaning protocol such that the membrane structure and integrity are not compromised affecting the life of membranes.

Electrodialysis (ED), Electrodialysis Reversal (EDR) and Electro Deionization (EDI) Processes

Electrodialysis (ED) is a special type of membrane separation process for demineralization of saline and brackish water. In this process dissolved ionic constituents in water are separated by application of an electrical field. ED cells contain a series of alternating cation and anion exchange membranes placed in parallel between an anode and a cathode electrode at the two ends of the cell as shown in Fig. 2.18. The cation exchange membranes selectively pass only the cations through the membranes

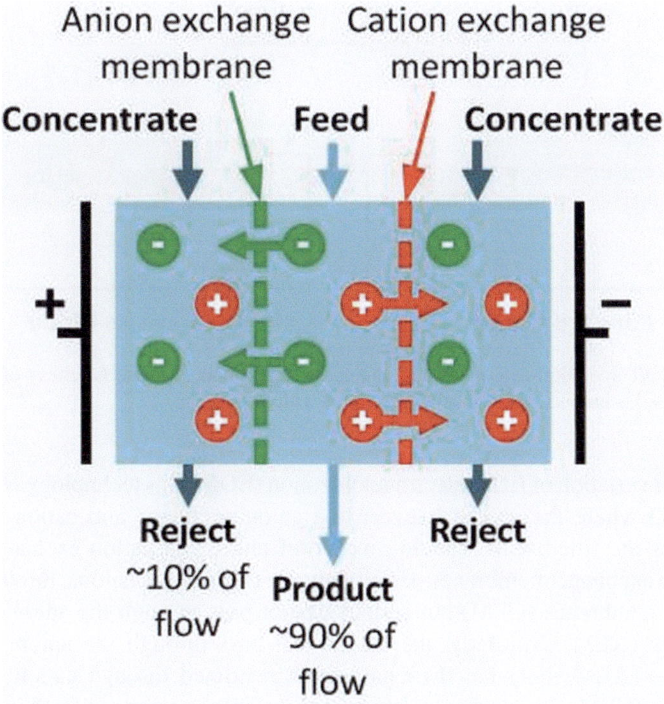

Fig. 2.18 Operating principle of electrodialysis (Photo Courtesy of Saltworks Technologies, Richmond, BC, Canada—Reprinted with Permission)

which migrate towards the cathode, but anions are obstructed. Similarly, only the anions pass through the anion exchange membrane and travel towards anodes while cations are arrested. This results in removal of ionic constituents from saline feed water which exits the ED unit as a concentrated reject stream. Purified water exits as the product stream.

ED is also a membrane-based process where separation of dissolved solids is achieved by application of electromotive force as opposed to RO, which utilizes pressure as the means of carrying out separation. Like RO, ED membranes are also subject to scaling and fouling. This difficulty is largely overcome by Electrodialysis Reversal (EDR) process which is a modified version of ED. In this mode of operation, the polarity of the electrodes is continuously changed at selected intervals of time. As shown in Fig. 2.19, with the alternating polarity of the electrodes the hydraulic channels 'reverse'. Thus, the direction of migration of anions and cations through the pores also continuously alternate, thereby preventing the opportunity of scalants and foulants to accumulate around the electrodes.

Unlike RO, EDR operation can avoid dosing of antiscalant and antifoulant chemicals. Also, this advantage allows EDR to process water containing much higher salinity compared to the levels typically treated by RO.

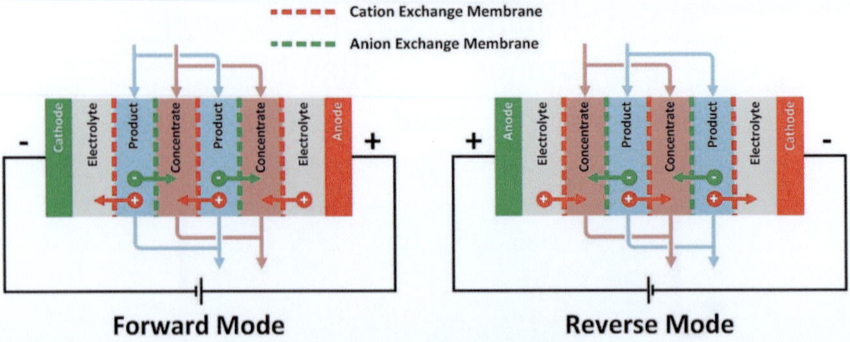

Fig. 2.19 EDR operation in forward and reverse polarity modes (Photo Courtesy of Saltworks Technologies, Richmond, BC, Canada—Reprinted with Permission)

Another variation of ED is electrodeionization (EDI). This technology is an extension of ED where the spaces between the anion exchange and cation exchange membranes are filled with random mixture of anion and cation exchange resins. The cation exchange membranes permit only the cations, not anions, through cation exchange membranes (CEM) but cations cannot pass through the anion exchange membranes (AEM). Similarly, the anions can pass through the anion exchange membranes (AEM) only but their passage is restricted through cation exchange membranes (CEM). Ion exchange resins in these spaces carry out further polishing of the dilute stream and help produce the treated effluent of even higher quality than that is achieved by ED.

Due to the very low ionic concentration of water in the dilute spaces the conductivity is very low. This, in combination with high electrical potential of ~ 100 to 600 Volt DC at 3 to 20 mA, electrolyzes water to hydrogen (H^+) and hydroxide (OH^-) ions. When the resins become exhausted over a period of time these hydrogen and hydroxide ions regenerate the resins in situ which precludes the necessity to add acid and alkali externally for regeneration of the resins as in the case of traditional ion exchange process. The internal components and working principle of an EDI system are presented in Fig. 2.20.

EDI is a technology for producing very high-quality water—typically for polishing RO product water. It can produce deionized water of multimegohm resistivity with up to 99% water recovery and >95% rejection of boron and silica, and > 99% rejection of sodium and chloride ions (Flynn 2009). To produce this quality of water EDI feed requirements are also quite stringent that needs to be fulfilled by pretreatment steps of UF, followed by RO. Table 2.10 lists the typical feedwater specifications for an EDI treatment system.

ED, EDR and EDI technologies are not commonly used in water treatment applications in petroleum industries unless there are specific needs in oil field produced water treatment for reuse, and use of seawater as the source of raw water.

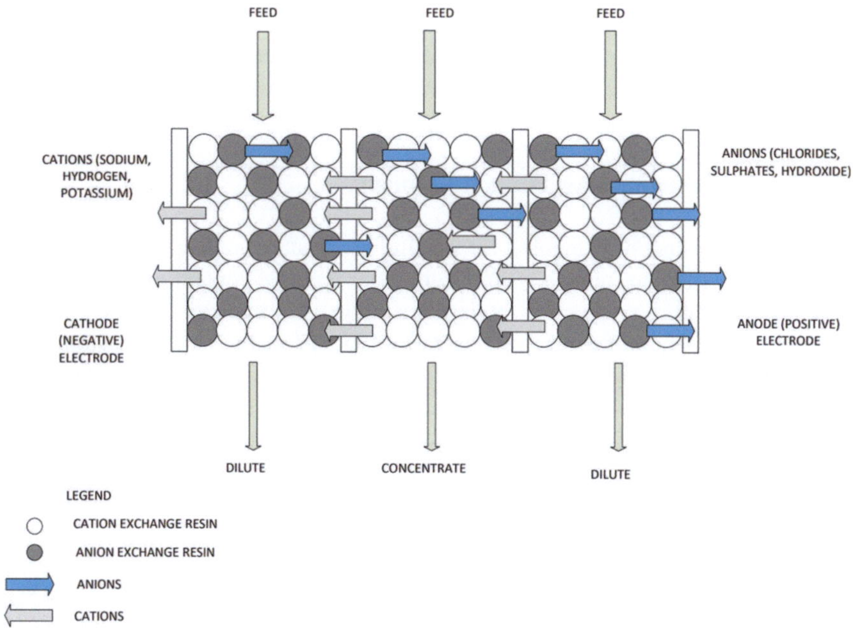

Fig. 2.20 A conceptual diagram depicting the working principle of an EDI system

Table 2.10 Typical feedwater specifications for EDI System (Adapted from Flynn 2009)

Parameter	Specification
Feed	RO product
Conductivity	<40 μS/cm
Total dissolved CO_2, HOC_3^- and CO_3^{2-}	<5 mg/L
Silica	<0.5 mg/L
Fe, Mn and H_2S	<0.01 mg/L
Total chlorine	<0.05 mg/L
Total hardness	<1.0 mg/L
TOC	<0.5 mg/L
pH	4–11
Temperature	5–45 °C
Pressure	~2–7 barg

2.1.2 *System Integration for Various Water Treatment Processes*

The sections above presented the details of the working principles of various unit operations and processes commonly used in water treatment. These are selected and integrated to develop water treatment trains for specific applications depending upon the feed water characteristics and the desired product quality. A few important applications and their treatment trains are discussed below.

2.1.2.1 Conventional Raw Water Treatment

Conventional treatment is applied for treating raw waters containing high concentrations of TSS but are relatively free from other types of contaminants e.g., hardness, dissolved iron, manganese, hydrogen sulfide, etc., that can adversely impact the downstream equipment and processes by fouling, scaling and corrosion. Both groundwater and surface water can be the source of this type of raw water. The conventional treatment trains are designed to handle a wide range of turbidity from 10 NTU to as high as 1,000 NTU. The unit operations and processes for conventional treatment are primarily focused on removing suspended solids by coagulation, flocculation, settling sedimentation and filtration. As shown in Fig. 2.21 the raw water is withdrawn by a set of feed pumps. The intake screens remove objects typically larger than 12 mm. The influent then goes through coagulation and flocculation steps. Most commonly in water treatment applications alum is the chemical of choice as a coagulant. Flocs of suspended solids settle by gravity in the settling tank. Clarified water is pumped through gravity or pressure filters to produce water of turbidity < 1 NTU that can be used as a feed for cooling water, boiler feed water and other advanced water treatment systems. If a refinery uses this water as a source of potable water for human consumption, then this water would require disinfection by gaseous chlorine or sodium hypochlorite solution to maintain a residual chlorine concentration ~ 1 mg/L. Sludge from the bottom of the settling tank is dewatered and disposed.

Softening of hard water and removal of dissolved iron, manganese, H_2S and VOCs from groundwater require an extension of the conventional treatment process described above. In each of these cases the solids formed in the chemical reactions are removed in the settling tank along with the incoming TSS in the raw water.

2.1.2.2 Hard Water Softening Treatment

Raw water from many surface and groundwater sources contains high levels of calcium and magnesium hardness that can cause scaling in downstream processes, equipment and pipes. Softening of such waters is essential prior to feeding them into treatment systems for boiler feed make up, cooling water make up and potable uses. Chemical reactions and treatment process train for hardeness removal are described in detail in Sect. 2.1.1.2 and Figs. 2.2 and 2.3. As stated above, the settling tanks for removal of solid products $CaCO_3$ and $Mg(OH)_2$ of softening reactions also co-precipitate the suspended solids (TSS) in the raw water. A conceptual treatment train for hard water softening process is presented in Fig. 2.22.

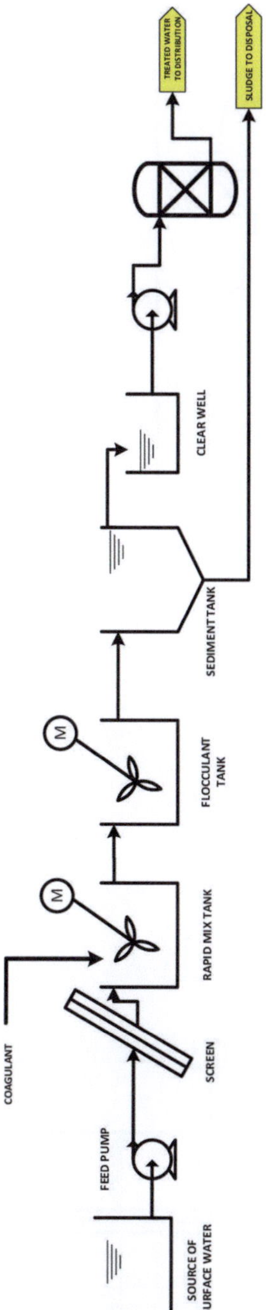

Fig. 2.21 Treatment train for conventional raw water treatment

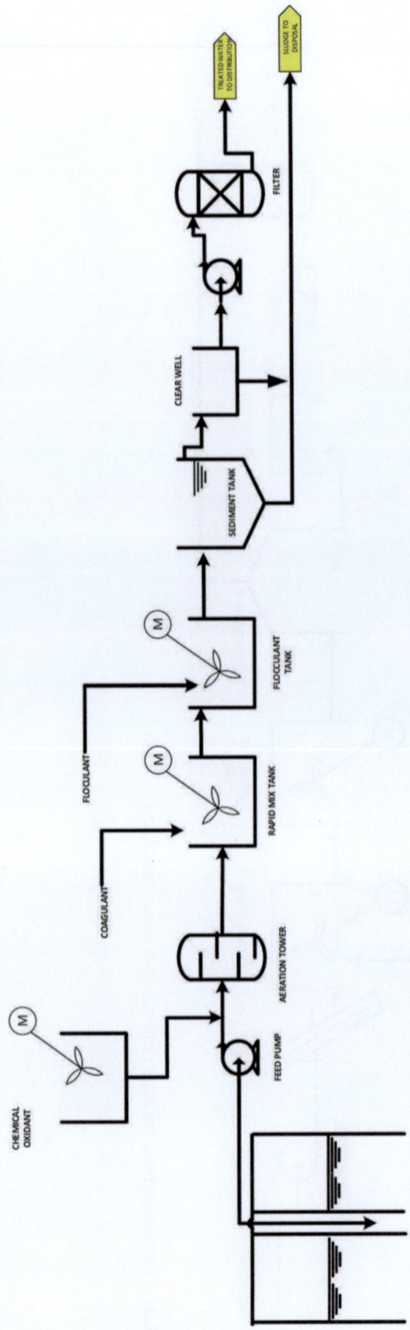

Fig. 2.22 Treatment train for water containing iron and manganese

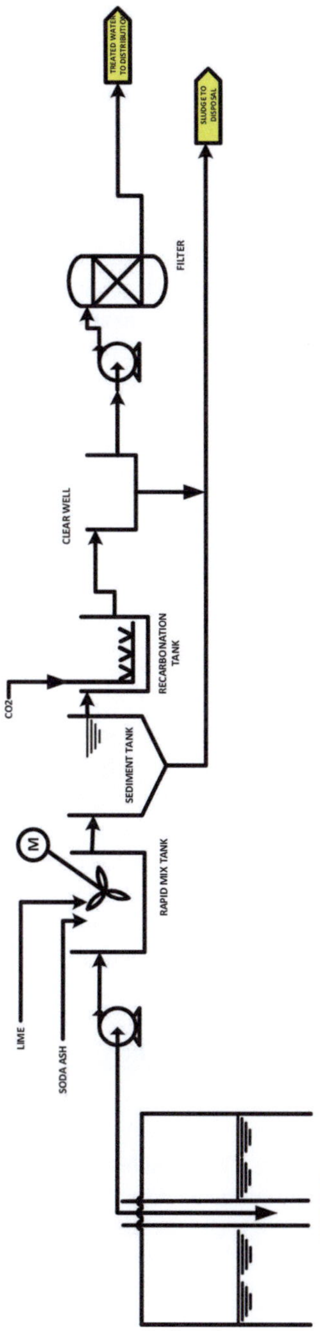

Fig. 2.23 Treatment train for water containing hardness

2.1.2.3 Groundwater Treatment for Iron and Manganese Removal

Dissolved iron and manganese in groundwater is removed by aeration as discussed in Sect. 2.1.1.2. For these applications typically cascade and spray aerator are used which induce air by natural draft to the flow of water. Air oxidizes the dissolved ferrous and manganese ions to $Fe(OH)_3$ and MnO_2 particles. Sometimes air is supplemented by dosing additional chemical oxidant e.g., chlorine dioxide (ClO_2) to achieve complete oxidation in cases where dissolved iron and manganese concentrations in the groundwater is high.

The water is then thoroughly mixed with a coagulant chemical in a rapid mix tank, followed by flocculation in a floc tank. A flocculant polymer may be added to enhance flocculation of suspended particles. In the next step the particles are removed by settling in a sedimentation tank. Incoming suspended solids in the raw water are also removed in this tank by co-precipitation. The clarified water is then pumped through a dual media filter to achieve a product of turbidity <1 NTU. This water is then treated further to advanced levels of treatment to produce high quality of water for cooling water, process water and boiler feed water make up in industrial applications. A typical treatment train for iron and manganese removal is presented in Fig. 2.23.

2.1.2.4 Sea Water Desalination

Availability of fresh water is inadequate for both human consumption and industrial uses in many parts of the world. Desalinated sea water could be a significant water resource for these regions. It has been used extensively in such regions as an effective and economical way of producing water for drinking and industrial use. Two main processes used in desalination of seawater are reverse osmosis (SWRO) and multi effect distillation (MED). They currently account for about two-thirds of the total installed seawater desalination capacity in the world (Henthorne 2009).

Design and operation of desalination plants and the selection of the best pretreatment is strongly dependent on the raw seawater quality and its variations.

<u>Desalination by Seawater Reverse Osmosis (SWRO)</u>

Sea Water Characterization

Seawater contains various mineral salts, organic substances, algae, bacteria, and other micro and macro size colloidal and particulate matters. Advanced analytical tools have been developed to allow thorough characterization of seawater samples. The physical, chemical, and biological characteristics of seawater are discussed below in further detail.

Physical Characteristics

- Density

The density of seawater generally increases with decreasing temperature, increasing salinity, and increasing depth in the ocean. The density at the surface of the ocean varies from 1,020 to 1,029 kg/m^3. It should be noted here that density is an important parameter in determining the power necessary for operation of the RO system.

- Silt Density Index (SDI)

Silt density index (SDI) is a parameter used for describing the fouling potential of water. It is one of the most important parameters for the design and operation of seawater reverse osmosis (SWRO) process. Definition of SDI was briefly introduced in Chap. 1 and discussed here in further detail in the context of SWRO process.

SDI analytical protocol is standardized in the ASTM D 4189–95, re-approved in 2002. It evaluates the amount of plugging of a 0.45 μm pore size filter that is caused by passing a sample of water through the filter for 15 min. SDI measurement is recognized as the standard test to estimate membrane fouling potential (Iwahori et al. 2003; Kim et al. 2006; Kremen and Tanner 1998; Mosset et al. 2008). It depends on the concentration of particles and is also representative of the types of foulants.

The SDI test is carried out at a pressure of 2.06 bars through a membrane filtration system with a pore size of 0.45 μm. Conventionally, the measurement is made over a period of 15 min (SDI 15) on the pretreated water. When the water has very high fouling properties, the measurement may be made over a period of 10, 5 or 3 min. It is expressed in %/min. Since the filter plugs over time, the rate of plugging (SDI) is important. SDI is then calculated according to the following equation:

$$\mathrm{SDI}_{\mathrm{Td}} = \frac{100}{T_d}\left(1 - \frac{T_i}{T_f}\right) \tag{2.52}$$

where:

T_d is the overall filtration time (3, 5, 10, 15 min).

T_i is the initial time (s) to filter 500 ml of water on a 0.45 μm membrane at 2.06 bar.

T_f is the final time (s) to filter 500 ml after 15 min of filtration.

The factors affecting SDI are pH and seasonal variation of water temperatures. An increase of SDI values has been observed when pH is increased from 7 to 8 and from 4 to 6. This is mainly due to the presence of dissolved substances (Ca, Mg, etc.), which precipitate with the increase of pH. As a rule of thumb, based on actual operating experience, if a 15 min SDI reading is not achievable pretreatment is necessary. The recommendation for reliable operation of seawater RO systems is to target a 15 min SDI of <3.0 with a maximum of 4.0.

It has been observed that, during the autumn and winter months, the average SDI is low with little variation. On the contrary, during the spring the SDI becomes high and varies greatly.

Chemical Characteristics

The chemical characteristics of the raw seawater must be determined to provide information for selection of the best pretreatment. These include:

- Total Dissolved Solids (TDS)

Most of the dissolved chemical constituents or salts found in seawater originate from mineral deposits and rocks on the sea floor. The TDS of sea water (usually 35 g/L) is made up by all the dissolved salts present in the water.

Salinity of seawater is typically defined using chlorine content. The following relation is used:

$$\text{Salinity, S (g/L)} = 1.80655 \text{ Cl (g/L)} \tag{2.53}$$

- Major Ions in Sea Water

Only six ionic species comprise about 99% of sea salts: chloride (Cl^-), sulphate ($SO_4{}^{-2}$), sodium (Na^+), magnesium (Mg^{2+}), calcium (Ca^{2+}), and potassium (K^+). Apart from these, iron and manganese are present as oxides and play an important role in the removal of various trace constituents from seawater. However, they can also cause fouling of membranes in seawater treatment. Iron and manganese oxides are co-settled with other settling particles and are reduced to dissolved ferrous and manganous ions during oxidation of organic matter. When ferrous and manganous ions diffuse upward through interstitial water they are transformed again to particulate iron (+3) and manganese (+4) oxides in contact with air at the surface. Thus, these two elements cycle between their dissolved and particulate forms as a part of their movement up and down the water column in the seawater.

Iron chemistry, such as inorganic speciation and organic complexes, in seawater is very complex and not yet fully understood. Iron is an essential micronutrient for phytoplankton growth and is an important component of such biochemical processes as photosynthetic and respiratory electron transport, nitrate and nitrite reduction, chlorophyll synthesis, and several other biosynthetic or degradative reactions (Weinberg 1989; Kuma, 1996; Geider and Roche 1994). The recommended upper limit for iron ions in seawater desalination feed system is 50 μg/L by membrane suppliers due to its possible oxidation on membrane surfaces which irreversibly damage the membrane (Kader 2011).

Manganese exists in the seawater in two main forms: Mn^{2+} and Mn^{4+}. The chemistry of manganese in seawater is largely governed by pH and redox conditions. Mn^{2+} ions dominate at lower pH and redox potential (LaZerte and Burling 1990). The oxidation of Mn^{2+} to Mn^{4+} occurs in a two-step process in which first Mn^{2+}

is converted to Mn^{3+} and then finally to Mn^{4+}. The Mn^{2+} ion is more soluble than Mn^{4+}, therefore, manganese will tend to become more bio-available with decreasing pH and redox potential. The presence of chlorides and sulphates increases manganese solubility (Schaanning et al. 1988).

Concentrations of manganese in open seawater range from 0.4 to 10 μg/liter. The limit recommended by the membrane suppliers for manganese is 20 μg/L, due to its possible oxidation on the membrane surfaces which will damage irreversibly the membrane surface. However, based on the above considerations, seawater in general does not pose a threat of membrane fouling, which may be the case with groundwater at certain locations.

Categories of Seawater Constituents and their Impacts on RO Membranes Fouling

Based on the above discussion of physical and chemical characteristics, seawater constituents can be broadly categorized as follows:

- Inorganic matter
- Organic matter
- Micro and macro biological species.

Fouling potential of individual candidates under each category is listed in Table 2.11 and discussed in the following section.

- Inorganic and Particulate Fouling

This type of fouling takes place due to accumulation of particles on the UF and RO membrane surfaces which were not removed from the raw seawater by pre-treatment. The indicators of sufficient reduction of suspended solids and particles are turbidity values of <0.2 NTU 95% of the time (<0.0.3–0.4 NTU max) and silt density index (SDI) values of <4 (as mentioned in the earlier discussion).

- Organic fouling

A large part of organic matter in seawater is Natural Organic Matter (NOM), mostly constituted of humic acids (HA) and fulvic acids (FA). Algal organic matters (AOM)

Table 2.11 Seawater constituents and potential membrane foulants

Inorganic and particulate fouling	Organic fouling	Biological fouling
Silica	Lipids	Algae
Quartz	Proteins	Plankton
Silt	Polysaccharides	Unicellular organisms
Carbonates/sulphates	N/A	N/A

consisting of extra cellular macromolecular and cellular debris are also another source of organic matter in seawater. The dissolved organic substances are mixtures of many aliphatic and aromatic compounds. However, 90% of them are represented by the humic materials among the total dissolved organic substance in seawater.

Humic and fulvic acids are adsorbed onto membrane surfaces. Adsorption of organic matters, particularly these organic acids on the membrane surfaces cause organic fouling. Organic fouling is very complex in nature. These foulants usually have functional groups such as carboxyl (-COOH) and phenolic groups (-OH), which also play key role in fouling.

Organic fouling can also be caused by hydrocarbon oils, naturally occurring, or resulting from spills during sea transportation, or pollution. These spills have been observed to be in the concentration range of 0 – 10 mg/L. These can pose an adverse effect on the desalination system. Emulsified oil and grease are the principal sources of immiscible liquid fouling in desalination facilities. Flotation or granular activated carbon are sometimes used to bring concentrations to acceptable levels upstream of the reverse osmosis membranes.

The rate and extent of fouling have been found to be directly proportional to ionic strength and the presence of divalent cations. These substances cause serious performance deterioration of SWRO membranes, including a gradual decline of membrane flux and an increase in TMP. These effects cause a decrease the permeate production rate requiring an increase in feed pressure and lead to a gradual degradation of membranes. Hence, pre-treatment of seawater is essential in order to control the concentration of organic substances prior to feeding the RO.

There are several measurement procedures for the characterization of the organic substances present in the seawater, including:

(a) Total organic Carbon (TOC) representing the total amount of NOM including particle content.
(b) Dissolved Organic Carbon (DOC) representing the amount of NOM dissolved in the raw water, which includes the aromatic content of NOM.
(c) Specific Ultraviolet Absorbence (SUVA) (ratio UVA @ $\lambda = 254$ nm/DOC) representing the part of the humic substances versus the non-humic substances.

- Biological fouling

Biological fouling happens when micro and macro biological populations build up on the membrane surface resulting in a slime layer. Biofouling is a special case of particulate fouling that involves living organisms that can cause loss of flux and degrade membranes.

Algae are a large part of microorganisms. They consist of varied groups of photosynthetic organisms that are found in oceans and other large water bodies. They exist in a wide variety of forms, from single cells and strings of cells to complex multicellular seaweeds and contain chlorophyll and other photosynthetic pigments. The best-known algae are red, brown, and green seaweeds, which are part of a group of large multicellular algae known as macroalgae. Algae are mostly single-celled species that float freely in the water column and are invisible to the naked eye and collectively in a group called phytoplankton. Excessive growth of phytoplankton can occur in coastal seawater and estuaries causing the seawater to appear colored - typically red, or brown, close to the surface of the seawater. This is commonly referred to as an algal bloom. Algal blooms may pose challenge to the operation of a desalination plant. High algal concentrations result in a high suspended solids load. Most marine algal blooms are harmless, resulting only in a discoloration of the water. Most algae are microscopic (ranging from 2 μm to 100 μm), several forms are macroscopic, with some species growing to lengths over 30–35 m (100 ft) (Brock and Clyne 1984).

SWRO Treatment Systems

Removal of foulants from seawater is necessary before it is fed into SWRO treatment system to prevent adverse impacts on the operation. Various pretreatment processes are used for the removal of:

- Fe, Mn, turbidity, suspended solids, and SDI
- Dissolved organic carbon
- Oil and chemical spills
- Algal blooms and red tides.

The pretreatment is predominantly conducted by following unit operations and processes:

- pH Adjustment
- Dissolved Air Flotation
- Filtration
- Granular Activated Carbon adsorption
- Disinfection by chlorination.

Following these steps, the pretreated seawater is fed into RO treatment system for removal of TDS and some organic contaminants (COD). A conceptual flow diagram for a typical SWRO system is presented in Fig. 2.24.

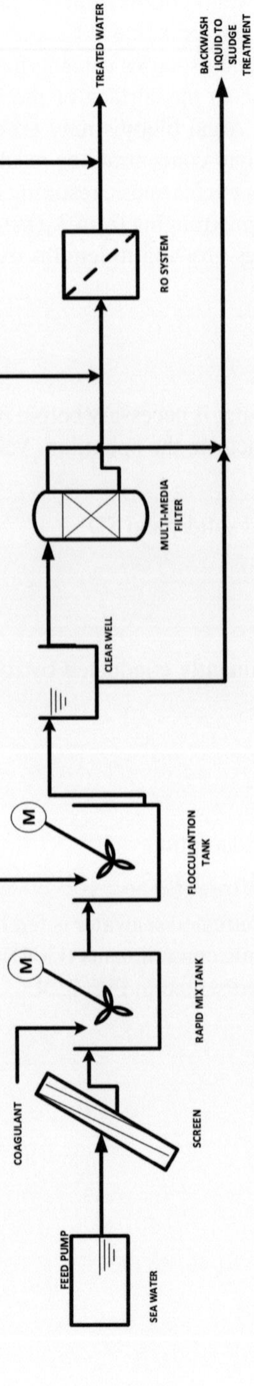

Fig. 2.24 Schematic view of a seawater RO (SWRO) treatment system

It is emphasized here that the membranes in SWRO applications are subject to far more aggressive operating conditions than those encountered by normal non-SWRO applications e.g., brackish water treatment applications (BWRO). As described above, the chemical and biological constituents and their concentrations in seawater vary widely based on the geographical location and seasonal conditions. Therefore, the success of SWRO systems depend largely on the performance of the pretreatment processes. While the pretreatment steps effectively remove the colloidal and suspended solids, the dissolved solids (TDS) are removed by the RO membranes. Salt concentration in seawater varies in the range of 32–40 g/L, which limits the recovery rate to <50% in a single stage. Additional stages can be used in the configuration as shown in Fig. 2.14 to improve recovery rates up to 60%. This renders the SWRO system operation very energy intensive. On the contrary, in BWRO applications the quality of RO membrane influent is much higher than that of seawater, and a high recovery rate of >90% can be achieved.

Control of algae is an essential step to minimize membrane biofouling and ensure smooth operation of SWRO systems. Chlorine is widely used as the disinfectant of choice, with shock dosing over continuous chlorination, for the control of biological fouling in the pretreatment system. Continuous chlorination followed by sodium bisulfite (SBS) dechlorination results in breakdown of large organic compounds to smaller molecules which then becomes food for slime forming bacteria, thereby resulting in a fouling problem.

Desalination by Multi Effect Distillation (MED)

MED units are a type of mechanical vapor recompression (MVR) evaporators. Separation of seawater into a high purity stream and a small, concentrated brine stream is driven by thermal energy. Seawater is pretreated first (as described in the section above), before feeding into a MED unit. Typically, conventional seawater desalination systems achieve about 50% recovery and produce permeate of TDS <500 mg/L (Sitterley et al. 2022). The principle of operation of this equipment is production of high-quality water by evaporation of seawater in multiple stages, each consisting of a plate type of heat exchanger.

Figure 2.25 schematically presents the treatment process for seawater using a two stage MED evaporator. Feed enters the preheater where it is heated by the outgoing brine and distillate streams. The pre-heated feed is next heated and evaporated in the first stage by high pressure steam from a boiler. Vapor from the first stage is recompressed to a higher pressure which elevates its boiling point. This becomes the heating steam for the second stage, which heats up and evaporates the liquid stream from the first stage. Liquid stream from the second stage returns to the preheater and leaves the MED system as concentrated brine. Condensates from the first and second stage are combined and returned to the preheater where the concentrated

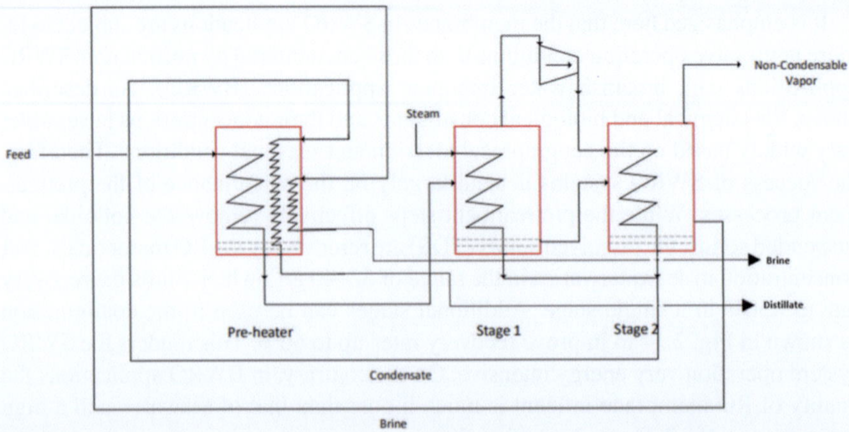

Fig. 2.25 Two stage multi effect distillation (MED) System

brine and the combined condensate streams give up their thermal energy to preheat the incoming feed in the preheater.

The high-quality product stream is collected and stored. However, this water may not be of adequate quality for direct use as a boiler feed water as it may need further treatment by Ion Exchange (IX) or reverse osmosis (RO) process for that purpose depending upon the feed water quality specifications for the specific application. The concentrated brine is rejected for proper disposal.

Multistage Flash (MSF) distillation is a later development of MED technology. The main difference is that the feed to the evaporator is sprayed so that the vaporization takes place without much contact with the heat exchange surfaces which helps avoid the potential of fouling and scaling. There have been several installations of MSF distillation systems primarily in the Middle East.

2.1.2.5 Brine Management

The concentrate stream from RO is referred to as 'brine' which has a high TDS ranging typically from 5 to 30%. The term "brine" is not restricted to the concentrate stream from SWRO alone and is equally applicable to high TDS water and wastewater streams from other petroleum production and processing operations also. Handling and disposal of brine require significant technical and regulatory considerations. This subject is discussed in Sect. 3.8 of Chapter 3 in the context with the handling, treatment and disposal of the brine stream generated in the upstream produced water treatment as a final step in in achieving the goal of Zero Liquid Discharge (ZLD).

Abbreviations and Glossary of Terms

AOM	Algal Organic Matter
AOP	Advanced Oxidation Processes
ASTM	American Society for Testing Materials
AWWA	American Water Works Association
CEB	Chemically Enhanced Backwash
CH	Carbonate Hardness
CIP	Clean in Place
CWA	Clean Water Act
DL	Double Layer
EBCT	Empty Bet Contact Time
ED	Electrodialysis
EDI	Electrodeionization
EDR	Electrodialysis Reversal
EU	European Union
FA	Fulvic Acids
FMA	Free Mineral Acidity
GAC	Granular Activated Carbon
HA	Humic Acids
HF	Hollow Fiber Membrane
IX	Ion Exchange
kDA	Kilo Dalton
MED	Multi Effect Distillation
MF	Microfiltration
MVR	Mechanical Vapor Recompression
NCH	Non-Carbonate Hardness
NF	Nanofiltration
NOM	Natural Organic Matter
NTU	Nephelometric Turbidity Unit
PAC	Powdered Activated Carbon
PES	Polysulfone/polyethersulfone
PVDF	Polyvinylidene fluoride
Re	Reynolds Number
RO	Reverse Osmosis
SAC	Strong Acid Anion
SBA	Strong Base Anion
SDI	Silt Density Index
SUVA	Specific Ultraviolet Absorbence
SWRO	Seawater Reverse Osmosis
TH	Total Hardness
ThMA	Theoretical Mineral Acidity
TKN	Total Kjeldahl Nitrogen
TMP	Transmembrane Pressure

UC	Uniformity Coefficient
UF	Ultrafiltration
US EPA	United States Environmental Protection Agency
UV	Ultraviolet
VOC	Volatile Organic Compounds
WAC	Weak Acid Cation
WBA	Weak Base Anion
WEF	Water Environment Federation
WHO	World Health Organization
ZLD	Zero Liquid Discharge
ZP	Zeta Potential

References

AWWA (2011) Water quality and treatment a handbook of community water supplies, 6th edn. McGraw Hill, pp 12.12 and 14.6

AWWA Committee Report (1980) The status of direct filtration. J AWWA 72(7):405

Basu S (1996) Catalytic oxidation and toxicity reduction of 2,4,6 trichlorophenol ion aqueous solution by Fenton's reagent, PhD Dissertation, Northeastern University, Boston

Bates W (2024) Personal communication, Nitto global membranes division

Brock TD, Clyne J (1984) Significance of algal excretory products for growth of epilimnetic bacteria. Appl Environ Microbiol 47:731–734

Delphos PJ, Wesner GM (2005) Mixing, coagulation, and flocculation. In: Baruth EE (ed), Water treatment plant design, American water works association, American society of civil engineers, Ch. 6, 4th edn. McGraw-Hill, Inc., New York

Flynn D (ed) (2009) The Nalco water handbook, 3rd edition. McGraw Hill, New York

Geider RJ, Roche JL (1994) The role of iron in phytoplankton photosynthesis, and the potential for iron-limitation of primary productivity in the sea. Photosynthesis Res 39:275–301

Henthorne L (2009) The current state of desalination. In: Proceedings of the international desalination association (IDA) world congress, Dubai, United Arab Emirates, November 7–12, 2019

Iwahori H (2003) Peculiar or unexpected behavior of Silt Density Index of pre-treated seawater for RO desalination, presented at the IDA Paradise Island Bahamas, November

Kader G (2011) Ning RY (ed) A large review of the pre treatment, expanding issues in desalination.. ISBN: 978-953-307-624-9

Kawamura S (1975) Design and operation of high rate filters. Am Water Work Assoc 67(12):710–710

Kim PH, Hong S, Choi S (2006) Variation and prediction of membrane fouling index under various feed water characteristics. J Membr Sci 284:248–254

Kremen SS, Tanner M (1998) Silt density index (SDI), percent plugging factor (%PF) their relation to actual foulant deposition. Desalination 119:360–372

Kuma K, Nishioka J, Matsunaga K (1996) Controls on iron (III) hydroxide solubility in seawater: the influence of pH and natural organic chelators. Limnol Oceanogr 41:396–407

LaZerte BD, Burling K (1990) Manganese speciation in dilute waters in the Precambrian Shield Canada. Water Res 24:1097–1101

Metcalf & Eddy/AECOM (2014) Wastewater engineering. In: Treatment and reuse, 5th edn. McGraw-Hill, New York

Mosset A, Bonnélye V, Petrya M, Sanz MA (2008) The sensitivity of SDI analysis: from RO feed water to raw water. Desalination 222:17–23
Qasim SR, Motley EM, Zhu G (2000) Water works engineering planning design and operation, 1st edn. Prentice Hall of India, New Delhi
Reynolds TD, Richards PA (1996) Unit operations and processes in environmental engineering. PWS Publishing Company
Sagle A, Freeman B (2004) Fundamentals of membranes for water treatment. Futur Desalination Tex 2(363):137
Sawyer CN, McCarty PL, Parkin GF (1991) Chemistry for environmental engineering, 4th edn, McGraw Hill, New York
Schaanning M, Naes K, Egeberg PK (1988) Cycling of manganese in the permanently anoxic Drammensfjord. Mar Chem 23:365–382
Sitterley KA, Cath TJ, Jenne DS, Yu Y-H, Cath TY (2022) Performance of reverse osmosis membrane with large feed pressure fluctuations from a wave-driven desalination system, Desalination 527
Skaggs BK (2012) Utilization of Ferrate as a disinfection technology for wastewater reuse.PhD Dissertation, Tulane University School of Public Health and Tropical Medicine
WEF (2006) Membrane systems for wastewater treatment, water environment federation. Alexandria, VA, p 284
Weinberg ED (1989) Cellular regulation of iron assimilation. Q Rev Biol 64:261–290

Chapter 3
Water Management in Upstream and Midstream Operations

Somnath Basu and Narayana Murthy Nadella

Abstract Water is intricately connected with all sectors of the petroleum industry. In the upstream sector it plays a critical role in the production from unconventional sources by hydrofracturing operation, and also in enhancing the secondary production from conventional sources. Another important consideration in the upstream business sector is the handling of large amounts of water produced from the exploration of crude. This chapter discusses the water use and water treatment in the upstream and midstream segments of the petroleum industry. Included are discussions of the sources and quality of water used and produced in these segments, and the methods of treatment practiced in this industry for appropriate disposal and/or reuse. To understand the subject of any industrial water and wastewater treatment process it is essential to develop an understanding of the types of chemicals that are present in those streams. With that in view a brief description of the chemical composition and characteristics of petroleum is presented in Appendix B.

Keywords Alkaline flood · Biological treatment · Brine · Completions · Dispersed · Disposal well · Divalent · Drilling · Fire flood · Hydraulic fracturing · Hydrocyclone · In-situ combustion · MPPE · NORM · Offshore · Oil/water separation · Onshore · Produced water · Production facility · Recovery · SAGD · Salt removal · SAR · Steam flood · Surfactant · TDS · Thermal enhanced oil recovery · TOC · TSS · Upstream · Water flood · Well stimulation · Zero Liquid Discharge (ZLD)

S. Basu (✉)
Jewel Innovations in Energy and Environment, Katy, TX, USA
e-mail: som.basu@jewelenergyenvironment.com

N. M. Nadella
Technip Energies USA, Houston, USA
e-mail: MurthyNadella59@gmail.com

S. Basu et al. (eds.), *Water Management in Petroleum Industries*, Water Resources Development and Management, https://doi.org/10.1007/978-981-19-3159-8_3

3.1 Introduction

Petroleum industry supply chain consists of four main segments—upstream, midstream, downstream, and marketing. A brief description of each sector and its operation has been introduced in the preface section of this book. This chapter is divided into the Upstream operations from Sects. 3.2 through Sect. 3.11 and the Midstream operations is covered in Sect. 3.12. These sectors of the petroleum supply chain are shown in the Fig. 3.1.

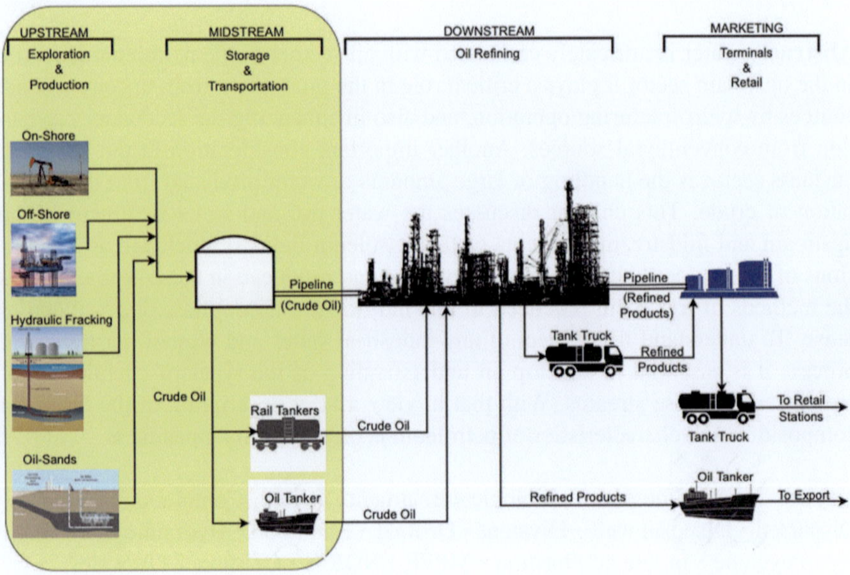

Fig. 3.1 Petroleum industry supply chain from production to marketing of refined products

3.2 Upstream Production Operations

Petroleum reservoirs have different subsurface conditions of temperature, pressure, density, viscosity, and composition based on their location. Therefore, the production methods must be tailored to maximize yield from a given reservoir based on the crude and reservoir characteristics. Even in the same reservoir, only about 25% of oil in place can be recovered by primary production and other methods are required for additional production. Different methods of petroleum recovery are described in Table 3.1.

Water is closely associated in the upstream operations both for production support, as in hydrofracturing, and enhancement as in water flooding, as well as in the form of produced water that flows out of the production wells along with crude oil and

Table 3.1 Methods of petroleum production

Methods of petroleum production	Description
Primary	Petroleum is produced using inherent reservoir pressure without any injected fluid for pressure enhancement
Secondary	A recovery improvement process such as water flooding, or polymer flooding, or CO_2 injection is implemented to boost up production. Hydraulic fracturing is considered a secondary recovery process
Tertiary or enhanced oil recovery (EOR)	EOR processes go beyond water or gas flooding. It may also involve steam, fire, polymer, or solvent

gas. Petroleum crude production sources are commonly classified as conventional and unconventional. Production from conventional sources is conducted mainly by the primary method where the crude flows aboveground from the reservoir without the application of any external stimulations. However, with time, as the reserves deplete production is supported by various means of stimulation e.g., water flooding, polymer flooding, CO_2 injection, etc. Conventional resources are underground rock formations that are porous and located at shallow depths underground. However, these reserves are dwindling fast, and production is increasingly being conducted from unconventional resources by application of advanced technologies from much more difficult terrains. A broad classification of different types of petroleum resources and water associated with production processes is presented in Table 3.2. Detailed discussions are presented in the following sections.

Table 3.2 Classification of upstream production resources

Production	Location and type	Water—production, processing and disposal/reuse
Conventional	On-shore	1. Produced water that needs to be treated for re-injection in water flood, or 2. Discharge to the surface water bodies after treatment
	Off-shore	1. Treated water suitable for reuse for secondary recovery, or 2. Discharge to sea after treatment
Unconventional	Hydraulic fracturing	1. Treated water suitable for reinjection into ground, or 2. Reuse for hydrofracturing and/or other beneficial reuses or 3. Discharge and disposal after treatment
	Thermal heavy oil production including oil sands	1. Treated water suitable for recycle and reuse as steam, or 2. Discharge to disposal wells after treatment

In produced water industry US units of measurement are used extensively for both technical literature and commercial purposes. Although crude production is reported in literature both in metric tons/day (MT/day) and US barrels/day (bbl/day), for produced water bbl/day is most widely accepted unit. Accordingly, US units have been primarily used in this chapter.

3.3 Conventional On-shore Production

The upstream facilities have process units and equipment to separate and remove water from hydrocarbons and store the oil, gas and water fractions for further processing. Many facilities, especially the larger ones are equipped with their own utility systems such as heating, cooling, and power generation units to support the facility operations.

The oil processing units are provided to:

- Separate the oil, gas, water, and solids.
- Treat the oil to meet transportation and sales specifications (e.g., Basic Sediment and Water (BS&W) content, salt content, vapor pressure).
- Measure and sample the oil to determine its value.
- Deliver it to transportation systems such as pipelines, trucks, ships/barges, or railroad cars.

The gas treating units may consist of:

- Processing to condense heavier components that can be transported as liquid.
- Removal of H_2S and CO_2.
- Gas dehydration to meet the sales and pipeline specification.
- Gas compression for sales or transportation.

Upstream water treatment consists of separating any oil, gas and solids and sending the water for disposal or for further treatment to recycle or reuse. A schematic of a typical upstream production facility is shown in Fig. 3.2 this is a general representation of an upstream water treatment facility but some of the equipment, especially desalter are typically used in downstream rather than upstream applications.

3.3.1 *Water Quality in Conventional On-shore Production*

During primary production, the free water and connate water that is present in the reservoir matrix flow along with the crude due to pressure depletion. At many locations water is more saline than seawater and must be separated from the oil and gas.

Water from different sources is used in upstream petroleum extraction. The type, quality and quantity of water used depend on the technology used in production,

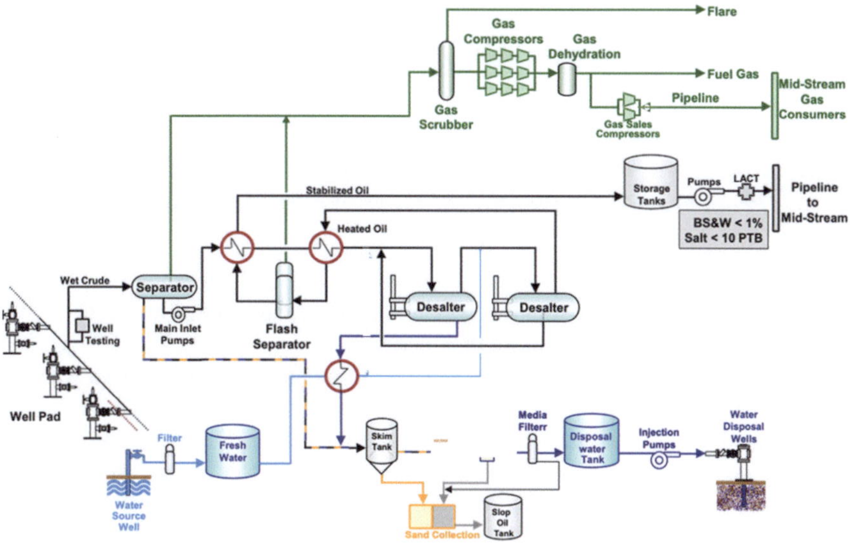

Fig. 3.2 Typical conventional upstream production facility

location of the facilities, as well as available water sources. A list of production operations and water requirements are shown in Table 3.3.

In addition to processing the oil and gas for distribution and sale, the produced water is treated for recycling and/or disposal. The type of produced water treatment

Table 3.3 Water sources and quality for conventional and unconventional production

Operation	Water required for	Sources	Quality required
Drilling and completions (see Sect. 3.8.1 for details)	Preparing drilling muds used during well drilling (see Sect. 3.8.1 for details)	Fresh water, brackish water, or sea water in limited quantities	Compatibility with drilling chemicals and reservoir matrix
Well stimulation (see Sect. 3.8.1 for details)	Dilution water for chemicals used during well workover (see Sect. 3.8.1 for details)	Fresh water, brackish water, or sea water in limited quantities	Compatibility with well material and reservoir matrix
Water flooding	Injecting into the reservoir	Large quantities of fresh water or brackish water or seawater or recycled produced water	Free of suspended solids, oxygen and be compatible with reservoir matrix
Desalting	Desalting	Small quantities of fresh water required	Low Total Dissolved Solids (TDS) (see 4.9 of Chap. 4 for the details of water quality specifications)

depends on its composition, and the type of final disposal or the expected reuse option. The processes can range from simple filtration to total deionization.

3.3.2 *Produced Water Characteristics and Treatment Processes in Conventional Onshore Production*

Produced water composition may vary widely. Table 3.4 presents a typical set of produced water characteristics that was used for the design of the treatment facility of an onshore production site in South America.

In onshore production facilities, oil and gas are separated from produced water in one or more stages operating at different pressures. The separated produced water is first treated in skim tanks and API/CPI separators, then further treated in flotation units, followed by media filters and/or nutshell filters to achieve the disposal water quality and then sent for deep well injection. Figure 3.3 is a block flow diagram of a typical onshore production facility produced water treatment. This represents the basic treatment train but may need additions if more stringent effluent qualities are required. Conventional onshore facilities with centralized produced water treatment are typically designed in trains processing up to 120,000 barrels per day (*Note: This is equal to 1900 m^3/day in SI units. However, in produced water industry US barrels are extensively used for units globally. The SI units used in crude production is tons/day. Throughout this document US barrel/day, or bbl/day is used as the rate of production as this is the most widely accepted unit internationally*). There may be multiple water treatment trains installed at a facility. Additional trains may be added as the produced water flows increase during the production life of the reservoir. The multiple train concept allows standardization of equipment sizes, sparing and maintenance philosophies, predictable transportation, and construction methods.

The flow rate of produced water treated in an onshore facility depends on the water to oil ratio during the production life of the reservoir. The design rates for produced water are about 2–3 times the crude oil production rates on an average (Veil et al. 2004). For example, if a facility is designed to produce 100,000 barrels of

Table 3.4 Typical produced water characteristics from an on-shore production site

Parameter	Unit	Influent composition	Treated effluent composition
Oil and grease	mg/L	1,000	<5
TSS	mg/L	200	<1
TDS	mg/L	50,000–70,000	–
Dissolved oxygen (DO)	mg/L	–	<0.2
Total bacteria count	cfu/mL	–	<1,000
Sulfur reducing bacteria	cfu/mL	–	<10–100

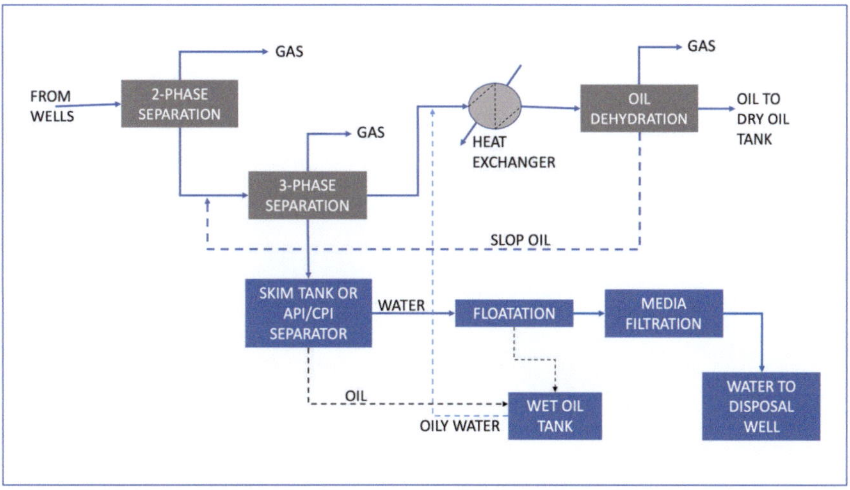

Fig. 3.3 Produced water treatment for conventional onshore production facility

crude per day, the produced water rates expected during the life of the facility will be 200,000–300,000 barrels per day.

3.4 Conventional Off-shore Production

Off-shore production technology has developed significantly over the last 80 years. Off-shore floating platforms have been built for production from wells at depths of tens to thousands of meters (less than a hundred feet to thousands of feet). These structures and pipelines to production wells are major undertakings involving thousands of tons of steel. As petroleum demand increases and onshore reservoirs get depleted, more offshore production facilities become economical and for some countries these are the only sources of crude oil. Figure 3.4 shows an offshore production platform and Floating Production Storage and Offloading (FPSO) vessel located at 10 km off the Gulf of Thailand.

The subsea portion of the offshore production involves production wells, pipelines connecting to the floating platforms or ships, and several equipment on the seabed depending on the production facilities. Figure 3.5 shows different subsea equipment and pipelines that are part of offshore production facilities.

A combination of high pressure, associated with low temperature in offshore fields, can cause formation of slugs of gas hydrate in transmission lines. Gas clathrates (commonly called hydrates) are crystalline structures that occur when water forms a cage-like structure around smaller molecules of gas. Gas hydrates are composed of water and any, or a combination of the eight molecules: methane, ethane, propane, isobutane, normal butane, nitrogen, carbon dioxide, and hydrogen sulfide. Hydrate

Fig. 3.4 Offshore production platform and FPSO in Gulf of Thailand (Rawlins 2017)

formation is possible in any place where water exists with one or more such molecules in natural or artificial environments and at temperatures close to the freezing point of water and at elevated pressure. For example, methane gas hydrates are found in cold waters off-shore Canada. These slugs of hydrates deposit on the internal surfaces of transmission pipelines and obstruct the flow. In these cases, anti-freezing agents like mono-ethylene glycol (MEG) and/or methanol are dosed into the well and pipes to prevent plugging of the pipelines. The function of these alcohols is to depress the freezing point of water so that it remains in liquid form and does not create solid slugs even at low temperatures. While MEG is dosed continuously during production process methanol is dosed intermittently and during start-up. Gas is then released from the production fluid by expansion and separated from the liquid phase. MEG is regenerated and recycled into the production process. The dosed methanol remains dissolved in the produced water which can potentially raise the COD levels to more than several thousand mg/L (Fig. 3.5).

Hydrate inhibitors MEG and methanol are detrimental since both promote scaling by accelerating nucleation of barite (Barium Sulfate). Therefore, addition of both chemicals needs to be optimized to avoid barite scaling. Baryte formation and its removal from produced water is discussed further in Sect. 3.9.2.

Fig. 3.5 Subsea production systems (SPE 2013)

3.4.1 Water Quality and Treatment Processes in Conventional Offshore Production

A conventional off-shore production facility will have the same objective as an on-shore facility to separate oil, gas, solids, and water. However, due to space limitations on offshore platforms, the equipment used is compact and disposal methods are different.

In offshore and deep-water production facilities, produced water is normally treated to remove the total oil and grease (TOG) and any toxic compounds before being disposed to sea or deep well injection. Compact treatment units like hydrocyclones and induced gas flotation (IGF) units are used. Figure 3.6 shows typical Deep-water Gulf of Mexico oil/water separation system (Bothamley 2004). The produced water rates depend on the water content in the production fluids. Due to space limitations, it is difficult to add more treatment trains as water content increases over production life. A common method in off-shore production is to use the spoke-and-hub structure. In this arrangement, multiple wells are connected to a centralized processing or hub platform, where preliminary oil–water separation takes place. The crude oil is then transported to onshore processing by pipelines or by ships. Compact, lightweight, and reliable technology to process the water near the source is the preferred solution to treat the produced water. As an example, the Bua Ban North field, which is about 10 km offshore Thailand, produces about 18,000 barrels

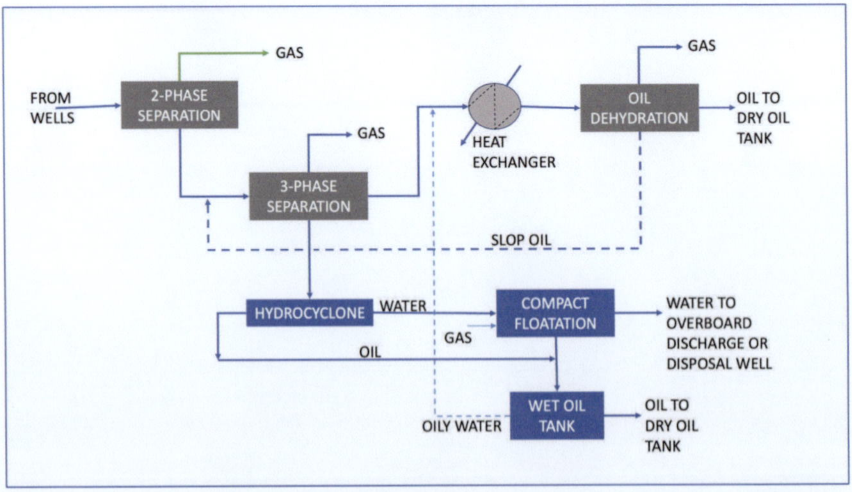

Fig. 3.6 Off-shore production facility block diagram

of oil per day from a total of 26 production wells. Water is about 90% which means total produced water treated is 162,000 barrels per day (Rawlins 2017).

Produced water from offshore facilities that use MEG and methanol as antifreeze chemicals need to conduct additional treatment beyond de-oiling as the dissolved methanol, that is dosed intermittently, creates spikes of high COD in tens of thousands mg/L of COD. This water is unsuitable for discharge into the ocean or injection. To mitigate this situation the de-oiled water is collected in an equalization tank to shave off the surges of COD and then fed to biological treatment to control COD. As methanol is readily biodegradable the BOD and COD values of this water are close to each other. A membrane bioreactor (MBR) system is suitable for locating on a platform as it requires a low footprint area. A detailed description of biological treatment processes with special reference to an MBR system is provided in Chap. 5.

3.5 Unconventional Production by Thermal Methods

Thermal enhanced oil recovery is a family of tertiary oil extraction processes defined as "any process in which heat is introduced into a subsurface accumulation of petroleum for the purpose of recovery".

The most common heating medium used is steam. Hot water and heated gasses have been tried but are not as effective as steam. According to 2014 *Oil and Gas Journal* survey (Kottungal 2014), steam enhanced oil recovery (EOR) projects accounted for more than 40% of the total for all tertiary enhanced recovery methods. Hydrocarbon gas injection and CO_2 gas injection are the only other significant contributors.

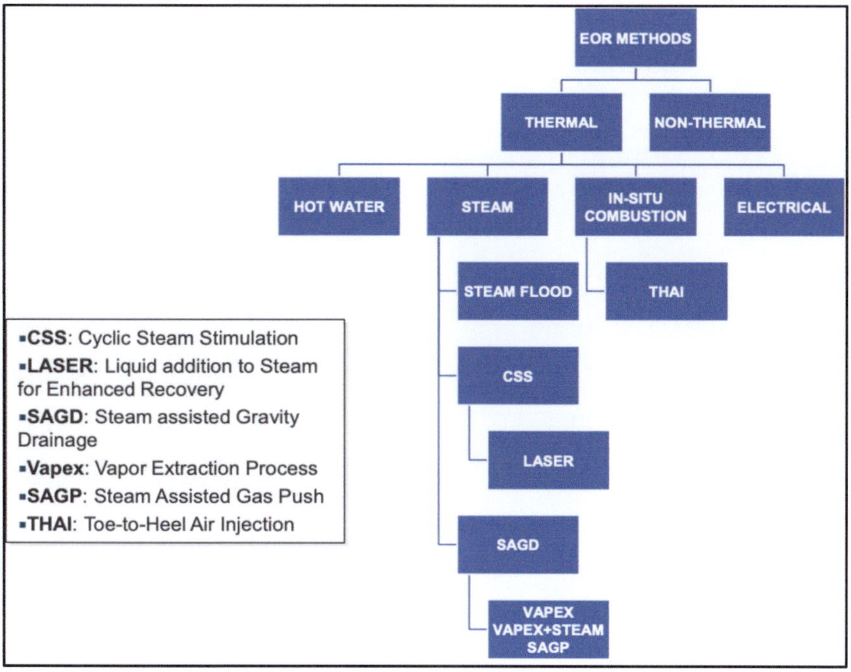

Fig. 3.7 Thermal EOR methods

Thermal methods aim at reducing the viscosity of oil to increase its mobility, through the application of heat. A 200 °C temperature increase reduces the viscosity of most oils to ~1 mPa S (cP). Many thermal processes have been developed and are shown in Fig. 3.7 (Thomas 2008).

Though all these thermal EOR methods have been piloted, only a few methods such as SAGD, CSS and Steam Flood are in commercial use. Some of these thermal EOR methods are described below (Nadella 2009).

- *Steam Assisted Gravity Drainage (SAGD)*: In this process, two horizontal wells are drilled into the bottom of the bitumen reservoir, the top well is called injector, and the bottom well is called the producer. High pressure (at about 7,000 kPa gage) saturated steam is injected through the top well which develops a steam chamber. The steam condenses and gives up the heat to melt and mobilize the bitumen. The bitumen flows by gravity to the bottom well and is extracted along with the steam condensate. Figure 3.8 shows the schematic view of SAGD process.
- *Cyclic Steam Stimulation (CSS)*: It is initially a "single well" injection-production process and is widely used in Canada and Venezuela because of its applicability in very viscous oil formations, and relatively quick payout. In Cold Lake, Alberta, the recovery is over 25% of the of the total reserve, and is expected to reach 35%, or higher. In Venezuela, recoveries as high as 40% (in Bolivar Coast area) have

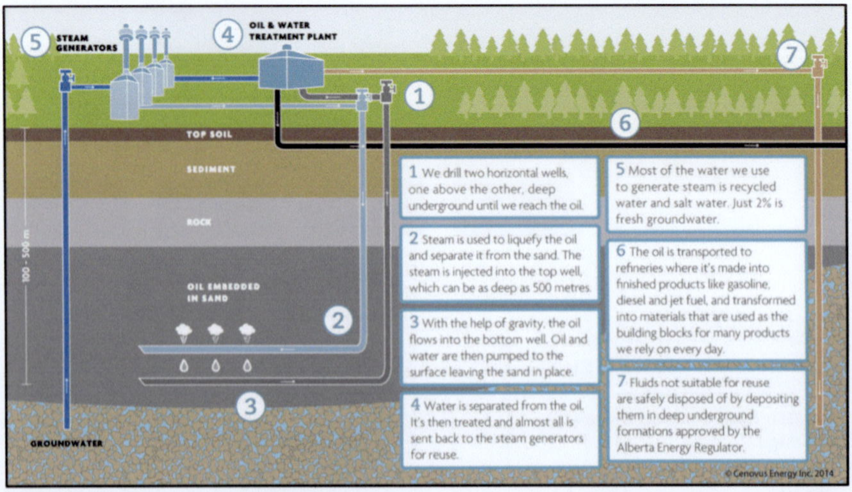

Fig. 3.8 SAGD process schematic (Cenovous 2022)

been noted. In California, cyclic steaming is the first stage in the development of a steam flood. See Fig. 3.9.

- *Steam Flooding*: The main advantage of this multi-well, pattern type method is the large aerial coverage and high recovery (50–60%), while the drawbacks are excessive heat loss, long payout times and high costs. This method is generally applicable to oils in the 100 mPa s viscosity range. A schematic of steam flooding process is shown in Fig. 3.10. This process has been used in Cold Lake, Alberta;

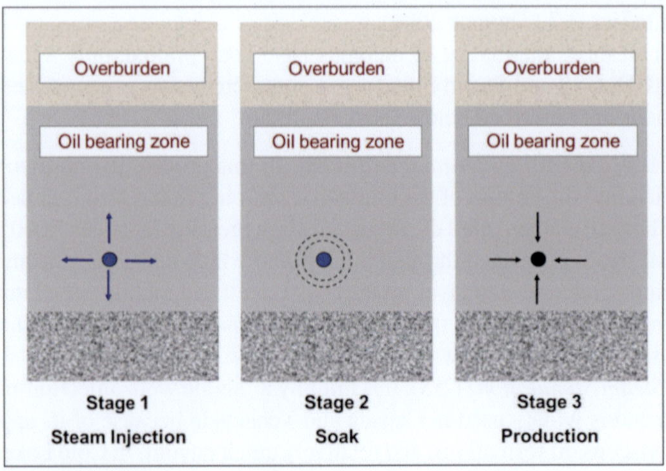

Fig. 3.9 CSS process schematic

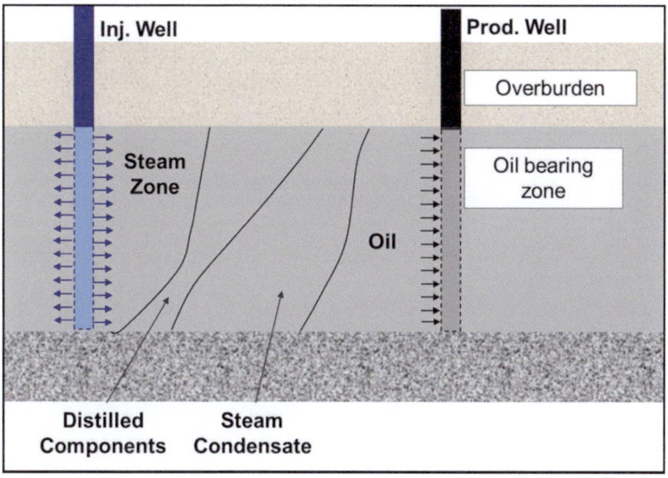

Fig. 3.10 Schematic of steam flood

Kern River and San Ardo, California; and Duri Steam flood project in Indonesia among several other locations.

The San Ardo, CA oil field comprises of a formation several hundred meters thick at ~550–600 m (1,800–2,000 ft) underground containing heavy oil of 10°–13° API crude (Webb 2009). Steam is used to reduce the viscosity and aid in the production. Water produced from this operation is desalinated using a treatment train consisting of degasification, ion exchange softening, multimedia filtration and cartridge filtration, followed by reverse osmosis in series. A part of the treated water is reused for raising steam and the excess amount is discharged into constructed wetlands and disposed through EPA Class II injection wells.

- *In-Situ Combustion or Fire Flood*: This process is said to be applicable for a wide range of oil gravities (8°–36° API), but commercial success has been possible only in oils that are sufficiently mobile at reservoir conditions. Forward combustion is theoretically the most efficient process and can yield high oil recoveries (30–50%). However, it suffers from several operational challenges, most notably fire-front control, high gas production rates and operational problems. Water injection has been used to improve the basic process. Oxygen enriched air injection and reverse combustion have certain advantages, such as in-situ upgrading of oil (in reverse combustion) but have been unsuccessful in the field. A schematic of the in-situ combustion is shown in Fig. 3.11. This process was piloted successfully as Toe-to-Heel Air Injection (THAI) process by Petrobank Energy in Alberta, Canada.
- *Hot Water Flooding*: It is seldom employed because of heat losses in the surface lines, wellbore formation, cause a large drop in temperature and reduce its effectiveness in decreasing the oil viscosity. Nevertheless, it is a process to consider for

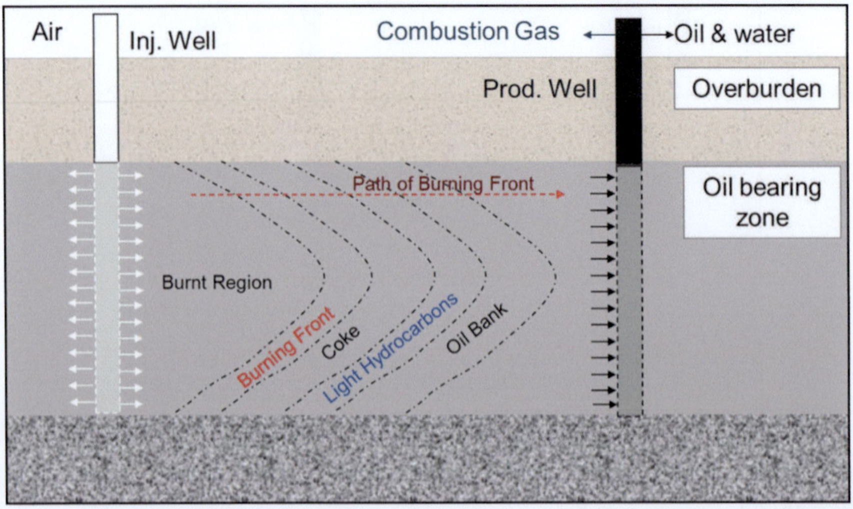

Fig. 3.11 Schematic of in-situ combustion process

deep heavy oil formations, where steam in not likely to be successful. An example of this is the Kuparuk field in Alaska.

3.5.1 Water Quality Required for Unconventional Thermal Production Methods

In the steam based thermal EOR operations, the water quality should meet the boiler water specification. These specifications are suggested by ASME Research and Technology committee and followed by the boiler manufacturers in providing the design and performance guarantees. Typical analysis of produced water and boiler feed water in SAGD and CSS operations is shown in Table 3.5.

3.5.2 Produced Water Treatment Processes in Thermal Production Methods

Produced water in thermal facilities is at about 200 °C when produced with oil. This temperature depends on the reservoir operating pressure at which steam condenses to give the heat of condensation to the reservoir matrix that mobilizes the heavy oil. The mobile oil and steam condensate are produced by either the operating pressure or by means of a pump or by using gas lift. Heat is recovered from this stream to preheat boiler feed water to conserve energy. The produced water is cooled to about 80 °C to treat and store in tanks at atmospheric pressure. Water is first separated from suspended solids and oil through skim tanks, Induced Gas Flotation (IGF) units, and crushed walnut shell media filters to bring suspended oil to less than 5 mg/L. The main impurities in this water are scale forming salts of calcium and magnesium, and silica. These are removed through hot or warm lime softening followed by ion exchange. Silica is brought below 50 mg/L and hardness is reduced below 0.5 mg/L as $CaCO_3$. This water can be recycled as boiler feed water through OTSG (Once Through Steam Generators) to generate 80% quality steam (80% steam and 20% water). The dissolved solids remain in water and are removed as blowdown. A block flow diagram of a thermal facility is shown in Fig. 3.12.

The water treatment train is a major part of a thermal facility that uses steam to produce oil. Typically, the steam to oil ratio varies from 3 to 4, which means for every barrel of oil produced steam injection of about 3–4 barrels of cold-water equivalent (amount of cold water that is required to generate steam) of steam is injected. Figure 3.13 shows the water treatment unit of a typical thermal production facility. Produced water from thermal operations is hot at about 180–200 °C from the wells. Due to high temperature, there are no bacteria present in produced water from thermal production facilities, and the water is cooled to 80 °C and stored in gas blanketed tanks to prevent contact with air as this water will be reused for steam generation.

Table 3.5 Produced water characteristics and treated water quality requirements for steam generation in thermal production method

Parameter	Produced water	Boiler feedwater (BFW) specifications (OTSG[a])	BFW specifications (drum boilers)
Generated steam pressure, psig (MPa)	–	900–1,500 (6.21–10.34)	900–1,500 (6.21–10.34)
Total dissolved solids (TDS), mg/L	7,100–10,000	<10,000	<2
Hardness, ppm $CaCO_3$	290	ND	ND
pH	7.4	9–11	8.8–9.6
Iron, mg/L	0.5	<0.02	<0.01
SiO_2, mg/L	200–250	<50	<0.04
Total Alkalinity, mg/L as $CaCO_3$	–	<100	–
Oil, mg/L	2000	<0.5	<0.2
Oxygen, ppbw	0	<7	<7

[a] OTSG is once through steam generator producing steam at 80–90% quality

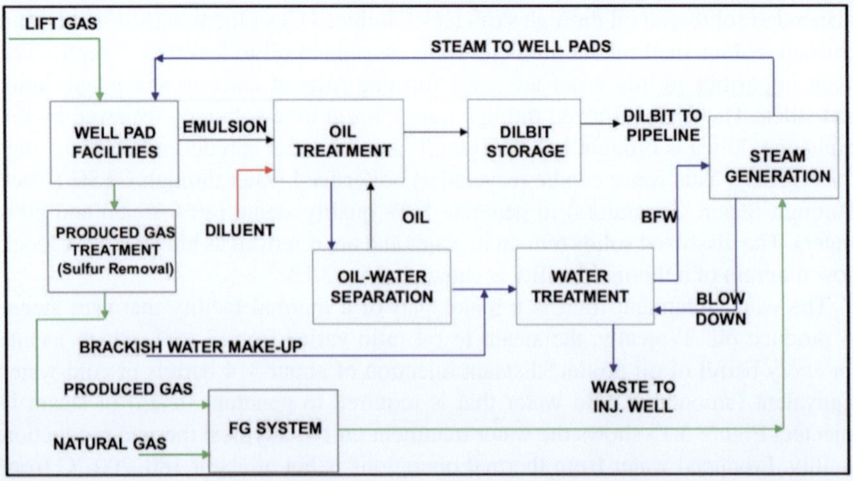

Fig. 3.12 Block flow diagram of a thermal EOR (SAGD/CSS/steam flood) surface facility

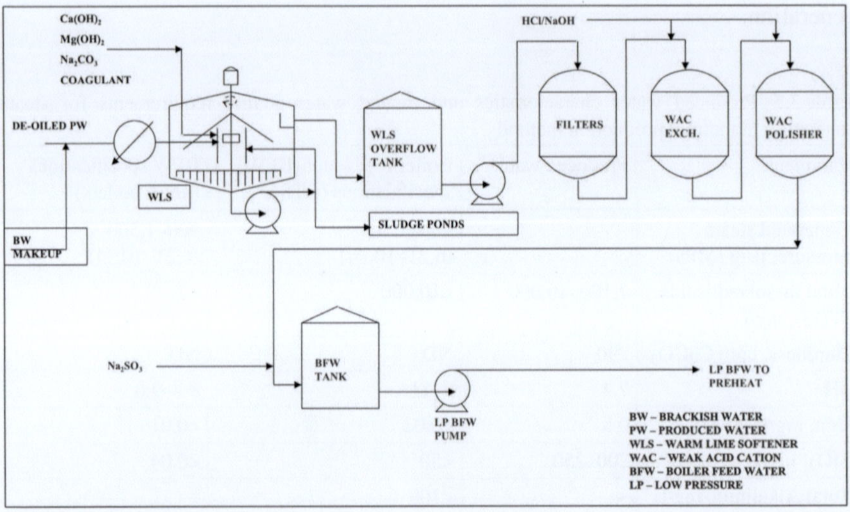

Fig. 3.13 Water treatment train in a thermal EOR central plant

3.6 Unconventional Petroleum Production by Hydraulic Fracturing

Exploration in difficult terrains at depths ranging from 1.5 to 5.5 km (5,000–18,000 ft) has been made possible by hydraulic fracturing or hydrofracturing technology. In principle, this involves injection of high-pressure water to fracture the rocks to release the oil and gas trapped in their tiny pores (which are in the size range of 100–1,000 nm. Porosity of a reservoir matrix directly impacts the movement of fluids in the matrix. With all other factors being equal, the reservoir with higher porosity has a higher rate of production. The rock structure and porosity of formations can be understood by comparing the permeabilities of different types of grains that make up these formations (King 2012):

- Unconventional Shale: 10^{-6} to 10^{-3} milli Darcy (mD)
- Tight Oil and Gas Formation: 10^{-3} to 10^{0} mD
- Conventional Production Domain: 10^{0} to 10^{3} mD

Formations in the permeability range of 10^{-6} through 10^{0} mD require hydrofracturing for production of unconventional oil and gas trapped in their structure. This technology, in combination with horizontal drilling, has enabled one single well to extend horizontally to greater than 4.5 km (15,000 ft) from the well-head and has made drilling for hydrofracturing a cost-effective operation. Pioneered by George Mitchell of Texas in the late 1990s this technology contributed enormously to the production of shale oil and gas from unconventional resources resulting in a sharp rise in the total U.S. production over the last decade.

A large part of hydraulic fracturing is driven by the shale plays in West Texas. Figure 3.14 (Source: US Energy Information administration (EIA)) compares conventional vertical wells with unconventional horizontal hydrofracturing wells.

Figure 3.15 shows a schematic view of the hydraulic fracturing process.

Up to 20% of injection water returns from the injection well as flowback water. Two types of water return during hydrofracturing operation—flowback water and

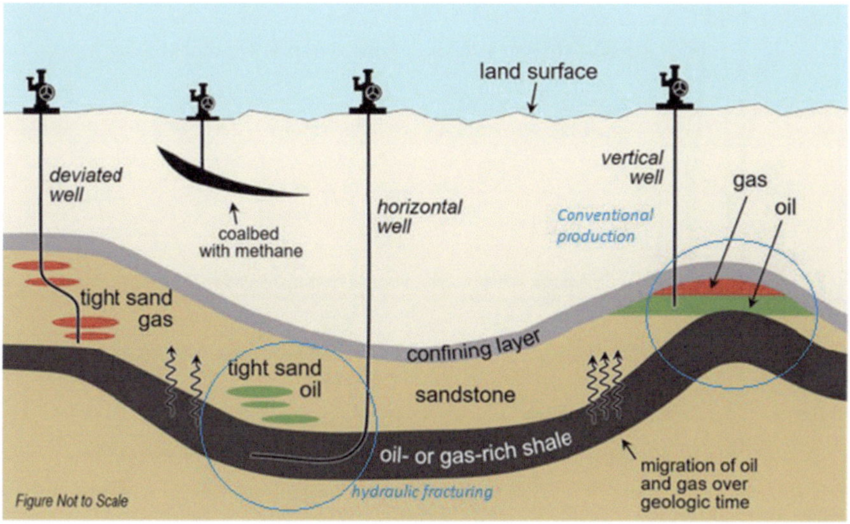

Fig. 3.14 Conceptual illustration of types of oil and gas reservoirs and production wells (Source: US Energy Information administration (EIA))

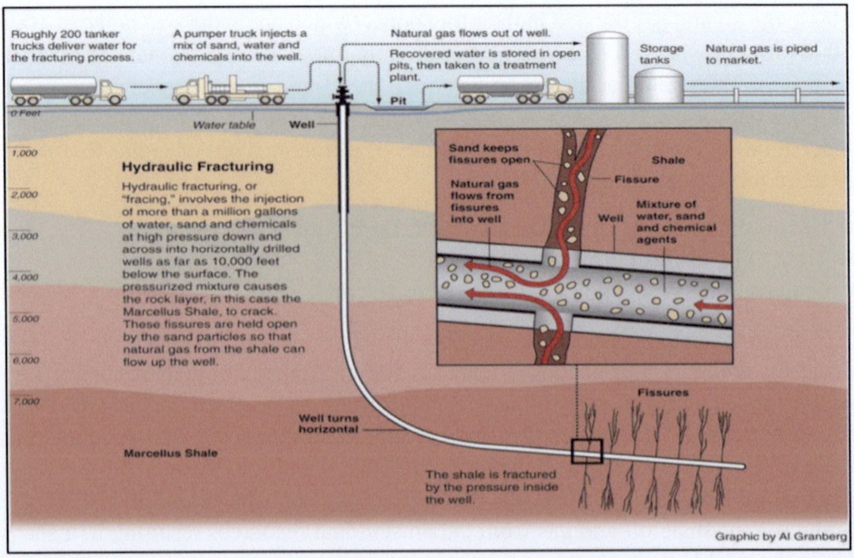

Fig. 3.15 Detail of hydraulic fracturing mechanism (Granberg and Schmidt 2010)

produced water. Figure 3.16 presents the conceptual flow profiles and durations of injected water, flowback water and produced water streams associated with the hydrofracturing of a production well.

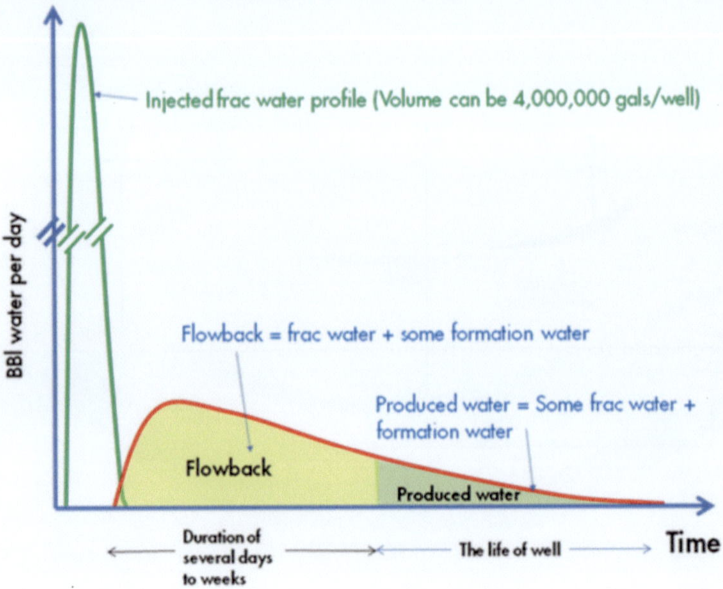

Fig. 3.16 Flow profiles and duration of flowback and produced water from hydrofracturing operations (Sun 2012)

Although the quantity varies with location, and even from well to well, approximately 3–20 barrels of water is produced for each barrel of oil production.

Two major roles of hydrofracturing water are:

1. To transmit pressure from surface to the well bottom to initiate fracture.
2. To maintain fracture permeability by suspending and transporting polymers, sand and other production chemicals as needed.

Hydraulic fracturing, also called hydrofracturing, or simply hydrofracking, can be conducted using matrix or continual fracturing methods. The fracturing fluid is water-based (>99% of water) with small amounts of additives—each of which plays a special role, as discussed in detail in the next section.

3.6.1 Chemistry and Role of Constituents of Hydrofracturing Fluid

Hydrofracturing is conducted by injecting a fracturing fluid consisting of sand and a small quantity of chemicals (anti-scalants, friction reducers and biocides) at high flow rates under pressure at depths of 450–2,000 m (~1,500–6000 ft). The type of fluid used for hydrofracturing depends on geology, rock structure and permeability of the formation. Large amounts of fluid, typically 15,000 m^3 (4 million gallons), are used for each well. Several chemicals are added to water to make it suitable for hydrofracturing operation. Table 3.6 (U.S. Department of Energy 2009) presents a typical composition of some of the more commonly used chemicals, their compositions, and functions. The total amount of additives ranges from 0.5 to 2.0% on a volume/volume basis.

Depending upon the rock structure, two main types of fracturing fluids are used—cross-linked gel fluid and slick water fluid. In certain applications, the viscosity of the fluid is increased by adding water-soluble gel polymers—primarily those which are guar gum-based—to ensure the uniform dispersion of proppants in the fracturing fluids. The purpose of proppants is to open-up the fractures and improve the fracture conductivity to facilitate the flow of trapped hydrocarbon molecules out of the rock structure. Fine sand particles are widely used as proppants. However, other proppants made of natural and synthetic resins are also used on a case-by-case basis when there is a need for high mechanical strength, because sand particles can break down into fines and potentially clog the small pores, obstructing the free flow of hydrocarbons.

Guar gum is a long-chain polysaccharide comprised of many repeating mannose sugar units as the backbone of its molecular structure, with galactose side chains attached to several mannose units (Fig. 3.17). Fracturing fluids also use several other components in addition to guar-based gels and proppants. Cross-linkers are among the most important components that are used to enhance the effectiveness of

Table 3.6 Chemical additives for hydrofracturing water (DOE 2009)

Main compound(s)	Volumetric composition (%)	Additive type	Purpose
Hydrochloric acid	0.123	Diluted acid (15% wt.)	Dissolves minerals to initiate fractures
Glutaraldehyde	0.001	Biocide	Controls bacteria
Ammonium persulfate	0.010	Emulsion breaker	Allows a delayed breakdown of gel polymer
N,N-dimethyl formamide	0.002	Corrosion inhibitor	Prevents pipe corrosion
Borate salts	0.007	Cross-linker for gel	Increases molecular weights of gel polymers by cross-linking, which increases elasticity and suspension properties
Polyacrylamide/ mineral oil	0.088	Friction reducer	Minimizes friction between fluid, pipe, and formation
Guar gum, hydroxyethyl cellulose	0.056	Gel polymers	Thickens the water to suspend the proppant, typically sand
Citric acid	0.004	Iron control	Helps control precipitation of metals
Potassium chloride	0.060	Brine carrier fluid	Stabilizes clays in the shale formation (reduce fines)
Ammonium bisulfite	0.002	Oxygen scavenger	Removes oxygen to help prevent corrosion
Sodium, or potassium carbonate	0.011	pH adjusting agent	Maintains effectiveness of materials such as cross-linkers
Silica, quartz sand	0.500	Proppant	Remains in fractures to prop them open and provide gas pathways
Ethylene glycol	0.043	Scale inhibitor	Helps prevent scale deposition and is a lubricant
Isopropanol	0.085	Co-surfactant	Helps reduce viscosity
Lauryl sulfate	0.001	Surfactants	Helps carry proppant, enhances water injection rate, cleans fractures, reduces friction, and breaks oil–water emulsions
Water	99.007	Aqueous base	Carrier fluid

gels by cross-linking multiple polymeric chains of guar gum molecules. The cross-linking process helps gel formation and enhances the viscosity that is necessary for the uniform dispersion of proppants. Among several others e.g., zirconium (IV), titanium (IV), aluminum (III) etc., borate ions are widely used as cross-linking agents. Molecular structures of guar gum polymers crosslinked by borate ions are illustrated in Fig. 3.18a, b (Menjivar 1986). If a pH above 10.5 and high borate concentrations are maintained, this cross-linked structure is very stable against thermal degradation up to a temperature of 166 °C.

Hydroxypropyl Guar (HPG) is another widely used gel derived from guar gum by a reaction with propylene oxide, resulting in the substitution of some of the Hydroxyl (OH) sites with –O–CH_2 –CHOH–CH_3, which removes some of the cross-linking sites. It is more stable at a high temperature and is therefore more suitable for applications where the well temperatures are above 150 °C.

Gel fracturing fluids use less amount of water but a large quantity of proppant. As explained above, the main function of gels is to ensure proper placement of proppants to maintain continuous production at desired rates. The gels also minimize loss of proppants through pores by formation of a cake.

Polymer breakers are introduced into the pores after hydrofracturing to break down the polymeric gels remaining in the pores of a rock structure after hydrofracturing and prevent the residual polymers from choking and fouling the pores that can cause

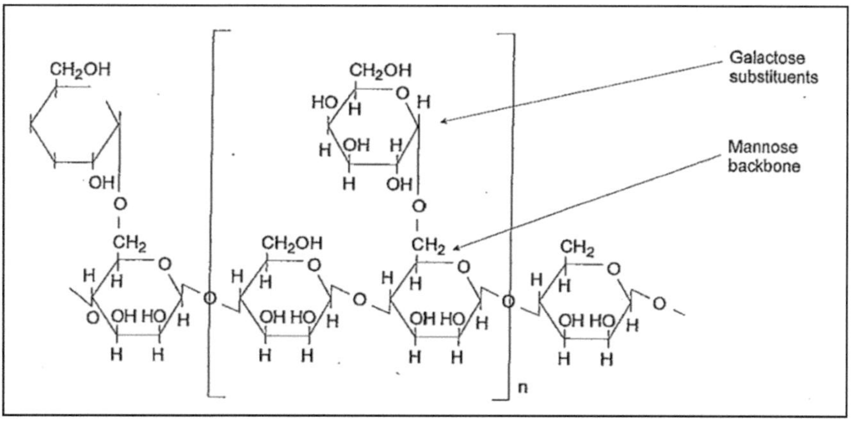

Fig. 3.17 Molecular structure of guar gum showing mannose backbone and galactose side chains (Gulbis and Hodge 2000)

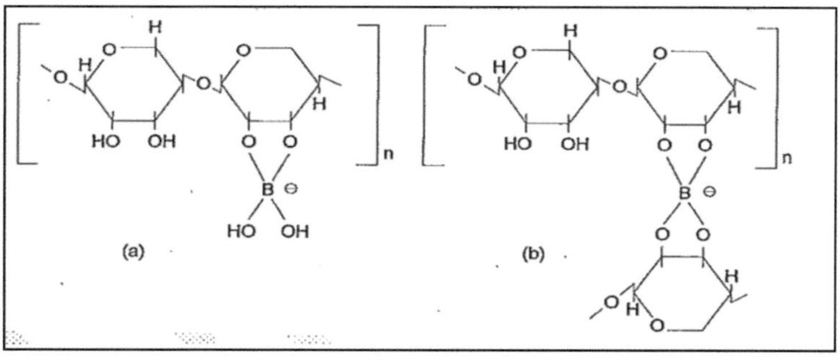

Fig. 3.18 **a** and **b** Structures of borate cross-linked polymeric chain of guar gum

obstruction to free flow of oil and gas. Breakers ensure the free flow of water back from a completed well. A well loaded with unbroken polymeric gels will not be able to produce oil or gas. Chemical oxidants, in particular chlorine dioxide (ClO_2) and sodium persulfate, are the most widely used breakers that break down the polymeric chain in situ.

Depending on the geology of the shale formation and the rock structure therein gel fracturing may not be necessary for production activities at all locations. Gel fluids have high viscosity and require very large pumping energy, and hence higher operating cost to deliver fluids at high rates necessary for hydrofracturing. An alternative is to avoid gel fluids at sites that permit use of slick water fracturing without compromising the production rates and efficacy. Slick water fluid uses large amounts of water mixed with a friction reducer, hydrolyzed polyacrylamide (HPAM) polymer to obtain commercial levels of production from low permeability tight reservoirs. Typically, slick water fracturing also requires lower amounts of proppants compared to gel fluid fracturing. These factors contribute to a significant reduction in the frictional resistance and the associated power consumption to maintain the high flow rate necessary to fracture the formation rocks. In order to optimize cost sometimes the operators utilize hybrid fracturing where slick water fracturing is followed by gel fluid fracturing.

Hydrofracturing methods are known as secondary recovery methods for oil production and recover about 50–70% of reserves. Flowback water is a chemically complex mixture since it contains biocides, polymers, polycyclic aromatics, salts, and dissolved metals. Spills of this water or release to the environment without treatment could result in significant health hazards. Many facilities re-use the water or dispose it off only after treatment. It has been observed that new well kick off, clean-outs, acid stimulations and hydrofracturing adversely affect produced water quality temporarily as the injected chemicals flow back in the beginning of well production. These fluctuations in the produced water quality normally even out when flows from individual wells are combined before processing and when wells are taken into service gradually.

3.6.2 Quality Requirements for Hydraulic Fracturing Injection Water

The inlet water for injection typically requires the following qualities:

1. Low salinity—High salinity causes an increase in cations which results in shielding of the negative ion groups on the polymer thereby reducing polymer's ability to coalesce droplets together and increase viscosity. Salinity of water can be reduced by treating the water with a reverse osmosis (RO) system or a nano filtration (NF) or a combination of Micro filtration (MF) and reverse osmosis (RO) systems.

2. Solids concentration—Low suspended solids concentration is required for continuous use of hydrofracking water since presence of high solids can lead to fouling.
3. Low chemical concentrations—Several species such as Iron (Fe^{2+}) and hydrogen sulfide (H_2S) in combination with oxygen produce free radicals which cause permanent damage to polymers in fracturing fluids by reducing their molecular weights and viscosities.
4. Oxygen Concentration: Presence of oxygen above 10 ppb(v/v) can cause corrosion of iron used in the well and connecting pipelines. Corrosion can cause leaks and spills, equipment failures and increase suspended solids that plug the flow lines.
5. Temperature—High temperature reduces polymer viscosity over days. The recommended temperature range is 40–120 °C to maintain hydraulic properties of the polymer for effecting sweep of the reservoir.
6. Bacteria Count—Presence of microorganisms in hydrofracturing water can cause biofouling and/or microbiologically induced corrosion (MIC) that affects the mechanical integrity of the metallic parts of production equipment (rigs, drill, etc.). The population of viable biomass in water is quantified by concentration of ATP in terms of ng/L (see definition in Chap. 1). Typically, the specification for bacterial count in hydrofracturing water is ATP concentration <50 ng/L.

3.6.3 Characteristics of Flowback and Produced Water

As flowback water is generated from hydrofracturing operation, it may contain some of the components present in the hydrofracturing water, although much of it is adsorbed in the formation rock underground. A part of the proppants, most commonly sand particles, return and exhibit as total suspended solids (TSS) in flowback water. Typically, produced water contains high total dissolved solids (TDS) in the forms of—hardness; cations of various metals, including high concentrations of sodium ions; and various anions, including high concentrations of chloride and sulfate ions. Apart from sodium, the other commonly encountered cations are those of calcium, magnesium, iron, boron, barium, strontium, arsenic, etc. Produced water also contains dissolved hydrogen sulfide resulting from the action of anaerobic bacteria on sulfate ions originating from gypsum in the underground rock structure.

Produced water from some unconventional resources contains very high levels of salinity that render them difficult to dispose or reuse. The three largest oil producing basins in the US at Bakken, Marcellus and Permian produce water containing TDS concentrations of up to 235,000, 166,000 and 154,000 mg/L, respectively.

Various organic constituents present in petroleum are also carried over into produced water in dispersed and dissolved phases which are quantified in bulk in terms of Chemical Oxygen Demand (COD) and Total Organic Carbon (TOC).

The characteristics of produced water at a shale play vary considerably with time, and even from well to well, depending upon the geologic formation of the reservoir and the mineral matter content that the water has been in contact with. Typical characteristics of Bakken and Permian basin water are listed in Table 3.7. It has been compiled using several sources based on the results of multiple sampling and analysis programs over the last several years.

Blauch (2009) conducted a study of the change of produced water characteristics with time based on a comprehensive sampling and analysis program at various locations in the Marcellus Shale. The progressive changes of chemical characteristics with the production volume from a well in southwestern Pennsylvania are shown in Table 3.8.

The concentrations of all constituent ions, except for those of sulfate, increased as the flowback water volume increased. This may be due to two possible reasons—(a) capture of sulfate by calcium ions and (b) biogenic reduction of sulfate to sulfide ions under the anaerobic condition prevailing at the large depths of shale oil and gas reserves. Part of the sulfide ions may also be immobilized in place by ferrous ions generated under the reducing conditions. The ratio of ferrous iron to total iron increased with progress of flowback water volume, which confirms the gradual conversion of iron to the ferrous ions by iron reducing bacteria, followed by precipitation in-situ as ferrous sulfide (FeS).

Table 3.7 Typical characteristics of produced waters from Bakken and Permian Basins

Parameter	Unit	Bakken Shale	Permian Basin
Total dissolved solids (TDS)	mg/L	235,000	154,000
Dissolved organic carbon (DOC)	mg/L	200	2,000
pH	su	5–7	6.3–6.5
Alkalinity	mg/L as $CaCO_3$	550	250
Total hardness	mg/L as $CaCO_3$	35,000	15,500
Turbidity	NTU	140	100
Ammonia	mg/L	N/R	530
Barium	mg/L	20	50
Boron	mg/L	750	60
Calcium	mg/L	33,450	5,000
Iron	mg/L	250	25
Magnesium	mg/L	2,000	850
Potassium	mg/L	13,500	850
Sodium	mg/L	120,000	45,000
Strontium	mg/L	3,000	1,100
Chloride	mg/L	17,000–200,000	120,000
Sulfate	mg/L	10–2,050	250

Note N/R = Not Reported

Table 3.8 Change of produced water composition with production volume at Marcellus Shale (Blauch 2009)

Parameter	Produced water volume m^3 (bbl)			
	1,908 (12,000)	2,067 (13,000)	2,226 (14,000)	2,544 (16,000)
pH, su	6.22	6.08	5.98	5.88
Alkalinity (bicarbonate only) as mg/L of $CaCO_3$	19,300	21,100	24,200	27,600
TSS, mg/L	144	175	498	502
Chloride, mg/L	54,500	59,000	62,000	67,800
Sulfate, mg/L	31	20	20	24
Sodium, mg/L	26,220	28,630	31,810	35,350
Potassium, mg/L	1,119	1,201	1,350	1,480
Calcium, mg/L	7,160	7,680	8,880	9,720
Magnesium, mg/L	341	463	488	805
Barium, mg/L	28.9	43.3	99.6	175.7
Strontium, mg/L	1,110	1,305	1,513	1,837
Iron (ferrous), mg/L	0.4	0.9	1.1	3.3
Iron (total), mg/L	63	66	72	78
Specific gravity	1,065	1,068	1,077	1,087
Conductivity, μmhos/cm	133,100	141,500	157,600	173,200
Langelier saturation index	1.02	0.84	0.72	0.55

The pH and alkalinity show a decreasing trend because of acidity due to the increased dissolution of carbon dioxide formed by anaerobic biological activity in-situ. This gradual drop of pH in turn results in increased dissolution of ionic species in the flowback water. Sodium and calcium ions are dominant in all samples. The sharp rise of strontium ions raises the possibility of formation of strontium sulfate, which forms a very hard scale on pipe and equipment surfaces that are hard to remove.

Produced water from many locations contains naturally occurring radioactive materials (NORM), e.g., radium, uranium, thorium ions, etc. Most common constituents are Ra 226 and Ra 228 cations. Radium ions are mobile in water and are present in very high concentrations up to the radioactive levels of 17,000–18,000 picocurie/liter (pCi/L) of the two isotopes combined together at various regions of Marcellus shale (Gusa et al. 2020). If a produced water is scale forming, NORM partitions into barite (barium sulfate) scale by co-precipitation in the form of a complex salt, as follows:

$$Ba^{2+} + Ra^{2+} + 2\,SO_4{}^{2-} \rightarrow Ba\,(Ra)(SO_4)_2$$

The solubility of barium sulfate decreases with decreasing temperature. The significant drop in pressure and temperature during the production process facilitates scaling potential of barium sulfate over calcium carbonate or sulfate in course of its travel from the reservoir to the ground.

NORM concentrations at Permian Basin are much lower. Radium 226 and Radium 228 median radioactivity levels are reported as 30 and 400 pCi/L, respectively. Gross α and Gross β radiations are reported in the ranges of 1,400–2,500 pCi/L and 1,400–2,200 pCi/L, respectively.

Note that Curie (Ci), the unit of radioactivity is defined as:

$1\ Ci = 3.7 * 10^{10}$ decays/s. This is approximately equivalent to the radioactivity of 1 g of Ra^{226}.

In addition to the chemical and radionuclide species, produced water also consists of microbiological population which is expressed in terms of ATP concentration. They proliferate even under adverse hypersaline conditions. ATP concentrations of 250–9,500 ng/L have been reported in samples from Permian basin. Absence of oxygen underground creates a highly reducing environment that fosters the growth of anaerobic and anoxic strains. Microbial Community Analysis (MCA) by 16S rRNA analysis of produced water samples indicate presence of nitrite reducing, sulfate reducing and phosphate accumulating organisms (PAO), all of which belong to the anaerobic type. Identification of reduced sulfide compounds and volatile fatty acids (VFA) confirm the presence of anaerobic bacteria that produce these species by biochemical reduction of sulfate ions and organic substrates associated with produced water.

Lipus et al. (2018) conducted a detailed microbiological study of produced water samples wells at Bakken shale. The communities they identified belong to the broad categories of *Bacillales*, *Halanaerobiales* and *Pseudomonadales* colonies with an abundance $0–10^4$ gene copies/mL as observed in other similar unconventional shale plays also.

Bacteria can exist both in sessile (attached) and planktonic (suspended) growth forms. Each type causes adverse impacts on the production equipment and their operation in various ways e.g., biofouling, corrosion, blocking the microporous rock structure downhole, and creating health and safety hazard by the formation of H_2S. These effects underscore the importance of disinfection of produced water for handling, storage and recycling for refracturing.

Knowledge of the characteristics of produced water, as described above, is essential in devising a treatment process for suitable disposal and/or reuse. A general outline of the overall treatment process is presented below.

3.6.4 *Produced Water Treatment Processes in Hydraulic Fracturing*

A treatment train typically includes a primary oil and solids separation by hydrocyclones or CPI separators depending on the location of well—onshore or offshore. Primary treatment also includes separation units like a dissolved air (or, gas) flotation (DAF or DGF) or an induced air (or, gas) flotation (IAF or IGF) unit if it is necessary to remove more oil and solids further depending upon treated effluent quality requirement for disposal or reuse. The sludge from both oil and solids separation units are transferred to a sludge storage tank for further disposal. In case the water contains significant amounts of divalent cations, e.g., calcium, magnesium, barium, and strontium then it will be necessary to conduct a secondary treatment process to remove these and prevent the possibility of scaling in downstream piping and equipment. It is a two-step process conducted in two completely mixed reactors in series. Solutions of soda ash (Na_2CO_3) and lime ($Ca(OH)_2$) are dosed in the first tank to react with and remove hardness in the form of calcium carbonate and magnesium hydroxide. Details of reaction steps are provided in Chap. 2. In the second reactor a solution of sodium sulfate is dosed which reacts with barium and strontium ions to form their sulfate salts and separate as solid precipitate. The solids subsequently are settled in a gravity clarifier and removed. If the dissolved COD of the water is high a biotreatment step may be included, followed by a secondary clarifier, to remove the organic constituents, although biological treatment is rarely reported for produced water due to the presence of high salt and other inhibitory substances in the water matrix. This is discussed in further detail in a later section of this chapter.

If the treated water is recycled to blend with fresh water for fracturing, it may need removal of dissolved solids. The alternatives for TDS removal are media filtration followed by membrane separation, or mechanical vapor recompression (MVR) evaporation. Reverse osmosis (RO) and evaporation can produce a concentrated brine of 150,000 mg/L TDS. If there is a requirement for Zero Liquid Discharge (ZLD) the suitable technologies are crystallization up to a TDS of ~250,000–300,000 mg/L, followed by disposal and evaporation in a solar pond and/or deep well injection. These are discussed in detail later in this chapter. In case of unavailability of sufficient land to site solar ponds or restrictions on deep well injection, the other alternative is disposal of crystallized solids in hazardous waste landfill. The treated water is passed through a deaerator before being reused for injection. Removal of oxygen prior to reuse is necessary to protect the materials of construction from corrosion.

Solids separated from primary treatment step are combined and dewatered in a centrifuge. The sludge cake is disposed in an especially permitted landfill.

Figure 3.19 presents a block flow diagram of a produced water treatment train for a hydrofracturing facility. A comprehensive list of all the individual unit operations and processes are presented Table 3.9. Details of each are presented in Sect. 3.8.

Management of produced water by proper disposal and/or reuse is a very important consideration in view of the large amount of water produced from the upstream exploration and production activities. The treatment process is selected based on

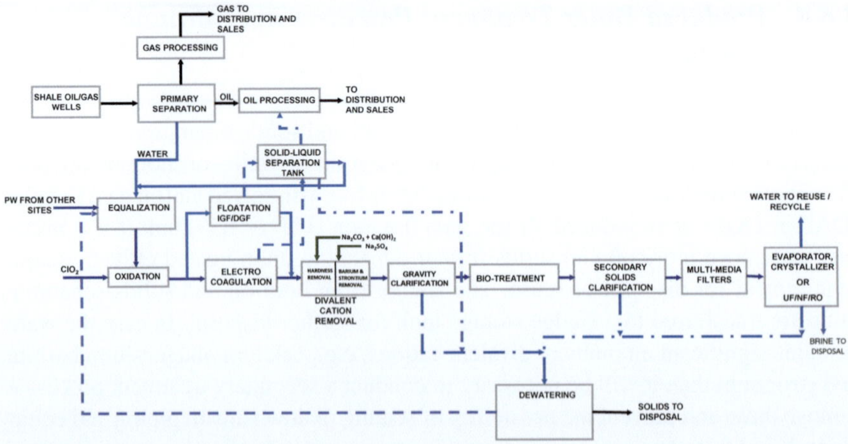

Fig. 3.19 Block flow diagram of a hydraulic fracturing water treatment train

Table 3.9 List of produced water treatment processes

Water treatment method	Technologies	Where used
Primary treatment	• Three phase separators • API separator • CPI separator • Skim tanks • Hydrocyclones • Oxidation • Induced gas flotation • Dissolved gas flotation • Flocculation, chemical coagulation and clarification • Macroporous polymer extraction (MPPE) system • Electrocoagulation • Crushed walnut shell filters • Media filters	• All locations • Onshore • Onshore and offshore • Onshore • Offshore • Onshore • All locations • All locations • Onshore • Onshore and offshore • Onshore and offshore
Secondary treatment	• Softening and divalent cation removal • Biological treatment (if necessary) • Disinfection	• Onshore and offshore • Onshore and offshore • Onshore
Tertiary treatment	• Membrane processes • Ion exchange • Evaporation • Crystallization	• Onshore • Onshore • Onshore • Onshore

the ultimate fate of produced water. Not all unit operations and processes shown in Fig. 3.19 are applicable in every situation. Some of these may be excluded depending on the feed composition and the treated water quality requirements. Each application is reviewed on a case-by-case basis during the planning and design stages. This discussion is further continued in Chap. 8 where the treatment processes for recycling, reuse and disposal of produced water have been discussed in detail.

3.7 Other Upstream Operations Consuming Water

3.7.1 Drilling, Completions and Well Stimulation

Petroleum producing wells are drilled from above ground to the oil-bearing reservoir matrix. The drilling operation is performed with mechanically driven drill bits and drilling fluids. The drilling fluids provide lubrication as well as ensure that the reservoir pressure is contained during the drilling operation. After drilling, the wells are completed with well casings, liners, and other devices to protect the well bore and prepare it for production. This operation is called well completion.

Water-based fluids (WBF) are most widely used for drilling and completions and are considered less expensive than oil-based fluids (OBF) or synthetic chemical-based fluids (SBF). Water for the base fluid may be fresh water, seawater, or brine. The type selected depends on anticipated well conditions and on the spacing between the wells being drilled. Water used needs to be free from suspended solids. Chemicals are added to improve the high temperature performance and to maintain the pH and density of the drilling mud. Drilling mud or drilling fluid is an important component of almost all drilling operations. The purpose of the drilling mud is to lubricate the wellbore, lifting drilling cuttings to the surface, cooling drilling bit, pressure control, signal transmission, add buoyancy and stabilize the formation near the well.

Production from a well will gradually drop from predicted rates over time, due to plugging of the reservoir, well bore or tubing in the well. This is when well stimulation is done to improve the well performance. Well stimulation is done by taking the well off-line and injecting acidic water to dissolve scale or deposits in and around the well bore and to restore the well productivity. Clean water is mixed with acids for well stimulation.

3.7.2 Enhanced Chemical Oil Recovery (cEOR) Method

For conducting a chemical EOR operation, a combination of high-molecular-weight and viscosity-enhancing polymers is added to the water to decrease the mobility of the flood water and, therefore, improve the sweep efficiency of the waterflood. The primary purpose of adding polymer is to increase the viscosity of the flood water; however, polymer addition to the flood water in many instances also imparts

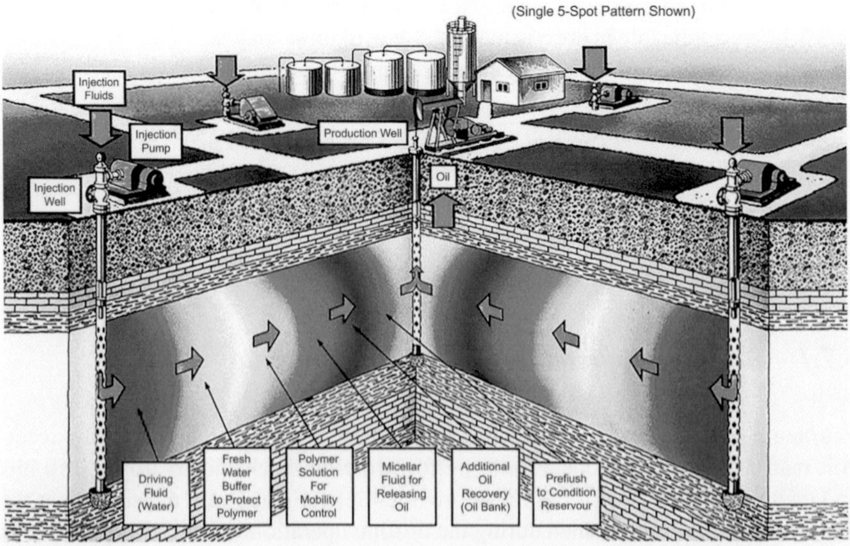

Fig. 3.20 Schematic of polymer waterflooding process (Lake 1989)

a permeability-reduction effect. Polymer waterflooding is normally applied when the waterflood mobility ratio (the ratio of mobility of water relative to oil) is high or the heterogeneity of the reservoir is high. Figure 3.20 shows the polymer waterflooding process (Lake 1989). Alkali Surfactant Polymer (ASP) flooding is one of the commonly used chemical EOR methods used in the industry. It is a combination of three types of chemicals:

- A surfactant that enhances the sweeping ability of the liquid to move crude from rock.
- An alkali to maintain high pH to minimize loss of surfactant adsorption on to rock.
- A polymer, typically hydrolyzed polyacrylamide (HPAM), to increase aqueous phase viscosity and decrease its mobility.

However, chemicals used in polymer flooding EOR also have a few adverse impacts:

- They can interact with divalent cations calcium and magnesium in produced water, form precipitate at the bottom of tanks, and scale equipment and pipe surfaces,
- Promote emulsion formation between water and oil phases making it difficult to separate them in downstream equipment,
- Chemicals return with the crudes and partition into the aqueous phase downstream that eventually end up in the wastewater treatment plant requiring breakdown and safe disposal.

It is necessary to consider these factors in the design of proper cEOR system and selection of chemicals.

3.8 Produced Water Characteristics and Treatment Processes

Produced water characteristics are functions of the production method and geochemistry of the reservoir. Production methods from conventional onshore, conventional offshore, unconventional thermal and unconventional hydrofracturing operations, which determine the produced water characteristics and treatment train in each case. Selection of a treatment train and individual equipment depend on the influent and specified effluent characteristics. The effluent characteristics are based on the intended use for the water either for disposal (e.g., surface discharge to sewer system or a body of water, deep well injection) or for reuse (e.g., recycling for hydrofracturing, land application, cooling water, industrial reuse, and use in a power plant). The sections below describe the treatment logistics and the unit operations used in the water treatment in upstream facilities.

3.8.1 Treatment Logistics

Onsite treatment is a viable alternative to truck-transportation and offsite disposal/management of produced water. There can be different spatial configurations of the treatment system, as follows:

Dispersed: These typically refer to truck-mounted packaged treatment systems. They are mobile and moved from region to region to serve clusters of wells. These are used mainly for smaller capacity treatment units for shorter well production life.

Stationary Centralized: This refers to a relatively large stationary and centralized treatment system that draws in the produced water from wells by a piping network and distributes the treated water to new well locations by another piping network. The piping systems are made flexible to accommodate the changing locations of wells. These are designed to treat larger flowrates of produced water from different well-pads. The treated water is then distributed for recycle and use as boiler feed water for thermal EOR methods.

Stationary Decentralized: The concept is like centralized treatment system, but the main difference is to set up a number (3 to 4) of treatment systems to serve wells at various regions within a given shale area. This may be more applicable for large shale plays or remote thermal well-pads where the produced water may not need to be recycled and can be disposed by deep well injection. This method avoids long and complex piping networks to and from one single treatment system. Decentralized plants have also been designed in thermal SAGD/CSS facilities to do primary separation at well pads and then send liquid and gas streams separately to a centralized plant (Fig. 3.21) where the water is treated to boiler feed water quality and sent to well pads where steam is generated and injected to the thermal wells. This avoids long steam pipelines and hence reduces heat losses.

Fig. 3.21 SAGD Central Plant Facility (MEG 2024)

The treatment logistics depend on the cost, capacity, and environmental considerations of the production operations. The treatment of produced water from hydraulic fracturing has evolved over years as shown in Fig. 3.22.

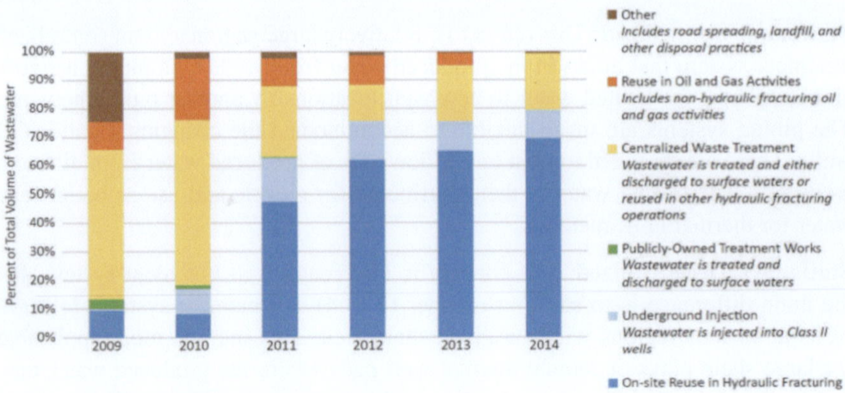

Fig. 3.22 Changes in produced water management practices over time in the Marcellus Shale Area of Pennsylvania (EPA 2013)

3.8.2 Unit Operations and Processes in Produced Water Treatment

Water treatment processes implemented in the upstream operations are standard processes adapted to the production methods used at a site—on-shore, offshore and unconventional. The treatment process trains comprise of integration of individual steps in series. These steps are determined on a case-by-case basis depending on the types of contaminants present in the water and the required quality of treated effluent, which, in turn, is determined by the ultimate destination for the treated water—reuse for hydrofracturing or steam generation for water flood, or discharge to a receiving body of water, or deep well injection. A comprehensive list of these individual processes is shown in Table 3.9.

While Table 3.9 presents a list of a broad range of technologies, for most sites, removal of TSS and Oil and Gas (O&G) are adequate. Secondary treatment utilized at sites requiring removal of hardness and other divalent cations to prevent the possibility of scaling of pipes and equipment. Dissolved solids (primarily arising from sodium chloride) removal by tertiary treatment is added to high TDS produced waters originating from Marcellus Shale, Bakken Shale and Permian Basin.

Some of these processes will be described in detail in the following sections. Tertiary treatment by evaporation and crystallization, and zero liquid discharge are not widely practiced at operating sites.

3.9 Technologies Used in Produced Water Treatment

The unit operations and processes listed in Table 3.9 are integrated appropriately to develop the produced water treatment train specific to a site. Not all the individual steps are necessary to treat the water at a site because those are determined by the water characteristics, end use of the treated water, economics of the treatment system and the underground formation chemistry. The technologies described below are individual unit operations and processes. Many of these are designed and sized using established industry standards e.g. API, ASTM, etc. while some others are offered by vendors based on proprietary in-house designs, and the reader should review the specific vendor literature to understand the details of their design and application.

3.9.1 Primary Water Treatment

Total suspended solids (TSS) are associated with produced oil and water. Sources are particulate matters carried over from the reservoir, ferric hydroxide and manganese oxide particles formed by chemical oxidation, and some ferric sulfide formed by the reaction of ferric ions and hydrogen sulfide in water. Flowback water also contains proppant particles carried over from hydrofracturing operations. These need to be removed before the treatment processes downstream for proper disposal or reuse.

3.9.1.1 Equalization

Equalization is an important step when multiple sources of produced water with varying compositions (pH, O&G, ionic content) are combined together into a single stream. It is very applicable to waters produced from hydrofracturing operation as their characteristics vary widely within the same geography from time to time and even from well to well. Given this high variability, it is necessary to collect the water into an equalization tank as it comes out of the ground, allowing 6 to 12 hours of retention time depending upon the level of fluctuation. Equalization is performed in large open basins or closed storage tanks depending on presence of dissolved gases and volatile organic compounds in the water. Typical arrangement for equalization process is depicted in Fig. 3.23.

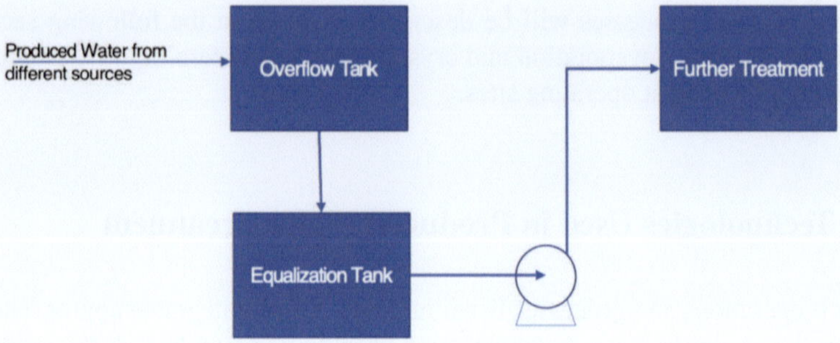

Fig. 3.23 Equalization process

This step may not be required for the produced waters from other methods such as water flooding, steam flooding, SAGD or CSS where the produced water compositions are in close range.

3.9.1.2 API and CPI Separators

The simplest process for oil–water-solids separation is gravity separation by a mechanical separator such as an American Petroleum Institute (API) separator, whereby solids settle down and oil floats up in a rectangular tank. Oil is skimmed off from the top surface and settled solids are removed from the bottom as sludge. Removed oil is returned to the crude tank after separation of water. Sludge containing emulsified oil, solids and water is treated using flash separator, centrifuge, or slop tanks where chemicals and heat are added to breakdown emulsions. The solids are disposed of in designated landfills while any recovered oil is added to the product oil stream.

In a CPI Separator the produced water flows through an inclined plate pack. The oil droplets coalesce and float up by gravity while water and solids settle downwards. The solids drop to the bottom and are drawn out and water overflows through a weir and is taken out. Any flashed gas is taken out from the top. The oil content of water stream drops to about 50 ppm at the outlet. API and CPI separators are also extensively used as primary oil–water separation equipment for wastewater treatment in downstream operations. A more detailed discussion of their working principle is presented in Chap. 5 and is not repeated here.

3.9.1.3 DGF and IGF Units

The working principles of DGF and IGF separators, and their comparisons, have been discussed in detail in Chap. 5. Both use natural gas as the means of flotation because it is available in abundance at the upstream production sites. The coalescence of fine and dispersed oil droplets is facilitated by the addition of suitable flocculant and demulsifying polymers.

DGF works on the principle of dissolution of gas (produced in the field) in water under high pressure followed by release of pressure in the tank. This liberates fine microbubbles of gas (10–100 μm range), which attach the oil droplets and TSS particles and float those up onto the surface as a blanket that is removed by skimming.

IGF also utilizes of gas bubbles for oil and TSS removal, but those are generated by dispersing gas into water by mechanical means (e.g., educator, impellers) that create fine-size bubbles (500–1000 μm size range) by shearing. The fine bubbles attach the oil droplets and particulate matter on their surface which float up and separate.

IGF is favored over DGF for upstream water treatment operations because of the following two reasons:

- Solubility of gas in water is largely diminished because of high TDS in produced water, adversely affecting DGF performance.
- Operation of IGF is simpler and more appropriate for field application than DGF.

Despite these disadvantages, many sites still use DGF where salt concentration is not extremely high. Dissolved gas flotation performs at a superior level if it is properly attended to from operation and maintenance points of view.

There are many manufacturer-specific designs of the IGF and DGF units that achieve this design objective. IGF reduces dispersed oil content of water to about 15–40 mg/L.

Like API and CPI separators, DGF and IGF are also used extensively in downstream wastewater treatment applications and discussed in detail in Chap. 5. Figures 5.10 through 5.15 in Chap. 5 show different views DGF and IGF units.

3.9.1.4 Hydrocyclones

Hydrocyclones are used mostly on offshore production platforms where their compact design saves space compared to flotation equipment and Free Water Knock-out (FWKO) drums. They are used both for solid–liquid separation and liquid–liquid separation. Figure 3.24 shows a liquid–liquid separation hydrocyclone unit with several individual cyclones installed in a pressure vessel.

In an oil–water hydrocyclone, the separation of oil droplets from water occurs as a result of large centrifugal forces which are generated by inducing a swirling motion in the fluid inside the device. The acceleration due to the swirling motion enhances the effect of density difference between the dispersed oil droplets and water. The inlet pressure to the hydrocyclone provides the driving force for separation. The oil rich phase (lighter phase) is at the lowest pressure and flows out as overflow from the top. The underflow or heavier phase (water rich phase) flows out from the bottom.

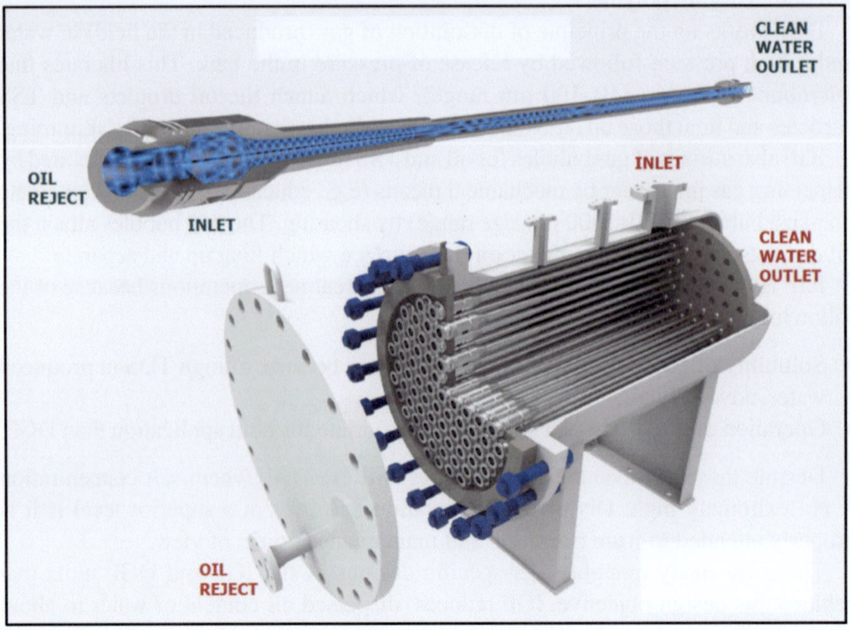

Fig. 3.24 Schematic of a hydrocyclone (Ditria 2014)

3.9.1.5 Macro-porous Polymer Extraction (MPPE)

On offshore production platforms, Macro-Porous Polymer Extraction, MPPE (Veolia 2022) units are used to remove dispersed and dissolved organics from produced water before downstream treatment or disposal. Figure 3.25 shows the schematic of this process.

In the MPPE process, water containing hydrocarbons is passed through a column packed with MPPE media. The polymeric media contain a specific extraction liquid (gel). The immobilized extraction liquid removes hydrocarbons from water. The types of hydrocarbons which have an affinity for the extraction liquid, are removed. The purified water can then either be sent downstream for further treatment or discharged.

Periodical in-situ regeneration of the extraction liquid is accomplished by heating the media with low-pressure steam. Hydrocarbons adsorbed in the extraction liquid are released as vapor and flow with the steam through a condenser to collect as a liquid and returned to the crude tank. The regenerated polymer gel particles are ready for use/reuse. The application of two columns allows continuous operation with simultaneous extraction and regeneration. A typical cycle is 1 h of extraction and 1 h of regeneration.

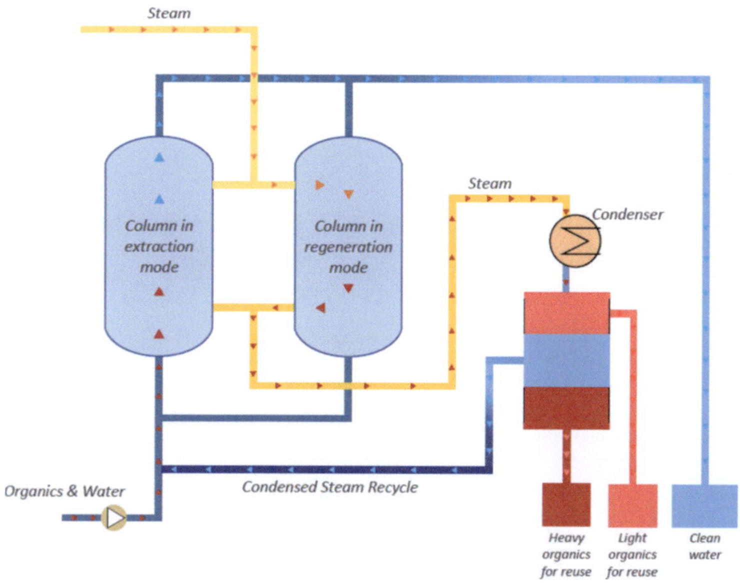

Fig. 3.25 Macro-Porous Polymer Extraction (MPPE) (adapted from https://www.veoliawatertechnologies.com/en/solutions/technologies/macro-porous-polymer-extraction-mppe)

3.9.1.6 Oxidation

This process is applicable to produced water when it is exposed to air. The environment inside the reservoir formation is highly reducing. There is also the presence of anaerobic bacterial population. Any iron and manganese in the geologic formation is present in a reduced state as dissolved ferrous (Fe^{2+}) and manganese (Mn^{2+}) ions. Once above ground, as water encounters air, these are oxidized into insoluble ferric hydroxide and manganese oxide, which settle in quiescent locations in the treatment process train and interfere with equipment performance. If H_2S is present, a chemical oxidation step with chlorine dioxide (ClO_2) is used, which also acts as a biocide. To prevent the accumulation of iron hydroxide and manganese oxide particles in the downstream treatment system, they are removed by Chemical Coagulation and Flocculation (CCF) followed by settling sedimentation in a gravity clarifier. It is described in detail in Chap. 2 and is not repeated here.

3.9.1.7 Chemical Coagulation and Flocculation

The colloidal particles in water and emulsions with repelling electric charges are neutralized by coagulant chemicals. These solids form flocs which are then settled in a sedimentation tank. A process schematic for Chemical Coagulation and Flocculation (CCF) is shown in Fig. 3.26. CCF process is discussed in detail in Chap. 2 and is not repeated here.

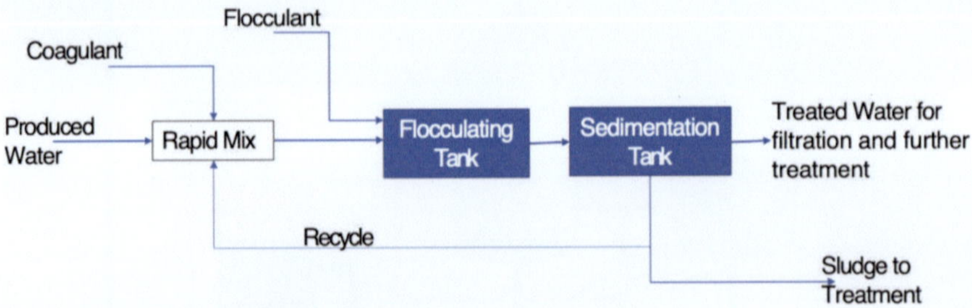

Fig. 3.26 Chemical coagulation and flocculation

3.9.1.8 Electro-coagulation (EC)

The electrocoagulation process can be defined as a chemical free enhancement of traditional coagulation process where the coagulant is generated in situ by application of electrical current. An EC reactor can be visualized as an electrochemical cell consisting of parallel anode and cathode plates immersed in an electrolyte, which is the water to be treated. The anode is made of the metal that is desired to serve as the sources of the coagulant cations, e.g., aluminum, iron, etc. For example when using aluminum as the cation in an application, the cell will consist of an aluminum anode. When two electrodes are externally connected, and electrical potential is applied, it will give rise to the following electrochemical reactions:

At the anode: $Al \rightarrow Al^{3+} + 3\ e$
At the cathode: $2H_2O + 2e \rightarrow 2OH^- + H_2(g) \uparrow$
In the liquid phase: $Al^{3+} + 3OH^- \rightarrow Al(OH)_3(s) \downarrow$

Precipitation of aluminum hydroxide facilitates removal of fine solid particles and dispersed oil by interparticle bridging mechanism. This explains the basic working principle of electrocoagulation process. A schematic view of the EC process is presented as Fig. 3.27.

Additionally, the aluminum ions participate in the reactions leading to the formation of various complex ions that serve as counterions to destabilize colloidal suspensions in the liquid medium. Hydrogen gas released at the cathode rises in the form of bubbles. This helps the destabilized fine particles and oily substances float up on the surface of water, which can then be removed by skimming. The heavier particles are removed by a subsequent flocculation step, followed by solids separation.

In addition to destabilizing colloidal and suspended particles, EC also helps in the removal of certain heavy metals including dissolved iron, chromium, organic compounds (TOC) and bacteria. The mechanism for removal of TOC is not clear at this time, but it is anticipated that the energy input to the electrochemical cell converts a part of the hydroxide ions (OH^-) produced at the cathode into hydroxyl

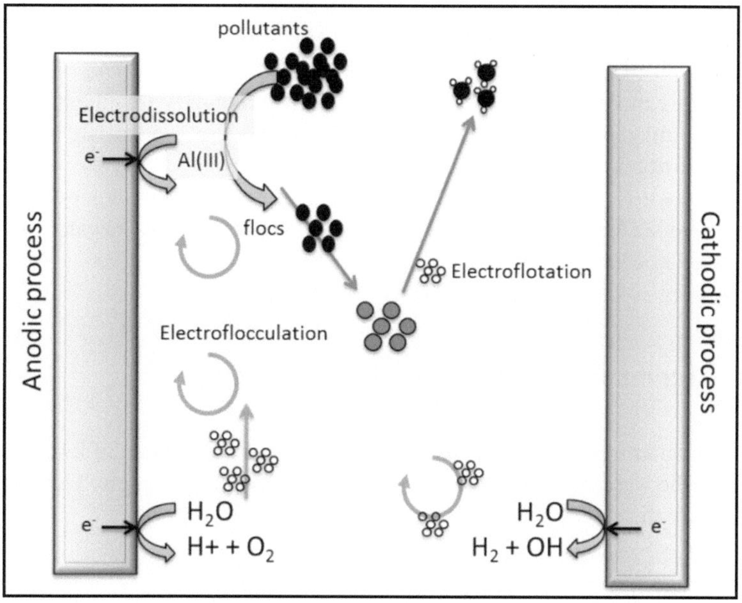

Fig. 3.27 A schematic view of the electrochemical coagulation (EC) process

radicals (OH^*). These are high energy radicals, with oxidation potential of 2.80 V which remove TOC by advanced oxidation of organic compounds to CO_2, H_2O and other byproducts of partial oxidation. Bacterial cells are also believed to be destroyed by the advanced oxidation of the cell contents following the same mechanism.

EC has advantages and disadvantages. As applied to flowback water and produced water in hydrofracturing operations, CCF has some limitations compared to electrocoagulation due to presence of polymers and residual guar gum in the flowback water. The principal advantages of EC over CCF include compactness, improved effectiveness over chemical coagulation as the fine particulate and colloidal matter are set into motion by electric charge, and simplicity (i.e., no moving parts). The principal disadvantage, on the other hand, is the passivation of electrodes which requires the periodic cleaning and replacement of sacrificial anodes and cathode plates. Occasional foaming has also been observed and can lead to a loss of efficiency and fouling of the electrodes.

The advantages outweigh the disadvantages and EC is gaining increasing popularity in full scale field services applications for the treatment of produced water. Pilot tests with water samples collected at a site of interest are recommended prior to the deployment of full-scale units.

3.9.1.9 Crushed Walnut Shell Filters

These filters are usually installed after IGF/DGF units. The suspended oil content is reduced to less than 5 mg/L from an influent concentration of 30 mg/L oil. In thermal EOR plants, produced water is also filtered in crushed walnut shell filters to bring the suspended oil content to less than 5 mg/L. This water is suitable for discharge to a disposal well or to open sea, but not suitable for discharge to surface water bodies or for steam generation. This water is then sent to secondary treatment to remove calcium and magnesium hardness and other divalent cations, primarily barium and strontium, if necessary.

Walnut shell filters are also used in downstream applications for polishing the effluent from secondary oil–water separation step in refinery wastewater treatment as discussed in Chap. 5.

3.9.2 Secondary Water Treatment

Secondary water treatment involves removing divalent cations that, if left untreated, build up in the water and cause scaling of the water handling systems. Their removal processes are discussed below.

3.9.2.1 Removal of Divalent Cations

The presence of high concentrations of divalent cations, especially calcium, magnesium, barium, and strontium, presents the challenge of scaling in the equipment, piping, tanks, and vessels of treatment systems. Because of this, removal of divalent cations is necessary to maintain stable operation of produced water treatment systems. If the treated water is reused for refracturing, then there is risk of scaling the pipes, resulting in high pressure drop and increases pump energy consumption. Scales can also block the pores in the shale structure. Based on the relative amounts of cations and anions present in water and their solubility products, calcium can form scales as calcium carbonate and sulfate, magnesium as magnesium hydroxide, and barium and strontium as their respective sulfate salts.

At certain shale plays in the US, especially at Marcellus Shale (Tieman et al. 2020; He 2014), produced water from upstream oil and gas operation contains divalent cations barium and strontium in significant concentrations in addition to the hardness ions calcium and magnesium. In such cases the effluent from the softener is treated by a solution of sodium sulfate to precipitate the cations as sulfate salts. When all of the major divalent cations (Ca^{2+}, Mg^{2+}, Ba^{2+}, and Sr^{2+}) are present in the water simultaneously the treatment is conducted in two completely mixed reactors in series—first one dosed with lime ($Ca(OH)_2$) and soda ash (Na_2CO_3) to remove calcium and magnesium by conventional or warm lime-soda process as described in Chap. 2, and the second reactor dosed with sodium sulfate (Na_2SO_4) solution to remove barium and strontium as their sulfate salts.

$$Ba^{2+} + SO_4{}^{2-} \rightarrow BaSO_4(s) \downarrow$$

$$Sr^{2+} + SO_4{}^{2-} \rightarrow SrSO_4(s) \downarrow$$

The solids formed in the two reactors are settled in a gravity clarifier, removed, and dewatered by centrifuge for disposal.

It is necessary to conduct a thorough sampling and analysis of the anions and cations in the produced water under consideration. Such information is very valuable in determining the scaling and corrosion potential of produced water in question, and in deciding the treatment steps that are required to make it suitable for reuse in refracturing and/or other purposes. The scaling and corrosion potential of water containing large concentrations of ionic species is determined with the help of mathematical modeling described in Chap. 6.

3.9.2.2 Biological Treatment Systems

In most cases the characteristics of produced water from conventional production are simple and may not warrant a multi-stage treatment train consisting of a number of unit operations and processes to remove a large number of contaminants. For

Table 3.10 Slop water characteristics

Constituent	Values
COD total	816 ± 94 mg/L
COD soluble,	530 ± 32 mg/L
TOC	390 ± mg/L
TPH	151 ± 3 mg/L
Chloride	24,000 ± 0.12 mg/L
pH	7.78 ± 0.62 su

on shore production, it may be adequate to remove only TSS and Oil and Grease by primary treatment for disposal or reinjection as shown in Fig. 3.2. For offshore production, if hydrate control by adding MEG and methanol is necessary, the COD of the produced water becomes high and warrants biological treatment before discharge into the sea. In such cases MEG is added continuously and methanol is added intermittently which causes short term spikes in COD to the tune of tens of thousands of mg/L. Such biological treatment facilities are package plants located on production platforms using treatment technologies with small footprint e.g., MBR, MBBR, etc., preceded by equalization tanks that prevent large shock loads of COD from entering the bioreactor.

On board slop wastewater treatment by aerobic granular sludge (AGS) technology was tested in pilot scale SBR reactors (Campo 2017). This slop water was collected from barges in Augusta Harbor in Sicily. The water characteristics are shown in Table 3.10.

Two sets of tests were conducted in parallel—one using granules already developed ex-situ under saline conditions, and the other granules were cultivated in the reactors in-situ using slop wastewater. Both reactors were started up by blending slop wastewater with synthetic water and increasing in steps the fraction of slop wastewater to 100%. In both cases initially the total COD removal efficiency of ~90% was observed but after the introduction of 100% slop wastewater it dropped to 30% in the first reactor and 67% in the second. The soluble COD reduction always remained at ~80%. The authors attribute the COD reduction to a combination of biological oxidation and bio-adsorption. The working principle of AGS process has been presented in Chap. 5.

Among the waters produced from unconventional resources, those from thermal production processes are not amenable to biological treatment due to their high temperatures that do not support microbiological population. The biological treatment of produced water from hydrofracturing operation has limited application because of the complex chemistry and refractory nature of this type of water. In some cases, reuse of produced water from hydrofracturing operation requires desalination. Reverse osmosis (RO) is the desalination process of choice for secondary treated produced water. However, RO requires pretreatment for removal of organics to prevent biofouling of membranes. Biological treatment provides a cost-effective pretreatment option. Apart from this application, biological treatment is seldom used because of high and variable TDS and salinity present in produced water from hydrofracturing operation. The negative impacts of salinity are:

- Rupture of cell walls due to high osmotic pressure of saline water leading to release of cellular material, and thus soluble COD spike.
- Interference of biological flocculation and settling in secondary clarifier by adversely affecting biopolymer formation mechanism on cell walls.
- Increased buoyant forces due to increased density of saltwater affecting settling for solid–liquid separation.

Dissolved solids in produced water contain large concentrations of monovalent sodium ions compared to divalent magnesium and calcium ions. According to Higgins and Novak (1997) when the monovalent (M) to divalent (D) cation ratio in water is greater than 2 the biomass in mixed liquor does not flocculate and settle in a clarifier creating high effluent turbidity and TSS. Produced water also contains many inhibitory or bio-refractory constituents which include, but are not limited to, high concentrations of heavy metals, metalloids, and refractory organic compounds. One key strategy to conduct microbiological breakdown of recalcitrant organic chemicals in adverse environment is to increase solids retention time (SRT) by retaining biomass within the reactor for long periods of time. This has been successfully tested by utilizing attached growth processes and membrane bioreactors (MBR). Pretreatment to breakdown refractory organics by advanced oxidation processes (AOP) can also make such chemicals amenable to biotreatment.

Despite the impediments described above there is still a significant interest among researchers and practitioners *in biological treatment* processes due to their simplicity and cost effectiveness. As a result, there has been continuous effort by researchers to innovate new biotreatment processes that can overcome the challenges stated above. A summary of the various bench, pilot and full-scale studies and their results are presented in the section below.

In a comprehensive review paper Jimenez (2018) presented the status of various biological treatment processes for produced water as reported by several groups of researchers. The results of several bench and pilot scale studies demonstrated a TPH (Total Petroleum Hydrocarbons) removal efficiency of 98–99% could be achieved by an activated sludge treatment unit with solids retention time (SRT) of 20 days. The authors concluded that leading mechanisms of biological removal of hydrocarbons are biodegradation and bio-flocculation.

Different mixtures of produced water and sewage were tested in a Sequencing Batch Reactor (SBR) with acclimated sewage sludge. In 45 and 35% (v/v) mixtures of produced water and sewage. COD removal efficiencies varied from 30 to 50%. Salinity was not found to have any significant effect on COD removal of mixed wastewater. Only the recalcitrant character of the produced water seemed to hinder the biodegradation. Other sources also corroborated the low effect of salinity. For instance, TOC removal was studied with 180 mg/L NaCl acclimated microorganisms. Produced water treatment studies were conducted in Sequencing Batch Reactors (SBR), trickling filters and chemostat reactors. It was observed that water containing as high as 200,000 mg/L TDS was still biodegradable although the SBR long term operation was not stable, biomass was gradually lost, and TOC removal performance worsened.

For barely treatable influents, immobilization of microorganisms can improve treatment efficiency. Experimental studies conducted with *Bacillus* sp. immobilized on polyvinyl alcohol (PVA) reported removal of more than 90% COD from oilfield produced water with initial COD value of 2600 mg/L. Also, positive results were obtained for PVA immobilization of B350 M and B350 commercial microorganisms in the shape of a biological aerated filter (BAF) with hydraulic retention time (HRT) of 4 h and a volumetric load of 1.07 kg COD/(m^3/day). The BAF removed 78% of TOC and 94% of oil, respectively. In addition to BOD and COD removal, BAF was also observed to remove ammonia and other nitrogenous compounds, iron, manganese, heavy metals, and hydrogen sulfide. However, as BAF retains solids, high concentrations of salts can decrease the biological activity of the system due to the toxicity effects caused by accumulation of salts.

Another study involving rotating biological contactors (RBC) demonstrated removal of BOD and Oil and Grease (O&G) up to 94% and 74%, respectively.

Lagoons have been studied as well for produced water treatment. A two-stage set-up was used with the first tank to oxidize the suspended oil and dissolved organic compounds, and a second one to oxidize dissolved ammonia compounds. Free water surface (FWS) and sub-surface flow (SSF) pilot lagoons were tested for oifield produced water. COD elimination was higher for SSF than for FWS wetlands. However, despite its cost-effectiveness the temperature dependence of these systems can be detrimental to effective treatment. In addition, groundwater contamination is a significant risk when lagoons are not properly lined and protected.

Reed beds have also been tested for hydrocarbon and heavy metals removal. For instance, a 800 m^2 *Phragmites Australis* pond treated 20 m^3/day of produced water which removed more than 98% of hydrocarbons. A 3000 m^2 reed bed system reduced 96% of total hydrocarbon content, while metal concentration decreased between 78 and 40% depending on the particular species. These kinds of systems are environment friendly and can be considered cost-effective methods, however the final effluents still need an additional polishing step and entail the requirement of a relatively large footprint.

Anaerobic degradation could also be a cost-effective alternative to aerobic biological oxidation of organics in produced water. Simulated produced water containing organic acids, especially naphthenic acid, was treated by anaerobic processes. Results showed that certain organic acids were depleted, nevertheless no degradation of naphthenic acids (NA) was obtained which demonstrated that NA is difficult to break down biologically. They can be partially degraded by advanced oxidation processes and removed by carbon adsorption and membrane separation processes as discussed in detail in Chap. 2.

A review and meta-analysis of biological treatment of oil and gas produced water by several groups were performed by Camarillo and Stringfellow (2018). They summarized the results of 59 biological treatment studies which included 16 pilot scale systems. The distribution of these treatment processes is shown in Fig. 3.28.

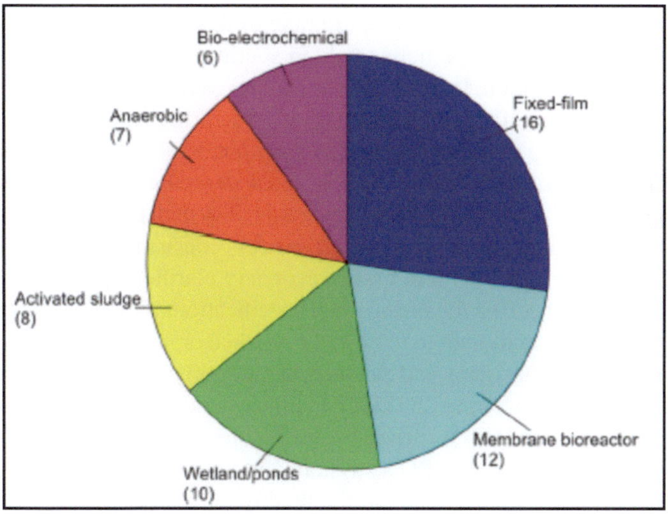

Fig. 3.28 Distribution of different types of biotreatment processes for produced water

The key takeaways of this review report are:

- About 73% of these studies were conducted on real produced waters at operating sites and 27% were synthetic wastes.
- Most of the tests (69%) were conducted in bench scale.
- The tests were conducted on many produced water samples of variable quality and characteristics—the mean TDS and COD were 28,000 and 1,125 mg/L respectively.

The general trend was that at TDS < 50,000 mg/L average removal of COD was ~73% which gradually dropped to 54% with increase of TDS to >50,000 mg/L.

The researchers reported that prior acclimation of microorganisms and their retention in the reactors contributed to the success of biological treatment. The main challenges to overcome are toxicity of the chemicals on microorganisms, biofouling, and scaling potential. Prior determination of toxicity by Microtox or similar tests, and scaling potential by thorough sampling and analysis of the major cations and anions, followed by modeling (see Chap. 6) will be helpful in planning for full scale treatment process developments and design. Notwithstanding the challenges the authors expressed hope in the future of biological treatment of produced water. The main drivers for biological treatment are lower cost compared to physiochemical treatment and shrinking options of deep well injection, especially in view of the environmental and public concerns of truck transportation and seismicity. This subject has been discussed in greater detail in Chaps. 7 and 8.

Advancements in understanding of biological treatment processes and different economic and feasibility drivers provide opportunities where it may be integrated as part of an overall treatment system, particularly for shale plays producing mid to lower

TDS waters. It is case specific. A good example of this is the produced water treatment facility in the Pinedale Anticline field in southwestern Wyoming in the Mesa Verde tight sands formation. Initially the water management practice for this facility was to truck fresh water from long distances for fracturing and produced water out for reinjection into the ground or disposal in spray ponds. However, these were inadequate to support the rapid growth of the field and water reuse was necessary. This led to the design and construction of an on-site treatment system that started operation in early 2010. The treatment train is a multistep process consisting of an API Oil/Water separator, anaerobic basin, aerobic basin, secondary clarifier, and a final polishing filtration. In order to meet the fracturing water specifications biological treatment is an essential part of the overall train to remove the residual organic compounds resulting from breakdown of guar gum and polyacrylamide polymers. Water in excess of that needed for hydrofracturing and other ancillary demands is further processed to fulfill the discharge standards to a river by an aerobic membrane bioreactor (MBR), followed by a reverse osmosis (RO 1) system and a boron removing ion exchange system in series. The improvements in the final quality of the discharge water are from 8000 to 100 mg/L for TDS, 15–30 mg/L down to below 0.75 mg/L for boron and TOC below the detection limit (Boschee 2012). To maximize recovery of water the RO-1 concentrate stream is further treated by electrocoagulation (EC), followed by an ultrafiltration (UF)—RO combination. The permeate from this second RO (RO-2) is combined with the mainstream treatment process prior to the boron ion exchanger and the concentrate sent to injection well for final disposal. Overall recovery of this treatment train is 84–91%. The chemical characteristics of raw produced water, the State of Wyoming Discharge Limits to the river, and the final analytical results from treated effluent samples in March 2010 are presented in Table 3.11. Limits on BTEX, methanol and boron were self-imposed and not mandated by the state. The results indicate that each of the chemical characteristics exceed the regulatory requirements. In addition, the effluent also fulfilled the Whole Effluent Toxicity (WET) test criteria (explained in detail in Chaps. 5 and 7) after remineralization by addition of calcium and magnesium which were removed by the treatment process. A conceptual block flow diagram of the treatment process is presented in Fig. 3.29.

An assessment conducted by the University of North Dakota (UND 2010) and published by the Journal of Petroleum Technology (Boschee 2012) estimated the average total cost of water in the range of $2.00–$16.80/barrel of water for operating exploration activities in the Bakken Shale. This includes the cost of acquisition of fresh water by trucking and the cost of disposal of the flowback and produced water, also by truck transportation, and deep well injection. The water treatment and recycling systems at Pinedale helped the operators overcome this economic burden, as well as the uncertainty associated with the transportation of fresh and produced water into and out of the site. Additionally, elimination of truck transport of water was beneficial to social and environmental causes.

In certain reuse applications, the refracturing water can be partially COD tolerant. In those cases, biological treatment may not be required to create water suitable for reuse in hydrofracturing and may be bypassed. Principles of biological treatment and the reactor configurations are discussed in detail in Chap. 5.

Table 3.11 Produced water influent, discharge limits and treated effluent characteristics (Shafer 2011)

Parameter	Produced water influent	State of Wyoming discharge limits to river	Final effluent (March 2010)
pH, standard units	6.5–8.5	6.5–9.0	7.34
Chloride, mg/L	3,600–6,750	230	18
Sulfate, mg/L	10–100	3,000	Non-detectable
Boron, mg/L	15–30	0.75	Non-detectable
TDS, mg/L	8,000–15,000	500	41
Conductivity, μS/cm	8,000–20,000	7,500	78
BTEX, mg/L	28–80	Non-detectable	Non-detectable
Methanol, mg/L	40–1,500	Non-detectable	Non-detectable
O&G, mg/L	50–2,400	10	Non-detectable

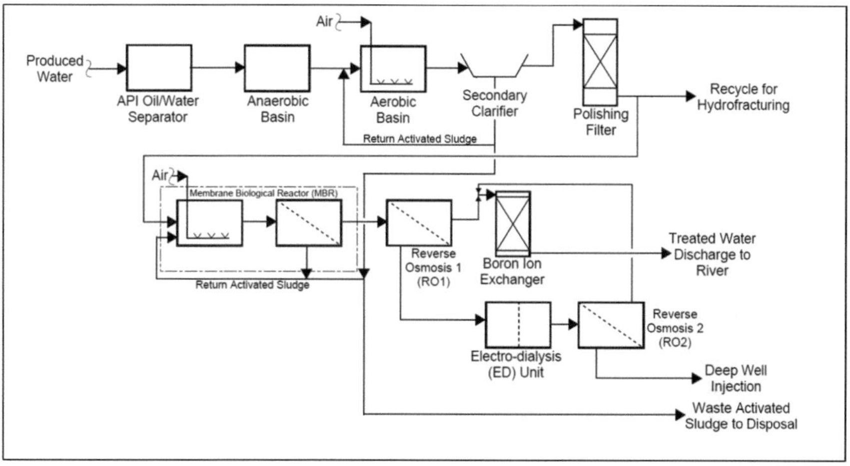

Fig. 3.29 Conceptual block flow diagram of the produced water treatment train at Pinedale Anticline Field, WY

3.9.3 *Tertiary Water Treatment*

When re-use of produced water needs to be implemented, tertiary treatment is used. In many cases produced water contains high salinity. For reuse applications, it is necessary to remove salt to an acceptable level. The treatment processes previously described do not remove TDS, and the removal by secondary processes is limited only to divalent cations. The bulk of the dissolved salts of sodium and other cations are removed by tertiary processes.

Reverse Osmosis (RO), Ion Exchange (IX), and evaporation are the commonly practiced processes to remove dissolved salt from water. However, all of these processes are quite energy intensive, and the recovery of treated water gradually diminishes with the increase of feed TDS. The ionic concentrations of water at the major shale plays in North America are high with TDS concentrations exceeding 100,000 mg/L.

Additionally, the presence of any salts with significant concentrations of divalent cations in the feed has the potential to deposit and scale on the equipment surfaces. Therefore, it will be necessary to remove divalent cations by the secondary treatment processes before treatment by reverse osmosis.

Ion exchange is another process to remove certain components of TDS and produces high-quality effluent. Ion exchange also requires pre-treatment to remove TSS and dissolved organic compounds to protect the active sites on the surfaces of ion-exchange resins. As a further complication, regeneration of the resin produces a highly saline brine that must be managed.

For produced waters with moderate levels of TDS ($\leq$40,000 mg/L), and pretreated by the primary and secondary treatment steps, RO and IX are good alternatives for TDS control and removal of certain inorganic species of concern. One such element is Boron that can be present in produced water from two potential sources—(a) connate water in the rock formation where the crude is trapped, and (b) breakdown of the cross-linked structure of the guar gum used in the fracturing water. At the normal pH range 6.0–8.5 of produced water concentration of dissolved boron can be 26–28 mg/L. Many states in the US regulate boron in water due to its associated health effects. Moreover, when it comes in contact with heavy metals Pb, Cu, Co, Ni, Cd, etc. in water it complexes with these metals and produces a synergistic effect on the toxicity of those metals. Therefore, boron control increasingly finds its place in the overall treatment scheme of produced water as in the case of the Pinedale Anticline field in Wyoming discussed earlier in this chapter. While IX and RO are the most widely used treatment technologies for boron control, electrodialysis (ED), adsorption and electrocoagulation also have been reported in the literature (Ezechi 2012) as effective technologies.

For high TDS concentrations above 40,000 mg/l, evaporation is a suitable process for tertiary treatment. At feed TDS above 300,000 mg/L, there is a potential for crystal formation and scaling on the heat exchanger tube surfaces, which also leads to high maintenance costs. As with reverse osmosis, scaling by salts of divalent cations is a concern here too, and prior removal of such cations is essential. Capital costs are also high because of the cost of alloy tube material that is required for handling of salt solutions at high temperatures.

A few other processes are also used for TDS control. These include forward osmosis, membrane distillation, capacitance deionization, mechanical vapor recompression (MVR), and humidification–dehumidification.

In Thermal Enhanced Oil Recovery (EOR) operations evaporators have been used to treat both the produced water and boiler blowdown water to recover maximum possible amount of water for reuse. A typical MVR system is shown in Fig. 3.30.

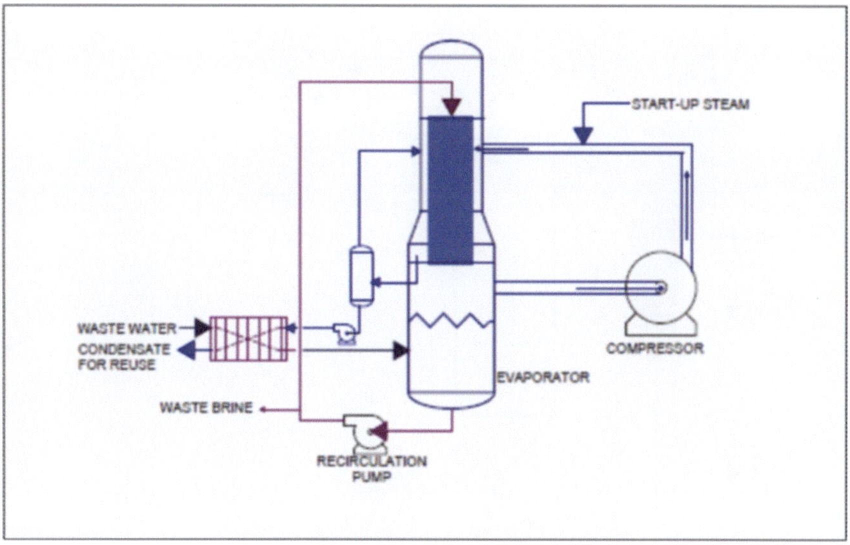

Fig. 3.30 Schematic of mechanical vapor recompression (MVR) unit

Distillate produced by evaporators is in highly pure state with a total dissolved solids (TDS) concentration of <500 mg/L and can be utilized for beneficial reuse.

There are several commercial designs of the evaporation units that are being deployed to maximize water reuse. In thermal production facilities, evaporators maximize the use of any surplus low-pressure steam by heat integration with steam generation and boiler blowdown water.

3.9.4 Brine Treatment to Achieve Zero Liquid Surface Discharge (ZLD)

This is the last step in the produced water treatment train if ZLD is desired or mandated. There are four avenues to achieve ZLD to surface water streams: (a) crystallization of salt by thermal crystallizers, (b) spray drying (c) disposal in solar evaporation ponds, and (d) deep well injection.

Drying high-concentration salt from reverse osmosis reject stream or evaporator blowdown stream in a solar evaporation pond is a viable alternative, but it has a few downsides. These include potential seepage of high concentration salts into the ground, with the possibility of groundwater contamination. This becomes even more concerning if the concentrated waste stream contains NORM (Naturally Occurring Radioactive Material). The other concern is its potential to contaminate surrounding water resources in the event of overflow during rainstorm events. Also, solar ponds are best suited for warm and dry climates.

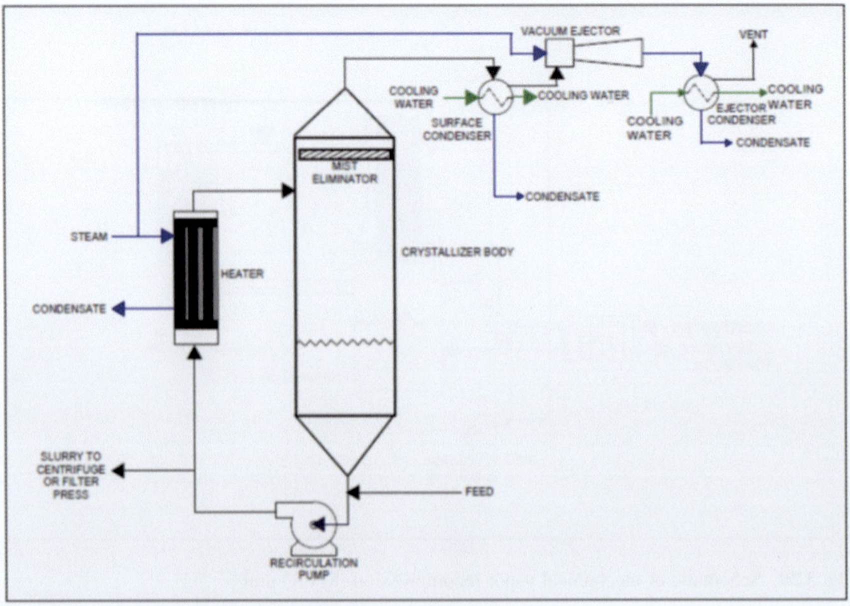

Fig. 3.31 Steam-driven forced circulation crystallizer

Thermal crystallization is frequently the preferred ultimate step to remove and recover the last drop of water in concentrated brine from reverse osmosis reject and evaporator blowdown streams. A steam-driven forced circulation crystallizer is shown in Fig. 3.31. Feed at TDS ~ 400,000 mg/L is introduced by blending it with circulating liquor. The solid–liquid slurry is withdrawn continuously and dewatered, the separated solids are disposed of, and the liquid is continuously recirculated through the crystallizer.

As in the case of evaporators condensate produced by thermal crystallizers is highly pure state with a TDS concentration of <500 mg/L and can be utilized for beneficial reuse. The concentrated stream is disposed by one or multiple means e.g., landfilling or discharge into a solar pond. However, it may require further processing to pass Paint Filter Liquid Test (EPA 2022) for the landfill to accept.

Deep well injection of produced water is practiced in the upstream petroleum exploration and production operations at places where suitable disposal wells to a receiving formation are available and permitted by the local authorities. Though this is not a strictly ZLD option, it is the most cost-effective solution available. Further details on Brine management are available in reference literature Chap. 5 of *Industrial Water Reclamation and Reuse to Minimize Liquid Discharge* published by Water Environment Federation (2021).

3.10 Reuse Opportunities

Water reuse is an important consideration given the water scarcity in many highly active regions of the United States in upstream oil and gas exploration and production, especially West Texas, New Mexico, Oklahoma, and California. It is increasingly gaining importance in other parts of the world also. This topic has been discussed in detail in Chap. 8.

3.11 Residuals Disposal

Every treatment process generates a residual stream that may be unusable and needs suitable disposal without creating any environmental or health hazards. Although upstream oil and gas exploration and production activities in the United States are exempt from regulation under the Resource Conservation and Recovery Act (RCRA), there are still certain requirements for disposal that must be met. In cases of produced water treatment, the sedimentation processes of solids from primary and secondary treatment will generate sludge that needs thickening and dewatering before disposal. Dewatered sludge from produced water treatment needs to be disposed of in suitable hazardous waste landfills because of the high concentration of heavy metals. Also, tertiary treatment processes generate concentrated streams that need proper and safe disposal.

3.12 Midstream Sector

The midstream sector is the second sector in the oil industry supply chain and includes multiple activities including:

- Transportation of crude oil and refined products
- Storage of crude oil and refined products
- Wholesale marketing of crude oil and refined products

Transportation of oil could be by pipeline, rail, barge oil tanker or truck. The movement by pipeline or oil tankers could be directly from the source of the oil extraction or tanks located at regionally central locations.

None of the operations involved in the midstream sector, however, use water or generate wastewater other than rainwater which does not come in direct contact with the hydrocarbons and therefore can be discharged without treatment.

Abbreviations and Glossary of Terms

Term	Definition	Explanation
CEOR	Chemically enhanced oil recovery	To further extract oil from aging fields using polymers or other chemicals
CSS	Cyclic steam stimulation	Intermittent steam injection for heating heavy oil for well stimulation. Also known as "huff and puff"
Darcy	Unit of permeability	Used in reservoir engineering to measure the permeability of a reservoir to fluids. Commonly specified as milli-Darcy or Darcy
EOR	Enhanced oil recovery	To further extract oil from aging fields using thermal, chemical, water and gas means
FPSO	Floating production storage and offloading	Floating oil platform used offshore
SAGD	Steam assisted gravity drainage	Continuous steam injection into wells to heat oil for heavy oil production
BS&W	Basic sediment and water	Specification for water and solids including sand in crude oil indicated in volume percentage (%)
BTEX	Benzene, toluene, ethylbenzene, and xylene	Aromatics usually present in crude oil measured in mg/l
TOC	Total organic carbon	Organic compounds dissolved in water measured in mg/l
VOC	Volatile organic compounds	Light organic materials typically consisting of 12 or fewer carbon atoms per molecule measured as mg/l in water

References

Blauch ME (2009) Marcellus Shale post-frac flowback waters—where is all the salt coming from and what are the implications? In: Society of Petroleum Engineers eastern regional meeting. Society of Petroleum Engineers, Charleston, West Virginia

Boschee P (2012) Handling produced water from hydraulic fracturing. Oil Gas Facil 22–26

Bothamley M (2004) Offshore processing options for oil platforms. In: SPE Annual technical conference and exhibition. Society of Petroleum Engineers, Houston, Texas

Camarillo M, Stringfellow W (2018) Biological treatment of oil and gas produced water: a review and meta-analysis. Clean Techn Environ Policy 20:1127–1146. https://doi.org/10.1007/s10098-018-1564-9

Campo RD (2017) Shipboard slop treatment by means of aerobit granular sludge: strategy proposal for granulation and hydrocarbons removal. In: Frontier international conference of wastewater treatment, Palermo

Cenovous (2022) SAGD operations. Accessed from Cenovous Corporate web page: www.cenovous.com

Ditria J (2014) Produced water treatment with deoiler hydrocyclone technology—misconceptions and corrections. NEL Produced Water Club. www.eprocess-tech.com
EPA (2013) Hydraulic fracturing for oil and gas: impacts from the hydraulic fracturing water cycle on drinking water resources in the United States. EPA, Washington DC
EPA (2022) SW-846 test method 9095B: paint filter liquids test. EPA, Washington, DC
Ezechi HE (2012) Boron in produced water—challendes and improvements: a comprehensive review. J Appl Sci 12(5):402–415
Granberg A, Schmidt KK (2010) Infographic: what is hydraulic fracturing? Water News
Gulbis J, Hodge RM (2000) Fracturing fluid chemistry and proppants. In: Nolte MJ (ed) Reservoir simulation, chapter 7. Wiley
Gusa AV, Tamani A, Zhang Z, Vidic RD (2020) Sulfate precipitation in produced water from marcellus shale for the control of natually ocurring radioactive material. Water Res 177
He CL (2014) Kinetics and equilibrium of barium and strontium sulfate formation in Marcellus Shale flowback water. J Environ Eng B4014001-1, ASCE
Higgins MJ, Novak JT (1997) Dewatering and settling of activated sludges: the case for usiong cation analysis. Water Environ Res 69(2)
Jimenez SM (2018) State of the art of produced water treatment. Chemosphere v:186–208
Veil, JA, Puder MG, Elcock D, Redweik RJ, Jr (2004) A white paper describing produced water from production of crude oil, natural gas and coal bed methane, Argonne National Laboratory, US DOE, NETL
King G (2012) Hydraulic fracturing 101: what every representative, environmentalist, regulator, reporter, investor, university researcher, neighbour and engineer should know about estimating frac risk and improving frac performance in unconventional gas and oil wells. In: SPE hydraulic fracturing technology conference. Society of Petroleum Engineers, The Woodland, Texas
Kottungal L (2014) Worldwide EOR survey. Oil Gas J
Lake L (1989) Enhanced oil recovery. Prentice Hall, Englewood Cliffs, New Jersey
Lipus D, Roy D, Khan E, Ross D, Vikram A, Gulliver D, Hammack R, Bibby K (2018) Microbial communitiesd in bakken region produced water. FEMS Microbiol Lett 365(12)
MEG (2024) Christina Lake operations. Accessed 15 May 2022, from MEG Energy Corporate website: https://www.megenergy.com
Menjivar JA (1986) Use of gelation theory to characterize metal cross-linked polymer gels. In: Glass EJ (ed) Water soluble polymers, chapter 13.
Nadella N (2009) Energy optimization in thermal EOR facilities. In: World heavy oil conference 2009. Society of Petroleum Engineers, Puerto La Cruz, Venezuela
Rawlins CH (2017) SPE-187109-MS, partial processing: produced water debottlenecking unlocks production on offshore Thailand MOPU platform. In: SPE annual technical conference and exhibition held in San Antonio, Texas, USA. www.spe.org
Shafer L (2011) water recycling and purification in the Pinedale Anticline field: results from the Anticline Disposal Project. In: SPE Americas E&P health, safety, security, and environmental conference. Society of Petroleum Engineers, Houston. https://doi.org/10.2118/141448-MS
SPE (2013) Alternative hydrocarbon production systems [OPL Subsea Production Wallchart]. Accessed 06 Nov 2022, from petrowiki.spe.org: https://petrowiki.spe.org/Offshore_and_subsea_facilities
Sun PM (2012) Presentation at WEFTEC Workshop 2012. Water Environment Federation, New Orleans
Thomas S (2008) Enhanced oil recovery—an overview. Oil Gas Sci Technol 63(1):9–19
Tieman ZG, Stewart BW et al (2020) Barium isotopes track the source of dissolved solids in produced water from the unconventional Marcellus Shale gas play. Environ Sci Technol 4275–4285

UND (2010) Bakken water opportunities assessment—phase 1. University of North Dakota Energy and Environmental Research Center
US DOE (2009) Modern Shale gas development in the United States: a primer. U.S. Department of Energy
Veolia (2022) Macro-porous polymer extraction. Veolia MPPE Brochure. Veolia
Webb CN (2009) Desalination of oilfield-produced water at the San Ardo water reclamation facility. In: SPE Western regional meeting. Society of Petroleum Engineers, San Jose

Chapter 4
Water Use and Wastewater Sources in Downstream and Marketing Operations

Mudumbai Venkatesh and Somnath Basu

Abstract Petroleum refineries are large complexes comprising of multiple process units that are integrated together to receive crudes from upstream operations and produce refined products for distribution and marketing. These are supported by the utility systems including water and wastewater treatment, and steam and power generation, which are critical to the smooth operation of a refinery complex as a whole. The quality and quantity of wastewater generated by any industrial chemical process is largely influenced by the feed and the products of that industry. With that in view, a brief overview of all the important unit operations and processes of a refinery has been presented in this chapter with the details of where water is used as well as discussions of the sources, flows and characteristics of wastewater generated. This includes a detailed discussion of water treatment for steam generation and cooling water circulation systems which are essential utilities supporting the successful operation of an integrated refinery complex.

Keywords API separator · Aquifers · Bottoms sediment and water (BSW) · Crude distillation unit (CDU) · Cooling water · Crude · Delayed coker · Desalination · Desalter · Diesel · Distillation · Ejectors · Fire water network · Flare · Fluid catalytic cracker (FCC) · Fuel gas · Fuel oil · Furfural extraction · Gasoline · Heating oil · HF alkylation · Hydrocracker · Hydrodesulferization · Hydrofinishing · Hydrotreating · Inside battery limits (IBL) · Jet fuel · MEK dewaxing · Merox · Mud wash · NESHAPS · Outside battery limits (OBL) · Petroleum coke · Power generation · Process water · Product blending · Purge treatment unit (PTU) · Raw water distribution · Reuse and recycle · Sanitary water · Sour water · Sour water stripper · Spent caustic · Steam generation · Storage pond · Stripping steam · Tail gas treating unit (TGTU) · TAN · Thermal

---

M. Venkatesh (✉)
Process and Energy Consultant, Andover, MA, USA
e-mail: m.venkatesh1@yahoo.com

S. Basu
Jewel Innovations in Energy and Environment, Katy, TX, USA
e-mail: som.basu@jewelenergyenvironment.com

S. Basu et al. (eds.), *Water Management in Petroleum Industries*, Water Resources Development and Management, https://doi.org/10.1007/978-981-19-3159-8_4

cracking · Utilities · Vacuum distillation unit (VDU) · Vacuum residue · Waste heat boiler

4.1 Introduction

The Petroleum supply chain consists of four main segments. Figure 4.1 illustrates these elements. This chapter provides a discussion of the water use and wastewater sources in the third and fourth segments of the oil supply chain.

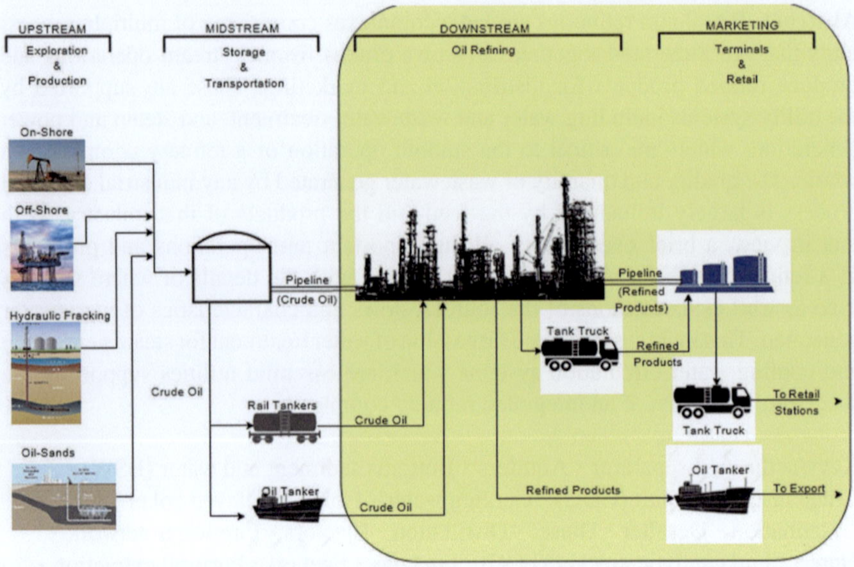

Fig. 4.1 Petroleum supply chain

4.2 Downstream Sector

This is the third sector in the petroleum industry supply chain. The primary function in this sector is oil refining.

Refinery installations are typically large integrated facilities with substantial infrastructure. These plants manage large quantities of raw materials and products and typically occupy a large area of several thousands of square meters and could include ship docking facilities, railroad yards, storage tanks and several reactors, distillation columns and ancillary facilities. In an oil refinery, crude oil is separated into various products using several technologies and processes.

The products produced at a typical refinery include:

- Fuel Gas
- Propane
- Butane
- Olefins
- Gasoline
- Jet Fuel
- Diesel
- Heating Oil
- Fuel Oil
- Lube Oil
- Petroleum Wax
- Sulfur
- Petroleum Coke
- Asphalt solids and other products

The plants also include utility systems as well as substantial support and infrastructure facilities. These include:

- Raw water Intake and distribution
- Steam generation and distribution
- Boiler house, steam turbines and power generation
- Cooling water distribution and Cooling towers
- Stormwater ponds
- Fire water storage and network
- Instrument air and service air network
- Nitrogen distribution
- Laboratories
- Extensive maintenance shops and facilities
- Storage Tank Farm

Figure 4.2 shows a block flow diagram of a typical refinery. It must be recognized that no two refineries are the same. The configuration and the process units in a refinery are dependent on the type of crude that is processed and the slate of products that will be produced. There are also local restrictions and regulations such as citing restrictions and limits on emissions and discharges that need to be taken into account in the configuration of refineries.

Oil refineries require a large amount of energy and water that are used in the process units to achieve the separation of crude oil into the various products. It is estimated that refineries use about 1.2 m^3 of water for each metric ton of crude processed. This water is used for the following purposes.

- Process water
- Utility water (used for washing at utility stations throughout the process units)
- Fire water
- Cooling water makeup

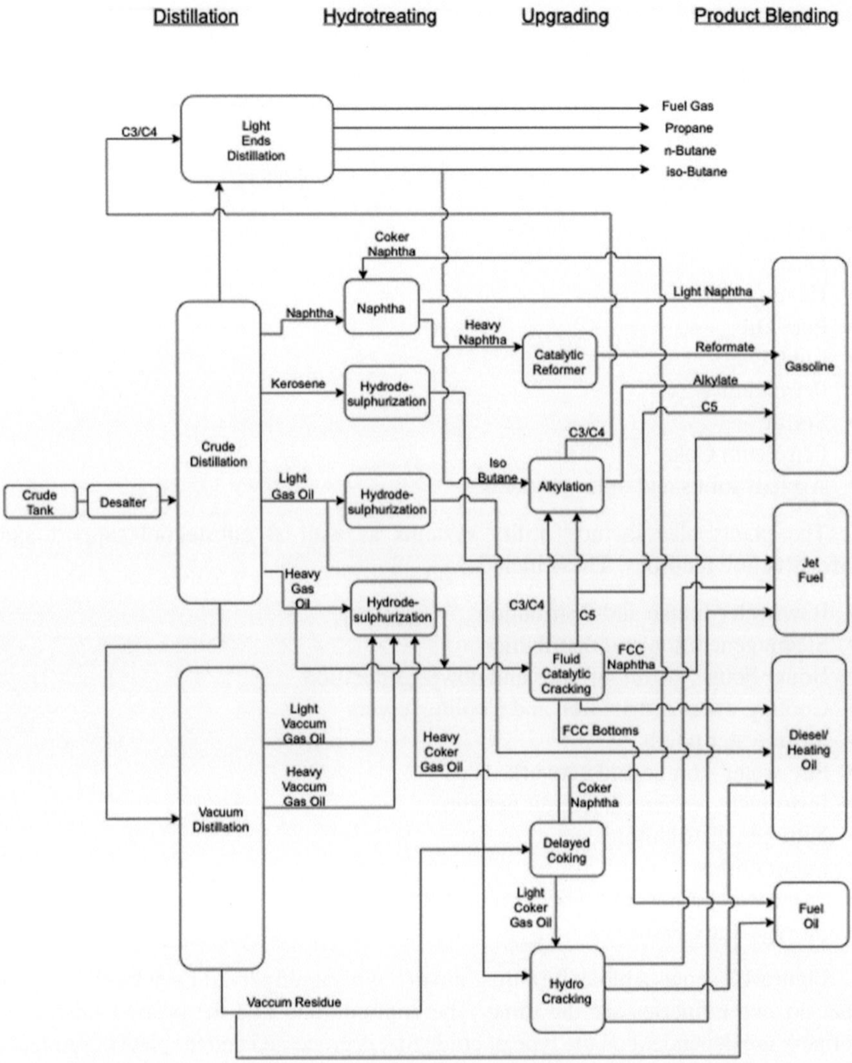

Fig. 4.2 Refinery block flow diagram (adapted from Giles 2023)

- Boiler feed water for production of steam
- Potable water
- Sanitary water

Figure 4.3 shows the approximate distribution of the water consumed in a refinery. It should be noted that these percentages could vary significantly depending on factors such as site location, plant operating practices and local climatic conditions. Because they are such large consumers of water, it is prudent to closely manage and monitor the use of water as well as look for opportunities to recycle and reuse water wherever

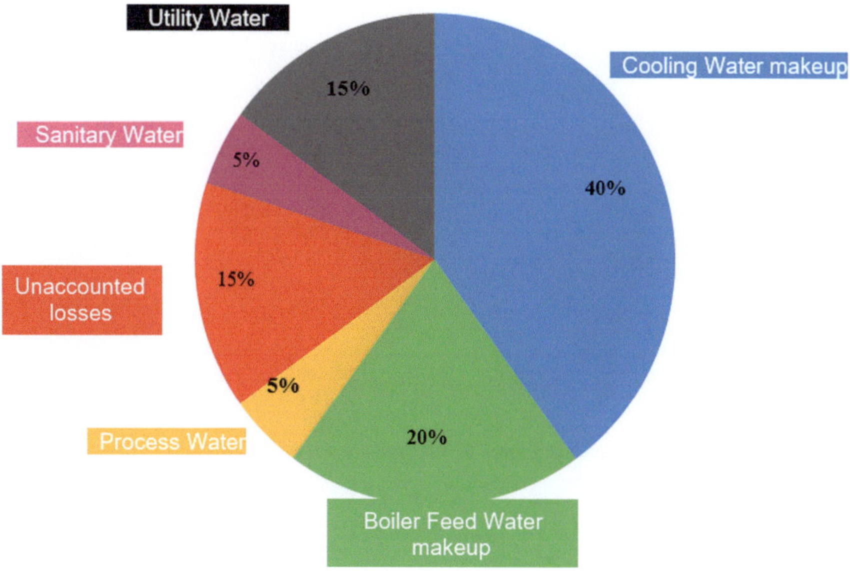

Fig. 4.3 Refinery water distribution

possible. It is well known that freshwater shortages exists in many parts of the world with some regions experiencing severe shortages.

Figure 4.4 illustrates the water resources distribution around the world. Refineries in many parts of the world are severely restricted in the amount of fresh water that can be imported into the plant. In some countries major industries are so severely restricted from taking water from inland sources such as rivers, lakes and aquifers that in many cases they desalinate sea water (which is an expensive option) to meet the water demand by the industries. Some refineries also use treated municipal wastewater for their makeup water needs.

In order to manage water used in the refinery it is important to understand the sources of wastewater so that control measures can be implemented where appropriate.

In the following sections of this chapter, the uses of water in the various operating units at the refinery are explained and the sources where wastewater is generated are discussed. This will enable the reader to gain an indepth understanding of water uses in the refinery so that opportunities for wastewater minimization and control can be explored and implemented.

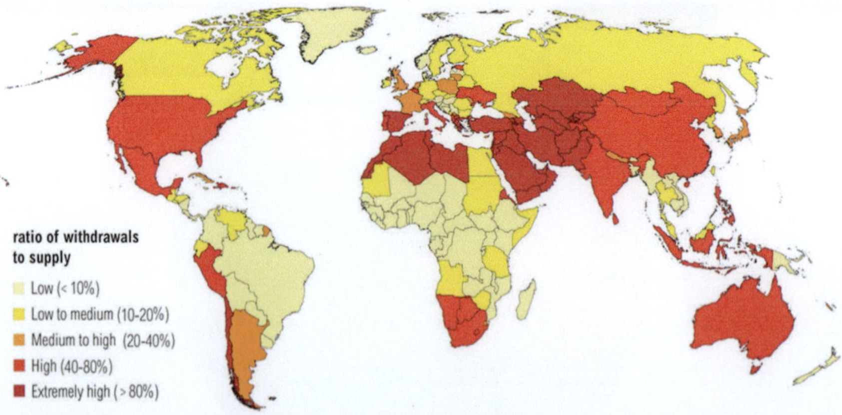

Fig. 4.4 Projected water stress by country in 2040 (World Resources Institute 2015)

4.2.1 Water Balance

The increasing pressure to reduce water use has increased the interest in understanding the "Water Footprint" of a refinery. The Water Footprint is defined as the total amount of water directly or indirectly used or consumed and the wastewater discharged as a result of manufacturing or other industrial operations. There are also considerations such as consumptive and non-consumptive use and the quality of the wastewater discharges that need to be taken into account when defining the Water Footprint of a facility. A good water balance is an important element in understanding the water uses in a refinery. A plant cannot reduce what they don't measure and monitor. Some of the potential hurdles typically encountered to put in the required time and effort in doing the water balance might be:

- We do not have time to do it properly
- We have always done it this way and don't want to change it
- Reduction of water use is not important enough to the plant
- Does not affect our day-to-day operation

These hurdles can be overcome by a commitment from the refinery management to reduce water use in the refinery. The following elements are very important when conducting a water balance at the refinery.

- Gather reliable information
- Evaluate the use of historical data versus snapshot data
- Review utility bills and inlet flow meters which can be important sources of reliable data
- Engage in discussions with plant personnel. It is important not to rely on anecdotal information but somehow check and verify the information that has been gathered
- Use engineering calculations

An illustration of a typical refinery water balance including a schematic is presented as Fig. 4.5.

For the purposes of the water balance, it is convenient to divide the refinery into three parts—Pretreatment, Utilities and the Process units. Many refineries have a storage pond where fresh water is brought in. Sources of water could be ground

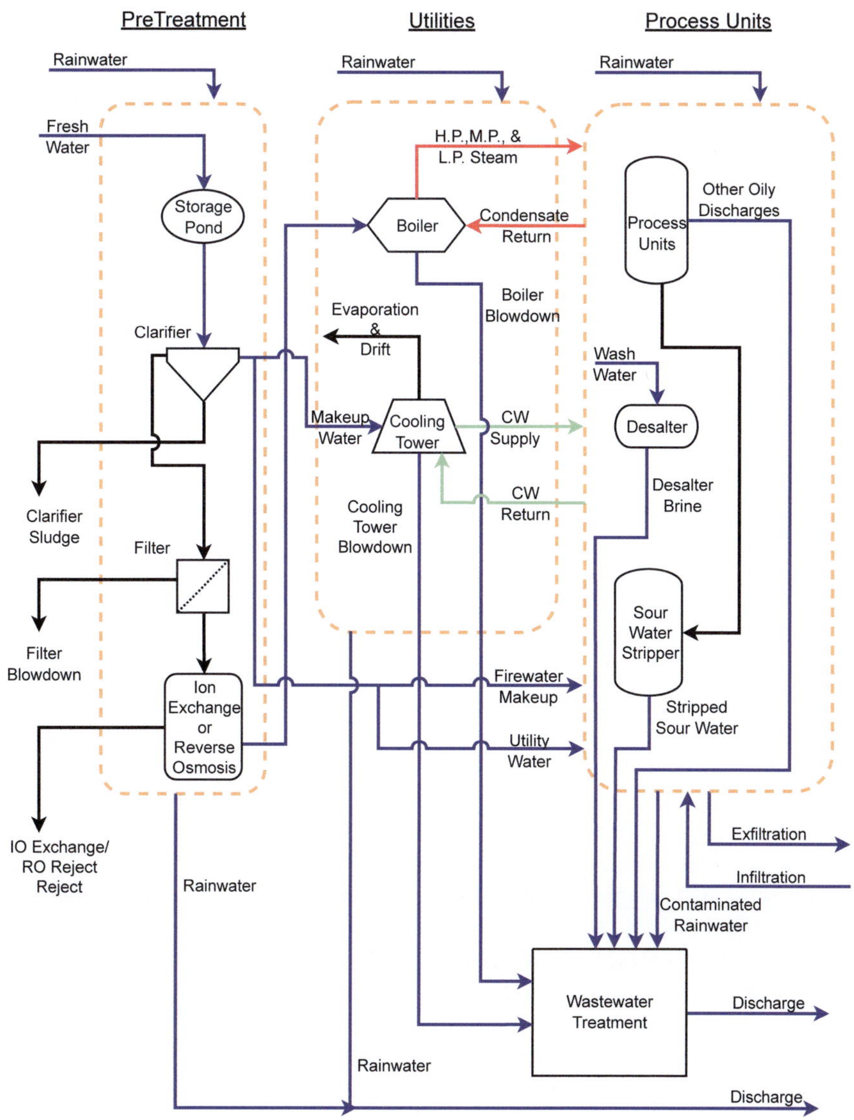

Fig. 4.5 Schematic of refinery water balance

water, rivers or lakes. In parts of the world where there is a scarcity of water, desalination of sea water is used as a source of water. Some refineries store stormwater in ponds and use it for firewater makeup and other uses. The water from the pond is clarified to remove suspended solids and the clarified water is sent to the cooling towers as makeup. This water is also used as makeup to the refinery firewater system as well as utility water that is used for various purposes at utility stations located throughout the process units. Some of the clarified water is also sent to further treatment which could include additional filtration and ion exchange (IX) or reverse osmosis (RO) to demineralize the water to make it suitable as boiler feed water. This additional treatment of raw water is discussed in more detail in earlier chapters of this book. Rainwater that falls on the pretreatment area is generally considered to be uncontaminated and therefore can be discharged without further treatment.

The boiler system in the refinery produces high pressure, medium pressure and low-pressure steam. The condensate from the refinery is returned to the steam plant. A certain amount of blowdown is discharged from the boiler to prevent buildup of dissolved solids in the system. The boiler/steam system in the refinery is discussed in greater detail later in this chapter.

Larger refineries typically have multiple cooling towers. Makeup water is added to these towers to account for losses via evaporation and drift as well as blowdown that is sent to wastewater treatment. The cooling system in the refinery including cooling towers are discussed in greater detail in later sections of this chapter.

In the refinery, steam and cooling water are used in the different process units. The cooling water and most of the condensate is returned to the utility system in the plant. Some of the low-pressure condensate is lost to the sewers in the refinery because it is not economical to recover all the condensate. Wash water is added to the desalter and the brine from the desalter is sent to the wastewater treatment plant. Sour water that is produced in the process units is sent to the sour water stripper for the removal of hydrogen sulfide and ammonia. The stripped sour water is sent to the wastewater treatment plant. The treated wastewater from the plant is discharged, typically, to either rivers or the sea. In many refineries, especially the older ones, there could be significant leaks due to the deterioration of the sewers. This results in exfiltration of wastewater into the ground water and infiltration from the ground water depending on the hydrology in the area. This is one of the reasons it is difficult to close the water balance in the refinery above 80–90%.

Following are examples of water balance for a refinery with an operating capacity of 15 million tons/day of crude. Tables 4.1, 4.2 and 4.3 present water balance in the pretreatment, utilities and the process units of a large refinery complex. Also included in the tables are the sources of the data as well as an explanation of how the data was obtained.

As can be seen from the tables above the imbalances range between −6% and + 18% which is not unusual and is within expected ranges.

Table 4.1 Water balance—pretreatment

Pretreatment			
	Source	Flow (m^3/h)	Data obtained from
Input	Raw water—source 1	1,431	From monthly totalizer data
	Raw water—source 2	494	From monthly totalizer data
	Total	*1,925*	
Output	Cooling tower makeup	627	Flow meter
	Demineralized water to steam plant	1,322	Flow meter at the steam plant
	Clarifier sludge	63	Estimated
	Raw water filter backwash	25	Estimated
	Demineralizer reject	38	Estimated
	Total	*2,037*	
	Input − output	*−112*	*Imbalance is = −6%*

Table 4.2 Water balance—utilities

Utilities			
	Source	Flow (m^3/h)	Data obtained from
Input	Demineralized water to steam plant	1,322	Flow meter at the steam plant
	Cooling tower makeup	627	Flow meter
	Total	*1,949*	
Output	Steam	870	Calculated based on blowdown and condensate return and losses
	Condensate return from refinery	579	Flow meter at steam plant
	Condensate losses	131	Based on internal refinery study
	Boiler blowdown	15	Based on internal refinery study
	Evaporation and drift	440	Calculated from cooling tower data
	Cooling tower blowdown	225	Calculated based on cooling tower operation
	Total	*2,259*	
	Input − output	*−310*	*Imbalance is = −16%*

Table 4.3 Water balance—process units

Process units			
	Source	Flow (m^3/h)	Data obtained from
Input	Steam	870	Calculated based on blowdown and condensate return and losses
	Fire water	135	From totalizer
	Utility water to Refinery	45	Estimated

(continued)

Table 4.3 (continued)

Process units			
	Source	Flow (m^3/h)	Data obtained from
	Rainwater	113	Estimated by internal refinery study
	Desalter 1 wash water	36	Based on 5% of crude flow
	Desalter 2 wash water	41	Based on 5% of crude flow
	Groundwater infiltration		Unknown
	Total	*1,239*	
Output	Condensate return	579	Flow meter at steam plant
	Desalter 1 brine	36	Based on wash water makeup rate
	Desalter 2 brine	41	Based on wash water makeup rate
	Sour water stripper—1 bottoms	42	Based on flowmeter
	Sour water stripper—2 bottoms	45	Based on flowmeter
	Sour water stripper—3 bottoms	46	Based on flowmeter
	Other oily water discharges	110	Rough estimate based on discussions
	Contaminated rainwater	113	Estimated by internal refinery study
	Groundwater exfiltration		Unknown
	Total	*1,012*	
	Input – output	*227*	*Imbalance is = +18%*

4.2.2 *Benchmarking*

Refineries use significant quantities of water for their routine operations. Strategies for effective management of water and wastewater could be an important criterion for refinery management. It is important to be able to compare the water use at one refinery to others in the industry. In order to do this we must first understand the distribution of water use in the refinery which is shown in the chart below. This is the same data that is presented as Fig. 4.3 in the form of a pie chart and is repeated here as a bar graph for the readers' convenience (Fig. 4.6).

- Cooling water makeup is the water added to cooling towers to compensate for the losses due to Evaporation and Drift from the cooling towers plus Blowdown from the system.
- Boiler feed water makeup is the water to compensate for losses due to Blowdown from boilers and Waste Heat Boilers, unrecovered condensate, and other condensate losses.
- Process water is the wash water used in the various process units. This water gets contaminated with hydrocarbons due to contact and is discharged to wastewater treatment.
- Unaccounted water includes the water used for other uses such as the laboratory etc. in the refinery.

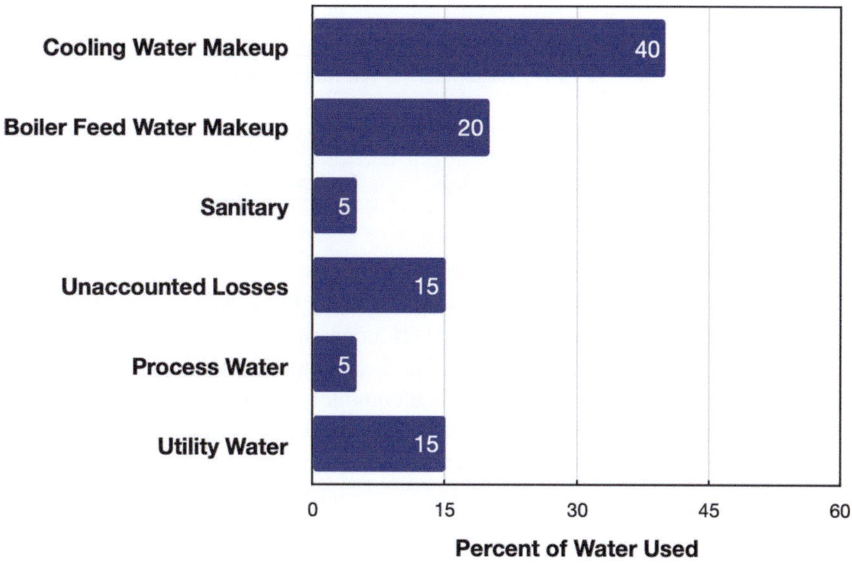

Fig. 4.6 Water use distribution

- Sanitary water is the water used in cafeteria and wash rooms in the refinery.
- Utility water is the water used in utility stations around the refinery.

As can be seen from the chart, cooling tower makeup is by far the largest consumer of water in a refinery. Evaporation from cooling towers is a strongly influenced by local climatic conditions such as temperature, pressure and humidity. Therefore, evaporation rates vary widely in refineries located in tropical areas when compared to plants located in cooler and drier climatic areas. In fact, some refineries located in warmer climates are forced to use supplemental refrigeration units in order to ensure that the cooling water supply temperatures are within acceptable limits. Additionally, some refineries do not use cooling towers but use once-through sea water or lake water for their cooling purposes. In these cases, the quantity of water used in the refinery can be considerably different from those at other refineries.

Therefore, it is not always meaningful to compare water use at various refineries and benchmark them against each other. The quantity of wastewater generated in a refinery is a good indicator of the water management practices at the refinery. Therefore, as an alternative to the comparison of water used in a refinery, a comparison of the wastewater generated offers a more meaningful evaluation of water use and management at refineries. Table 4.4 provides the benchmarks for wastewater generated as well as three other important parameters in refineries.

The benchmark numbers are based on the tons of crude oil that is processed in a refinery and compiling data from more than 30 refineries worldwide.

- Table 4.4 presents category of wastewater and the corresponding flowrate of wastewater discharged from the refinery.

Table 4.4 Benchmarking data

	Category	Units	Benchmark
1	Wastewater	m^3/ton of crude	0.27–0.7
2	Sour water	m^3/ton of crude	0.03–0.18
3	Spent caustic	m^3/1,000 tons of crude	0.05–4.4
4	Wet slops	% of crude processed	0.5

- Sour water is the quantity of sour water that is generated by all the process units in the refinery. It must be understood that this number has a wider variation because it is dependent on the type of crude and sulfur content of the crude being processed in the refinery as well as whether the refineries include catalytic crackers or cokers as these units are generators of significant amounts of sour water.
- Spent caustic is the water generated by all the Merox units in the refinery including the prewash water. This is a strong function of the sulfur content in the crude being processed as well as the effectiveness of the operating practices of the Merox units in the refinery. Merox is the short form of a process for mercaptan removal by oxidation from petroleum products, a proprietary process developed and marketed by UOP. It is described in detail later in this chapter.
- The wet slops is the amount of oil recovered and recycled from the API separator in the wastewater treatment plant. This is also a measure of the effectiveness of all the oil/water interface controllers in process vessels various units in the refinery and is a measure of the quantity of inadvertent discharges of oil to the wastewater treatment plant. Production of more slops results in a larger recycle of hydrocarbons to the crude unit that require reprocessing and consumes more fuel, steam and power in the refinery.

4.2.3 Wastewater Composition

The composition of wastewater streams in the refinery varies widely depending on the type of crude that is being processed as well as operating practices. The ranges of contaminant composition from various refinery wastewater streams based on data from many refineries is shown in Table 4.5.

There are many factors that influence the generation of wastewater as well as its composition in a refinery as discussed below.

- *Origin of Crude*—The composition of crudes vary widely depending on the country and region of origin. Crude oil includes many water-soluble compounds such as salts, sulfides, ammonia, and various hydrocarbons. These contaminants get transferred to water up to the solubility limit of the contaminants when the fresh water or steam condensate comes in contact with crude. Additionally, various metals and metalloids such as copper, selenium, iron, vanadium and arsenic that are present in the crude are also transferred in varying amounts. Sulfur and acid (naphthenic acid) contents of the crude can also significantly affect the wastewater composition. The amount of bottom sediment and water (BS&W) present

Table 4.5 Wastewater composition

Wastewater source	COD (mg/L)	Oil and grease (mg/L)	Suspended solids (mg/L)	Phenol (mg/L)	Sulfide (mg/L)
Refinery wastewater to wastewater treatment	300–3,500	200–120,000	200–30,000	5–200	1–200
Stripped sour water	500–1,000	5–10	10–25	25–100	≤10
Desalter effluent	400–1,600	400–10,000	500–15,000	10–120	50–80
Spent caustic[a]	15,000–70,000	50–1,000	50–150	100–7,000	100–20,000
Cooling tower blowdown	100–150	2–5	50–200	–	–
Boiler blowdown	50–100	2–5	10–25	–	–
Stormwater	50–100	2–6	50–100	–	–

[a] Spent caustic compositions vary widely depending upon the type—sulfidic, phenolic and naphthenic. See Tables 4.22, 4.23 and 4.24 for details

in the crude is dependent on various factors such as the amount of water present in the crude reservoir from which it was extracted, type of crude transportation (pipeline or crude container ship) and the duration for which the BS&W was allowed to settle in the crude tank before the mixer in the crude tank was turned on. The amount of slops (off spec products) that are recycled within the refinery and added to the crude tank for reprocessing could also influence the composition of the wastewater that is generated in the refinery. A detailed discussion of the origin and characteristics of crude is presented in Appendix B of this book.

- *Process Unit Configuration*—The type of units in a refinery also have an impact on the quantity and composition of the wastewater produced in a refinery. Plants operating with high sulfur crudes are likely to have more hydrodesuferization units and thus will produce more sour water. These refineries will also tend to have more Merox units resulting in the production of more spent caustic. Refineries operating with high TAN crudes (greater than 1% naphthenic acid which results in a higher TAN number) will likely have Merox units that treat the jet fuel or kerosene produced in these refineries. The spent caustic that is produced from these Merox units include compounds that are refractory to biological treatment. Refineries that are focused on maximizing gasoline or diesel production will include Catalytic Cracking and Cokers as part of the process units. Both these types of units produce a significant quantity of sour water that is high in phenolic compounds which are difficult to treat biologically and therefore might require special handling. Alkylation units that use hydrofluoric acid as the catalyst produce a waste which is referred to as acid soluble oil (ASO). This ASO is normally neutralized with caustic and discharged to the wastewater treatment plant. This wastewater will contain sodium fluoride which will pass through the treatment plant to the discharge. This can be a significant source of fluoride from the refinery which is usually regulated in the final discharge from the wastewater treatment plant.

- *Refinery Maintenance Events*—Refinery process equipment usually are on a routine schedule that requires decommissioning and maintenance periodically. Additionally, there are instances in which some equipment will need to undergo emergency shutdown due to unexpected failures. In both of these type of events, the hydrocarbon liquids in the equipment is pumped forward to the extent possible. Any remaining heel in the equipment is drained to the sewer which increases the hydrocarbon load to the wastewater treatment plant. After the equipment is drained it is steamed out to remove residual hydrocarbons. A significant portion of this steam condenses and is drained to the sewer resulting in an increased load to the treatment plant. Some newer units in the refinery may be equipped with a closed decommissioning drum (which is vented to the flare) that is used to capture all the decommissioning waste and minimize discharges to the treatment plant during maintenance events.
- *Utility Systems in the Refinery*—The cooling water and steam systems in the refinery contribute significantly to the hydraulic load of the wastewater treatment plant. If there are any leaks in the tubes of the process heat exchangers that are in service, this will result in the contamination of the cooling water in the system. If the leak is significant, the blowdown from the cooling tower will need to be increased significantly in order to purge any build-up of hydrocarbons in the cooling water circuit. This could result in a significant increase in both hydraulic and contaminant load to the wastewater treatment plant.
- *Rainwater*—The climate where a refinery is located has a substantial impact on the quantity of stormwater that is sent to the wastewater treatment plant. The volume of water generated is dependent on the type of real estate occupied by the refinery and the area of the paved and non-paved parts of the refinery. If the refinery has segregated sewers that separate the rainwater from paved and non-paved areas it will help in reducing the flows to the treatment plant. Another strategy that can be implemented to minimize the quantity of rainwater that requires treatment is to treat only the first flush (25–50 mm of rainfall) that falls on the paved parts of the refinery. Refineries that are located in cold climates will also have to contend with the melting snow that occurs in the spring.
- *Local Temperature*—Refineries that are located in colder climates also need to winterize the lines in the plant which is done by either steam or electrically tracing the lines. However, it might be impractical to winterize all the utility stations in the refinery. Frequent freezing is prevented by continuously bleeding a small flow of water from the utility stations. This could contribute to increased hydraulic load depending on the number of utility hoses that are left partially open. At some refineries located in warm climates it is not uncommon to apply external cooling on heat exchangers by directing hoses to spray water continuously on the external surfaces of the exchangers to compensate for undersized or fouled heat exchangers. This water ends up in the sewers and then to the treatment plant.

- *Fire Water*—At some refineries the fire water system is tested weekly by flushing the lines for brief periods. In general, firewater lines are larger in diameter and therefore could result in relatively large discharges of water to the sewer during the flushing periods.

4.2.4 *Refinery Process Units*

This section provides water use and wastewater information pertaining to the process units in the refinery. For each unit a brief description and purpose of the unit is provided together with the information regarding the wastewater generated by the unit.

4.2.4.1 Crude Tanks

In a refinery crude oil is typically brought in either by pipeline or large ship tankers or rail cars. The crude is transferred to large crude tanks that can be more than 30–40 m in diameter and up to 20 m tall. Refineries usually have multiple crude tanks in the tank farm. The incoming crude contains connate water which is water that is present in the oil formations, sea water and drilling muds that are entrapped with the oil during the drilling operation. The water and muds are allowed to settle out to the bottom of the tank and is referred to as bottoms sediment and water (BS&W).

The typical arrangement of a crude oil tank drain in refineries is shown in the Fig. 4.7.

Crude oil tanks are equipped with floating roofs to minimize losses of light end products from the oil. The tanks are usually located inside an earthen berm/dike in order to contain the oil in the event of a catastrophic leak of oil from the tank. The walls of the berm are usually 1–2 m high. The tank has an internal sump (multiple sumps in the case of very large tanks) which is equipped with a drain line that is routed to a sump located adjacent to the tank. Any rainwater that is collected in the

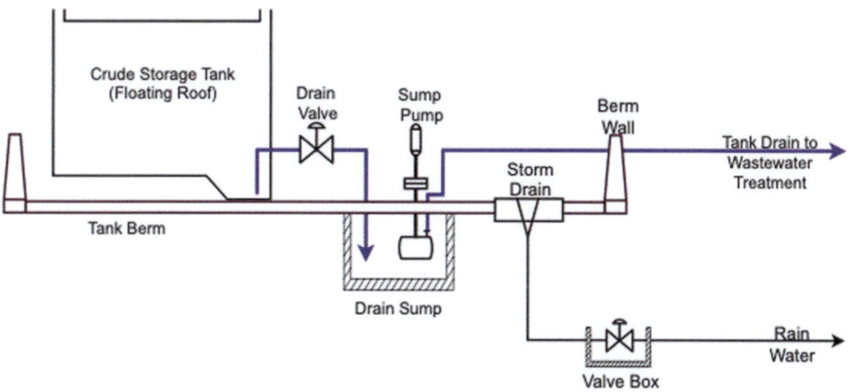

Fig. 4.7 Crude tank drain arrangement

Table 4.6 Crude tank drain data

Contaminant	Units	Value
Chemical oxygen demand	mg/L	Up to 1,000
Sulfides	mg/L	Up to 500
Ammonia	mg/L	Up to 500
Total suspended solids	mg/L	Up to 3,000

berm is either allowed to evaporate over time or drained off using the storm drain located inside the berm. The draining is controlled by a valve contained in a valve box that is usually located outside the berm. The destination of this rainwater depends on whether it is contaminated or not. The level of contamination can be determined by testing the water.

After the BS&W in the crude is allowed to settle the draining operation is started. The drain sump is equipped with a drain pump, the discharge of which is routed to the refinery wastewater treatment plant. During the tank draining operation, the drain valve is opened, and the water/mud is allowed to flow into the sump. The sump pump is activated simultaneously so that the drained water flows to the treatment plant. In many refineries the draining operation is monitored manually to ensure that free oil does not drain into the sump. In some refineries, the crude tanks are equipped with interface probes/controllers to automate the tank draining operation which typically takes several hours. Since the tanks are very large not all the mud/sludge gets drained and some of it accumulates in the tank. The accumulated sludge is cleaned out when the crude tanks are decommissioned, typically once in 10 years.

The crude tank drain is one of the most highly contaminated wastewater streams that is generated in the refinery and has a significant impact on the operation of the refinery wastewater treatment plant. Table 4.6 shows the expected levels of contaminants in the crude tank drain wastewater.

The draining of crude tanks is an intermittent operation and the flow can be estimated using the capacity of the sump pump and the duration of operation.

4.2.4.2 Other Storage Tanks

Typically, refineries have many storage tanks that are located in a dedicated tank farm. The types of tanks include:

- Slop Tanks
- Intermediate Tanks
- Gasoline Tanks
- Jet Fuel Tanks
- Diesel Tanks
- Heavy Oil Tanks
- Other Tanks

Some of these tanks such as slops tanks, gasoline tanks and jet fuel tanks may contain water that settles out in the tanks. These tanks are periodically drained using sumps similar to the ones shown in the crude tank section discussed above and sent

Table 4.7 Composition of wastewater—other storage tanks

Contaminant	Units	Value
Sulfides	mg/L	Up to 300
Ammonia	mg/L	Up to 200
COD	mg/L	Up to 500

to wastewater treatment the Table 4.7 shows the typical composition of wastewater from these types of tanks.

4.2.4.3 Desalters

Crude oil from underground formations contains various contaminants including—basic sediment and water (BS&W) water soluble salts such as chlorides and sulfates of sodium, magnesium and calcium; and various heteroatomic organic compounds. The adverse effects of these contaminants are listed in Table 4.8.

Desalting is a process where wash water is used to solubilize the salts in the crude and then remove them using gravity separation with the aid of an electrical field. The function of the desalters is to remove salts and sediments from the crude to the best possible extent and make it suitable as feed to the crude distillation unit (CDU) for fractionation into different product streams. Oil and water separate into two phases in the desalter. The dissolved salts and suspended solids in the crude are transferred to the aqueous phase in the desalter and leave as the brine stream. Desalted crude from the top layer is fed into the crude distillation unit (CDU) column and brine from the bottom layer leaves for the wastewater treatment plant. Desalters are operated under high alternating (AC) electrical current. This causes the water droplets to coalesce together. This effect, in combination with demulsifying chemicals, facilitates formation of larger drops of water that can separate easily.

Desalters are part of the preheat train in the crude distillation unit and are located upstream of the crude furnace and the crude distillation tower. Many refineries carry out desalting using two horizontal vessels in series (two-stage desalting) although

Table 4.8 Contaminants in petroleum crudes and their adverse effects

Contaminant	Effect
Particulate matter	1. Equipment fouling 2. Equipment erosion and abrasion 3. Promote emulsion formation leading to poor oil–water separation in desalters
Water	1. High energy consumption in preheating 2. Reduction in desalter capacity 3. Impose high hydraulic load to wastewater treatment
Salt	1. Carryover with crude and cause catalyst deactivation downstream 2. Create potential for corrosion and scaling
Organics	1. Create potential for scaling 2. May lead to low pH and corrosion 3. Create potential for emulsion formation leading to poor oil-water separation in desalter

there are some older refineries that still have single stage desalters. Typical salt removal efficiency of a well-designed and operated single stage desalter is ≥90% and that of a two stage desalter is ≥95%. Figure 4.8a and b show the configurations of a one-stage and a two-stage desalter system respectively.

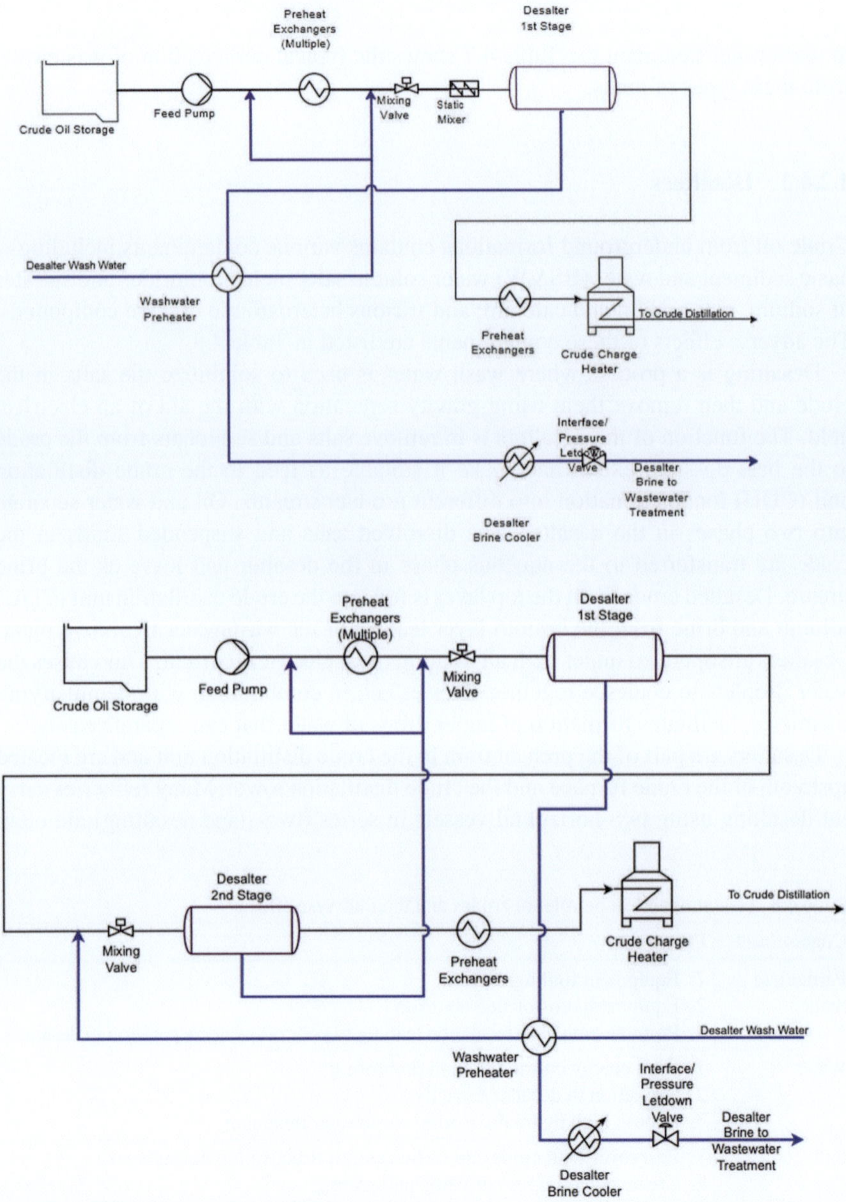

Fig. 4.8 **a** Process flow scheme for a one stage desalter, **b** process flow scheme for a two-stage desalters

For a single-stage unit the crude is pumped from the crude tank via preheat exchangers to the desalter. Typically, there are multiple heat exchangers in the preheat train in order to get the crude heated up to the required temperature. Wash water is added both upstream of the heat exchangers as well as the mixing valve followed by a static mixer that are located ahead of the desalter. The purpose of the mixing valve and static mixer is to thoroughly mix the water with the crude so that the salt in the crude gets dissolved in the water. In the case of a two-stage desalter, the crude oil from the first stage goes to the second stage desalter. Fresh wash water is added upstream of the second stage desalter. The wash water flows counter-current to the crude with the separated water from the second stage being sent upstream of the first stage. The water, sediment and drilling muds present in the crude separate out in the desalter and are sent to the wastewater treatment plant. The crude oil from the second stage is sent to the crude distillation unit.

Desalting can be viewed as a three-step process—(a) preheat the crude to reduce its viscosity and facilitate mixing, (b) mixing with wash water aided by mixing valve and static mixer, and (c) dissolution of salts in water and separation of oil-water layers. The wash water used in the desalting process is usually about 5% of the volumetric flow of crude through the desalter although in some cases the amount of wash water used is limited to the maximum available hydraulic capacity of the desalter vessel. Following are some common sources of wash water:

- River water after clarification
- City or town water supply
- Stripped sour water from the refinery

Stripped sour water (SSW) is the most desirable source of wash water as it reduces the consumption of fresh water and allows capture of the organics in the SSW stream by transferring those to the crude. Regardless of the source, wash water needs to fulfill the set of specifications as listed in Table 4.9. Typical ranges of parameters representing desalter operating conditions are listed in Table 4.9.

Table 4.9 Desalter wash water specifications

Parameter	Limits	Effects
pH	Between 7 and 8 at 40 °C	Minimizes scaling or corrosion potential
TSS	≤50 mg/L	Reduces equipment fouling
DO	≤50 μg/L	Reduces corrosion potential
$NH_3 + NH_4^+$	≤150 mg/L	Reduces emulsion formation potential in desalter as high ammonia elevates pH which leads to emulsion and rag layer formation
$H_2S + HS^-$	≤50 mg/L	Reduces corrosion potential, H_2S hazard and high sulfur in crude
Cl^-	≤100 mg/L	Prevents corrosion of stainless-steel equipment and piping
Calcium hardness	<150 mg/L as $CaCO_3$	Avoids scaling potential

Table 4.10 Typical ranges of desalter operating parameters

Operating parameter	Range
Wash water rate	5–8% (v/v) of feed
Operating temperature, °C	125–150
Applied voltage	Up to 30 kV
Demulsifier rate	Up to 100 ppm of feed

Apart from the above parameters it is also important to make sure that the presence of amines and surfactants are at minimum levels both in wash water and in the crude. Both of these ingredients cause foaming and promote emulsion formation. Both are present in certain crudes. Surfactants are used as agents in chemically enhanced oil recovery (cEOR) process. Amines are used to tie up hydrogen sulfide in sour crudes and suppress the associated odor and health hazard. Amines transferred from crude to desalter brine also contribute to the wastewater treatment plant influent TKN load and results in increased aeration demand for nitrification. Typical operating parameters of a desalter are listed in Table 4.10.

Desalter operating temperature is chosen based on crude oil properties. Operating pressure should be such that it is above the boiling point of the fluid in the desalter vessel as boiling will create turbulence and disturb clear separation of oil and water phases. Therefore, the operating pressure should be maintained higher than the sum of the vapor pressures of oil and water at the operating temperature.

The aqueous phase (brine stream) from the desalter is the most highly contaminated stream that is sent to the wastewater treatment plant and also contributes the highest contaminant load since it is a continuous stream. It is worth examining and understanding the operation of the desalter. Figure 4.9 shows the internal details of a typical desalter and Fig. 4.10 depicts the view of a model EDGE^{TM} desalters marketed by Forum Energy Technologies of Houston, Texas, USA.

The desalter vessel is usually a large horizontal vessel—about 5 m in diameter and about 15 m long although the size is dependent on the crude throughput of the refinery. Most refinery desalters are equipped with electrodes through which electric current at high amperage is passed to help coalesce water droplets present in the oil phase and thus aid in their easier separation from oil. The incoming crude is introduced between the electrodes. An oil outlet header is located at the top of the desalter which removes the oil from the desalter. The bottom of the vessel is equipped with a mud wash header which is equipped with spray nozzles. This header is used to periodically (usually once a day) spray the bottom with high pressure effluent water (using a mud wash booster pump) to wash out accumulated muds from the bottom of the vessel. This washing operation is carried out for several minutes and it increases the contaminant discharged to wastewater treatment during this period.

Desalters are operated at elevated temperatures (up to 150 °C) in order to lower the viscosity of the crude oil which makes oil/water separation easier. Some of the heat in the desalter brine is removed by exchanging it with the incoming wash water. The

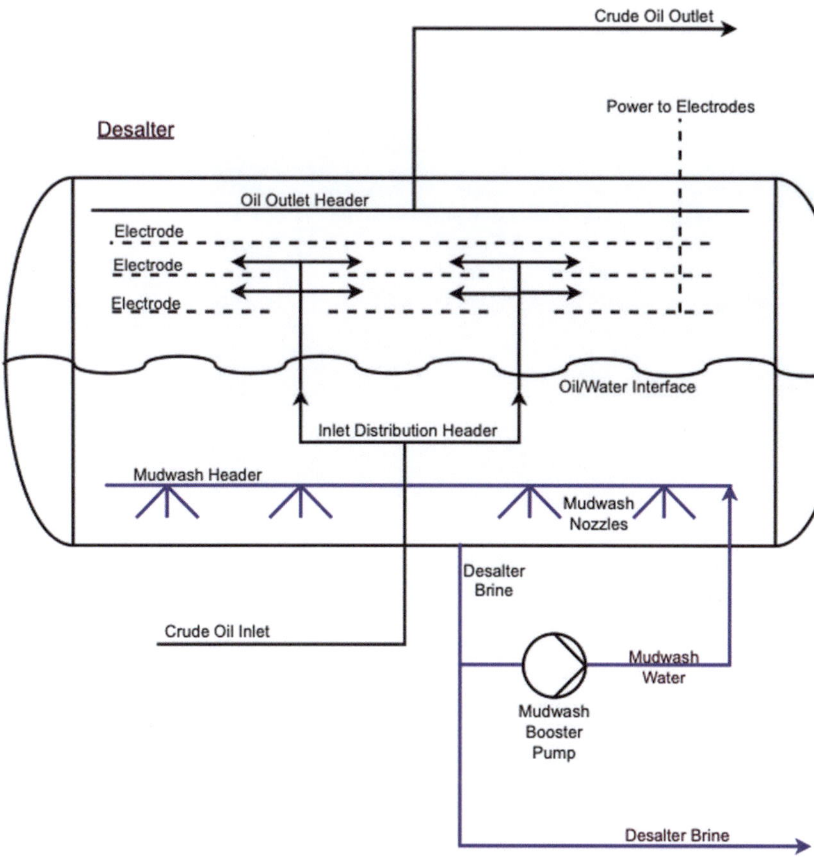

Fig. 4.9 Desalter internal details

brine stream is cooled further to about 35 °C prior to discharge. Downstream of the brine cooler the interface/pressure letdown valve is used to control the interface in the desalter vessel. Proper control of the interface in desalters can be quite challenging. Success of oil/water separation in a desalter depends upon a distinct separation of the individual layers of oil and water. However, frequently a rag layer forms in the desalter due to the presence of emulsions, and this interferes with the proper operation of the interface controller and can cause inadvertent discharges of free oil with the brine.

Proprietary demulsifiers from chemical vendors are introduced into the desalter feed to break emulsion and improve desalter performance. There are two types of demulsifiers—water soluble and oil soluble. Oil soluble chemicals dissolve and remove the thin oil layers surrounding the water droplets thus allowing them to coalesce into larger drops. The water-soluble demulsifying chemicals dissolve and remove water layers surrounding the fine oil droplets in the brine thus freeing up the

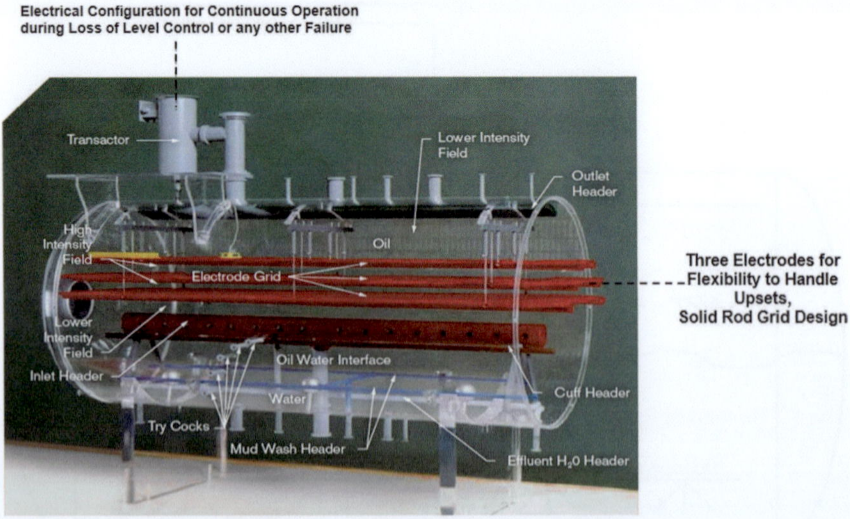

Fig. 4.10 Internal view of EDGE™ desalter (Photo courtesy of Forum Energy Technologies, Houston, TX)

fine oil droplets to flocculate together into larger drops. The right type and dosage of these chemicals are determined by pilot and/or field scale tests.

Desalter Operating Challenges and Remedies

As discussed above, one of the common operating challenges of desalters is their ability to create and maintain a distinct separation between oil and water layers. Instead of a sharp interface, the two layers are often separated by a rag layer consisting of oil-water emulsions made up of fine droplets of water and entrained solids dispersed into oil. A thick emulsion layer indicates poor oil-water separation in the desalter and results in the escape of water in oil phase and under carry of oil in the water phase. Emulsions of oil in water result from entrapment of micron size droplets of oil into the water layer, and water in oil results from the presence of micron size water droplets in the bulk oil layer. These droplets remain in suspension for an indefinite period due to their surface charges which prevent them from coalescing together and forming larger droplets. Escape of oil with brine stream imposes increased loads of COD and performance challenges in the operation of the wastewater treatment plant. On the other hand, carryover of excessive water into the crude column causes high thermal load and potential corrosion and/or fouling arising from associated inorganic salts dissolved in water. Many refineries have installed AGAR probes to properly detect the oil/water interface, and this results in more effective interface control that prevents carryover of water with the desalted crude and discharges of free oil in the wastewater. The working principle of AGAR probes is based on the detection of water concentration in an oil-water emulsion by measuring the energy absorption of the liquid surrounding the probe. These are manufactured and marketed by Agar Corporation, Houston, Texas. Sonar probes are another type of probe used in desalters for oil-water interface monitoring and control. Sonar probes work on the principle of ultrasonic detections of the liquid-liquid interface. Transducer USA of Houston, Texas manufactures and markets these probes under the brand name of InterFazer. Nuclear probes are a less commonly used type of probe for desalter oil-water interface monitoring and level control. Tracerco of Pasadena, Texas markets a product under the brand name Profiler which utilizes gamma rays to establish a vertical density profile to detect and control liquid-liquid phase interface level. Details of their working principles are available in respective manufacturers' literature.

Crudes contain chloride salts of sodium, calcium and magnesium dissolved in connate water. Uncontrolled carryover of water from desalter also helps unintended passage of such salts into the CDU column. Both magnesium and calcium chloride hydrolyze under the prevailing temperatures in the crude column according to the following reactions:

$$MgCl_2 + 2H_2O \rightarrow 2HCl + Mg(OH)_2 \text{ at} \sim 120\,^{\circ}C \quad (4.1)$$

$$CaCl_2 + 2H_2O \rightarrow 2HCl + Ca(OH)_2 \text{ at } 200\text{–}230\,^{\circ}C \quad (4.2)$$

The HCl produced by hydrolysis reaction moves up the column with the crude. It forms hydrochloric acid when it comes into contact with water at the top of the column and causes corrosion of column internals, ultimately leading to the mechanical damage and failure of top trays in the column. Sodium chloride, however, is a stable salt and does not hydrolyze under the column operating conditions. Therefore, caustic soda (NaOH) solution is dosed into the desalted crude to convert dissolved calcium and magnesium chloride into sodium chloride and prevent HCl formation by hydrolysis inside crude distilling column.

So called Opportunity crudes (see Appendix B) are a group of crudes from various parts of the world that are difficult to process due to their chemical constituents. Yet, those are increasingly being used by refiners because of the economic advantage they offer over the traditional crudes. Some of the challenges in handling and processing of opportunity crudes are listed below:

- Low API Gravity and high viscosity resulting in higher pumping costs, inadequate mixing in desalters resulting in low salt removal from crudes.
- High sulfur content creating potential health and safety risks, and environmental pollution in transportation, handling, and processing.
- High concentration of filterable solids (FS) leading to deposition and accumulation in equipment and piping, and promoting formation of a rag layer in desalters.
- High TAN leading to stabilization of rag layer in desalter, corrosion, difficulty of biological treatability of wastewater, and effluent toxicity.

- High amine content leading to high pH and rag layer formation in desalter, high operating cost for wastewater treatment due to increased COD and nitrification loads.

As can be observed above, most of the undesirable characteristics of opportunity crudes directly impact desalter performance. Since desalters significantly influence the overall performance of a petroleum refinery, it is imperative that the desalters be operated and maintained with sufficient care to ensure the smooth operation of the refinery. One of the key parameters to ensure high desalter performance is proper oil-water separation and avoidance of stable rag layer.

As the crude slates in a refinery frequently change due to availability of crudes and economic considerations it is a common practice to test new crudes on a bench scale using a Portable Electric Desalter (PED) prior to running it in full scale operation. Results of the PED tests are used in making decisions regarding selection of the right type and dosage of demulsifier, amount of wash water to be used for desalting, operating temperature, and applied voltage to the desalter to produce the best result. Details of PED test set up and procedure are provided in reference literature (Basu 2018).

Desalter Brine Handling and Disposal

Brine from the desalter is the most highly contaminated stream that is sent to the wastewater treatment plant and also contributes the highest contaminant load since it is a continuous stream.

Table 4.11 shows the typical contaminant concentrations in the desalter brine.

At some refineries the effluent from the desalter is sent to a tank for further separation of oil and sludge. This tank is typically located near the desalter in the process unit. A diagram of this type of tank follows (Fig. 4.11).

This tank is a floating roof tank that is used to minimize emissions and odors. The skimmed oil from the tank is sent to the slops system. The sump in the tank where the sludge collects is pumped out and sent to the API Sludge treatment system in the refinery wastewater treatment plant. The separated brine is sent to the API separator in the wastewater treatment plant. Some refineries have reported limited success in operating this type of tank.

Table 4.11 Desalter brine composition

Contaminant	Units	Value
Chemical oxygen demand	mg/L	Up to 1,600
Sulfides	mg/L	Up to 80
Phenol	mg/L	Up to 120
Ammonia	mg/L	Up to 200
Oil and grease	mg/L	Up to 10,000
Total suspended solids	mg/L	Up to 15,000

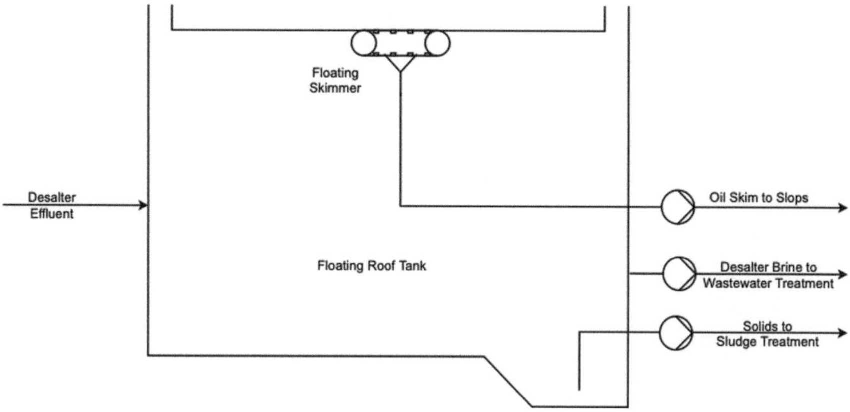

Fig. 4.11 Desalter effluent tank

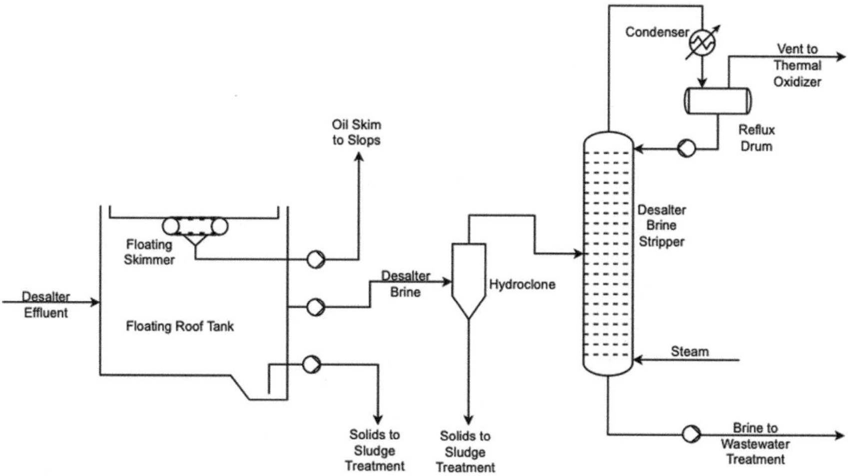

Fig. 4.12 Desalter brine stripper

In order to capture benzene emissions in response to the NESHAP (National Emission Standards for Hazardous Air Pollutants) regulations in the United States, some refineries in the US have installed strippers for the desalter brine streams to strip the benzene contained in the stream. Refer to Chap. 7 for a detailed discussion of the Benzene NESHAPS regulation of the United States that is monitored by the US EPA. Figure 4.12 shows a desalter stripper system.

Brine is sent to a hydroclone for further solids separation and removal. The stripper is equipped with high open area trays to minimize fouling. Live steam is used as the stripping medium to minimize fouling. The overhead stream from the stripper is condensed and the water is refluxed back to the stripper. The vent from the reflux drum is sent to a thermal oxidizer. The stripped brine is sent to the refinery wastewater treatment plant. Refineries that have installed these systems report that it needs close monitoring and frequent maintenance.

4.2.4.4 Crude Distillation Unit (CDU)

The crude distillation unit (CDU) is the first process unit in petroleum refineries. In this unit, the crude oil is separated in various distilled products such as naphtha, kerosene, diesel etc., by fractional distillation. A flow diagram of the crude distillation unit follows (Fig. 4.13).

The crude oil from the desalter is heated further in the preheat train and sent to the Preflash Drum where light gases are flashed off and sent to the Crude tower. The bottoms from the drum is sent to the Furnace and then to the Crude Tower. In this tower the oil is separated into different fractions of different boiling ranges which are processed further in other units in the refinery. The system consists of the crude

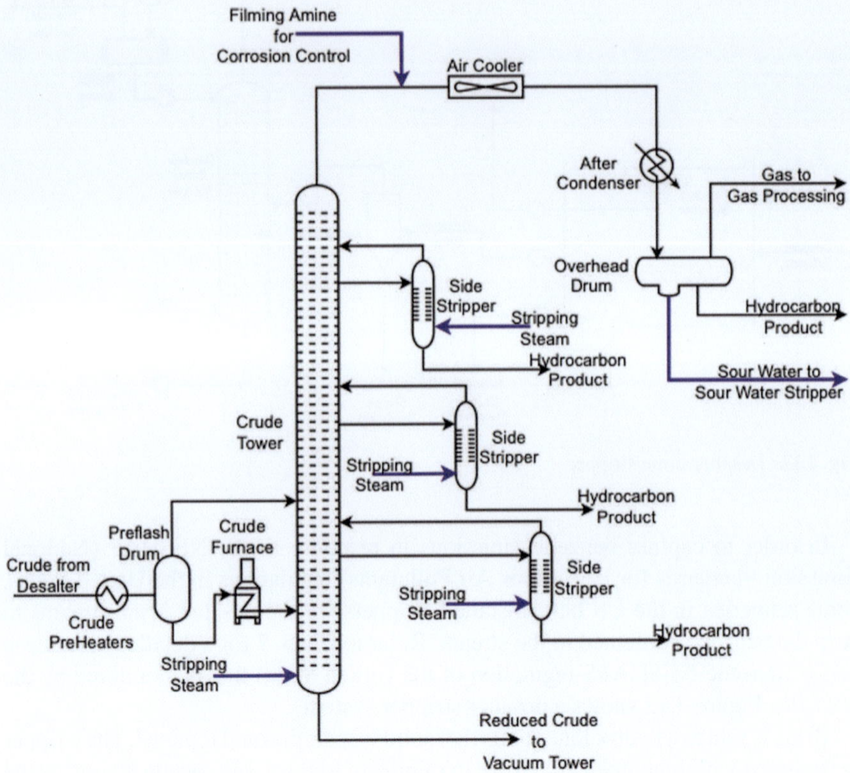

Fig. 4.13 Crude distillation unit

Table 4.12 Crude tower sour water composition

Contaminant	Units	Value
Sulfides	mg/L	Up to 800
Ammonia	mg/L	Up to 700

tower and multiple side strippers which are used to adjust the end boiling points of products from the tower by adding stripping steam.

Stripping stream is added to the main tower as well as the side strippers. The vapors from the side strippers are sent back to the crude tower. All the steam is condensed in the overhead system by means of air cooler and condenser and sent to the overhead drum. In this drum the condensate is separated from the hydrocarbons and leaves the system as sour water. Filming amines are also added to the overhead line of the crude tower to control the potential corrosion due to the presence of hydrogen sulfide in the overhead line as well as the condensers. These amines also end up in the sour water. The bottoms from the crude tower is sent to the Vacuum Tower in the refinery.

The quantity of sour water generated in the Crude Distillation Unit is dependent on the type of crude being processed as well as the sulfur content of the crude. The sour water flow can be up to 60 m^3 per 1,000 ton of feed to the unit. The flow of sour water from the CDU can be calculated using the data from the flow meters of the stripping steam added to the crude tower and the side strippers. Some refineries reuse the sour water from the CDU as wash water in the desalter, but most refineries route this stream to the sour water stripper.

The typical contaminants in the sour water from the CDU is shown in the Table 4.12.

The only other wastewater that is generated in the CDU is contaminated rain water which is normally sent to the refinery wastewater system via the sewer system. It should also be noted that it is the case with all the process units discussed below in Sects. 4.2.4.5 through 4.2.4.17.

4.2.4.5 Vacuum Distillation Unit (VDU)

Once the first cut of distillation is carried out in the crude distillation unit (CDU), the heavy components present in the residue from the crude tower is sent to the VDU. The system consists of the vacuum tower, steam driven ejectors, condensers, collection drum and associated pumps. A typical Vacuum Distillation Unit is shown in Fig. 4.14. In the vacuum tower the distillation is carried out under vacuum conditions in order to help vaporize heavier components.

The CDU bottoms stream first enters the vacuum furnace where it is heated to about 450 °C and then sent to the flash zone in the tower. Typically, the tower is equipped with a combination of trays as well as high efficiency packing in order to minimize the pressure drop across the tower. In most cases two or three side stream products are produced in the vacuum tower. To aid in the fractionation the tower is equipped with pump-around internal condensers. Stripping steam is added at the

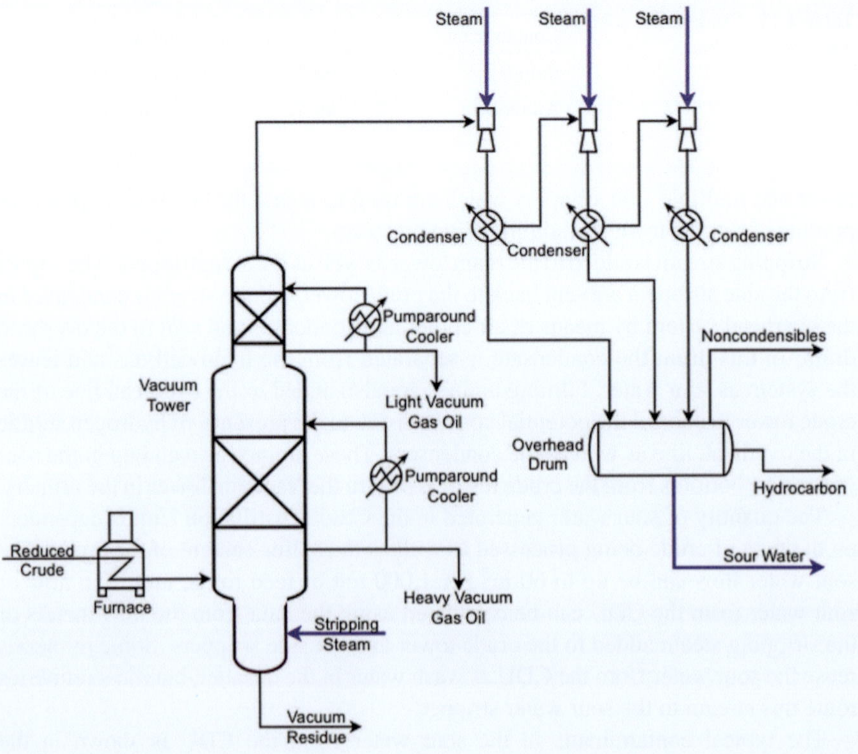

Fig. 4.14 Vacuum distillation unit

bottom of the tower to further aid in the stripping of lighter components. The overhead vapors are sent to multiple steam ejectors that are used to generate vacuum conditions in the vacuum distillation system. Typical operating pressures in the column is 25–40 mmHg although lower pressures might be required in some cases. The level of vacuum depends on the type of crude being processed as well as the pressure drop in the system. A significant quantity of medium pressure (10 bar) steam is used to generate the vacuum in the system.

The discharge of the ejectors are sent to condensers where steam as well as hydrocarbons are condensed and sent to the overhead drum where hydrocarbons and water are separated from the non-condensable gases. The separated water is considered sour water because of the presence of sulfur compounds in the system.

The quantity of sour water produced from the Vacuum unit can be up to 70 m^3 per 1,000 tons of feed to the unit. The flow of sour water from the VDU can be calculated using the data from the flow meters of the stripping steam added to the vacuum tower. The steam used in the ejectors can be estimated based on the operating pressure on the vacuum tower as well as the pressure of the steam being used. Most refineries route sour water stream to the sour water stripper.

Table 4.13 Vacuum distillation sour water composition

Contaminant	Units	Value
Sulfides	mg/L	Up to 250
Ammonia	mg/L	Up to 150
Phenols	mg/L	Up to 100

The typical contaminants in the sour water from the VDU is shown in the Table 4.13.

4.2.4.6 Catalytic Cracker (Cat Cracker)

As the name implies, catalytic cracking is a process that takes large hydrocarbon molecules and cracks them using a catalyst to make products such as gasoline and distillate fuels. There are three types of cracking processes—fluid catalytic cracking (FCC), moving bed catalytic cracking and thermal cracking units. FCC is a very flexible process and is used in most refineries although thermal cracking units are used in some refineries. Figure 4.15 is the diagram of a typical FCC unit.

The feed to the cat cracker is introduced into the catalyst circulation loop which then goes to the reactor. The reactor is equipped with internal cyclones to separate the catalyst fines from the product gas. The catalyst leaves the reactor from the bottom and is sent to the catalyst regenerator where the coke that was formed in the reactor is burnt off. The flue gas from the regenerator which contains sulfur dioxide, carbon monoxide and other components is first sent to the catalyst separator. The flue gas from the separator passes through an expander and then to the carbon monoxide boiler where the CO in the gas is converted to carbon dioxide and some of the sulfur dioxide is converted to sulfur trioxide. The gas from the boiler then passes through an electrostatic precipitator where more catalyst fines are removed. Regenerated catalyst is returned to the catalyst circulation loop.

The vapor product from the FCC reactor is sent to the Main Fractionator where the hydrocarbons are separated into different fractions. Although only one side stripper is shown in the diagram, some FCC units could have more than one depending on the number of products that are made. Wash water is added to the overhead line of the fractionator to minimize corrosion due to the presence of hydrogen sulfide. The overhead stream from the fractionator is condensed and sent to the accumulator where sour water is separated out. The uncondensed gases are sent to the refinery gas processing plant for recovery of propane and butane. The bottoms from the Fractionator are sent to the slurry separator where a hydrocarbon product is separated. The slurry from the separator is sent back to the FCC reactor.

The amount of sour water generated is dependent on the type of feed being processed in the unit, the sulfur content of the feed as well as local operating practices. The sour water flow from this unit can be up to 160 m^3 per 1,000 tons of feed to the unit. The flow of sour water from the catalytic cracker unit is significant and is

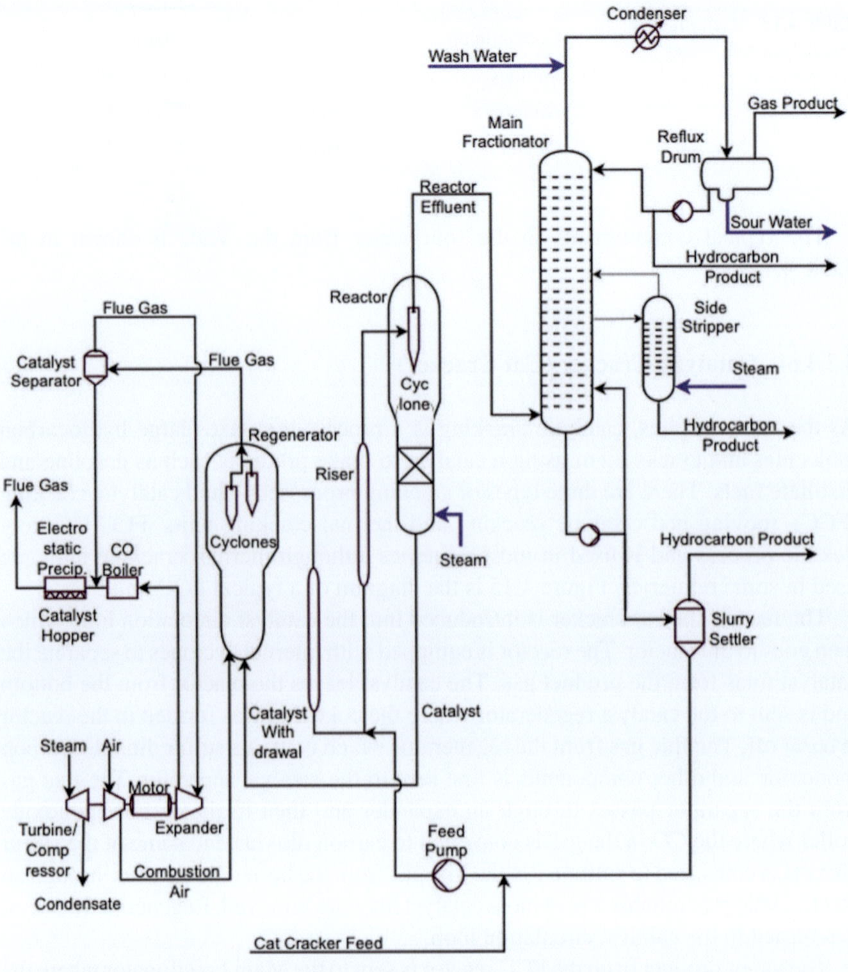

Fig. 4.15 Fluid catalytic cracking unit

therefore monitored closely using flow meters by most refineries and this data should be readily available.

The catalytic cracker produces relatively large amounts of sour water containing significant concentrations of phenols. The typical contaminants in sour water from the FCC is shown in the Table 4.14.

The flue gas from the electrostatic precipitator still contains a significant amount of sulfur dioxide, nitrogen oxides and catalyst fines. It is therefore sent to a purge treatment unit (PTU) at many refineries. A diagram of a typical FCC purge treatment unit follows (Fig. 4.16).

In this unit, the flue gas from the precipitator is sent to a scrubber where it is contacted with water to knock down the catalyst fines in the flue gas. The flue gas

Table 4.14 FCC sour water composition

Contaminant	Units	Value
Sulfides	mg/L	Up to 4,500
Ammonia	mg/L	Up to 4,000
Phenols	mg/L	Up to 500
Cyanide	mg/L	Up to 25

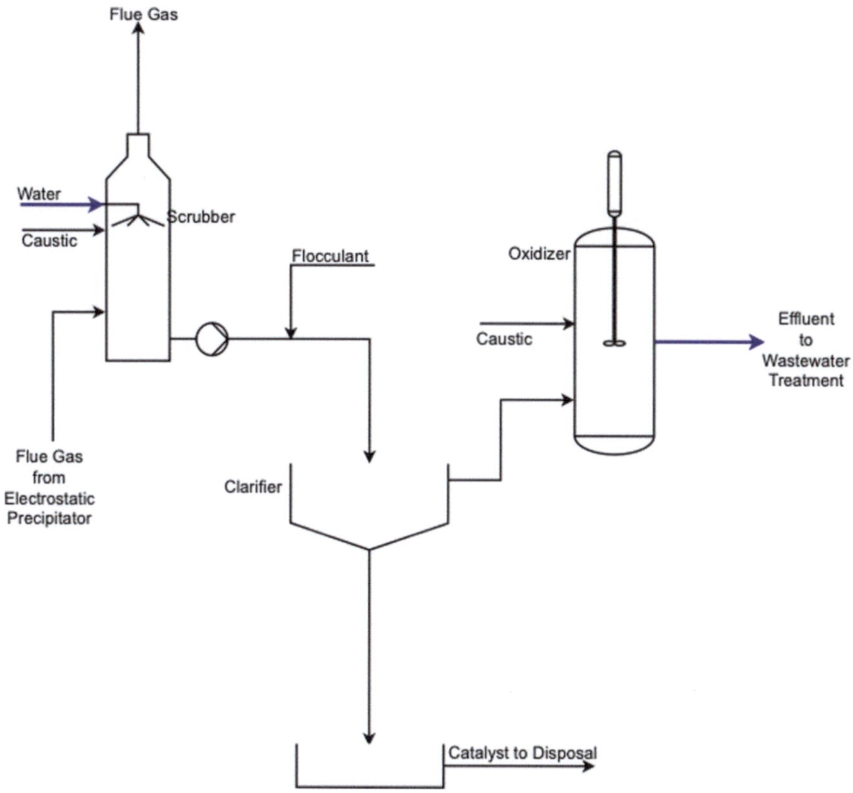

Fig. 4.16 FCC purge treatment unit

from the scrubber is discharged and the effluent from the scrubber is sent to a clarifier. Flocculant is added upstream of the clarifier to aid the catalyst settling. The settled catalyst is sent to disposal and the liquid is sent to an oxidizer where caustic soda is added. In the oxidizer the dissolved sulfur oxides in the effluent is neutralized prior to discharge to the refinery wastewater treatment plant.

In some cases, the effluent from the PTU could impart a significant load of both COD and suspended solids on the refinery wastewater treatment plant.

The typical contaminants in the water from the Purge Treatment unit is shown in the Table 4.15.

Table 4.15 FCC purge treatment unit WW composition

Contaminant	Units	Value
COD	mg/L	Up to 1,500
Suspended solids	mg/L	Up to 5,000
Total dissolved solids	wt%	Up to 10

4.2.4.7 Delayed Coker

This is a type of coking process that is used in many refineries. In this process heavy residue that is produced in upstream units in the refinery is subjected to very high temperatures (up to 500 °C) to thermally crack the hydrocarbons into lighter components and petroleum coke. Steam is added upstream of the furnace to prevent the formation of coke in the furnace which is upstream of the coke drum and is therefore referred to as delayed coking. Figure 4.17 is a diagram of a typical Delayed Coking Unit.

In the delayed coking unit the heavy residue feed is added to the bottom of the Main Fractionator. From there it is pumped to the Coker Furnace. Since the coking process is a semi-batch process the unit is typically equipped with multiple coke drums. From the furnace the residue is sent to one of the drums where over a period of up to 24 h the coke drum fills up with petroleum coke that is formed in the coking process. At this time the feed is switched to the second drum and the first drum switches to the quench cycle. During the quench cycle the vapors from the drum are sent to the Quench tower, where it is cooled, for recovery of hydrocarbons and sour water. After the quench cycle, the top and bottom heads of the coke drum are opened and the solid coke is drilled and cut out of the coke drum using large quantities of high-pressure water. The cutting water together with the cut coke drops into movable rail cars located at the bottom of the drums. The water that drains out of the rail cars is collected and sent to a clarifier for separation and recovery of coke fines. The water from the clarifier is recovered and reused as cutting water. Typically, no cutting water is sent to wastewater treatment.

The hot vapors from the coke drum that are sent to the Main Fractionator where they are partially quenched by the incoming feed which minimizes coke formation at the bottom of the Fractionator. Steam is added to the bottom of the Fractionator. A side-stream stripper is used to recover an intermediate cut product. Steam is also added to the side stripper to aid the stripping process and also control the end boiling point of the product from the side stream stripper. The overhead vapors from Fractionator are sent to a condenser to recover hydrocarbons and sour water. Uncondensed vapors are sent to the Gas Recovery Unit in the refinery for recovery of light gases such as propane and butane.

The quantity of sour water produced from a Delayed Coker unit can be highly variable and is dependent on various factors including the type of feed to the unit as well as the sulfur content of the feed. The sour water flow can be up to 200 m^3 per 1,000 ton of feed to the unit. The quantity of sour water generated from the system

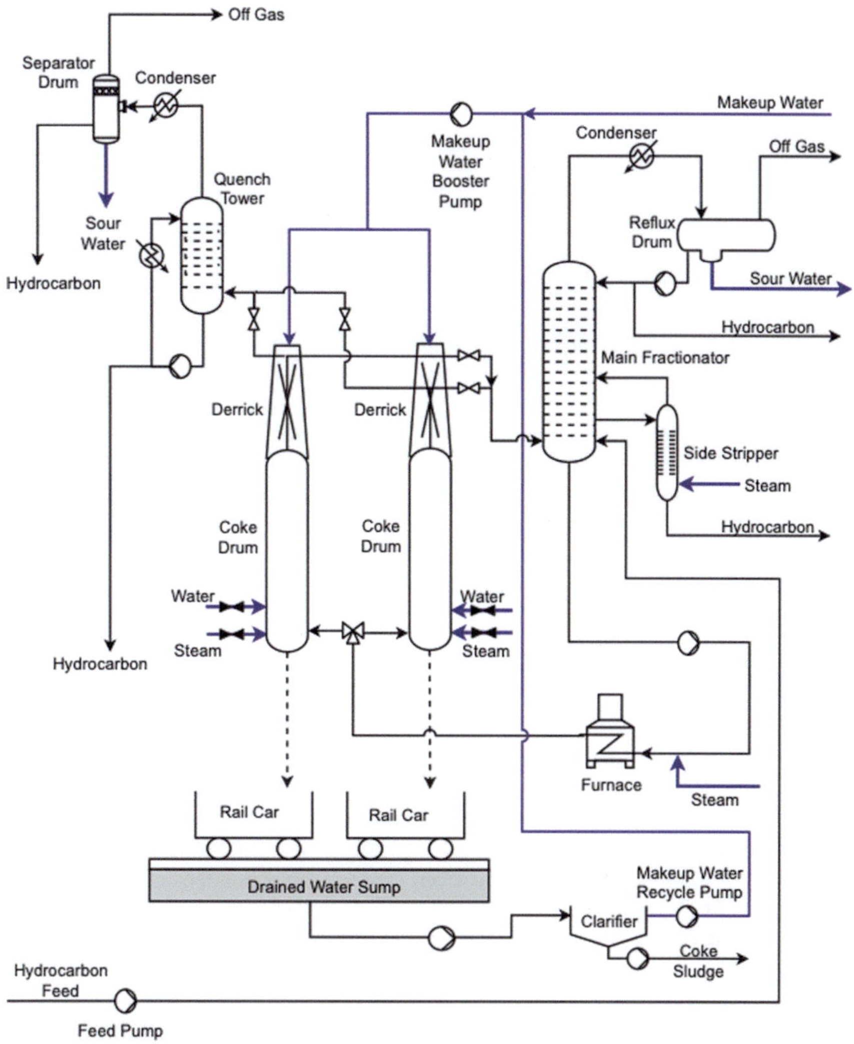

Fig. 4.17 Delayed coking unit

can be calculated using the flow meters on the steam that is added upstream of the furnace and also the fractionator as well as the sidestream stripper.

The typical contaminants in the sour water from a Delayed Coking unit is shown in the Table 4.16.

Table 4.16 Delayed coking unit sour water composition

Contaminant	Units	Value
Sulfides	mg/L	Up to 1,000
Ammonia	mg/L	Up to 600
Phenols	mg/L	Up to 400
Cyanide	mg/L	Up to 100

4.2.4.8 Hydrodesulfurization (HDS) Unit

Removing sulfur compounds from the refined products in a refinery is very critical. Hydrotreating, also referred to as hydrodesulfurization (HDS) is a catalytic process in which sulfur and nitrogen contaminants are removed from refined products. For example, this is the process used to treat diesel in order to make ultra-low-sulfur diesel. It is also used to remove sulfur from naphtha since sulfur is poisonous to the noble metal catalysts that are used in the downstream reforming process. Many refineries have multiple HDS units to treat different products. A block flow diagram of a typical HDS unit follows (Fig. 4.18).

The hydrocarbon feed is sent to the fixed bed reactor via the feed furnace where if heated to temperatures up to 400 °C. The products from the reactor are cooled and depressurized to an intermediate pressure and sent to the high pressure separator from which the gas is sent to the amine scrubber. Wash water is used upstream of the high pressure separator. The scrubbed gas is compressed and sent back as recycled hydrogen to the reactor along with makeup hydrogen.

The liquid from the HP separator is cooled further and sent to the Low Pressure Separator where more hydrocarbons are separated from the sour water. The hydrocarbons are sent to a stripper for further separation. Steam is added to the stripper as the heating medium. The overhead from the stripper is condensed and produces sour water and non-condensable gas. Condensed hydrocarbons are refluxed back to the stripper. The stripped product is sent to a dryer which produces some more sour water and a slop hydrocarbon stream. The bottoms stream from the dryer is the desulfurizer product.

The sour water produced from a hydrodesulfurization unit is dependent on the sulfur content of the feed to the unit and can be up to 60 m^3 per 1,000 tons of feed to the unit. The flow of sour water from the HDS can be calculated using the data from the flow meter of the stripping steam added to the stripper tower. Steam used in the ejectors can be estimated based on the operating pressure of the system as well as the pressure of the steam being used. The wash water rate for the water that is used upstream of the HP separator can be estimated based on the throughput of the system. Most refineries route the sour water from the HDS to the sour water stripper.

The typical contaminants in the sour water from the HDS is shown in Table 4.17.

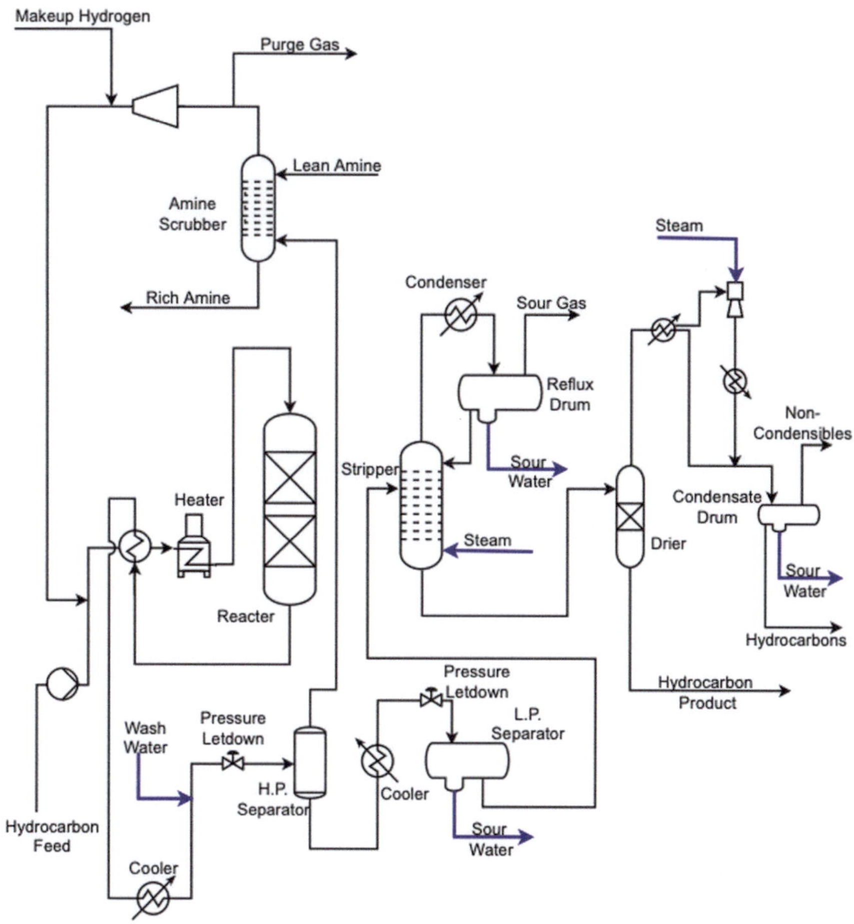

Fig. 4.18 Hydrodesulfurization unit

Table 4.17 Hydrodesulfurization unit sour water composition

Contaminant	Units	Value
Sulfides	mg/L	Up to 5,000
Ammonia	mg/L	Up to 2,400
Phenols	mg/L	Up to 35

4.2.4.9 Hydrocracker

Hydrocracking is a process in which heavier fractions of crude oil that contain the large molecules are cracked into lighter compounds using a catalytic reactor in the presence of hydrogen. The difference between catalytic cracking and hydrocracking is that in the catalytic cracking process cracking is carried out by breaking down heavier molecules and rejecting excess carbon molecules as coke fines whereas, in hydrocracking it is carried out by hydrogen addition. Moreover, catalytic cracking is an endothermic process whereas hydrocracking is an exothermic process. There are multiple configurations of the hydrocracking process such as single stage and two stage process. Figure 4.19 shows the diagram of a two-stage reactor system.

Distillate streams from the Crude Distillation Unit, Vacuum distillation Unit, Delayed Coking units and Catalytic Cracking Units could be used as feed to the Hydrocracking unit. The feed is introduced into the first stage reactor via preheaters and a furnace where it is heated to temperatures up tp 460 °C. The reactor is operated at pressures up to 200 bar. The reactor has multiple catalyst beds and hydrogen is introduced between the beds. The effluent leaves the bottom of the reactor and is cooled and sent to the first high pressure separator. Wash water is added upstream of the separator to wash out hydrogen sulfide and ammonia. Sour water is removed from this separator. The vapors from the separators are sent to the amine scrubber

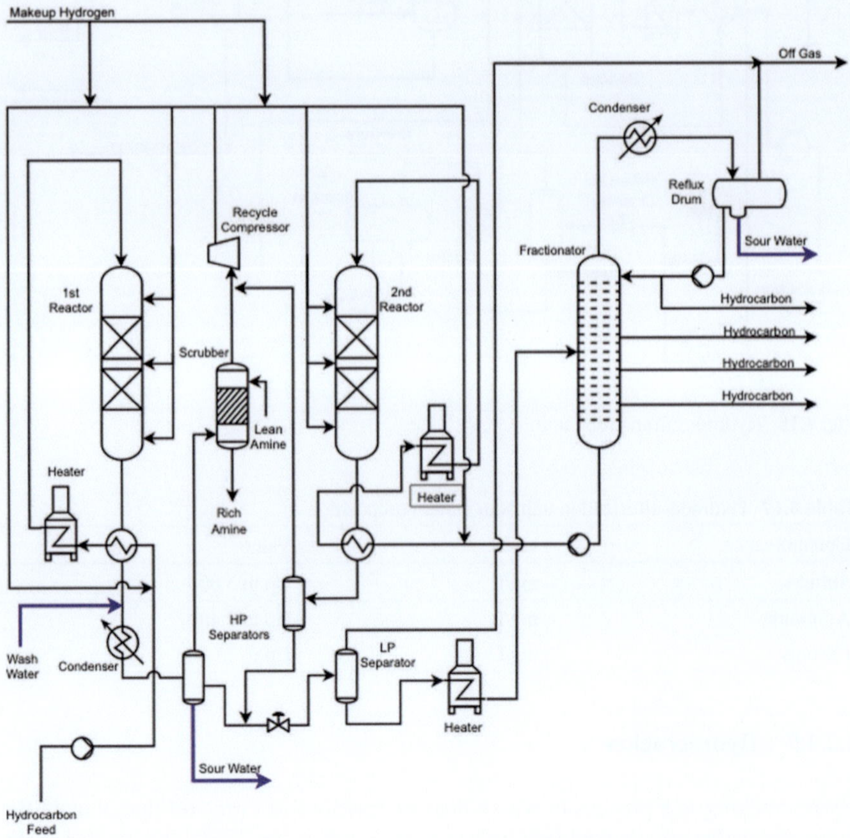

Fig. 4.19 Hydrocracker unit

Table 4.18 Hydrocracker unit sour water composition

Contaminant	Units	Value
Sulfides	mg/L	Up to 40,000
Ammonia	mg/L	Up to 17,000

where it is contacted with lean amine into which sulfur compounds are absorbed. The vapors from the absorber are compressed and mixed with makeup hydrogen and distributed to both the first stage and second stage reactors.

The hydrocarbon from the HP separator is letdown further and sent to the low pressure separator where the separated gas is removed as offgas. The liquid from this separator is heated in a furnace and sent to the Fractionator from where multiple streams are removed as products. The overhead from the tower is condensed and hydrocarbons are separated from the sour water. Condensed hydrocarbons are removed as a product and also refluxed back to the tower. Uncondensed gases are removed as off gas.

The bottoms from the fractionator are sent to a heater and then to the second stage reactor where hydrogen is added between the catalyst beds. The effluent from the second reactor is cooled and sent to a second high pressure separator. The vapors from this separator are sent to the recycle compressor and the liquid is mixed with the bottoms from the first HP separator.

The Hydrocracker unit produces a significant amount of sour water and many refineries are equipped with a dedicated sour water stripper since the stripped water is reused as wash water in the Hydrocracker. This wash water needs to be protected from contamination with chlorides which could poison the reactor catalyst. Quantity of sour water produced in hydrocracking process is dictated by sulfur content of the feed which can reach up to 200 m^3 per 1,000 of feed to the unit.

Table 4.18 shows the typical concentrations of contaminants in the sour water from the unit.

4.2.4.10 Thermal Cracking

Thermal cracking is a process used in the refinery where heavy residue is subjected to high temperatures to break down long-chain, high boiling hydrocarbons into short-chain lower boiling components. This process is also referred to as vis breaking, Thermal cracking is not very common since many refineries are equipped with fluid catalytic crackers which offers more operational flexibility. A flow diagram of a typical Thermal Cracking unit is shown in Fig. 4.20.

In this unit, residue from the Vacuum Distillation Unit is routed through a furnace, where it is subjected to a temperature of up to 900 °C, to the reactor. Steam is added to the discharge of the furnace in order to minimize the formation of petroleum coke. The effluent from the reactor is sent to the fractionation tower. Any tar that is formed in the reactor leaves the bottom of the reactor. Steam is also added to the fractionator to aid in the stripping. Additional heat is added to the fractionator

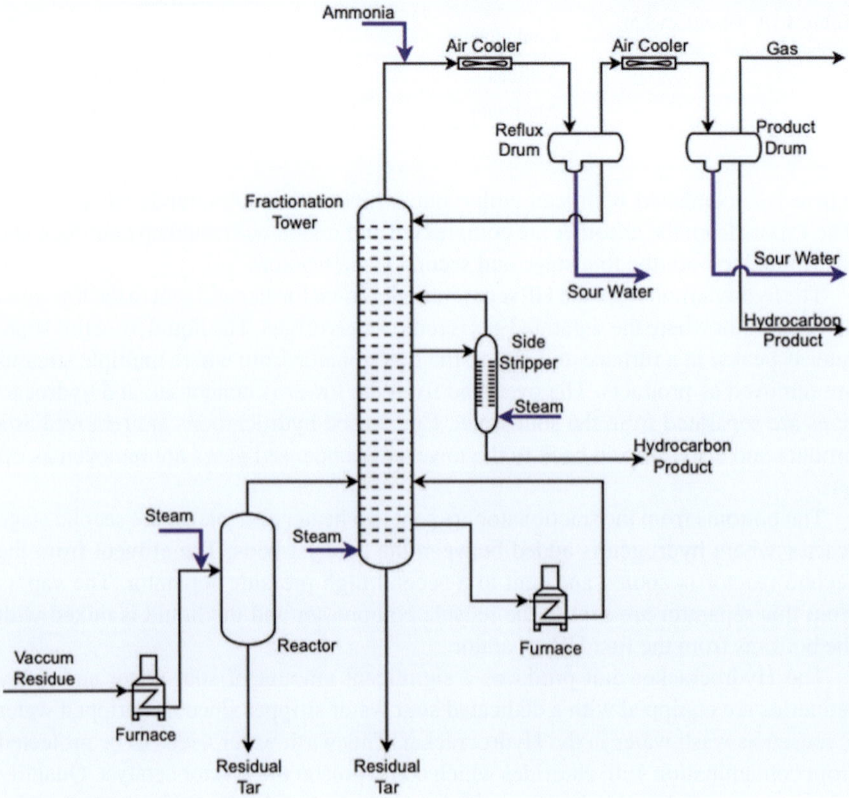

Fig. 4.20 Thermal cracking unit

using a second furnace. A side-stream stripper is used to remove an intermediate product from the tower. The overhead stream from the tower is sent to an air cooler where hydrocarbons and water are condensed. Ammonia is added upstream of the cooler for corrosion control. The condensed hydrocarbons are separated from the sour water and refluxed back to the fractionator. The sour water is removed from the system. The uncondensed hydrocarbons are routed to a second air cooler and then to a condensate drum. Any condensed sour water is removed from the unit and hydrocarbons are removed as product. The uncondensed gas is sent to the refinery gas processing system.

The quantity of sour water generated in a Thermal Cracking unit can be up to 25 m^3 per 1,000 tons of feed to the unit. The quantity of sour water generated from the system can be calculated using the flow meters on the steam that is added upstream of the furnace and also the fractionator as well as the side-stream stripper.

The typical contaminants in the sour water from a Thermal Cracking unit is shown in the Table 4.19.

Table 4.19 Thermal cracking unit sour water composition

Contaminant	Units	Value
Sulfides	mg/L	Up to 800
Ammonia	mg/L	Up to 800
Phenols	mg/L	Up to 400
Cyanide	mg/L	Up to 20

4.2.4.11 Sulfur Plant

The sulfur plant is an important process unit in a refinery where all the hydrogen sulfide that has been separated from the various distillates and products gets converted to elemental sulfur. The sulfur product can be shipped out as a powdered product or can be pelletized. In some cases, sulfur is also shipped out as molten sulfur. Figure 4.21 is a diagram of a typical sulfur unit in a refinery.

The sulfur plant typically has two parts—the Sulfur Unit and the Tail Gas Treatment Unit (TGTU). In the sulfur unit the sour gases from the refinery are sent to the reaction furnace where it is subjected to high temperatures of up to 1500 °C. Some of the hydrogen sulfide is oxidized to sulfur dioxide which in turn reacts with hydrogen sulfide to produce sulfur in accordance with the following reactions.

$$2H_2S + 3O_2 \rightarrow 2H_2O + 2SO_2 \tag{4.3}$$

$$4H_2S + 2SO_2 \rightarrow 4H_2O + 6S \downarrow \tag{4.4}$$

$$\text{Overall, } 2H_2S + O_2 \rightarrow 2H_2O + 2S \downarrow \tag{4.5}$$

The reaction products are sent to the condenser where heat is recovered in the form of steam. Condensed sulfur is drained into a pit from where it is further processed and converted to the final sulfur product. Residual gases from the condenser are sent to a reheater and then to a catalytic converter for further conversion of hydrogen sulfide to sulfur. The converted product is sent to a condenser for additional recovery of sulfur. Residual (tail) gas which contains sulfur dioxide is sent to the Tail Gas Treatment Unit. Some refineries are equipped with two or more catalytic converters/condensers to increase the overall recovery of sulfur.

The tail gas at the Tail Gas Treatment Unit flows through a burner and then through a hydrogenation reactor for conversion of sulfur dioxide to hydrogen sulfide. Other compounds such as carbon sulfide and carbon disulfide are also converted to hydrogen sulfide. The reactor gases are then cooled and sent to a quench tower. Condensed water is removed as sour water and sent to the sour water stripper. The quenched gases are sent to the amine absorber where the hydrogen sulfide is absorbed into the circulating amine solution. Off-gases from the absorber are sent to an incinerator. The rich amine is routed to a Regenerator where the hydrogen sulfide is

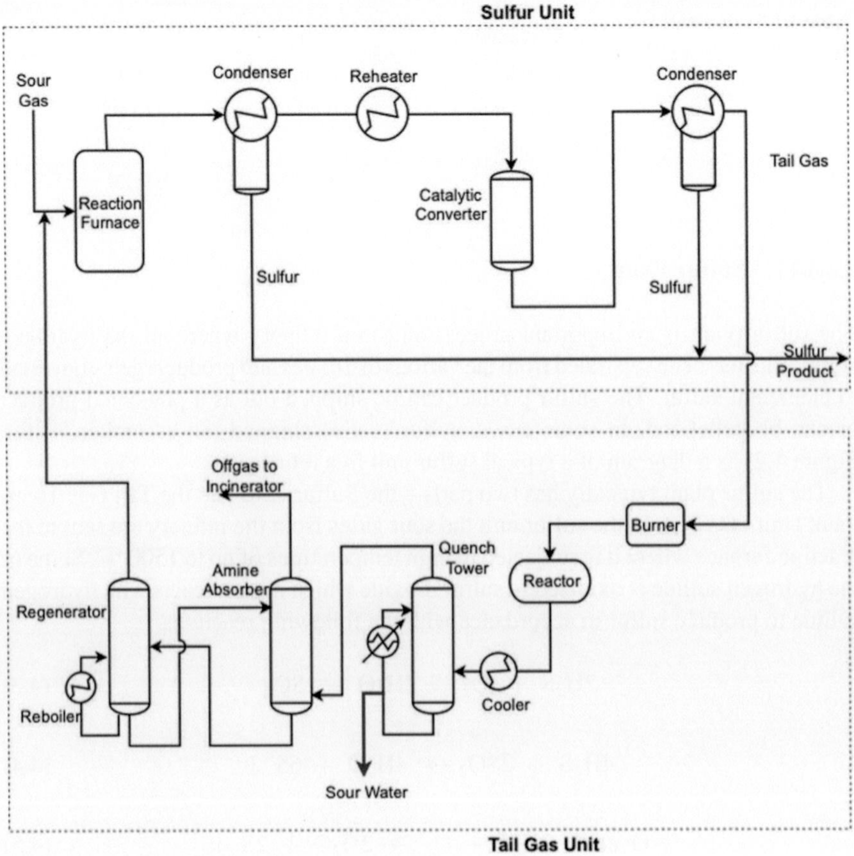

Fig. 4.21 Sulfur plant

separated and recycled back to the sulfur unit for conversion. The lean amine from the regenerator is sent back to the absorber.

Water is produced in the TGTU by the following reaction.

$$SO_2 + 3H_2 \rightarrow H_2S + 2H_2O \tag{4.6}$$

The hydrogen required for the above reaction is usually supplied by the excess hydrogen in the Sulfur unit offgas. In some cases supplemental hydrogen is added to the unit. The quantity of sour water produced in the TGTU can be variable and depends on various factors such as the type of crude that is being processed in the refinery, the sulfur content of the crude and sulfur plant feed gas throughput. The sour water produced can be up to 450 m^3 per 1,000 tons of feed to the sulfur plant.

Table 4.20 shows the typical composition of sour water from TGTU.

Table 4.20 Sulfur unit sour water composition

Contaminant	Units	Value
Sulfides	mg/L	Up to 1,000
Ammonia	mg/L	Up to 500

4.2.4.12 Alkylation

Alkylation is a process that is used in the refinery to covert gases such as iso-butane and olefins (propylene and butenes) to alkylate which is added to the gasoline pool in the refinery as an octane enhancer. There are two types of catalysts used in alkylation processes—sulfuric acid alkylation and hydrofluoric acid (HF) alkylation. Sulfuric acid alkylation has minimal impact on the refinery wastewater treatment plant. HF alkylation on the other hand produces a waste stream that could have an impact on the treatment plant. Figure 4.22 is a flow diagram of the HF Alkylation process.

The feed which consists of iso-butane and olefines that are produced in the refinery gas processing unit are sent to the reactor where hydrofluoric acid (catalyst) is added. The reactor products are sent to an acid settler where the HF is separated and sent to the HF regenerator. The HF is removed from the regenerator and recycled. The acid soluble oils are sent to neutralization where they are mixed with caustic soda and discharged to the refinery wastewater treatment plant (WWTP).

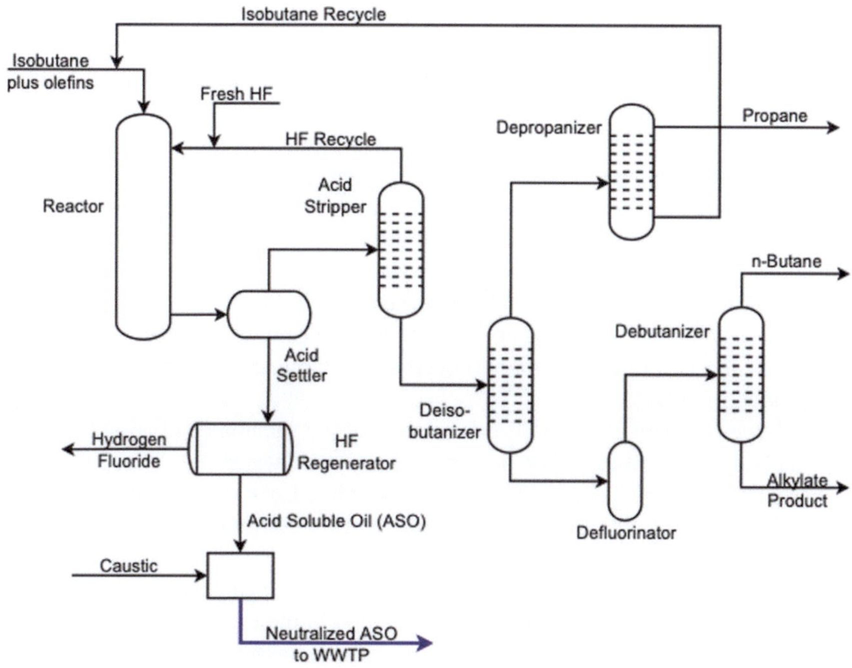

Fig. 4.22 HF alkylation unit

Hydrocarbons from the acid settler are sent to the distillation section of the unit where propane and n-butane are separated as by-products from the alkylate. The isobutane that is recovered in the depropanizer is returned to the reactor as recycle.

The neutralized effluent that is sent to the WWTP contains sodium fluoride which will pass through the treatment plant to the discharge. Many refineries have tight limits on the quantity of fluoride in the discharge and could create non-compliance issues. An alternative would be to neutralize the acid soluble oil (ASO) with calcium hydroxide. This results in calcium fluoride salt which is significantly less soluble in water than sodium fluoride and therefore can be precipitated out prior to discharge to the WWTP. Another alternative would be dispose the ASO offsite if this option is cost effective and feasible.

4.2.4.13 Lube Oil Plants

Some refineries include lube oil plants as part of the process units in the plant. These plants are used to produce lube oil products that is used for automobiles and trucks. A flow diagram of a typical lube oil plant follows although there could be other configurations and units (Fig. 4.23).

The feed to a lube plant is the residue from the crude distillation unit. This is fed to a vacuum tower that is similar to the vacuum tower in the vacuum distillation unit that has been described in a previous section. Steam is used in the ejectors that produce the required vacuum in the tower. Stripping steam is also used in the vacuum tower. All the steam is condensed and is separated from the hydrocarbons in the overhead drum and sent to the sour water stripper. The vacuum residue is sent to the Propane Deasphalting unit where light hydrocarbons are extracted using propane as the solvent. The extract is sent to the Furfural Extraction unit. The tar from the Propane Deasphalting unit is removed as a bitumen product.

The side stream products from the vacuum tower are also sent to the Furfural Extraction unit. Furfural is the solvent that is used to selectively remove aromatics and undesirable compounds containing oxygen, nitrogen, sulfur and metals. The products from the unit are sent to MEK Dewaxing unit. MEK (methyl ethyl ketone) is used to separate the wax using crystallization. The dewaxed product is sent to the Hydrofinishing unit in which non-hydrocarbon contaminants are removed from the final product.

In a Lube Oil plant all the units downstream of the vacuum distillation tower are essentially dry operations and do not produce any wastewater.

The sour water produced in the vacuum tower in a Lube oil unit can be up to 70 m^3 per 1,000 ton of feed to the unit. For the readers convenience the table shown before is repeated below (Table 4.21).

The flow of sour water from the vacuum tower can be calculated using the data from the flow meters of the stripping steam added to the tower. The steam used in the ejectors can be estimated based on the operating pressure on the vacuum tower as well as the pressure of the steam being used.

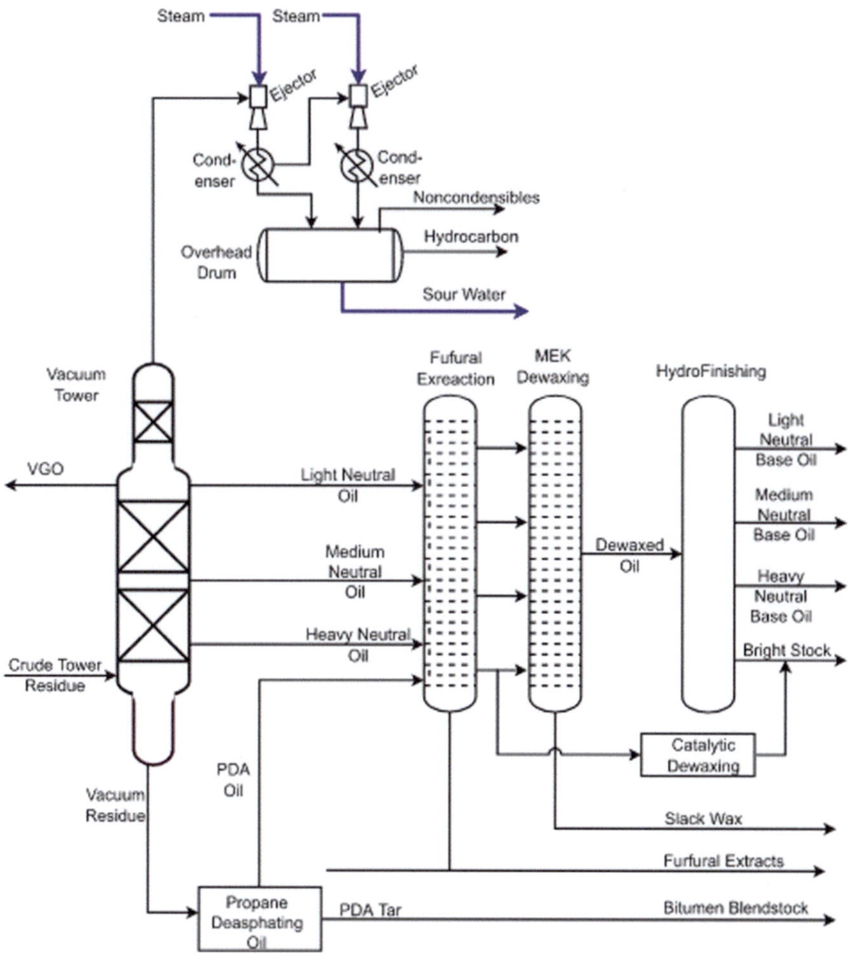

Fig. 4.23 Lube oil plant

Table 4.21 Lube oil plant sour water composition

Contaminant	Units	Value
Sulfides	mg/L	Up to 250
Ammonia	mg/L	Up to 150
Phenols	mg/L	Up to 100

4.2.4.14 Other Process Units

Other process units in the refinery such as the Gas Processing Units where light gases such as Ethane, Propane and Butane are separated are considered dry units where no water or steam is used in the process. The steam that is used as the heating medium in the distillation tower reboilers do not come into contact with the hydrocarbons. No wastewater is generated in these units other than potentially contaminated rainwater from the unit.

Due to changes in operating conditions (pressure and temperature) small quantities of water (referred to as drips and drains) come out from the tower overhead drums and are occasionally drained to the sewer.

Some of the other units such as Catalytic Reforming are also dry processes and do not generate wastewater other than contaminated rainwater.

4.2.4.15 Sour Water Strippers

In refineries, water is used for a number of purposes including scrubbing, washing, quenching, ejector steam for vacuum generation and steam-stripping. During these operations the water comes into close contact with the hydrocarbon stream so that contaminants such as hydrogen sulfide, ammonia, hydrocarbons and suspended solids get extracted and are mixed with water. The contaminated wash water and condensate from these unit operations which are referred to as sour water can be toxic and emit unpleasant odors and therefore it is unsafe to discharge to the refinery sewers. Typical contaminants in sour water include hydrogen sulfide, ammonia, phenols and hydrogen cyanide. Sour water is treated in a stripper prior to reuse as desalter wash water and/or discharge to wastewater treatment. The objective of the sour water stripper is to remove volatile organic compounds (VOC), hydrogen sulfide and ammonia from the sour water using steam as the stripping medium. For this discussion hydrogen sulfide is defined as the sum of dissolved H_2S, bisulfide (HS^-) and sulfide (S^{2-}) ions; ammonia as the sum of dissolved ammonia (NH_3) and ammonium (NH_4^+) ions; carbon dioxide as dissolved CO_2, bicarbonate (HCO_3^-) and carbonate (CO_3^{2-}) ions; and cyanide as dissolved hydrogen cyanide (HCN) and cyanide (CN^-) ions.

Although sour water strippers are located within the battery limits in a refinery and are considered a process unit it has a significant impact on the refinery wastewater treatment plant because of the relatively large hydraulic load it imparts on the treatment. There are two configuration options for sour water strippers—single stage and dual stage strippers and are shown in Fig. 4.24.

In the first option the sour water is fed to a single stripper in which the hydrogen sulfide and ammonia contained in the feed are stripped and sent to the sulfur plant.

In the two-stage option the feed is passed through a feed/bottoms heat exchanger against the bottoms from the second stripper and then fed to the hydrogen sulfide stripper where the H_2S is stripped out and sent to the sulfur plant. Mineral acid is added to this stripper to enhance the stripping of H_2S. The bottoms from this stripper is sent to the Ammonia stripper where ammonia is taken out as the overhead stream. In this stripper the pH is increased by the addition of caustic soda to enhance the recovery of NH_3. This refers to the discussion in Chap. 1 which shows that, at low pH (at $pH < 7.1$) all forms of H_2S (H_2S, HS^- and S^{2-}) exist together in the form of H_2S dissolved in water, and at high pH ($pH > 9.25$) ammonia and ammonium ions

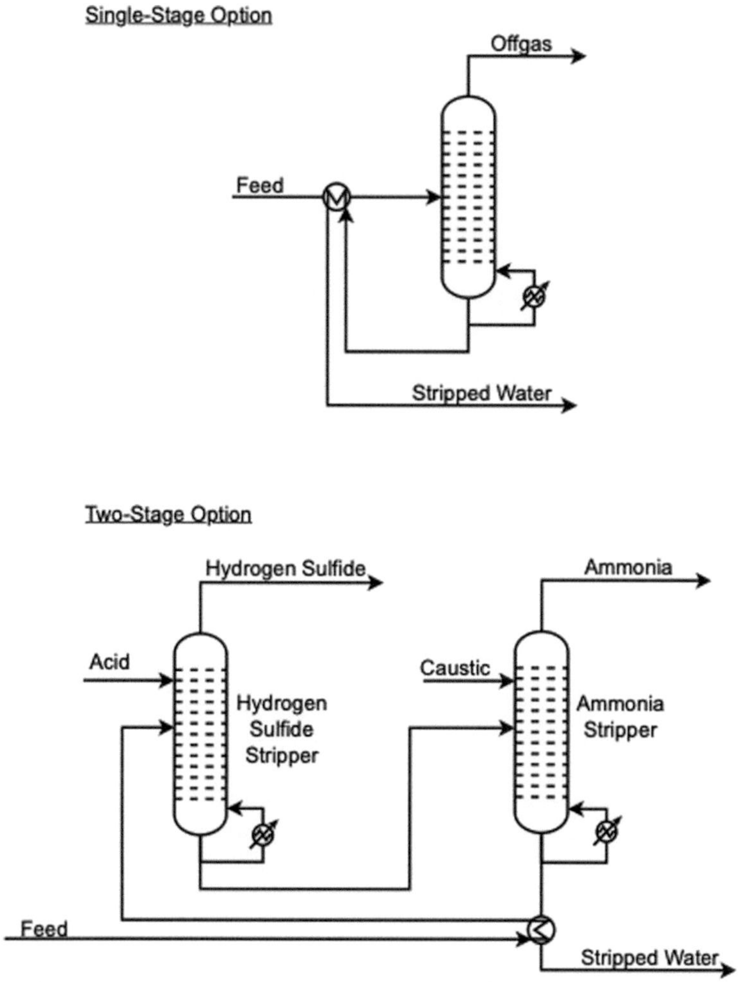

Fig. 4.24 Sour water stripper options

together exist in aqueous solution as NH_3. The bottoms from this stripper can be recycled within the refinery as required or sent to the WWTP. One advantage of this option is that the no ammonia is sent to the sulfur plant. It has been reported that H_2S concentrations of less than 1 ppm in the bottoms can be achieved in the H_2S stripper and the NH_3 concentration of 30–50 ppm can be achieved in the ammonia stripper. The two-stage sour water stripper was first developed by Chevron and is now licensed by Bechtel Corp.

The single stripper configuration is more common in many refineries and warrants a more detailed discussion.

Stripper Design Considerations

For the design of sour water strippers there are a few design options that need to be considered:

- Reboiler versus live steam as heating medium
- Packing versus trays for column internals
- External versus internal condensers

Reboiler Versus Live Steam as Heating Medium

A significant design consideration that needs to be taken into account is the use of reboilers as opposed to using live steam as the heating medium in the stripper.

- Sour water stripper is subject to fouling. Addition of live steam to provide the heat required for vaporization in the stripper minimizes/eliminates the fouling issues in the stripper.
- Use of a reboiler results in additional capital costs. The reboiler is also subject to fouling and will require to be cleaned out periodically. To accommodate this event, either a spare reboiler will need to be installed on the stripper or the feed tank will need extra capacity to deal with the shutdown. Both these scenarios also result in increased capital costs.
- The amount of steam live required can be significant (more than 1 kg of steam per kg of feed). All this steam will condense in the column and more than double the flow of stripped water. This imposes a significant additional hydraulic load on the wastewater treatment plant.
- Many refineries operate strippers equipped with reboilers as opposed to using live steam to provide the required heat input to the tower. Using live steam as the heating medium in the stripper more than doubles the water traffic in the column because all that steam is condensed and gets sent to the wastewater treatment plant.
- Since all the steam fed to the stripper is lost this will require constant additional makeup of boiler feed water to the refinery boiler system resulting in increased operating costs. On the other hand all the steam used in the stripper reboiler is recovered so no additional hydraulic load is imposed on the wastewater treatment plant.

Most of the refineries that are relatively new or refineries that have revamped their sour water stripper have installed reboilers as opposed to using live steam as the heating medium.

Packing Versus Trays for Column Internals

Another design consideration that needs to be taken into account is the use of packing instead of trays in the stripper.

- Packing provides more theoretical stage per unit of column length than trays. This could result in a smaller overall column height. However, vapor and liquid

maldistribution could be a problem with packed columns. Extra packing height is usually provided to compensate for maldistribution.
- Packed columns provide lower pressure drop per theoretical stage resulting in a lower column operating pressure. Lower pressure favors more effective stripping.
- Packed columns do not require minimum vapor flow for proper operation. However, they require minimum liquid flow to ensure proper wetting of the packing in the column.
- In general entrainment is lower in packed columns than in trayed towers but they are more susceptible to fouling than trays.
- In packed columns, visual inspection of internals is not possible whereas trayed column internals can be inspected visually.
- Trays are usually less expensive than packing but they provide less theoretical stage per unit column height than packing.
- Maldistribution of vapor and liquid in trayed columns is not an issue but they require minimum vapor flow in the tower to prevent liquid dumping.
- Trayed columns are subject to only minimal fouling.

In practice most refineries use trays in their sour water strippers as opposed to packing.

External Condensers Versus Internal Condensers

A third design consideration is the use of internal condensers for the sour water stripper instead of using a conventional external condenser.

- The overhead vapor from a sour water stripper is concentrated with H_2S and NH_3 which can be very corrosive and results in the rapid deterioration of the tubes in the condenser. With an internal pump-around condenser, there is no heat exchanger that is exposed to these concentrated vapors.
- Use of internal condenser results in increased capital costs due to the additional column height required in the stripper to accommodate the extra trays for the condenser section plus the pump required for the pump-around condenser.
- The tubes in the external condenser can be subject to severe fouling and therefore the stripper will need to be periodically shutdown for replacement of the tubes in the condenser. This will require a larger stripper feed surge tank to store the sour water during the shutdown or a spare sour water stripper. Both these scenarios result in increased capital costs.

Refineries with modern sour water strippers tend to have internal pump-around condensers.

Figure 4.25 represents the typical components of a sour water stripper system (SWS).

Sour water is first sent to a Flash Drum where it is depressurized, and the gases are sent to the flare. Any hydrocarbons in the drum are skimmed off as necessary. Water from the drum is sent to the Feed tank prior to being sent to the stripper. In the tank, a layer of diesel oil is maintained on the liquid surface of the tank in order minimize emissions and odors. Any gases that are vaporized in the tank dissolve into the diesel

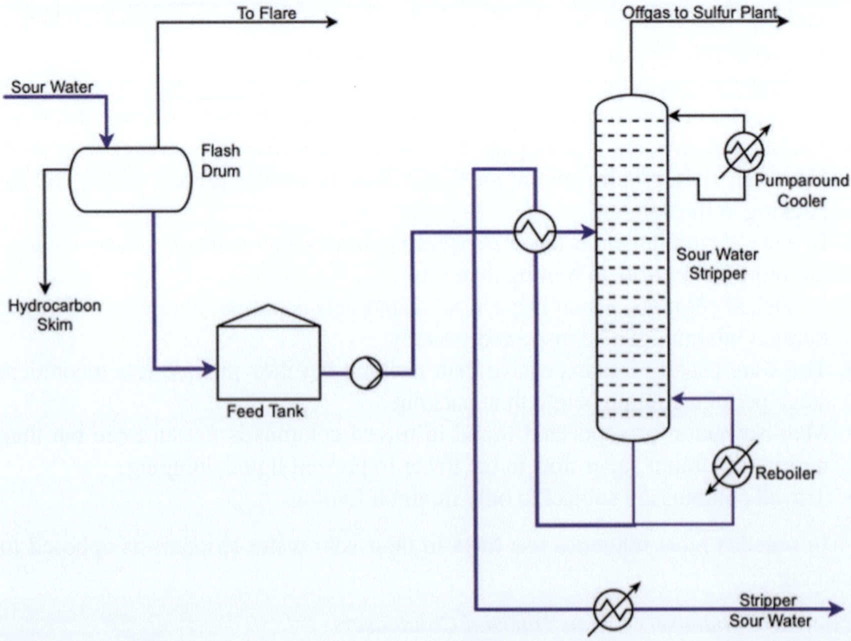

Fig. 4.25 Sour water stripper

layer and remain in the tank. The diesel layer is periodically purged and replaced with fresh diesel. Sour water from the tank is pumped through a Feed/Bottoms heat exchanger in order to preheat the feed prior to being introduced into the stripper. This is done to minimize the usage of steam in the stripper reboiler. The sour water stripper is usually a trayed column with about 40 trays. Packing is not typically used in this service due to significant corrosion and fouling issues. Operating pressures in the column are usual in the range of 0.1–3 bars and the temperature is usually between 38 and 130 °C. Even though this is called a stripper, the feed to the column is introduced several trays below the top of the column and the stripper is operated like a conventional distillation column with a rectification section. In many refineries the top few trays are operated as an internal condenser with a pump around cooler instead of a conventional external condenser due to the high corrosion potential in the external condenser. The overhead stream typically has a very high concentration of hydrogen sulfide which is highly corrosive. These gases are sent to the sulfur plant for recovery of sulfur.

The bottoms from the stripper is cooled using the Feed/Bottoms exchanger and trim cooled with cooling water and then sent to wastewater treatment plant.

Table 4.22 shows the typical composition of the stripped sour water that is sent to the wastewater treatment plant.

Table 4.22 Stripped sour water composition

Contaminant	Units	Value
Sulfides	mg/L	Less than 10
Ammonia	mg/L	Less than 50
Phenols	mg/L	Up to 100
Cyanides	mg/L	Up to 20
Temperature	°C	Less than 35

It should be recognized that to have an ammonia concentration of about 50 mg/L in the stripper bottoms is potentially beneficial to the wastewater treatment plant because this can satisfy the nutrient nitrogen requirement in the treatment plant.

Phenols are a significant contaminant in the sour water produced in certain units in the refinery such as the FCC, Delayed Cokers etc. but cannot be stripped out in sour water strippers. It is also difficult to effectively treat phenols biologically and this results in the risk of occasional exceedances of tight phenol limits that are usually imposed on many refineries. For this reason, it is beneficial to operate multiple sour water strippers with one dedicated to the treatment of phenolic sour water (from the cat cracker and delayed coker). The stripped sour water from this stripper is recycled and used as wash water in the desalter. By doing this, up to 90% of the phenol contained in the stripped water is extracted into the crude oil in the desalter thus significantly minimizing the phenol load to the wastewater treatment plant. The phenols that are extracted into the crude oil get distributed into the distillate streams in the crude distillation unit. Figure 4.26 illustrates an example of the segregated sour water system.

In existing refineries one of the issues that will need to be addressed is that the segregated sour water loads will need to be matched with the available capacities in the existing strippers. If necessary, some of the non-phenolic sour water can also be processed in the phenolic stripper.

Additionally, some refineries with Hydrocracker units choose to operate a dedicated stripper to treat the sour water from this unit. The stripped water from this tower is recycled back to the Hydrocracker as wash water. This is done to prevent introduction of chlorides into the wash water which could potentially poison the catalyst used in the Hydrocracker unit.

Operating Challenges of Sour Water Strippers

Corrosion

Several acid gases are formed at different stages of crude processing which become associated with the respective sour water streams and enter the sour water stripper (SWS). These include HCl, H_2S, HCN and carbonic acid. SWS internals, especially the top trays of the column, the shell of the column, and condenser become vulnerable to corrosion by aqueous acid solutions.

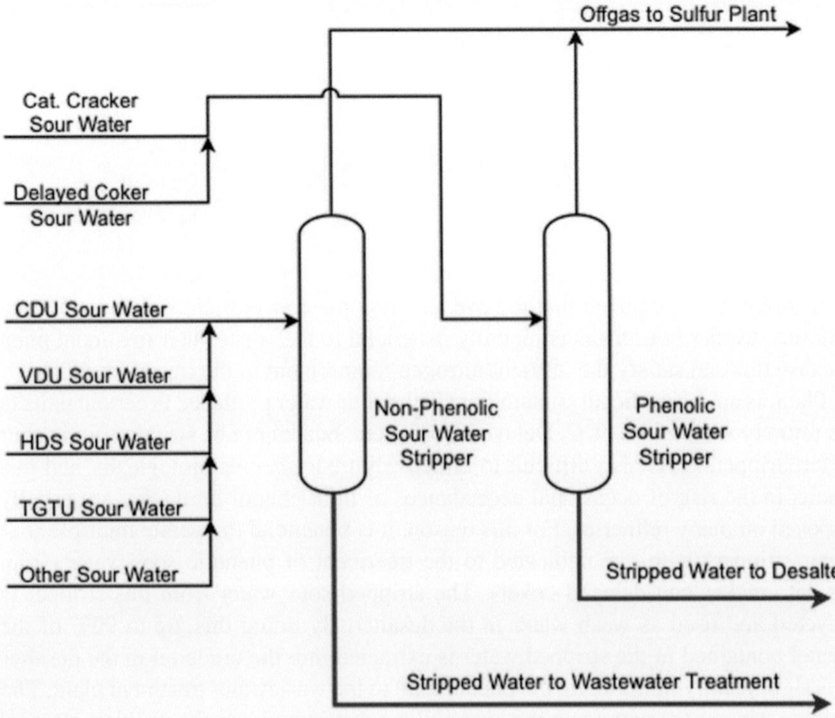

Fig. 4.26 Segregated sour water stripping

At the top of the tower ammonia combines with the acids to form ammonium bisulfide, chloride, cyanide, carbonate and carbamate. These are corrosive salts. This explains the fact that most of the incidents of corrosion occur at the top of the column. This is followed closely by column internals e.g., trays, liquid distributors, feed/effluent exchangers, reboilers, etc. Corrosion mitigation of stripper and reflux condenser internals require careful consideration of materials, coatings and routine maintenance.

Fouling

Strippers are prone to fouling due to deposition of fine solids carried over from crude into the sour water streams. Fouling of column internals can lead to maldistribution of liquid, channeling of liquid and inadequate liquid-vapor contact leading to poor mass transfer. Additionally, overhead section of the strippers is also very susceptible to fouling by deposition of acid salts. Particularly, solid ammonium chloride (NH_4Cl), ammonium bisulfide (NH_4HS) and ammonium carbamate ($NH_4{\cdot}COO{\cdot}NH_2$) salts desublimate and deposit under certain temperature and pressure conditions. Commercially available computer simulation programs can predict desublimation conditions for the salts which helps plants avoid operating close to those conditions. Steam jacketing of overhead gas line from sour water strippers is a good practice, especially at refineries in cold climates, to avoid the problem of desublimation. Typically, SWS column internals—trays/packing, liquid distributors, nozzles, etc. lead the locations of fouling, followed by overhead section and feed/effluent exchangers.

Foaming

Foaming causes operational problems leading to poor vapor-liquid interaction and mass transfer within the stripping column. Foaming is caused by presence of chemicals that alter the surface properties. Presence of amines in sour water can cause emulsions which can lead to foaming in the column. Occurrence of foaming is sudden and cannot be predicted ahead of the time. There is no specific mitigation measure that can be taken to avoid the foaming issue. At the time of occurrence of this operational problem it is best to reduce flow to the SWS until it stops.

In addition to the above, other factors also contribute to the operating problems of sour water strippers. These include lack of proper operational control of key parameters e.g., Steam:Feed Ratio, column operating temperature and pressure profiles, reflux ratio, etc. Also, as refiners have been increasingly using high sulfur crudes, coupled with increasing demand for desulfurization for products to comply with tightening air emissions regulations, existing sour water strippers are constantly under pressure to handle high NH_3 and H_2S loads and increasing feed flows. Many of those strippers were originally not designed for such high flows and loads and underperform. Eventually, this affects the operation of wastewater treatment plants in terms of performance and operating cost. Considering that stripped sour water contributes to approximately 50% of the total wastewater treatment operating cost it is essential that SWS units be monitored closely and maintained periodically to avoid unexpected failures and associated costs.

4.2.4.16 Spent Caustic Treatment and Disposal

Although the spent caustic produced in a refinery might not be considered a wastewater, it has a significant impact on the wastewater systems in the refinery because it is routinely discharged to the sewer. Therefore, a good understanding of the issues associated with spent caustic is imperative for effective management of the wastewater system and wastewater treatment in the refinery. The sulfur balance in a typical refinery is shown in the Fig. 4.27.

Recognizing the environmental trends and the refinery sulfur balance, it is clear that over the long run, more emphasis will be placed on increasing the recovery of the sulfur contained in the crude as elemental sulfur. This will potentially reduce H_2S and other reduced sulfur compounds emissions to the environment and the SO_x emissions via the combustion of refined products (gasoline, diesel, furnace oil, heavy fuel oils).

As refineries process heavier crudes that usually have higher sulfur content, there is more environmental pressure to decrease the sulfur content in refined products as well as emissions from refineries. More hydrotreating will be required in the refinery to remove sulfur from gasoline, diesel etc. as well as more amine treating of gaseous

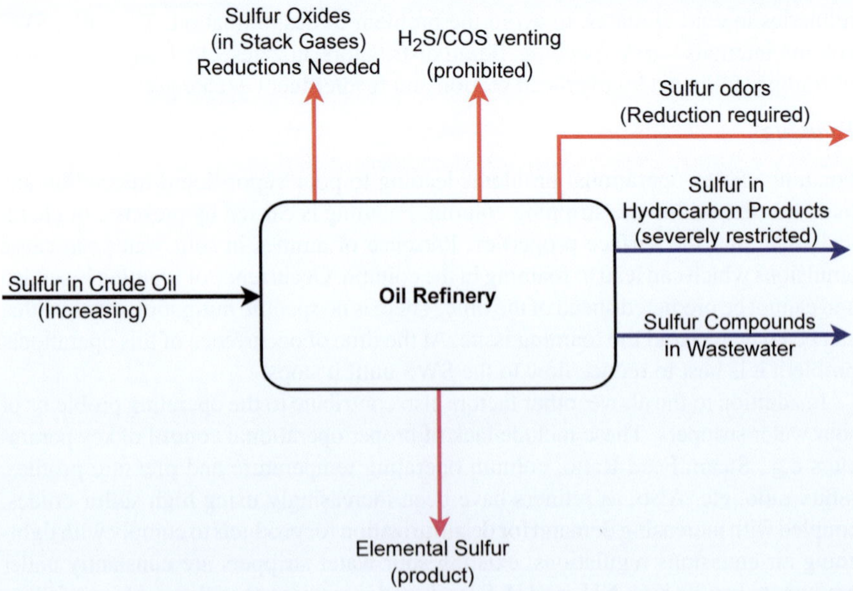

Fig. 4.27 Refinery sulfur balance

streams to remove sulfur compounds will be required. Additionally, caustic treating is used extensively to remove residual sulfur compounds from refined products and this treatment results in the generation of spent caustic.

The intermediate/product streams most frequently treated with caustic in a refinery are shown in the Table 4.23.

The technical considerations associated with handling, treatment and disposal of spent caustic are:

- High odor potential
- Sulfide content (in sulfidic spent caustics)
- Phenols, Cyanide, Soluble organics (in phenolic spent caustics)
- BOD/COD Content necessary for biological treatment
- Disulfide Oil Content

Table 4.23 Stream impurities

Intermediate/product	Usual contaminants/impurities				
	H_2S	RSH	Phenol	HCN	Other
Straight run LPG	√	√			√
Light straight run naphtha		√			√
FCC C3 + C4 LPG	√	√	√	√	√
FCC gasoline		√	√	√	
Coker C3 + C4 LPG	√	√	√	√	√

- Volatile Organic Compound (VOC) Content (Benzene, Toluene, Ethylbenzene and Xylenes)
- High pH (due to large fraction of unreacted NaOH)

In general, there are three types of spent caustic streams that are produced in refineries. Table 4.24 provides details regarding the definitions of the types of spent caustic streams.

At many refineries Merox systems are used to remove residual sulfur compounds from refinery products such as LPG, light naphthas, gasoline, kerosene, and jet fuel. The Merox (an acronym for mercaptan oxidation) system is a proprietary process offered by UOP, in which they are converted to disulfides. A block diagram of a typical Merox system follows (Fig. 4.28).

The hydrocarbon feed is first sent to a Prewash Tank where it is washed with caustic to remove the hydrogen sulfide contained in the stream in accordance with the following equation.

$$NaOH + H_2S \rightarrow NaHS + H_2O \quad (4.7)$$

The wash water is drained off from this tank as spent caustic. The hydrocarbons are then sent to the oxidation reactor, where the mercaptans are converted to disulfides. The reaction is as follows:

$$4RSH + O_2 \rightarrow 2RSSR + 2H_2O \quad (4.8)$$

The hydrocarbons are then subjected to a water wash and the water is drained from this tank. The hydrocarbon stream then passes through a salt bed for removal of any residual water and then though a clay bed for removal of other impurities. This is the description of a Merox at a concept level, and the details could vary depending on the specific product being treated.

The following tables show the typical composition of spent caustic streams that originate from various sources (Tables 4.25, 4.26 and 4.27).

Table 4.24 Spent caustic classifications

Unit of origin	Type of product treated	Type of spent caustic generated
Crude distillation	Saturated LPG	Sulfidic spent caustic
	Straight run naphtha	
	Kerosene	Naphthenic spent caustic
	Jet fuel	
Fluid catalytic cracker	Unsaturated LPG	Phenolic spent caustic
	Light gasoline	
	Full range gasoline	
Delayed coker	Coker LPG	
	Coker naphtha	

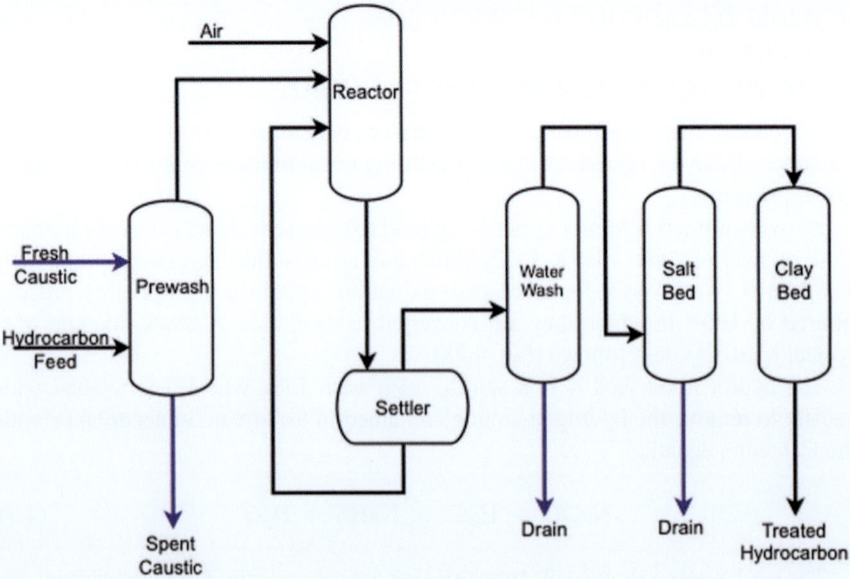

Fig. 4.28 Typical Merox unit

Table 4.25 Sulfidic spent caustic composition

Sulfidic spent caustic		
Contaminant	Units	Value
Sulfides	mg/L	Up to 20,000
Phenols	mg/L	Up to 1,000
COD	mg/L	Up to 35,000
Free hydrocarbon	mg/L	Up to 900
Suspended solids	mg/L	Up to 100

Table 4.26 Phenolic spent caustic composition

Phenolic spent caustic		
Contaminant	Units	Value
Sulfides	mg/L	Up to 20,000
Phenols	mg/L	Up to 7,000
COD	mg/L	Up to 35,000
Free hydrocarbon	mg/L	Up to 1,000
Suspended solids	mg/L	Up to 150

Table 4.27 Naphthenic spent caustic composition

Naphthenic spent caustic		
Contaminant	Units	Value
Sulfides	mg/L	Up to 1,000
Phenols	mg/L	Up to 6,000
COD	mg/L	Up to 70,000
Free hydrocarbon	mg/L	Up to 900
Suspended solids	mg/L	Up to 100

While the flow of spent caustic tends to be small—less than a few m^3/h in most cases, obtaining real flow data for spent caustic streams in the refinery can be quite challenging because flow meters are not typically installed on these streams. It is best estimated based on discussions with plant engineers and operators.

The following options can be considered for the disposal of spent caustic.

In-process Abatement

The following in-process options have the common objective of minimizing quantities of spent caustic that is generated.

- Minimize the production of spent caustic by exercising tighter control of caustic treating operations.
- Monitor hydrocarbon/caustic phase separation in the caustic treater to prevent carryover of caustic with the product to tankage. Carryover will require draining the water/caustic from product tanks.
- Consider pre-washing hydrocarbons with stripped sour water to reduce the quantity of acidic compounds in the feed to the caustic treater that can lead to increased consumption of feed caustic soda solution.
- Consider recycling water/caustic that is drained from product tanks to the caustic treater for reuse.

End-of-Pipe Treatment

The following are examples of options for dealing with spent caustic streams after it has been generated:

- Bleed rather than batch dump spent caustic into an appropriate point in the wastewater treatment system. This will help minimize odor emissions and upsets in the wastewater treatment system.
- Before discharge into the wastewater treatment system, neutralize spent caustic with sulfuric acid to an acidic pH to "spring" and facilitate extraction of hydrogen sulfide, phenols, and cyanide.
- Traditionally, spent caustic has been disposed of in a number of different ways. Dumping to the sewer system is common. Discharging phenolic spent caustic to wastewater treatment plant will cause odor problems at the treatment plant. Some refiners have attempted to use the phenolic spent caustic in the Purge Treatment

Unit of the Catalytic crackers in the plant with limited success because of odor problems. At some locations offsite firms will accept "phenolic spent caustics" for the value of organic compounds contained in these streams.

- There is little interest in offsite disposal of "sulfidic spent caustics" (often the largest portion of refinery spent caustic), because transportation costs to "recyclers" can be high and there is minimal recoverable value in this stream. In most cases sulfidic spent caustic can be sent to the wastewater treatment plant. This is best done by piping the spent caustic separately to the treatment plant and introducing it directly to the biotreater as close to the aerators as practical. This will facilitate the quick oxidation of the sulfides and minimize the potential odors in the wastewater treatment plant. In this context the reader is referred to the section *Treatment of Reduced Sulfur Compounds* in Chap. 5.
- Many refineries process crudes with a high Total Acid Number (TAN). TAN is a measure of naphthenic acid in crude. Refer to Appendix B for a discussion on TAN and its desirable range of values. An acid number greater than 1% is considered a high TAN crude. When kerosene or jet fuel that is produced while processing high TAN crudes is Merox treated, it produces naphthenic spent caustic. Unfortunately, this type of spent caustic is refractory to biological treatment and therefore cannot be treated in the wastewater plant. One option is to dispose of it offsite. At certain refineries there are companies that will take this spent caustic and recover the naphthenic acid which is a valuable product. If the production of this type of caustic at the refinery is large enough it might be cost effective to treat this stream onsite. The treatment can be done with either high pressure wet air oxidation or an alternative acidification/deep neutralization process. The effluent from either treatment process can be safely added to the refinery wastewater treatment plant for tertiary treatment.

4.2.4.17 Refinery Flares

All refineries have flare systems that is used to burn gases in the event of inadvertent vapor releases or over pressurization or fires in the refinery. Some refineries could have multiple flares depending on the size of the refinery, number of process units and physical layout of the refinery plot. Figure 4.29 is a diagram of a typical refinery flare system.

The flare header in a refinery is first routed to the Flare Knockout Drum where any entrained liquids in the flare gas are separated from the gas.

Hydrocarbons are allowed to separate in the drum and are drained out and sent to refinery slops. Any separated water is pumped out to the refinery sour water system. The gases from the knockout drum are sent to the Flare Seal Drum. The purpose of the seal drum is to prevent flame flashback from the flare into the flare header. In the seal drum a constant level of water is maintained to provide a liquid seal against flashback. A constant flow of makeup water is added to the seal drum to prevent accumulation of hydrocarbons. A constant overflow is maintained from this drum and is pumped to the refinery sour water system. The flare gas enters the flare at the

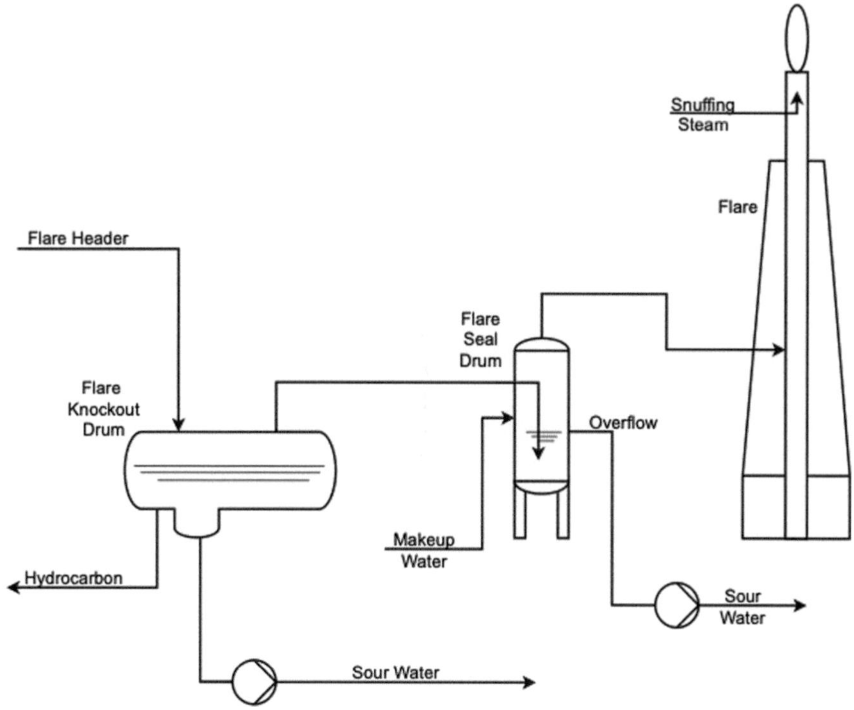

Fig. 4.29 Refinery flare system

Table 4.28 Flare sour water composition

Contaminant	Units	Value
Sulfides	mg/L	Up to 600
Ammonia	mg/L	Up to 300

bottom and is burned at the tip of the flare. A constant flow of low pressure steam is added at the top of the flare as snuffing steam to minimize smoke generation.

Table 4.28 shows the typical composition of sour water from the flare system.

4.2.5 Refinery Utility Systems

Utility systems are part and parcel of an overall refinery complex. They support the smooth and integrated operation of a petroleum refinery starting from crude storage and handling, processing, and finally the storage and delivery of the finished products. Main parts of the utility systems are as follows:

- Raw Water Treatment System
  - Preliminary Treatment
  - Primary Treatment
  - Advanced Water Treatment

    Boiler Feed Water Treatment
    Steam and Power Generation, and Condensate Polishing System
    Cooling Water Treatment

- Refinery Process Sewer System
- Refinery Process Wastewater Treatment System
- Refinery Complex Sanitary Water and Wastewater System
- Stormwater Management System
- Firewater Storage and Management System
- Tank Farm

The following sections provide a discussion of a few of the important utility systems and the issues related to their operation within a refinery complex.

4.2.5.1 Raw Water Treatment

Water that is imported into the refinery from various sources—groundwater, surface water and city water usually contains sediments, suspended and dissolved solids. Surface water, especially from rivers and sea, contains large amounts of debris that are removed by screens located at the intake stations. Sediments and suspended solids are removed by passing the water through a Clarifier. The raw water enters the clarifier where the sediments are settled to the bottom by gravity. The separated solids are removed as sludge from the clarifier. The overflow from the clarifier is sent to the refinery.

Additionally, raw water, especially from groundwater resources contains dissolved constituents e.g., hardness, reduced iron, manganese, and natural organic matter (NOM). It is essential to remove these constituents in preparation for additional treatments to produce cooling tower make up water, boiler feed water and process water for use in the refinery. The unit operations and processes applied to the preliminary, primary and advanced level raw water treatment have been discussed in detail in Chap. 2. Advanced treatment processes for cooling water, boiler feed water and steam cycle are discussed in the following sections.

4.2.5.2 Boiler Feedwater Treatment

Boiler feed water (BFW) is used to generate steam in boilers that are utilized for various purposes in a refinery. These include—process steam that takes part directly in hydrocarbon processing, stripping steam, ejector motive steam and turbine drive steam. BFW is the highest quality water in a petroleum refinery requiring extensive and advanced levels of treatment. Due to losses from the boiler feedwater system, make-up water is continuously added into the boiler water loop to maintain a steady inventory in the system. BFW quality requirements become increasingly stringent with increasing pressure of steam. Typical specifications for BFW make-up water for boilers at >690 kPa (>100 psi) are as follows (GWI 2012):

- Conductivity <1 μS/cm (μS is micro Siemens)
- Hardness <0.3 mg/L
- Chlorides <0.05 mg/L
- Sulfates <0.05 mg/L
- Total Silica <0.01 mg/L
- Sodium <0.05 mg/L
- Dissolved Oxygen (DO) <0.007 mg/L

BFW treatment is a three-step process—pretreatment, external treatment and internal treatment. Pretreatment of raw water is discussed in Sect 4.2.5.1 and described in detail in Chap. 2. Pretreated water is suitable for demineralization, which is the external treatment to remove dissolved solids by ion exchange (IX) and/or reverse osmosis (RO) processes. Finally, the internal treatment step conducts chemical treatment of demineralized water in the boiler feed system to minimize the possibility of scaling or corrosion of boiler tubes.

Ion exchange is the most widely used demineralization process. The principle of ion exchange has been discussed in detail in Chap. 2. Figure 2.7 depicts a simple demineralization system consisting of a cation exchanger and an anion exchanger in series to remove the major cations and anions from pretreated water to make it suitable to feed into a boiler for steam generation. In petroleum refineries steam is generated at high pressures (HP) for production of power in-house. For production of HP steam it is necessary to remove dissolved impurities to a very high degree. In this context it is necessary to discuss the trace impurities that escape the standard cation and ion exchange processes.

SAC (Strong Acid Cation) and SBA (Strong Base Anion) resins have preferential affinities for the cations and anions respectively in the following decreasing order:

For SAC:

$$Al^{3+} > Ca^{2+} > Mg^{2+} > K^{+} > Na^{+} > H^{+}$$

For SBA:

$$SO_4{}^{2-} > NO_3{}^{-} > Cl^{-} > HCO_3{}^{-} > SiO_3{}^{2-} > OH^{-}$$

As the feed water flows down the SAC bed the cations in the feed water will replace the hydrogen ions and attach on to the resin structure. Based on the affinities described above these ions will form layers of Al^{3+}, Ca^{2+}, Mg^{2+}, K^+ and Na^+ from top to bottom of the cation exchanger vessel in the order of their ionic affinities with the resin. Similarly, the anions will also form layers of $SO_4{}^{2-}$, $NO_3{}^{-}$, Cl^{-}, $HCO_3{}^{-}$ and $SiO_3{}^{2-}$ in that order from top to the bottom of the anion exchange vessel. Therefore, when the ion exchangers become exhausted the first ions to breakthrough are Na^+ and SiO_3^{2-} from cation and anion exchanger beds respectively. These two beds may not simultaneously exhaust, and their individual times of exhaustion are indicated by a sharp rise in sodium ions concentration for cation exchanger and silicate ions for anion exchanger beds, respectively.

Due to the position of the sodium and silicate ion layers in cation and anion exchangers, some continuous leakage of these two ions in the treated water is unavoidable. This affects the boiler water chemistry, scaling and corrosion of boiler tubes. Passage of silica into the boiler causes adverse effects since silica vaporizes at high temperatures and entrains into the steam as a contaminant. As steam enters the power generation turbine silica vapor de-sublimates and deposits onto turbine blades, thus affecting the efficiency of power production. If the deposition is uneven, it can create imbalance in the turbine rotor resulting in vibration that may prevent the turbine from operating at full load.

Removal of trace cations and anions, including silica, requires addition of a mixed bed exchanger downstream the anion exchanger to prevent those from escaping into the steam generation boiler. As the name suggests, the mixed bed exchanger consists of a bed of both cation and anion exchange resins mixed randomly in the same vessel. Mixed bed exchanger produces very high-quality water suitable for make-up in a recirculating steam drum boiler water system. Reader is directed to literature on steam generation (Stultz and Kitto 2005) for a detailed discussion on boiler operating principle and the function of steam drum.

Regeneration of a mixed bed exchanger is a multi-step process. Since they carry out both cationic and anionic exchange reactions slippage of any ionic species in the effluent triggers the need for regeneration. First a thorough backwash with demineralized water is conducted to remove any particulate matter. Upon completion of backwash the resin particles are allowed to settle in a quiescent condition. The cationic resins with higher specific gravity (1.26–1.30 g/mL) settle forming a bottom layer and lighter particles of anionic resins (1.06–1.08 g/mL) settle in the top layer. The cationic layer is regenerated by a 2–5% sulfuric acid solution and the anionic layer is regenerated by a 4% sodium hydroxide (caustic soda) solution. The two layers are regenerated sequentially, and the internal piping arrangement prevents the regenerant solution from one layer to cross over to the other layer. After regeneration, these two layers are rinsed by demineralized water. Following the rinse, air is introduced from the bottom to mix up the cationic and anionic resins thoroughly so that the vessel is ready to function as a mixed bed exchanger again.

Since the cation exchanger effluent is highly acidic the carbonate and bicarbonate ions in the influent water exist as carbonic acid (H_2CO_3), which is in equilibrium with CO_2 (g) dissolved in water (refer to Fig. 1.3 in Sect. 1.1.2.6). It is advantageous to remove this dissolved CO_2 by stripping with air prior to entry into the anion exchanger. This relieves the anion exchangers from the load of removing carbonate and bicarbonate ions, extends their service run between consecutive regenerations and also reduces the required amount of regeneration chemicals. Air stripping of CO_2 from cation exchanger effluent is conducted in a vertical trayed stripper, called Decarbonator (also called Degasser). Water is sprayed from the top and air from a blower is introduced from the bottom of the tower. The stripped water is collected in a drum and fed to the anion exchanger. Figure 4.30 presents a comprehensive view of the boiler feed water demineralization process. It depicts the IX and RO as two alternative treatment processes for boiler feed make-up water. The right selection depends upon the feed water characteristics, treated water quality goals, and capital and O&M costs.

Treated effluent from mixed bed exchanger is of high quality and stored in a Demineralized Water Tank for feeding into the boiler water system. Boiler water recirculates in a closed loop and the demineralized water is added into the loop as make up.

Rapid development of membrane separation technologies has enabled reverse osmosis (RO) as a viable process for BFW treatment in place of, or in conjunction with ion exchange (IX) process. Placement of RO can be ahead of IX system, or vice versa. Each has its own advantages. If a RO system follows an IX softener, bulk of the dissolved contaminants are removed by the IX. The RO permeate serves as the boiler feed water and the concentrate from the RO may be of adequate quality for use as make up for an evaporative cooling tower system or other process water systems. However, in certain other applications, power generation utilities install UF and RO membrane systems for bulk of the BFW demineralization, followed by IX system for polishing. This configuration allows substantial reduction in the cycles of regeneration of ion exchange resins and hence the cost of regenerant acid and alkali solutions is also lower. The preferred options by petroleum industries (GWI 2012) are:

- Single pass UF-RO combination, followed by IX, and
- Double pass RO followed by polishing by EDI. It is necessary to feed EDI a greater purity water than IX, and thus requires installation of a two pass RO.

In 2018 Sarroch Refinery on the island of Sardinia in Italy started up a new 12,000 m^3/day SWRO treatment system. Seawater, screened offshore from an intake structure, is pretreated by 0.125 μm self-cleaning strainers followed by UF membrane module prior to feeding into a two pass RO system. The effluent is further polished by an electrodeionization (EDI) system for boiler feed make-up water preparation.

Recently, stand-alone RO systems have also been used for demineralization of boiler feed make-up water. A few such installations at refineries are listed in Table 4.29 (GWI 2012).

Table 4.29 Installation sites using stand-alone RO systems for demineralization of BFW make-up

Name/Location	Capacity (MGD)	Membrane process	Industry
PEMEX, Mexico	6.9	UF (submerged) + RO	Refinery
Conoco Phillips, UK	3.8	UF (submerged) + RO	Refinery
Chennai Petroleum Corporation India	2.5	UF (pressurized) + RO	Refinery

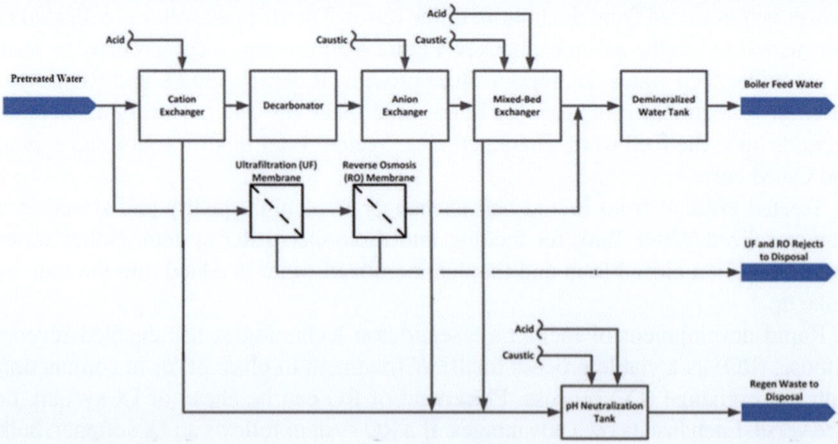

Fig. 4.30 Demineralization process for boiler feed make-up water treatment by IX and RO processes

4.2.5.3 Refinery Steam and Power Generation, and Condensate Polishing Systems

Steam at various pressures is used extensively in refineries for various purposes including:

- Turbine for power generation
- Dilution medium in reactors
- Motive fluid in ejectors
- Stripping steam in distillation towers
- Stripper steam in deaerator stripper
- Heating medium in reboilers
- Steam tracing of pipelines
- Utility stations

Steam is generated in the main boiler plant in the refinery as well as in waste heat boilers where heat is recovered from process streams and converted to steam.

Steam Generation in Fuel Fired Boilers

Natural gas and/or liquid fuels are used to raise steam in fuel fired boilers. Figure 4.31 shows a diagram of the feed water, steam, and condensate system at refineries.

Typically, there are three levels of steam pressures in a refinery—high pressure (HP), medium pressure (MP) and low pressure (LP). An example of the HP, MP and LP steam pressures are 42, 10 and 3.5 bars respectively. However, these vary depending upon the needs of the refinery. HP steam is sent to turbine(s) where the

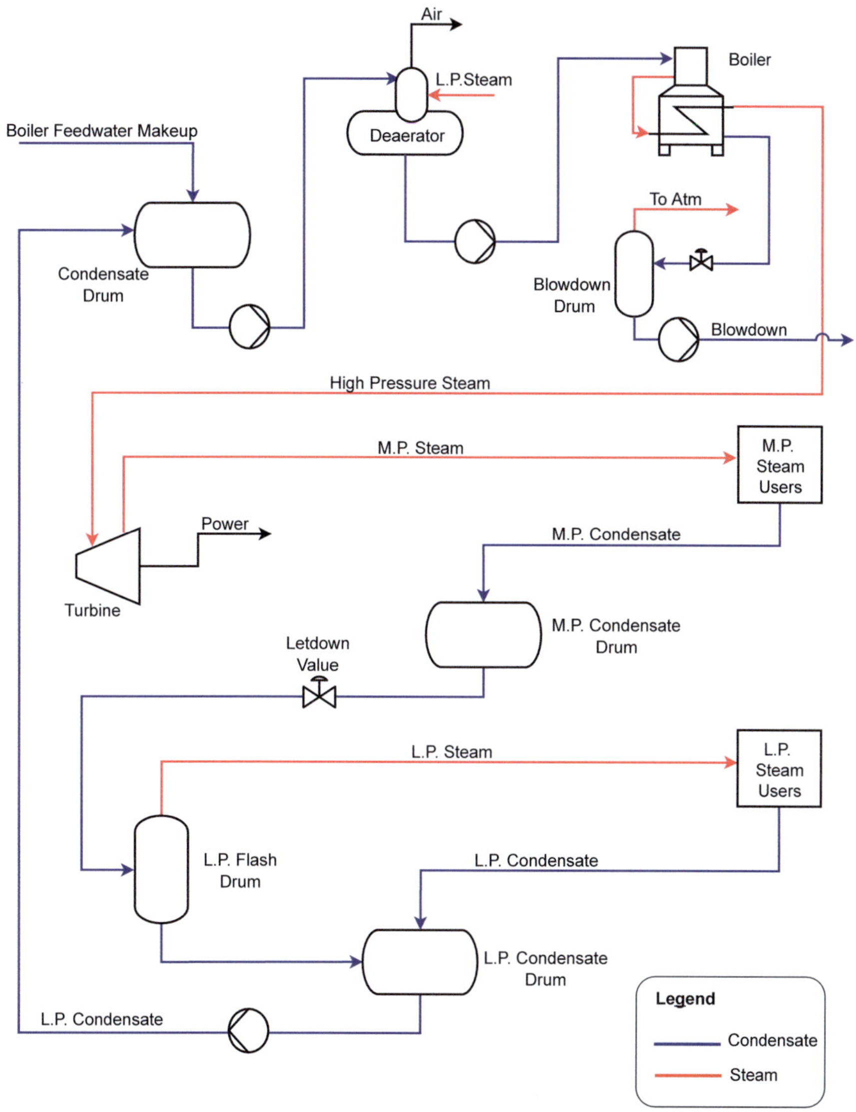

Fig. 4.31 Refinery steam generation system in a fuel fired boiler

pressure is let down to medium pressure (MP) and power is generated for use in the refinery. MP steam is then sent to various users in the refinery complex, which may include another set of power generation turbines. Condensate from MP steam users is sent to the MP Condensate drum for collection. This condensate is then let down to low pressure and sent to the LP Flash drum. Steam from the drum is sent to the LP steam users in the refinery, condensate is recovered and sent to the LP Condensate drum.

Table 4.30 Low pressure steam users

User	Disposition of condensate
Stripping steam	Condensed in the tower and sent to wastewater treatment via sour water stripper
Reboiler steam	Recovered using pump and returned to condensate drum
Snuffing steam to flare	Lost to atmosphere
Steam tracing	Recovered at some locations using pumping traps and lost at other locations to the sewer because recovery is considered impractical/uneconomical
Steam at utility stations	Not recovered and lost to the sewer

LP steam is used extensively in the refinery for various purposes. Table 4.30 shows the typical users and the disposition of the condensate.

In addition to generating power some refineries also use steam turbines at all pressure levels to drive process gas compressors. In some cases, the LP turbine exhaust steam is condensed under vacuum. Operation under vacuum helps the turbine to extract maximum amount of energy from steam to produce the maximum output. Also, it removes the non-condensable gases from the condenser.

The above is a general description of a typical refinery steam system. There could be many variations to this depending on local conditions and preferences.

Condensate Collection and Recovery

Condensate streams from various steam users and other parts of the circulating steam system collect in the LP Condensate Drum. Return condensate stream from this drum and the boiler feed water make up stream collect in the Boiler Feed Water Collection Drum which feeds the combined BFW to the Deaerator. The make-up is about 20–25% of the total boiler feed water. Overall condensate recovery at most refineries is in the range of 80–90% with the remainder being lost to the sewer and atmosphere.

The Deaerator is a tray type steam stripper with water entering from top and LP steam fed from the bottom. Air is stripped off from the BFW to remove oxygen that can cause corrosion of boiler tubes and other metallic parts in the boiler water and steam system. Water collects in a sump that feeds the economizer section and then the radiant section of the Boiler for raising steam under high pressure. Although this is high quality water there are still incidents of boiler tube failure due to scaling and corrosion. Especially, the condensate return stream is prone to pick up of various dissolved gases e.g., O_2, CO_2 and ammonia into the system, and metal ion resulting from corrosion.

Ingress of oxygen and CO_2 occurs from leakage of air into the return condensate stream originating from a LP stage turbine condenser operating under vacuum. Additionally, CO_2 also originates from degradation of any dissolved organic compounds in the make-up water that may pass through untreated in the water demineralizer system upstream. Dissolved oxygen creates pitting corrosion of iron producing iron hydroxide ($Fe(OH)_3$) as shown below:

$$4Fe + 6H_2O + 3O_2 \rightarrow 4Fe(OH)_3(s) \tag{4.9}$$

Iron hydroxide breaks down into ferric oxide (Fe_2O_3) particles and accumulate in the condensate stream. Presence of CO_2 accelerates the corrosion of iron by 10–40% compared to corrosion by each of them alone (Flynn 2009). The proposed chemical reaction is:

$$Fe(OH)_3 + 3CO_2 \rightarrow Fe(HCO_3)_3 \tag{4.10}$$

This is mitigated by internal treatment which consists of chemical addition to remove traces of contaminants that promote corrosion, scaling, and fouling.

Internal Treatment of Boiler Feed Water

One of the main constituents that cause corrosion of boiler tubes is dissolved oxygen (DO). The first step in the removal of DO from BFW is conducted by LP steam stripping in the deaerator. Trace amounts of residual DO is removed by dosing oxygen scavenging chemicals into the deaerator sump. Two commonly used chemicals are sodium sulfite and hydrazine. Their chemical reactions are as follows:

$$Na_2SO_3 + 0.5O_2 \rightarrow Na_2SO_4 \tag{4.11}$$

$$N_2H_4 + O_2 \rightarrow N_2 \uparrow + 2H_2O \tag{4.12}$$

These two chemicals are applied under different conditions. Sodium sulfite is recommended for boilers operating below 60 bars pressure due to the potential of formation of breakdown products. On the other hand, hydrazine is suitable above 60 bars pressure since the kinetics of its reaction with oxygen is slow at low pressure and temperature. However, hydrazine breaks down to ammonia above 230 °C (saturation pressure of steam at this temperature is 2800 kPa or 28 bars) in steam cycle as follows:

$$3N_2H_4 \rightarrow 4NH_3 + N_2(gas) \uparrow \tag{4.13}$$

Thus, the effect of hydrazine as an oxygen scavenging agent is limited depending upon the steam generation pressure. Moreover, ammonia, the breakdown product, promotes corrosion of copper as explained below.

Another corrosion control measure undertaken by steam plants is chemical dosing to raise the pH of the water to the alkaline range. Most commonly aqueous solution of ammonia is dosed to maintain a pH in the range of 9.0–9.3, which is optimum for both copper and iron. These are mainly the two metals that boiler water comes in contact with as the cooler and condenser tubes are made of copper alloys and boiler tubes are made of ferritic steel materials. Neutralizing amines are also used for pH control due to their propensity to absorb carbon dioxide. As explained before, CO_2 is an acid which may be entrained due to incomplete removal of carbonate or decomposition of organic compounds in the external treatment steps. Most commonly used amines are cyclohexamine and morpholine. However, their application is limited to temperatures below ~300 °C since they decompose at higher temperatures.

It should be noted that ammonia dosed into the condensate stream for pH control and that produced by thermal breakdown of hydrazine carried over to the steam cycle, promote corrosion of copper in the cooler and condenser tubes in presence of dissolved oxygen. The mechanism involves dissolution of the protective copper oxide layer to form cupro-ammonium complex ions:

$$2Cu + O_2 \rightarrow 2CuO \tag{4.14}$$

$$CuO + 4NH_3 + H_2O \rightarrow Cu(NH_3)^{4+} + 2OH^- \tag{4.15}$$

In addition to corrosion control, scale control on boiler tube surfaces is also very important. Presence of dissolved solids in the system because of slippage through ion exchangers and/or introduction of treatment chemicals can cause scaling. Phosphate treatment is effective in preventing solids from formation of hard scale on metal surfaces by keeping them in suspension while maintaining a high pH. The phosphate treatment program consists of dosing a combination of di- and trisodium phosphate (Na_3PO_4 and Na_2HPO_4) in an approximate molar ratio of 1:1, which is suitable as a solid dispersant. A total phosphate (PO_4^{3-}) concentration in the range of 3–8 mg/L with the desired di- to trisodium phosphate molar ratio of 1:1 fulfills the requirements for both scaling control and corrosion control by maintaining the pH in the range of 9.0–9.3.

Condensate streams from refinery vacuum stripper overhead condensers and vacuum ejector condensers contain hydrocarbons and are therefore sent to wastewater treatment. Blowdown from recirculating steam systems contain relatively high concentrations of dissolved solids. Therefore, the condensate streams from steam utilized by these systems are considered contaminated and cannot be returned to the boiler water loop for reuse. These streams are discharged to the wastewater treatment plant.

Handling and Management of Boiler Blowdown Streams

As in any recirculating water system boiler water also accumulates solids in the recirculation loop which is purged out from the system in the form of continuous and intermittent blowdown. Continuous and intermittent blowdowns are purged from the boiler in order to prevent a buildup of dissolved solids in the feed water. The blowdown is sent to a blowdown drum where steam is flashed off and the liquid is sent to wastewater treatment as blowdown.

Water balance for the deaerator can be expressed as:

$$F = M + C + D - V \tag{4.16}$$

Water and steam balance for the steam drum can be expressed as:

$$S = F + B \tag{4.17}$$

Dissolved solids balance for the steam drum can be expressed as:

$$F * TDS_F = B * TDS_B \tag{4.18}$$

where,

F = Feed Water to Steam Drum (kg/h)
M = Make up Water (kg/h)
B = Blowdown (kg/h)
C = Condensate Return (kg/h)
S = Steam Flow from Drum (kg/h)
D = Deaerator Stripping Steam (kg/h)
V = Loss through Stripper Vent (kg/h)
TDS_F = TDS of Feedwater (g/m^3)
TDS_B = TDS of Blowdown (g/m^3)

All the flow terms in the above quantities are in terms of mass flow rate (kg/h).

Steam drum blowdown (B) is the fraction of feedwater (F), and expressed in terms of percentage as

$$\% B = B * 100/F \tag{4.19}$$

Cycles of Concentration (COC) is the ratio of the concentration of dissolved solids (TDS) in the boiler water to that of the make-up water. COC is inversely proportional to blowdown rate and can be increased or decreased by adjusting the blowdown. COC can be expressed as follows.

$$COC = TDS_B/TDS_F = F/B \tag{4.20}$$

The boiler is operated at an optimum COC corresponding to a make-up rate to avoid scaling of the tubes and also making sure that there is only a limited amount of energy loss from the system through blowdown.

Table 4.31 shows the typical composition of the boiler blowdown in refineries. It must be recognized that the composition can vary widely depending on operating practices of the refinery.

Table 4.31 Boiler blowdown composition

Contaminant	Units	Value
Total dissolved solids (TDS)	mg/L	Up to 800
COD	mg/L	Up to 100
Free hydrocarbon	mg/L	Up to 5
Suspended solids	mg/L	Up to 25

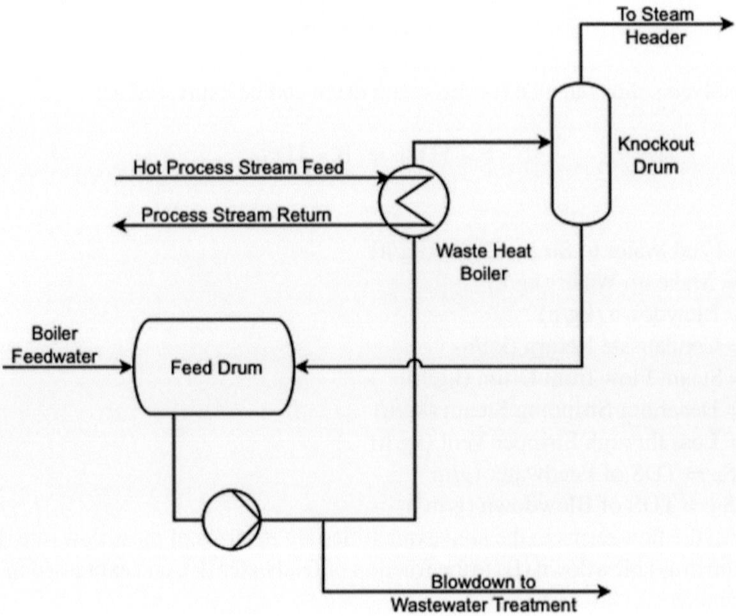

Fig. 4.32 Waste heat boiler

Steam Generation in Waste Heat Boilers (WHB)

In a refinery there are several relatively hot process streams that need to be condensed/cooled. Instead of rejecting this heat to the air (using air coolers) or cooling water using exchangers, the heat is recovered by generating steam in waste heat boilers. Figure 4.32 is a diagram of a Waste Heat Boiler which is also referred to as a Steam Generator.

Boiler feedwater is sent to the Feed Drum and then pumped to the Waste Heat Boiler. Steam from the boiler is sent to a knockout (KO) drum from which the separated steam is sent to the steam header. Any water in the KO drum is returned to the feed drum. A continuous blowdown is taken from the WHB feed drum in order to prevent a buildup of dissolved solids in the system. Typically, WHB are installed to recover medium pressure steam. Heat from process stream that are capable of generating only low pressure steam are generally not recovered because at most refineries there is an excess of low pressure steam. This excess steam from the LP steam header is normally vented to the atmosphere on an as required basis.

Calculation of Blowdown rates

The blowdown rate from a WHB can be calculated using the following equation:

$$\text{Water Balance: } F = B + S \tag{4.21}$$

Solids Balance: $F * TDS_F = B * TDS_B$, (all solids in feedwater are concentrated and discharge with blowdown)

Therefore,

$$(B + S) * TDS_F = B * TDS_B \tag{4.22}$$

Or,

$$B = S * TDS_F / (TDS_B - TDS_F) \tag{4.23}$$

where,

F = Feed Water to Steam Drum (kg/h)
B = Blowdown Rate (kg/h)
S = Steam Flow from Drum (kg/h)
TDS_F = TDS of Feedwater (g/m^3)
TDS_B = TDS of Blowdown (g/m^3)

Sample Calculation

Steam rate, $S = 10{,}000$ kg/h
$TDS_F = 100\ g/m^3$
$TDS_B = 600\ g/m^3$
Blowdown, $B = 2{,}000$ kg/h

4.2.5.4 Refinery Cooling Systems

In an oil refinery, various products are made using reactors and distillation columns. In the case of reactors the reactants are heated up to a high temperature in many instances and then the effluent from the reactors are condensed and cooled. In distillation systems the liquid mixtures are separated into various components by heating, boiling, condensation and cooling. All this cooling is accomplished using a combination of air coolers and heat exchangers that use cooling water as the cooling medium.

There are three types of cooling water systems used in refineries:

1. Once-through cooling water system.
2. Closed loop cooling water circulation system.

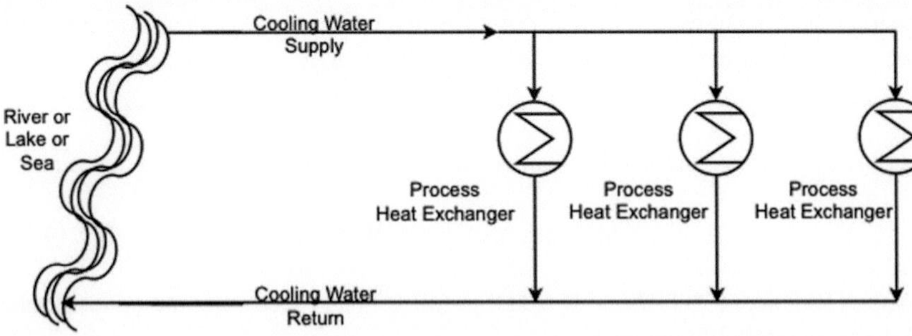

Fig. 4.33 Once-through cooling water system

3. Cooling water system with an evaporative cooling tower.

Once-Through Cooling Water System

Figure 4.33 is a diagram of a Once-through cooling water system.

In this type of system water is taken from one of the following three types of sources:

- Sea
- Rivers
- Lakes

The water is circulated through the cooling exchangers and condensers in the refinery. Warm water from the exchangers is the returned downstream of the source from where it was taken. Sea water is used most frequently in these types of system. River water is used in some refineries and lake water is rarely used. The cooling water return temperature is usually restricted to 5–7 °C more than the intake temperature due to environmental considerations. The maximum permitted discharge temperature is 60 °C (EPA 1999). This results in relatively large circulation rates of cooling water. An advantage of this type of a system is that there is no blowdown from this type of system and does not impose a hydraulic load on the refinery wastewater system. All the cooling exchangers in the refinery are designed so that the process side pressure in the exchangers are greater than the cooling water side to ensure that the process fluid does not get contaminated with cooling water in the event of inadvertent tube leaks in the exchangers. The system needs to be monitored to ensure that in the event of leaks of process fluids into the cooling water, the cooling water return does not contaminate the receiving waters.

Operations Issues

The primary operational challenges in these types of systems is Scaling, Fouling and Corrosion.

Scaling

Likelihood of scaling is less than recirculating cooling water system since there is no buildup of dissolved solids due to evaporation. Also, at the operating maximum temperature limit of 60 °C calcium carbonate and calcium sulfate, the main species of concern remain dissolved in water.

Fouling

Once through cooling water systems are prone to fouling due to the presence of debris, aquatic macro and microbiological species, mud and silt. The raw water is withdrawn from river or sea, passed through the intake screens, and directly pumped to the cooling water distribution system. The likelihood of fouling is less in the once through systems where the water is pretreated to remove suspended solids and related turbidity. The types of fouling caused by seawater and their remedial measures are described in detail earlier in Chap. 2 in the section on seawater desalination.

One additional type of fouling that is unique to once through cooling water system is macrobiological fouling. Various vertebrate and non-vertebrate macrobiological species include various types of fish, shellfish, oysters, snails, etc. Most of these are effectively removed by the influent screens but smaller organisms pass through and create fouling inside the coolers and condensers. Snail infestation within heat exchanger tubes resulting in severe loss of their performance has been reported at refineries using once through seawater cooling systems. Young population of these organisms are small in size and pass through the intake screens. The tubes in heat exchangers provide an ideal environment for them to thrive and proliferate, eventually blocking the cooling water from passing through the tubes. An example of snail infestation in the condenser tube bundle at a refinery in the Philippines is shown in Fig. 4.34. Shock dose of chlorine dioxide as a biocide, coupled with periodic mechanical cleaning of tube bundles has proven to be an effective measure to combat this fouling.

Fig. 4.34 Image of snail infestation with the magnified view of a snail in the condenser tube bundle of a refinery using once through seawater cooling system

Corrosion

The potential for corrosion is similar to that in recirculating cooling water system. The added concern here is the high salinity (>3% dissolved solids) of seawater, which is corrosive to metals. Selection of heat exchanger tube materials in seawater cooling service is very important. 304 and 316 stainless steels are not very effective in corrosion resistance to chloride ions. Copper alloys, especially admiralty brass containing 65–70% copper and 25–29% zinc offer good resistance to salinity and are widely used in the industry for sea water service. Other materials e.g., titanium and tantalum also have high corrosion resistant properties to salinity but are not economically viable options.

Closed Loop Cooling Water Circulation System

A closed loop cooling water system is shown in Fig. 4.35.

In this type of system, the cooling water is stored in a Surge tank or pond in the refinery and is circulated throughout the refinery and passed through process heat exchangers as required. The warm water from the exchangers is passed through a cooling water cooler where the heat is rejected to either sea water or river water on a once through basis. The sea/river water is sent back to the source of the water. There is no continuous blowdown from this system. The process heat exchangers are designed so that the process side pressure is always higher than the cooling water pressure so that in the event of any exchanger tube leaks the process side fluid always

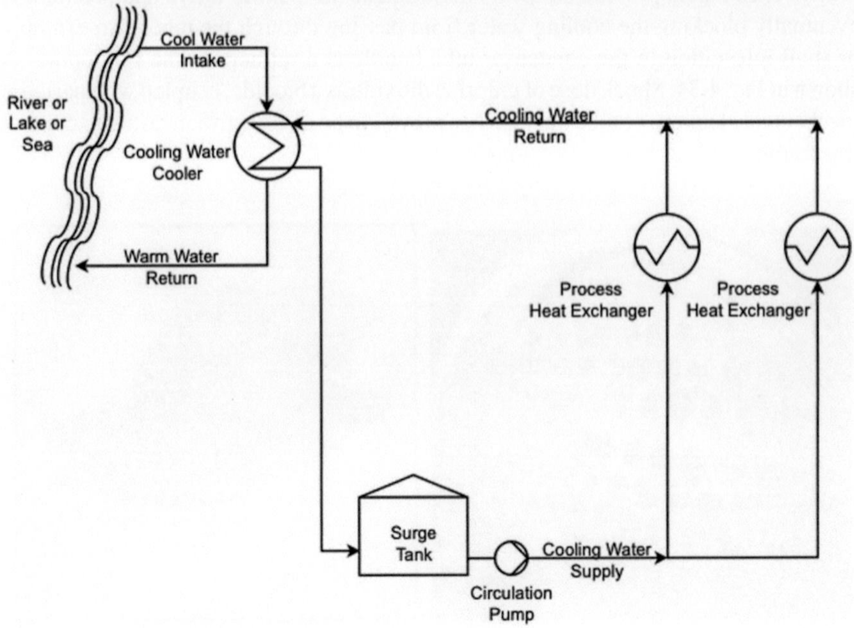

Fig. 4.35 Closed loop cooling system

leaks into the cooling water. In this event some of circulating cooling water will need to be purged to the refinery wastewater system on a batch basis to prevent build up of contaminants in the system.

Evaporative Cooling Towers

Figure 4.36 is a diagram of a cooling water system with an evaporative cooling tower.

In an evaporative cooling tower system, the circulating cooling water is passed through the process heat exchangers in the refinery and then sent to the cooling tower where the heat is rejected by evaporation of the circulating water. The return water is sprayed on the top of the tower and is contacted with air that passes counter currently through the tower either by forced draft or induced draft. The water drains into a sump that is located at the bottom of the tower. Makeup water is added to this sump to account for the losses from the tower due to evaporation and drift (entrainment). A significant quantity of blowndown is also taken from this sump to prevent the buildup of dissolved solids such as calcium and magnesium in the system. This blowdown is sent to the refinery wastewater treatment system and imposes a significant hydraulic load on the wastewater treatment system. In some rare instances where the ambient conditions are tropical and very humid, it might be necessary to include a supplemental refrigeration system into which some of the heat is rejected in order to maintain acceptable cooling water supply temperatures.

Since evaporative cooling towers are prevalent at most refineries a more detailed discussion of this system is warranted. From a refinery water use and wastewater

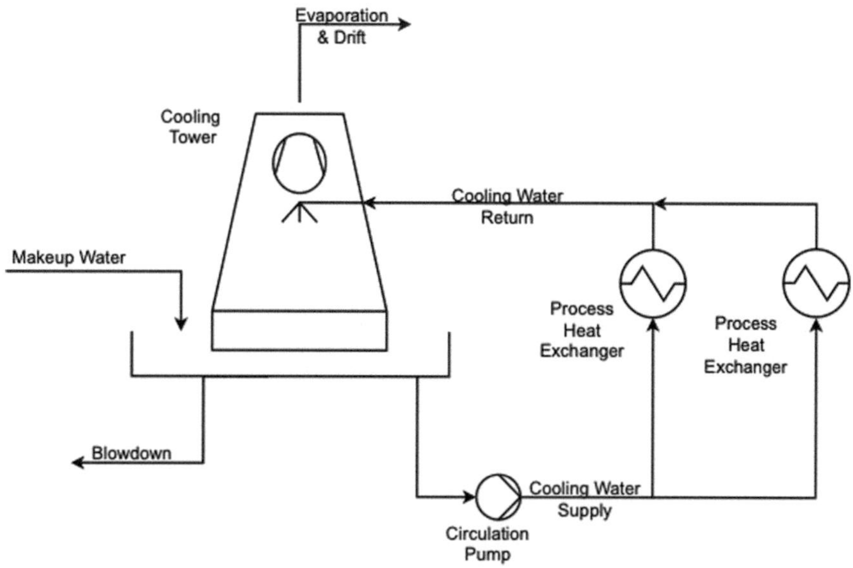

Fig. 4.36 Evaporative cooling towers

standpoint it is important to understand the issues that govern the makeup and blowdown rates of water from the cooling water system.

The make up water rate (M) is the sum of the evaporative losses (E), drift losses (L) and the blowdown (B).

$$M = E + L + B \tag{4.24}$$

The evaporative loss is a function of the heat rejected by the cooling tower and the drift loss is a function the circulation rate.

The quantity of dissolved solids (TDS) coming into the system with the makeup water has to go out of the system with the blowdown stream to prevent TDS buildup in the circulating cooling water. Cycles of concentration (COC) in the cooling tower is calculated by:

$$COC = \frac{\text{TDS concentration in circulating water drift(y)}}{\text{TDS concentration in makeup water (x)}} \tag{4.25}$$

Therefore, from solids balance

$$y = COC * x \tag{4.26}$$

And, since the TDS concentration of the blowdown is the same as that of the circulating water, and the evaporative losses can be assumed to have no TDS, the TDS in the make up water is equal to the combined TDS of the drift losses and blowdown.

Therefore, from solids balance

$$M * x = (B + L) * y \tag{4.27}$$

Combining Eqs. 4.26 and 4.27:

$$M * x = (B + L) * COC * x \tag{4.28}$$

Therefore,

$$M = COC * (B + L) \tag{4.29}$$

Combining Eqs. 4.24, 4.26 and 4.27:

$$\begin{aligned} B &= M - E - L \\ &= COC * (B + L) - E - L \end{aligned} \tag{4.30}$$

Therefore,

$$B = \frac{E - (COC - 1) * L}{(COC - 1)} \quad (4.31)$$

$$M = \frac{COC * E}{(COC - L)} \quad (4.32)$$

where

M = Amount of makeup water (kg/h)
E = Amount of evaporated water (kg/h)
L = Amount of water loss by drift (kg/h, specified by tower manufacturer)
B = Amount of blowdown (kg/h)
COC = Number of cycles of concentration
x = TDS of the makeup water (g/m^3)
y = TDS of circulating Water (g/m^3)

When determining the water balance around the refinery it is important to be able to calculate the makeup and blowdown rates from the cooling tower.

The following example illustrates the calculation of these parameters (Table 4.32).

Operations Issues

The operating challenges of evaporative cooling water systems are scaling, fouling and corrosion. These effects, and their mitigation methods are discussed below.

Scaling

Scaling adversely impacts the heat transfer capability of cooling water. Moreover, scaling constricts the flow of water and results in increased pressure drop. Therefore, scaling has to be routinely removed from heat exchangers and cooling water lines. The principal reasons for scaling are:

1. Temperature rise: Cooling water temperature rises due to heat rejected from process fluids and condensing steam. Solubility of certain dissolved solids,

Table 4.32 Sample calculation for cooling tower water and heat balance. In these calculations all the flows for Cooling Tower Circulation Rate, Evaporation Rate, Drift, Blowdown and Makeup have been expressed in terms of m^3/h instead of kg/h. These have been based on a conversion of kg/h to m^3/h using the density of water = 1,000 kg/m^3

Cooling tower circulation rate (assumed for this calculation)	4,000 m^3/h
CW temperature difference (assumed for this calculation)	9 °C
Calculated duty (circulation rate x temperature difference)	36 MM Kcal/h
Heat of vaporization (HV)	576 kcal/kg
Evaporation = $\frac{Duty}{HV}$	62.5 m^3/h
Drift (assumed to 0.1% of circulation)	4 m^3/h
Assumed cycles of concentration	5
Blowdown (by Eq. 4.31)	11.6 m^3/h
Makeup (by Eq. 4.32)	78.1 m^3/h

specifically carbonate, sulfate and phosphate salts of calcium, which are commonly present especially in groundwater, decreases with increasing temperature and tend to form scale. It is necessary to routinely monitor the recirculating and make up water analyses to make sure that the concentrations of ionic species are within the limits of the operating cycles of concentration. For calcium carbonate Langelier Index (discussed in Chap. 1), provides good guidance on the potential for scaling. When multiple cations and anions are involved, it is best to input the data into a model to predict scaling versus corrosion potential. This has been discussed in detail in Chap. 6.

2. Presence of scale forming ions in the source water: As discussed above, calcium is the most common scale forming ion in the source waters. The softening processes, as discussed earlier in this chapter, are very effective in removing hardness and overcoming this operational difficulty.
3. Oversaturation of silica bearing compounds: Silica can be present in water in colloidal form as SiO_2, which is amorphous, and as silicate ions dissolved in water. Colloidal silica can deposit on metal surfaces as a foulant. Silicate ions, however, form hard scales in combination with calcium. Colloidal silica can be removed by coagulation-flocculation process, and also by micro or ultrafiltration. Removal of silica that is dissolved in water requires the use of ion exchange or RO process.

If not routinely controlled, scaling can lead to shutdown of affected equipment requiring significant downtime and maintenance. Mitigation of scaling can be done by adopting appropriate pretreatment processes. Additional measures include routine water analysis for pH, alkalinity, hardness, and other ionic and non-ionic species of concern; maintaining the desired cycle of concentration; and on-line chemical dosing. Acid dosing is commonly practiced keeping Langelier Index in control if calcium carbonate is expected to be the main scaling chemical. Certain commercial products consisting of surfactants and solubilizing agents of metal ions are offered by chemical vendors. These include inorganic and organic phosphorus-based chemicals. Phosphorus contained in these compounds finds its way to wastewater treatment plant with the cooling water blowdown stream. This is beneficial because phosphorus is an essential nutrient for microbiological growth in biological treatment and refinery wastewater is typically deficient in phosphorus. Polyacrylates, which modify the crystal structure of calcium carbonate, is another type of scale control organic chemical for cooling water treatment. It prevents scale from adhering to the metal surfaces and leaves the system with blowdown stream.

Fouling

Fouling creates similar effects as scaling. It also reduces the heat transfer in the heat exchanger tubes, and increases pressure drop in the cooling water system. Fouling can be of physicochemical or biological in nature. Physicochemical foulants include debris carried over with make-up water, sand and silt with make-up water, cooling tower wood decomposition particles and metallic corrosion products. The best means to control fouling is effective pretreatment to remove materials associated with make-up water. The foulants that are formed in the recirculation system can be prevented from accumulation by adding surfactant chemical that keep such materials in suspension and remove those with the blowdown stream. Also, larger objects are prevented from entering the recirculation loop by the suction strainers of the cooling water pumps into the distribution network.

Biological fouling is caused mainly by the growth of algae and other organisms in the recirculating water. The presence of plentiful sunlight in combination with CO_2 from open air in the cooling towers creates an ideal environment for algal growth by photosynthesis. Circulating cooling water carries these as debris that are deposited on heat exchanger tube surfaces and in the pipes of the cooling water system. Biocide chemicals need to be shock dosed routinely to maintain algal growth under control.

Corrosion

Corrosion potential of cooling water can be both from chemical and biological sources. As in the case of scaling, corrosion potential can be controlled by adjusting the Langelier Index (LI) for calcium carbonate saturation with the help of acid or alkali dosing appropriately. This measure, in conjunction with commercially available corrosion inhibitors, ensures good corrosion control. When several cations and anions are present in significant concentrations then modeling must be done to determine scaling or corrosion potential of the recirculating water as detailed in Chap. 6. Proper chemical dosing program must be instituted for effective scale or corrosion control.

Microbiologically induced corrosion (MIC) is caused by various aerobic and anaerobic microorganisms in the recirculating cooling water and make-up water. Corrosion of metals increases by several orders of magnitude by the presence of microbes. Aerobic iron related bacteria (IRB) thrive in the cooling tower decks which utilize iron in the cooling water pipes as electron donor and dissolved oxygen as electron acceptor. This sets up the electrochemical reactions leading to corrosion of iron pipes as shown by the simple half-cell reactions below:

$$4Fe^0 \rightarrow 4Fe^{3+} + 12e \text{ (anodic reaction)} \tag{4.33}$$

$$3O_2 + 6H_2O + 12e \rightarrow 12OH^- \text{ (cathodic reaction)} \tag{4.34}$$

$$\text{Overall reaction}: 4Fe^0 + 3O_2 + 6H_2O \rightarrow 4Fe(OH)_3 \tag{4.35}$$

The typical anodic and cathodic reactions presented above demonstrate the mechanism of corrosion of iron pipes in cooling water distribution systems in presence of oxygen. Inside the cooling water loop as oxygen is depleted, sulfate ions are reduced to sulfide by anaerobic action of sulfate reducing bacteria (SRB) according to the reaction:

$$SO_4{}^{2-} + 8H^+ + 8e \rightarrow S^{2-} + 4H_2O \tag{4.36}$$

They proliferate in attached films on the heat exchanger tube surfaces on the cooling water side. IRB and SRB populations work together to promote corrosion of cooling water internal distribution network and exchanger tube materials and ferric sulfide particles are formed according to the reaction 4.37.

$$2Fe^{3+} + 3S^{2-} \rightarrow Fe_2S_3 \quad (4.37)$$

Two types of IRB—*Leptothrix* filaments and *Gallionella* ribbons are depicted in Figs. 4.37 (400×), and 4.38 (400×). These were identified in the attached biofilms on the corrosion coupons placed in the makeup water pretreatment clarifier clearwell and in the cooling water return header in a petrochemical plant in the US Midwest. The SRBs identified belong to the Genus *Desulfotomaculaum* and *Desulfovibrio* (not shown in the photomicrographs below). Figure 4.38 demonstrates an example of the action of both IRB and SRB bacteria on iron pipes. The first action is by *Gallionella* IRB which oxidizes iron to ferric ions by the biochemical reaction 4.33. Subsequently, ferric ions are consumed by SRB.

Algal blooms in cooling tower decks also cause corrosion due to pH swings. Photosynthesis reaction, which is the principal metabolic mechanism of algae during the day, consumes CO_2 and causes localized increase in alkalinity and pH. However, in the darkness at night, algae respire by taking up dissolved oxygen and releasing CO_2, thereby decreasing pH. This diurnal pH swing further amplifies the microbiologically induced corrosion potential of recirculating cooling water systems.

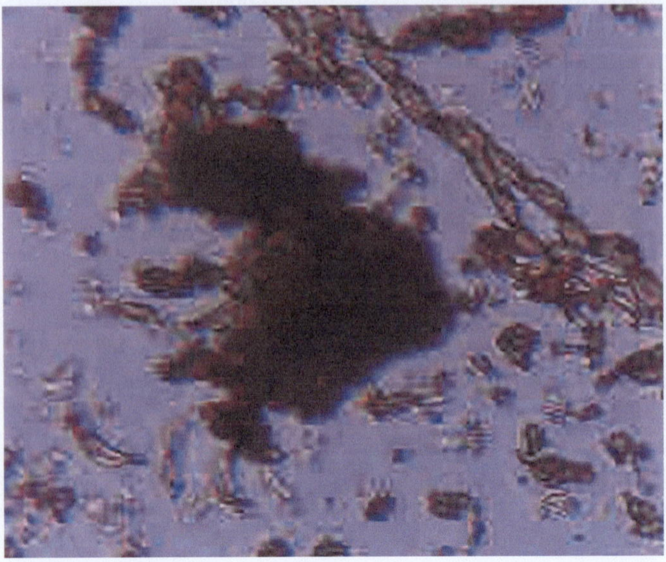

Fig. 4.37 Gallionella ribbons attached on corrosion coupon in make-up water treatment clarifier clearwell (magnification 400×)

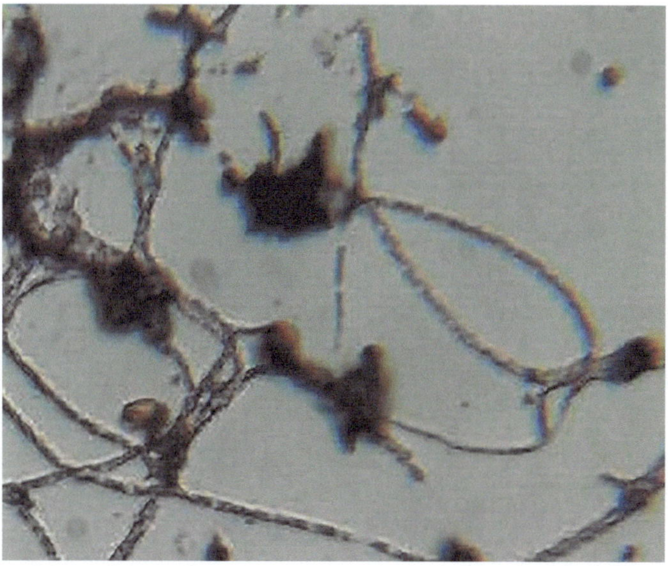

Fig. 4.38 Leptothrix filaments attached on corrosion coupon in the cooling water return header (magnification 400×, black patches represent deposit of ferrous sulfide, Fe_2S_3 particles within the cell structures)

Disinfection in open recirculation cooling water system is critical to protect the mechanical integrity of heat exchangers, coolers and condensers. Typically, this is done by dosing biocides. Both oxidizing and non-oxidizing biocides are used. Gaseous chlorine, sodium hypochlorite solution and chlorine dioxide solution are the most widely used oxidizing biocides, and glutaraldehyde and quaternary ammonium salts are used as non-oxidizing type biocides. Typically, the cooling water treatment systems at petroleum refinery sites use a combination of disinfectants and dispersant chemicals. As the microbial population exists in the system primarily in attached growth form it is essential to loosen those films, so they become exposed to the disinfecting chemicals. Depending upon the extent of deposition some special programs are sometimes instituted. These include shock dosing and targeted dosing of specific heat exchangers, or group of exchangers in the affected areas the plant. When a large amount of chemical is applied to the entire recirculating water system the effect of the chemical is diluted. Targeted and shock dosing avoids such dilution.

Cooling water treatment systems are analyzed and recommended by specialized companies that offer the combination of chemicals on a service contract basis. Some of those chemicals, particularly the biocides, dispersants, surfactants, antiscaling and corrosion control agents contain formulations that are proprietary to these companies.

The typical contaminants in the blowdown from typical cooling towers is shown in the Table 4.33.

Table 4.33 Cooling tower blowdown composition

Contaminant	Units	Value
Total dissolved solids (TDS)	mg/L	Up to 1,000
COD	mg/L	Up to 150
Free hydrocarbon	mg/L	Up to 5

4.2.6 Refinery Sewer Systems

Sewers are used in refineries to convey wastewater produced in the plant to the wastewater treatment system. Most refineries have multiple categories of sewers conveying different types of effluent streams, such as:

- Sewer hubs connected to oily water sewer
- Process unit paved area drains combined with oily water sewer or piped separately to wastewater treatment plant
- Area storm water drains sent to discharge

Figure 4.39 shows the routing of the different categories of sewers in a refinery.

It must be understood that the number of sewers and the routing of the sewers can vary widely depending on the location of the refinery, local environmental regulations, and refinery operating philosophy among other factors.

The blowdown from the cooling towers and the boiler usually come from the Utilities area of the refinery and are conveyed separately to the wastewater treatment plant. The brine from the desalters and the crude tank drains and the stripped sour water are in many cases conveyed separately to the treatment plant. The oily sewer is collected from the different process units and routed to the treatment plant. These are Continuously Oil Contaminated (COC) streams and are major sources of pollutants that will need treatment. Streams contributing to COC sewer system are:

- Crude and product tank bottom drainage
- Process Effluent (Process Sumps)
- Laboratory sump
- Desalter Effluent
- Stripped Sour Water

Refineries that receive ballast water from oil tankers (although this is becoming increasingly rare since most modern tankers do not use ballast water anymore) route this water to a storage tank in the refinery. The wastewater from this tank is bled into the wastewater treatment plant at a controlled rate. In some refineries the contaminated stormwater collected in the various process units are conveyed separately to the treatment plant while in others it is combined with the oily sewer. In these instances, it makes it more challenging to handle the large variations in hydraulic flow during rain events and could require larger equalization tanks/ponds in the wastewater treatment plant. The rainwater that is collected from the stormwater drains located adjacent to the process unit area should be uncontaminated and therefore can be discharged without treatment. These are considered Entirely Oil Free (EOF) streams and are segregated from the contaminated COC streams. Examples of these streams include:

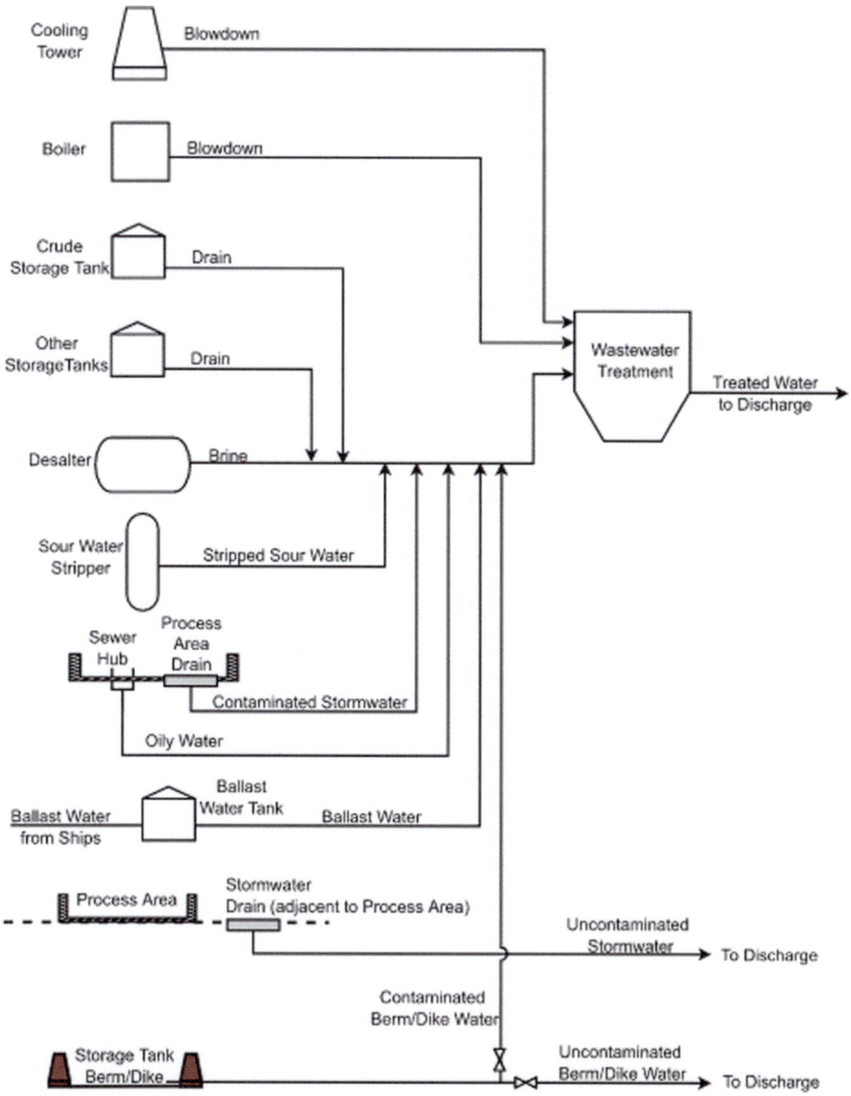

Fig. 4.39 Refinery sewer categories

- Oil free Cooling Water Return (CWR) or Sea Cooling Water Return (where applicable)
- Roof draining
- Oil-free surface run-offs from Process Area (Plant Water)

The following paragraphs provide additional details regarding the sewer hubs and area drains in typical process units. In the refinery there are number of process units that are usually divided into various sectors depending on the process scheme at the refinery. The process units could be spread out within an area that is designated as ISBL (inside battery limits). Other parts of the refinery such as the Utilities area, Storage tank farm, Wastewater treatment, Laboratory and other infrastructure such as maintenance and transportation are considered OSBL (outside battery limits) and are located away from the ISBL areas. In each of the process units, all the equipment is typically located in a paved area that is usually curbed from adjacent areas. Figure 4.40 shows a typical paved process unit.

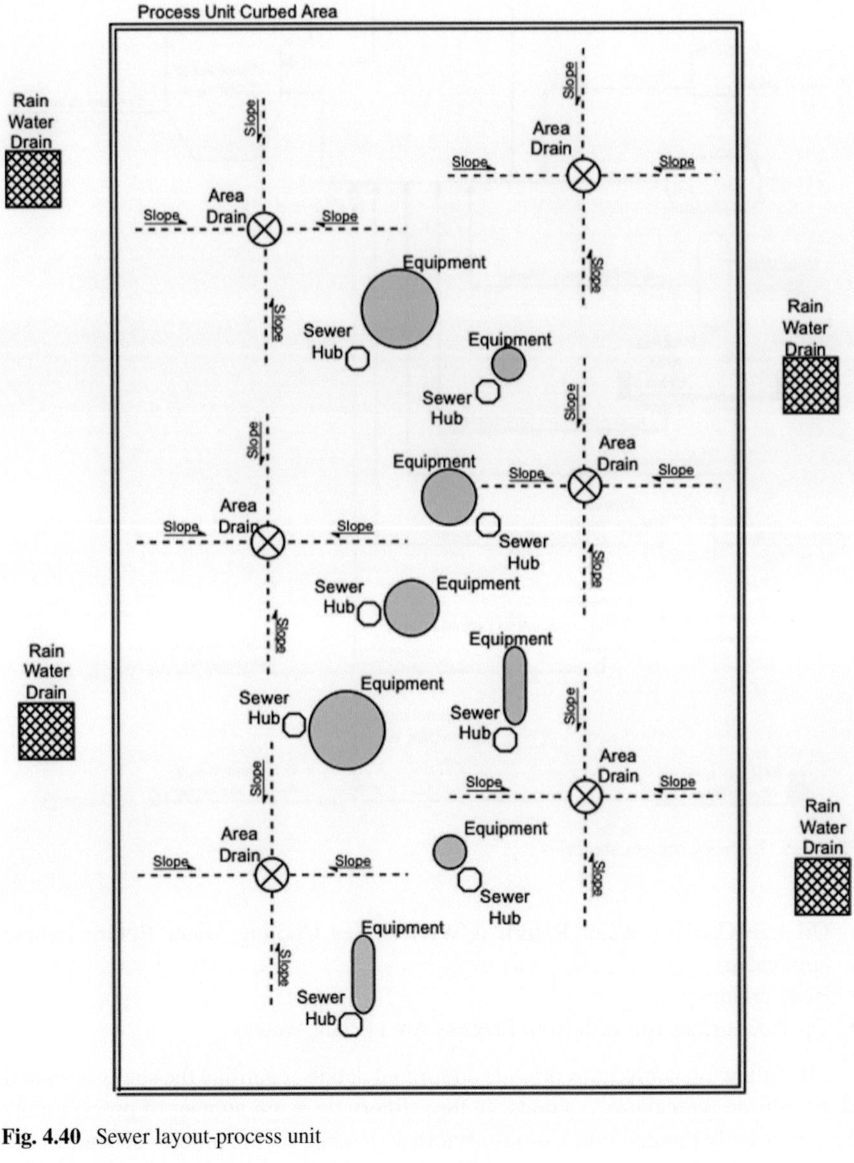

Fig. 4.40 Sewer layout-process unit

The typical layout of process equipment in the paved area in the process unit is shown in the diagram. Adjacent to each piece of equipment in the unit is a raised sewer hub (see the Fig. 4.41) which is used when the heels left in equipment are drained when the equipment needs to be decommissioned. Normally, these hubs are connected to the oily water sewer in the plant. Additionally, there are area drains located in the unit that collect and drain any rain water that falls on the unit. This rainwater could be contaminated by any hydrocarbons that may have been inadvertently spilled during the normal course of operations. Many refineries have a separate contaminated stormwater sewer system to which these drains are connected to prevent the wastewater treatment system from being hydraulically overwhelmed during a storm event.

Outside the paved area, drains are located to convey the rain water away from the area. Typically this rain water should be clean since it has not been in contact with hydrocarbons and therefore can be discharged without treatment.

Details regarding the process drains and sewer hubs are discussed below. Figure 4.41 is a diagram of typical oily water sewer hub.

The figure shows a typical arrangement of a process vessel with a a drain at the bottom. The sewer hub is located adjacent to the vessel so that the heels of fluid left in the vessel after decommissioning can be drained out. The hub is usually raised,

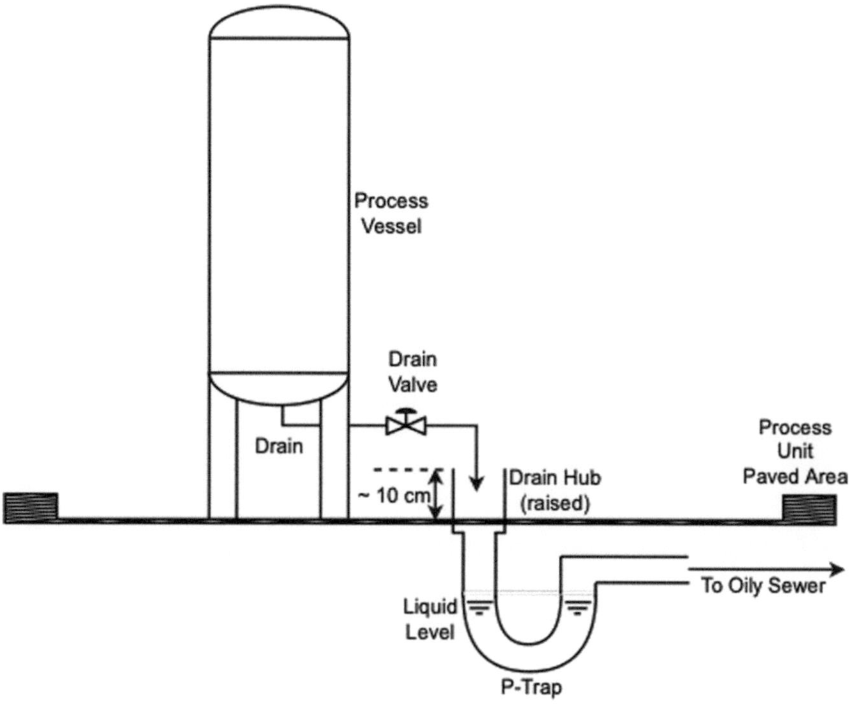

Fig. 4.41 Sewer hub—process unit

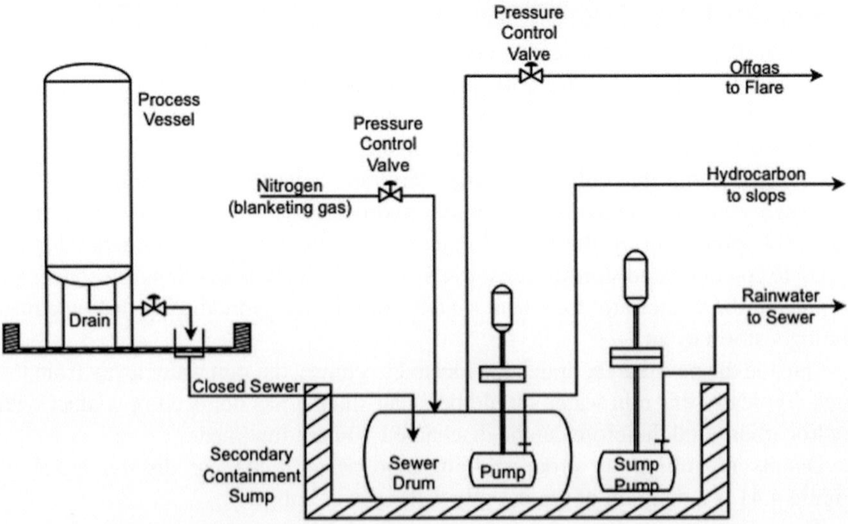

Fig. 4.42 Closed sewer system

up to 10 cm above the pavement in the unit. This is done to prevent rain water from the paved area being drained to the oily water sewer. Each sewer hub is equipped with a p-trap located underground and the liquid in the trap prevents vapors from the sewers from being backed up in the unit.

Some units in newer refineries or newer units in older refineries have installed closed sewer systems in order to contain and recover hydrocarbons when equipment is decommissioned and drained for maintenance purposes. Figure 4.42 is a diagram of a closed sewer system.

The heels of fluid that are left in process vessels during decommissioning are drained to sewer hubs which are connected to a drum that that is located in an underground sump. The sump is installed as secondary containment to catch any leaks from the drum. Any rainwater that collects in the sump is pumped away to the sewer. The drum is nitrogen blanketed to prevent any emissions from the drum and is vented to the refinery flare. Water and hydrocarbons that collect in the drum are pumped away to the slops tank for reprocessing.

4.2.7 Rainwater Management

Management of rain water at refineries is very important for the proper operation of the wastewater treatment plant (WWTP). During a storm event instantaneous hydraulic flow from rain water runoff can be very significant and overwhelm the WWTP if not managed properly. There are strategies that can be implemented to manage and control the amount of rainwater that needs to be treated while the remainder can be discharged without treatment.

Rainfall data for different locations can be obtained from a variety of sources. For example, in the United States rainfall data is for various cities and towns is available from the National Oceanic and Atmospheric Administration (NOAA). Most meteorological stations (frequently located at nearby airports) also collect rainfall data and this data can be used to estimate the rainfall pattern at any particular site. Figure 4.43 shows the way rainfall data is usually reported.

The x-axis shows the duration of the rainfall event in minutes and hours. The y-axis shows the rainfall intensity in millimeters of rainfall per hour. The curves represent the different return period for rainfall events in years. The curves that are shown are for return frequencies of 2, 5, 10, 25 and 100 years. This means, for example the 10 years curve is the expected rainfall intensity for a 1 in 10 year rainfall event based on statistical data.

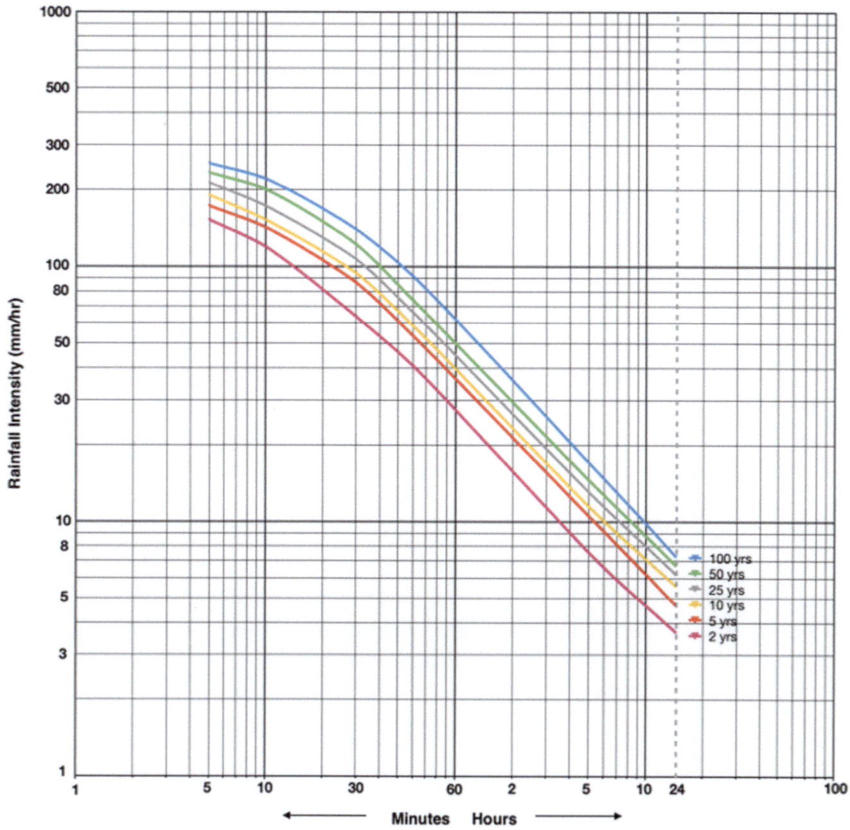

Fig. 4.43 Typical rainfall data

Table 4.34 Runoff coefficients

Description of surface	Runoff coefficient
Lawns	0.05–0.35
Unimproved areas	0.1–0.3
Residential areas	0.3–0.75
Business areas	0.5–0.95
Industrial areas	0.5–0.9
Asphalt streets	0.7–0.95
Brick streets	0.7–0.85
Roofs	0.75–0.95
Concrete streets	0.7–0.95

The rainwater management system at a refinery is based on choosing a rainfall return period for which the system will be designed. The higher the frequency event the greater the rainfall volumes will be that require treatment. Most refineries choose to design their systems for a 1 in 10 year event although in some jurisdictions by regulation it is required that rainfall management systems be designed for a 1 in 25 year event. Once the rainfall return event is chosen the volume of rainfall is calculated and the volume of rainwater storage required is determined.

The calculation of the rainfall volume is based on the use of rainfall runoff coefficients. The runoff coefficient defines the percentage of rainwater that runs off as opposed to be absorbed into the ground. For the storm water calculations, different sets of runoff coefficients are considered; based on the type of land cover. For example, for paved areas, a runoff coefficient of 0.95 is used and for non-paved areas, a run-off coefficient of 0.4 is used.

Typical runoff coefficients for other types of areas is shown in the Table 4.34.

The following example illustrates the calculation of rainfall flows. For this example the following hypothetical plot plan for typical refinery is used (Fig. 4.44).

In this diagram the areas in gray represent the unit paved areas in the refinery and the yellow represent the areas around the refinery that are not paved. It is important to delineate these areas because the rainfall runoff estimates from each of these categories is very different and will impact the design and management of the stormwater system. The areas of each of the numbered sections in the refinery plot is calculated and shown in the Table 4.35. It also shows the runoff coefficient used for each area as well as the calculated effective area for rainwater runoff.

The storm water runoff is a function of the following factors:

- Drainage area
- Rainfall intensity
- Runoff coefficient (fraction of rainfall that runs off and is not absorbed into the soil)
- Time of concentration (time it takes for the rainfall to travel to the farthest point of the area under consideration to the point of collection)
- Duration of storm event

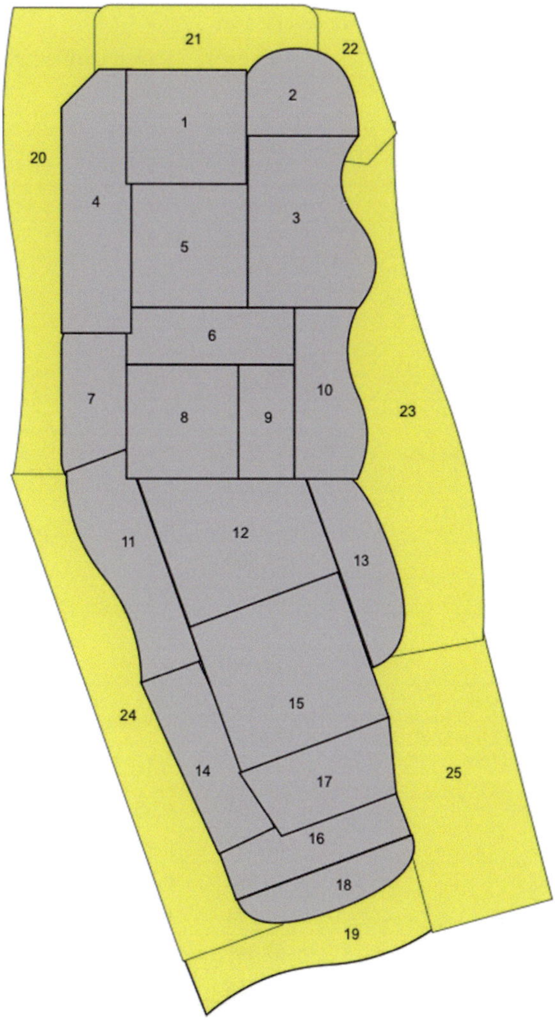

Fig. 4.44 A hypothetical refinery plot

The peak flow of storm water can be calculated using the Rational Method (*developed by* ***Kuichling*** *in 1889*) which is described by the following equation

$$Q_d = c * i * A \tag{4.38}$$

where

Q_d Peak flow (m^3/h)
c Rational method runoff coefficient (dimensionless)
i Rainfall intensity (mm/h)
A Drainage area (m^2)

Table 4.35 Refinery process unit area

Plot designation	Area (m^2)	Runoff coefficient	Effective area for runoff (m^2)
1	32,067	0.9	28,860
2	12,827	0.9	11,544
3	41,687	0.9	37,518
4	41,687	0.9	37,518
5	35,274	0.9	31,746
6	25,654	0.9	23,088
7	20,844	0.9	18,759
8	32,067	0.9	28,860
9	12,827	0.9	11,544
10	16,034	0.9	14,430
11	35,274	0.9	31,746
12	51,307	0.9	46,176
13	27,257	0.9	24,531
14	28,860	0.9	25,974
15	57,721	0.9	51,949
16	14,430	0.9	12,987
17	24,050	0.9	21,645
18	10,582	0.9	9,524
Total for paved area	*520,447*	–	*468,403*
19	41,687	0.45	18,759
20	64,134	0.45	28,860
21	27,257	0.45	12,266
22	10,582	0.45	4,762
23	83,374	0.45	37,518
24	67,341	0.45	30,303
25	73,754	0.45	33,189
Total for non-paved area	*368,129*	–	*165,658*

The rational method was originally developed to estimate peak discharges for sizing drainage structures, such as storm drains and culverts. The modified rational method (MRM) is an extension of the rational method to produce simple runoff hydrographs. The MRM is often called the rational hydrograph method. Application of the MRM produces a runoff hydrograph and runoff volume in contrast to application of the rational method, which produces only the peak design discharge flow. The modified Rational method which is an extension of the conventional Rational method uses the peak flow calculating capability of the Rational method paired with assumptions about the inflow and outflow hydrographs to compute the approximate storage volume required to contain the rainwater.

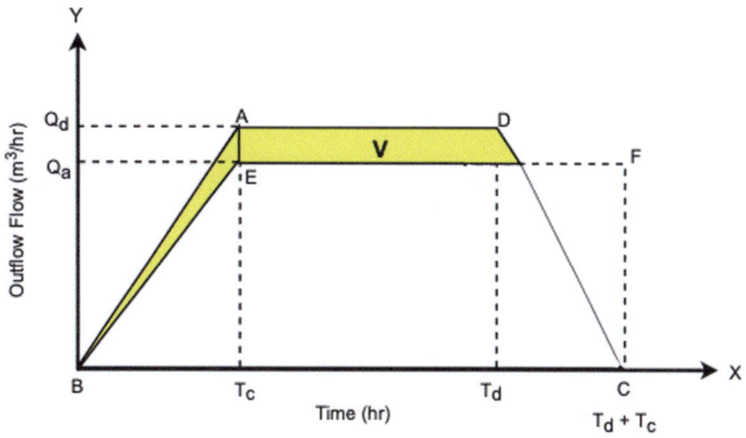

Fig. 4.45 Rainwater runoff

This approach assumes the storm water runoff hydrograph for a design storm event is trapezoidal in shape. It is assumed the peak outflow hydrograph falls on the recession limb of the inflow hydrograph and the rising limb of the outflow hydrograph is approximated by a straight line.

Figure 4.45 schematically represents the situation during a rainfall event.

Q_a Outflow required from storage volume (m^3/h)
Q_d Peak stormwater flow (m^3/h)
T_d Stormwater duration (h)
T_c Time of concentration (h)
V Volume of the storage (m^3)

The following equation relates the required pump out rate to the required storage volume for the rainwater in order to prevent overflow.

$$Q_a \times (2T_d + T_c) = 2 \times (Q_d T_d - V) \tag{4.39}$$

Solving for Q_a

$$Q_a = 2 \times (Q_d T_d - V) / (2T_d + T_c) \tag{4.40}$$

and Solving for V

$$V = Q_d T_d - \left[\{ Q_a \times (2T_d + T_c) \} / 2 \right] \tag{4.41}$$

The stormwater calculations in this example are based on the following rainfall data (Table 4.36).

Table 4.36 Sample rainfall data

Duration (min)	Duration (h)	10-year storm (mm/h)	25-year storm (mm/h)
5	0.08	68.6	81.85
10	0.17	48.75	60.15
15	0.25	40.15	49.95
30	0.5	27.6	35.35
60	1	18	23.1
360	6	4.7	5.95
720	12	2.85	3.3
1,440	24	2.1	2.4

This type of data is usually available from local meteorological stations. Typically, stormwater calculations are based on cumulative rainfall for a storm of one-in-25 year return storm or a one in 10-year return events.

The time of concentration (the time is takes for the rain falling on the furthest point on the refinery plot to the destination) for the drainage areas in the refinery is dependent on the layout of sewer lines, the size of the sewer lines, the material of construction of the lines and the condition of the lines (affects friction factor) and requires detailed analysis of the system. For the purposes of this example, the time of concentration in refinery drainage systems is assumed to be 30 min.

The peak stormwater flow Q_d can be calculated using Eq. 4.38 and the data from the Table 4.36.

Using Eq. 4.41 the storage volume V can be calculated for different outflow rates. For the purposes of this example outflow rates of 50 and 150 m^3/h are used (Tables 4.37, 4.38 and 4.39).

The storage volume that can be provided at the refinery will need to be evaluated based on factors such as available hydraulic capacity in the wastewater treatment plant (WWTP) to handle stormwater flow, available real estate for stormwater storage, capital cost for building the required storage volume, etc.

Table 4.37 Peak stormwater flows

Duration (h)	Peak flow 10-year storm (m^3/h)	Peak flow 25-year storm (m^3/h)
0.08	32,132	38,339
0.17	22,835	28,174
0.25	18,806	23,397
0.50	12,928	16,558
1.00	8,431	10,820
6.00	2,201	2,787
12.00	1,335	1,546
24.00	984	1,124

Table 4.38 Storage volume calculation—50 m^3/h

Outflow rate (Q_a)	50 m^3/h	
Duration (h)	10-year storm (m^3)	25-year storm (m^3)
0.08	2,661	3,178
0.17	3,785	4,675
0.25	4,677	5,824
0.50	6,426	8,242
1.00	8,369	10,758
6.00	12,896	16,409
12.00	15,407	17,936
24.00	22,395	25,767

Table 4.39 Storage volume calculation—150 m^3/h

Outflow rate (Q_a)	150 m^3/h	
Duration (h)	10-year storm (m^3)	25-year storm (m^3)
0.08	2,628	3,145
0.17	3,743	4,633
0.25	4,627	5,774
0.50	6,351	8,167
1.00	8,244	10,633
6.00	12,271	15,784
12.00	14,182	16,711
24.00	19,970	23,342

Another rainwater management strategy that can be implemented is what is commonly referred to as the "first flush" concept. The first few millimeters of rain—say 25–50 mm—is captured and treated while the remainder is discharged without treatment. This concept assumes that any hydrocarbons that are present on the paved areas get washed off with the first few millimeters of rain and therefore the remainder would be largely uncontaminated. It must be recognized that the topography of the refinery plot might require collection of more water from areas of lower elevation so that the target of 25–50 mm of rain is collected from the areas of higher elevation thus ensuring that the hydrocarbons are adequately washed off from both higher and lower elevation areas. Figure 4.46 illustrates how such a system could be implemented.

The rainwater from the paved area is collected and sent to storage, either a pond or an above-ground tank. Tanks are used in those locations where the lack of adequate real estate precludes the installment of ponds which occupy a relatively large surface area. In either case the total water collected can be monitored using a totalizer. When the total volume reaches the equivalent of 25 or 50 mm of rainfall, the flow is shutoff using a solenoid valve and the remainder is diverted to discharge using the diversion solenoid valve. The rainfall that is collected in the pond or tank is worked off as quickly as possible on a controlled basis to ensure that it does not overwhelm the

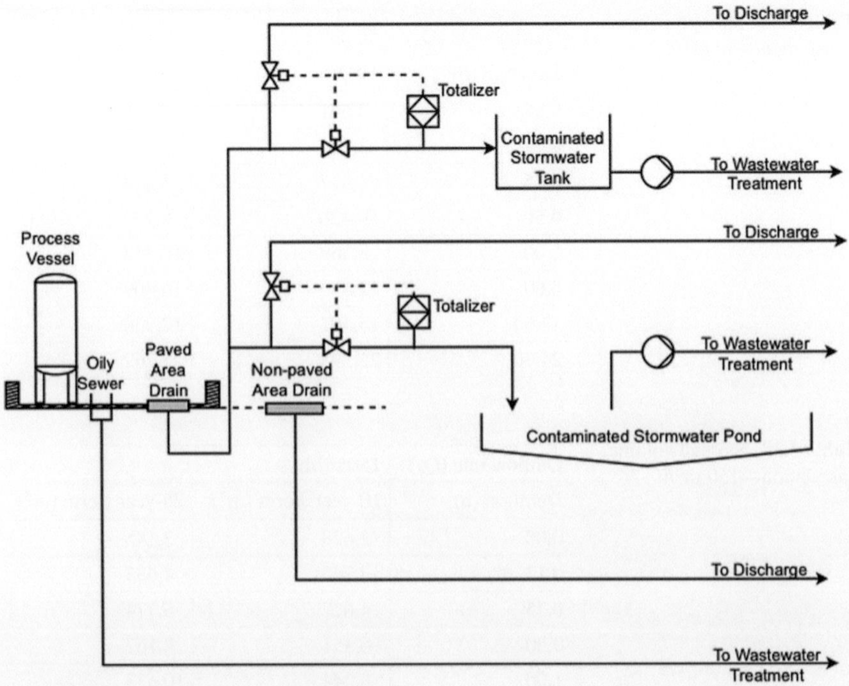

Fig. 4.46 First flush rainfall management

WWTP. For this strategy to be successful, the pond/tank must be emptied as soon as practical to ensure that it is ready to receive rainwater from the next rainfall event.

At some refineries, depending on the location and local regulations it might be possible to recover rainwater for recycling. This can be done using ponds/basins when there is enough real estate available, and the topography allows for gravity flow. Figure 4.47 illustrates this concept.

This example shows a pond/basin that is divided into the following chambers—Inlet flow channel, Peak flow basin, First flush basin and Outflow channel. The flow into each of these chambers is controlled by means of weirs and baffles. The rainwater from the paved areas flow into the inlet channel. Then first flush from the paved areas in the refinery flows into the First flush basin from the inlet channel by means of a weir. The volume of this basin has been calculated to collect a controlled volume of rainwater from the refinery. Any excess rainwater beyond the first flush flows into the Peak flow basin for collection. When this basin gets full it overflows into the outflow channel from where it is sent to discharge.

The water collected in the First flush basin is sent to wastewater treatment and the water from the Peak flow basin is sent back to the refinery for recycle. Given that oil refineries use significant quantities of water, recycle and reuse of wastewater produced in the refinery is an important consideration. This issue is discussed in detail in Chap. 8 of this book.

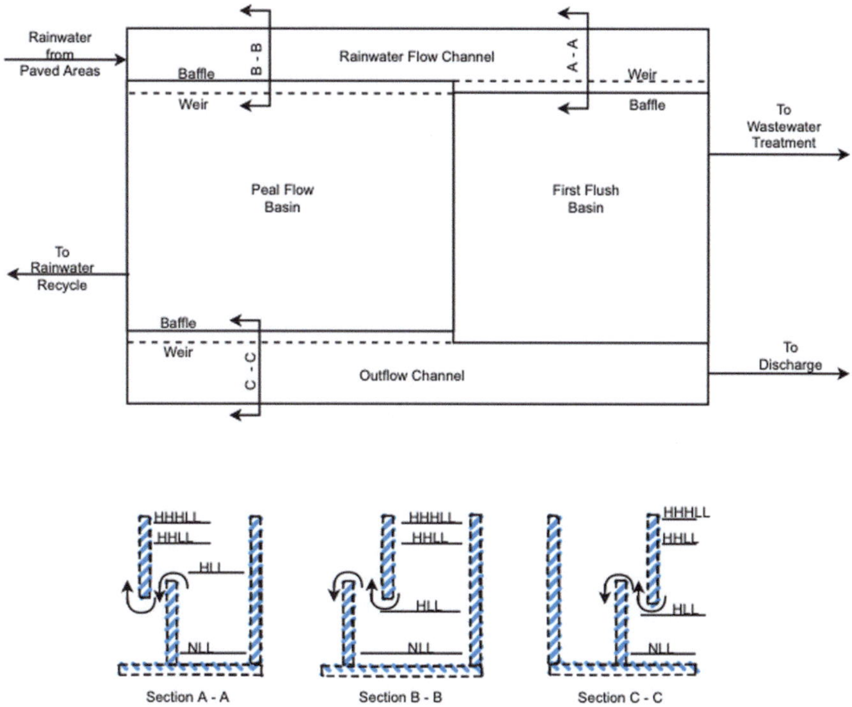

Fig. 4.47 Recycle of rainwater

4.2.8 Firewater Storage and Management System

Firewater system is a critical component of any petroleum industrial operation, including refineries, to ensure the safety and security of staff members and communities, and to prevent the potential loss of assets. It consists of the components:

- Source and Storage
- Pumping and Distribution
- Management of Operation and Maintenance

Firewater can be sourced from natural sources of water e.g., lakes, rivers and streams. Although these are unlimited sources, the potentials for change in water levels during flooding or draught conditions, freezing and clogging situations in cold climate operations, and fouling by debris and suspended solids need consideration during design. Some refineries that possess large land areas, store stormwater from non-process paved areas, and uncontaminated post first flush rainwater from process paved areas, in ponds or reservoirs to utilize it totally or partially for fire water consumption. These are examples of beneficial water recycle and reuse. Rainwater is relatively free from suspended solids and may not need clarification for use as firewater.

Table 4.40 Guidelines for fire water pump selections based on refinery crude processing capacity

Crude processing capacity, cu m/d (bpd)	Minimum pumping capacity, Liters/min (GPM)/pump	Minimum no. of pumps
≤1,600 (10,000)	11,360 (3,000)	2
1,600–3,200 (10,000–20,000)	15,100 (4,000)	3
3,200–8,000 (10,000–50,000)	22,710 (6,000)	3
8,000–16,000 (50,000–100,000)	37,850 (10,000)	4
≥16,000 (100,000)	37,850–75,700 (10,000–20,000)	6

The pumping capacity is specific to a site and determined on a case-by-case basis. A general rule of thumb in the practice of firewater system design is presented in Table 4.40.

The water source should be capable of supplying the amount of water for a minimum duration of four (4) hours for a small to medium size refinery. It should be higher for larger refineries based on 150% capacity of all the operating pumps combined together for 4 hours. The pumps should be horizontal centrifugal type with positive suction during the entire operation or vertical centrifugal type. At least 50% of the operating pumps should be diesel driven such that fire water pumping is not interrupted in the event pf power failure during a fire event.

Considering the critical nature of fire water system the entire pumping and distribution network should have a routine monitoring and preventive maintenance program.

4.2.9 Odors and Emissions

At refineries odors are caused due to emissions of light hydrocarbons as well as miscellaneous sulfur compounds. These emissions come from various sources such as:

- Leaks from floating roof storage tanks roof seals
- Process sewer hubs
- Process drains
- Sewer vents
- Crude tank drain sumps
- Other storage tank drain sumps
- Wastewater Treatment Equipment including
  - Lift Stations and Collection Sumps
  - API Separators
  - Oil Collection Tank
  - Dissolved Air Flotation

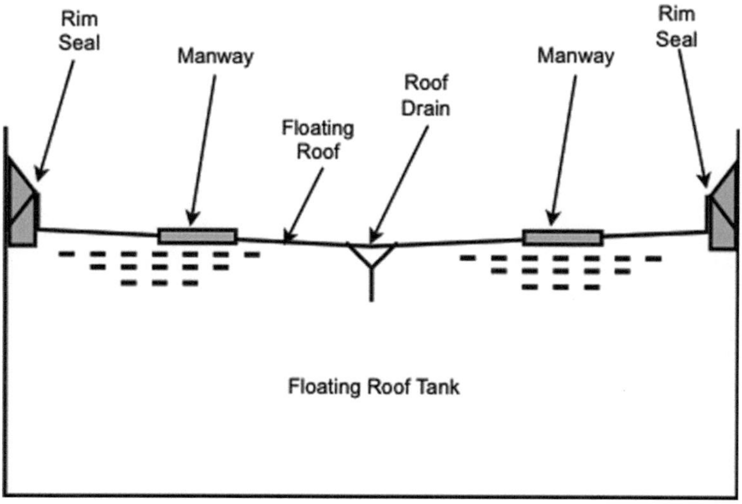

Fig. 4.48 Floating roof tank roof seal

Odor Abatement Strategies

Vents from process equipment in the refinery are normally connected to the flare header in the refinery thereby eliminating potential odors from these sources.

Floating Roof Tanks

Leaks from floating roof storage tank roofs can be minimized by monitoring and maintaining floating roof rim seals. Figure 4.48 is an illustration of a typical floating tank including the rim seals.

As the floating roof in the tank moves up and down the walls of the tank is coated with a thin film of hydrocarbon. When this gets exposed to the atmosphere the light components will vaporize and could contribute to emissions in addition to the emissions due to leaky rim seals. If necessary, these emissions can be eliminated by converting the tank to an internal floating roof tank which is illustrated in the Fig. 4.49.

In this type of tank an external cone roof is installed on the tank to contain any vapors. However, this introduces another level of complexity in that it could lead to the presence of explosive mixtures under the roof due to the presence of air. This situation can be eliminated by nitrogen blanketing the vapor space and controlling the vent. However, this is deemed to be an impractical solution especially with large crude storage tanks which can be 40 m in diameter or more and therefore rarely implemented.

Process Sewer Hubs and Process Drains

Emissions from sewer hubs and drains in the process units can be minimized by ensuring that the drain line from process equipment is introduced below the level of the pavement. Although this will not eliminate odors it could significantly minimize it.

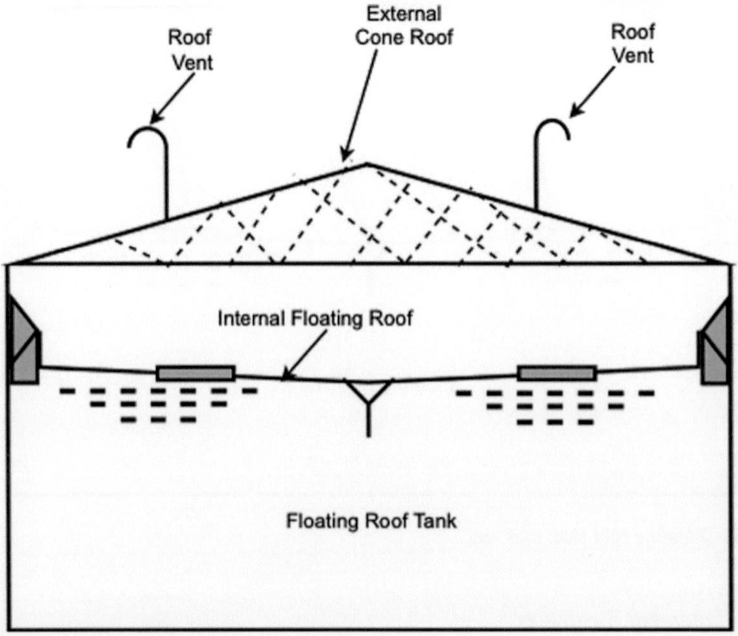

Fig. 4.49 Internal floating roof tank

Sewer vents

Some refineries in the United States have installed vents on their sewer systems in order to capture benzene emissions in response to the NESHAPS (National Emission Standards for Hazardous Air Pollutants) regulations. These vents are routed through canister that contains granular activated carbon (GAC) to adsorb any volatile organic compounds (VOCs). The carbon canisters need to be monitored regularly for breakthrough which occurs when the carbon in the canister gets saturated and is unable to adsorb any more VOCs. Care must also be taken to design the system such that the back pressure imposed by the vent control equipment on the sewer system is minimal so that proper gravity flow in the sewers can be maintained.

Crude Tank Drain Sumps

These sumps are normally open to the atmosphere and can be a significant source of emissions during the draining operation. In many parts of the world these sumps are required to be covered and vapor controlled. Figure 4.50 shows how this can be accomplished.

In this setup, the tank drain is sent to a vapor tight covered sump. In order to prevent the presence explosive mixtures in the sump, it is blanketed with nitrogen. The vent from the sump is sent under pressure control to a canister that is filled with activated carbon into which can adsorb the VOCs. The vent from the canister is sent to atmosphere. The vent will need to be monitored regularly to ensure that there is no break-through of the carbon. When this happens the canister will need to be filled with new carbon or replaced with a new canister.

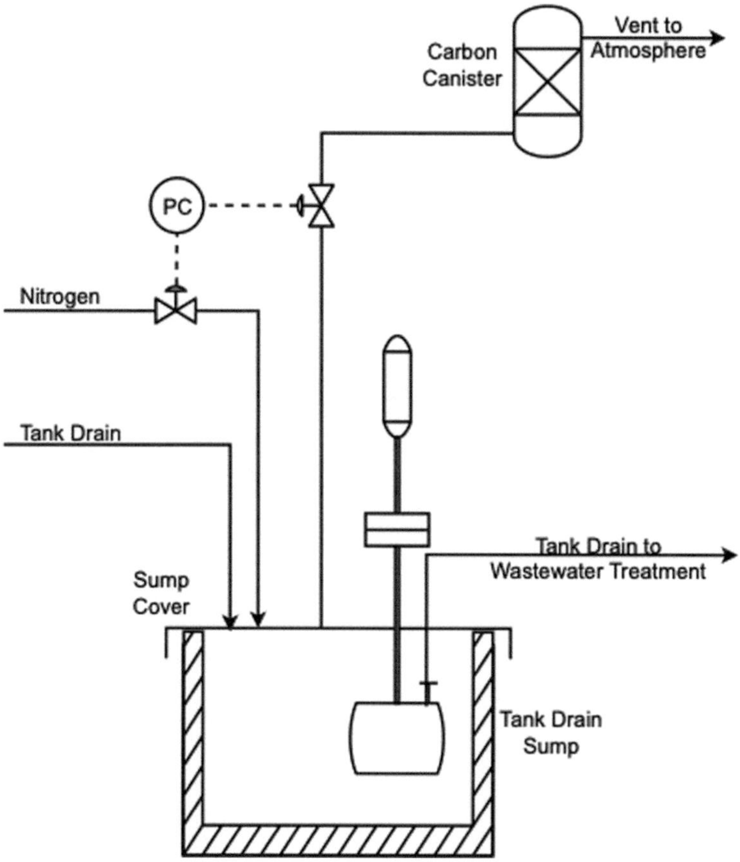

Fig. 4.50 Covered tank drain sump

Other Storage Tank Drain Sumps

Emission and odor control from these sumps is similar to that shown above with the Crude tank drain sumps. The emissions/odor issues associated with the wastewater treatment equipment is discussed in Chaps. 6 and 7.

4.3 Marketing

The fourth sector in the oil industry supply chain is marketing. The activities in this sector include storage, transportation and distribution of various refined products. The terminals are located at various urban locations as well as airport in major cities.

Typical distribution terminals consist of a number of storage tanks where products such as gasoline, diesel and heating oil are stored for further distribution. Additives are usually added to the products before they are sent out to marketing stations. The tanks in the terminals are usually supplied by pipelines from refineries. There could be multiple terminals located in large urban areas. Terminals are also located at airports for the supply of jet fuel. The products from the tanks are loaded on to trucks at loading stations located at the terminal. Figure 4.51 shows a diagram to a typical loading station.

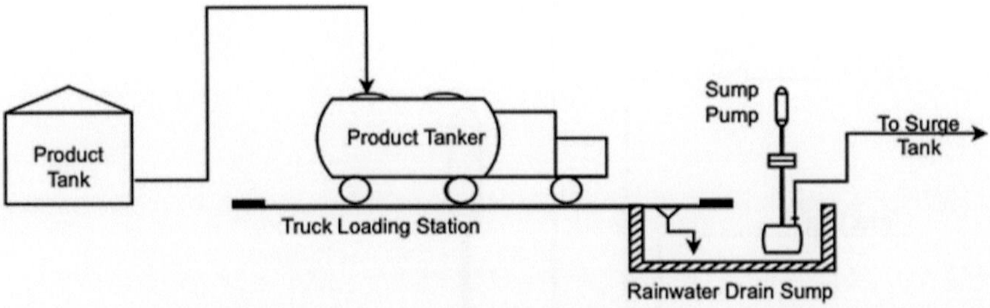

Fig. 4.51 Terminal loading station

The figure shows a truck being loaded with a product such as gasoline. There could be multiple loading stations at a terminal. Each loading stations is typically equipped with a drain that collects rainwater as well as any inadvertent spills that might occur. The drains are collected in an underground sump which is usually located adjacent to the loading station. The sump is equipped with a pump that is used to remove any collected water.

Gasoline and jet fuel are generally stored in floating roof tanks to avoid loss of product due to vaporization and also prevent emissions of light hydrocarbons to the atmosphere. The floating roof tanks are equipped with roof drains to remove any collected rainwater. Figure 4.52 shows the arrangement of the roof drain in a floating roof tank.

The tanks are located inside a berm/dike to contain any potential leaks. The roof drain is routed through the inside of the tank by means of a flexible hose or articulated pipe and sent to a sump that is usually located close to the tank. A sump pump is used

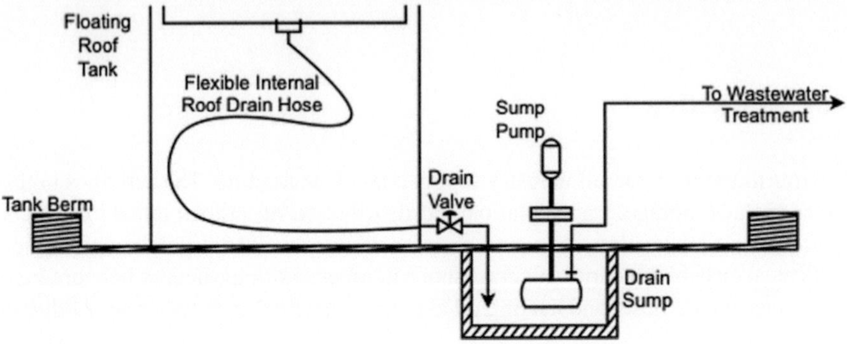

Fig. 4.52 Floating roof tank roof drain

Table 4.41 Expected contaminants—terminal wastewater

Contaminant	mg/L
Chemical oxygen demand	Up to 3,000
Biological oxygen demand	Up to 2,200
Total suspended solids	Up to 200
Free hydrocarbons	Up to 200

to remove the contents of the sump. If there is any deterioration of the flexible hose or articulated pipe it will result in the leakage of tank product into the drain line and will get mixed in with the collected rainwater which will need to be treated. Table 4.41 shows the expected contaminants in the wastewater from marketing terminals.

Abbreviations and Glossary of Terms

API	American Petroleum Institute
ASO	Acid Soluble Oil
BFW	Boiler Feed Water
BOD	Biochemical Oxygen Demand
BS&W	Bottoms Sediment and Water
CDU	Crude Distillation Unit
cEOR	Chemically Enhanced Water
COC	Cycles of Concentration
COD	Chemical Oxygen Demand
CWR	Cooling Water Return
DO	Dissolved Oxygen
EDI	Electrodeionization
EOF	Entirely Oil Free
EPA	Environmental Protection Agency
FCC	Fluid Catalytic Cracking
FS	Filterable Solids
GWI	Global Water Intelligence
HDS	Hydrodesulfurization
HF Alkylation	Hydrofluoric Acid Alkylation
HP	High Pressure
IRB	Iron Reducing Bacteria
ISBL	Inside Battery Limit
IX	Ion Exchange
KO Drum	Knockout Drum
LP	Low Pressure
LPG	Liquified Petroleum Gas

MEK	Methyl Ethyl Ketone
Merox	Mercaptan Oxidation
MIC	Microbiologically Induced Corrosion
MP	Medium Pressure
MRM	Modified Rational Method
NESHAP	National Emissions Standard for Hazardous Air Pollutants
NOAA	National Oceanic and Atmospheric Administration
NOM	Natural Organic Matter
O&M	Operation and Maintenance
OSBL	Outside Battery Limit
PED	Portable Electric Desalter
pH	$-\log_{10}[H^+]$
PTU	Purge Treatment Unit
RO	Reverse Osmosis
SAC	Strong Acid Cation
SBA	Strong Base Anion
SRB	Sulfur Reducing Bacteria
SSW	Stripped Sour Water
SWS	Sour Water Stripper
TAN	Total Acid Number
TDS	Total Dissolved Solids
TGTU	Tail Gas Treating Unit
TKN	Total Kjeldahl Nitrogen
UF	Ultrafiltration
UOP	Universal Oil Products
VDU	Vacuum Distillation Unit
VOC	Volatile Organic Carbon
WHB	Waste Heat Boiler
WWTP	Wastewater Treatment Plant

References

Acids and caustics from petroleum refining. https://www.petroleumhpv.org
https://ceng.tu.edu.iq/ched/images/lectures/chem-lec/st4/c2/lec.21&22.pdf
Basu S (2018) Impact of opportunity crudes on refinery desalter and wastewater performance, hydrocarbon processing, August 2018
EPA (1999) Phase I final rule and technical development document of uniform national discharge standards (UNDS), seawater cooling overboard discharge: nature of discharge, EPA-842-R-99-001

Flynn D (ed) (2009) The Nalco water handbook, 3rd edition. McGraw Hill, New York
Giles KA (2023) Fundamentals of petroleum refining. PDHengineer.com
Global Water Intelligence (2012) Industrial desalination and water reuse, GWI report, August 2012
Stultz SC, Kitto JB (2005) Steam: its generation and use, 41st edition. The Babcock & Wilcox Company
World Resources Institute, 26 August 2015

Chapter 5
Wastewater Treatment—Downstream and Marketing Operations

Mudumbai Venkatesh, Andrew R. Shaw, Somnath Basu, Saurabh P. Deshpande, and Arvind D. Kannan

Abstract An important part of water management in the downstream sector of the petroleum industry is proper handling, treatment and disposal of wastewater generated from refining operations. Refinery effluent streams contain chemical constituents that may be harmful to human health and toxic to aquatic life. Proper treatment of such streams is essential for safe disposal into receiving water bodies and/or for beneficial reuse. This chapter includes a discussion of wastewater treatment in the downstream sector of the petroleum industry. This includes the various unit operations and processes that are typically used in a refinery wastewater treatment plant and their working principles. Emissions/odor issues related to each of the relevant unit operations are also discussed.

Keywords Activated carbon · Activated sludge · Aerated lagoon · API separator · Attached growth · Belt filter press · Centrifuge · Chemical oxidation · CPI separator · Diffuser · Dissolved air flotation (DAF) · Dissolved oil · Downstream · Emulsified oil · Equalization · F:M ratio · Floating roof tank · Free oil · GAC ·

M. Venkatesh (✉)
Process and Energy Consultant, Andover, MA, United States
e-mail: m.venkatesh1@yahoo.com

A. R. Shaw
Black and Veatch, Houston, TX, USA
e-mail: shawAR@bv.com

S. Basu
Jewel Innovations in Energy and Environment, Katy, TX, USA
e-mail: som.basu@jewelenergyenvironment.com

S. P. Deshpande
Black and Veatch, Toronto, ON, Canada
e-mail: DeshpandeSP@bv.com

A. D. Kannan
Carollo Engineers, Dallas, TX, USA
e-mail: akannan@carollo.com

S. Basu et al. (eds.), *Water Management in Petroleum Industries*, Water Resources Development and Management, https://doi.org/10.1007/978-981-19-3159-8_5

IFAS · Induced air flotation (IAF) · Lift station · Membrane bioreactor (MBR) · Microfiltration (MF) · Mixed liquor suspended solids (MLSS) · Mixed liquor volatile suspended solids (MLVSS) · Moving bed biological reactor (MBBR) · Oil & grease · Oil/water separation · PACT · RAS · Reed bed · Rotating biological contactor (RBC) · Sand filtration · Sequencing batch reactor (SBR) · Stoke's law · Surface aerator · Suspended growth · Tertiary treatment · Thickener · Trickling filters · WAS · Wetlands

5.1 Introduction

Characteristics and treatment of industrial wastewaters depend upon the type of industry discharging such wastes. This section is focused on introducing the basic elements of wastewater treatment from petroleum industries as it relates to the downstream sector (oil refineries) and marketing sector (terminals). Detailed discussions on design and operation of the important parts of refinery wastewater treatment systems and terminal wastewater treatment systems have been presented in the following sections of this chapter.

Major unit operations and unit processes in the wastewater treatment plants include preliminary treatment, primary treatment, secondary treatment, tertiary treatment, and solids handing. A broad overview of these processes is summarized in Table 5.1.

5.2 Preliminary Treatment

Preliminary treatment is the very first step of a wastewater treatment plant. Its main purpose is to protect the equipment and the overall operations of the treatment plant. It facilitates the removal of large debris and heavy inorganic materials like metal cans, rags, sand, etc. that may damage mechanical equipment and interfere with the operations of a treatment plant.

Preliminary treatment mainly consists of screening, shredding, and grit removal to separate coarse materials from the wastewater being treated. These treatment units are located at the front end, also called headworks, of the treatment plant and protect downstream treatment processes. A preliminary treatment system is typically not utilized in the petroleum industry and hence this unit operation is not discussed in detail. The section below highlights only some essential features of this treatment step.

Table 5.1 Unit operations and processes in wastewater treatment

Treatment	Unit operation/unit process	Application
Preliminary treatment[1]	Screening	Unit operation which is used to remove large objects e.g., debris, rags, plastic, and metal cans, etc.
	Grit removal	Unit operation to remove heavy materials like sand, grit, gravel etc.
Primary treatment	TSS and oil/water separation by settling or floatation using primary oil/water separation	Unit operation to remove settleable solids, oil, and floatable materials. Mainly this step is intended to remove free phase oil and particulate matters
	Secondary oil/water separation by floatation	Mainly this step is intended to remove emulsified/dispersed oil and fine solids
Secondary treatment	Biological treatment	Unit process to remove dissolved and colloidal organic contaminants
	Secondary solid/liquid separation	Unit operation for solid–liquid separation by gravity settling or floatation to remove solids produced by the biological treatment and produce high quality effluent suitable for disposal and/or reuse
Tertiary treatment	Sand filtration, Membrane separations, Granular activated carbon (GAC) adsorption, chemical oxidation, nitrification and denitrification[2], other advanced treatments as needed, e.g. selenium control	Unit processes to remove fine particles, recalcitrant organics, ammonia, and Total Nitrogen (TN),[2] selenium (if required by the discharge regulations), etc.
Disinfection[3]	Chlorination, ultraviolet radiation	Unit process to reduce water borne pathogens and other microorganisms from the wastewater
Residual handling, treatment, and disposal	Thickening, Stabilization, Dewatering	Unit operations and processes to reduce the sludge volume and produce stabilized and dewatered sludge solids for a suitable disposal

Notes

[1]—An important unit operation in domestic wastewater treatment. In petroleum refineries it is used in open circulating cooling systems that use river or seawater as cooling water

[2]—Not common for petroleum refinery wastewater treatment, but for new installations at many locations this is increasingly becoming a regulatory requirement

[3]—Typically not used for petroleum refinery wastewater treatment unless sanitary waste is present. In petroleum refineries it is preferrable to segregate the sanitary waste and treat it separately. Disinfection is an important unit process in cooling water treatment to control algae growth and microbiologically induced corrosion (MIC)

5.2.1 Screening

Screening removes objects such as rags, paper, plastics, and metal cans or containers to prevent damage and clogging of downstream equipment and piping. Screens can be classified as coarse screen and fine screen based on the screen opening size, or the solid's particle size. Some wastewater treatment plants use both coarse and fine screens to achieve enhanced solids removals for the protection of downstream equipment (Table 5.2).

Table 5.2 Types of screens in wastewater treatment Facilities

Broad classification of screen	Screen type	Opening size	Description
Coarse screen	**Manually cleaned screen**	**>6 mm**	These screens are typically installed at small treatment facilities. These are generally used as a standby screen during peak flow conditions or during maintenance of mechanically cleaned screens
	Mechanically cleaned screen	**>6 mm**	These screens consist of parallel bars or stepped plates placed at an angle in the path of the wastewater flow and are used to remove debris and large solids entrained in wastewater. Mechanical rakes clear debris from the bars. The screenings are washed and compressed to remove excess water and are ultimately disposed of by burial in a landfill. Refer to Fig. 5.1 for the front view of a mechanically cleaned screen
Fine screen	**Drum screens**	**<6 mm**	Wastewater either enters these screens from one side (single entry) or both sides. The wastewater is separated and collected within the screen and the solids are released to a hopper. The debris are washed by a series of water jets
	Step Screens	**<6 mm**	These screens consist of two sets of step shaped plates in which one is movable and other is fixed. The movable steps move in vertical motion; this motion separates the solids from the wastewater moving them up the steps. The solids are then collected in a hopper at the top of the screen
	Traveling band	**<6 mm**	Multiple screens are attached together to form screen panels or bands. These move vertically upwards trapping solids inside the screens. The trapped solids are then washed and collected from the top

5.2.2 Grit Removal

A variety of inorganic solids such as sand, silt, pebbles, glass, metal pieces, etc., combined with organic solid objects, found in wastewater is known as grit. Accordingly, grit characteristics vary significantly. Grit removal systems utilize separation by settling of grit particles. Traditionally, grit removal equipment has been designed to remove up to 75% of particles with the settling velocity of sand or silica of specific gravity 2.65, and 0.15 mm (0.006 in) size. However, in field applications, it has been observed that the performance of grit removal equipment sometimes falls short of the design expectation due to the presence of surface active agents (SAA) which comprise of oil, grease, and other organic matters. These materials coat the surface of the grit particles that render them lighter than sand such that the settling velocity of a grit particle is typically smaller than that of a clean sand particle of the same size. The term sand equivalent size (SES) is defined to relate the settling velocity of wastewater grit to that of clean sand. SES of a grit particle is the size of a clean sand particle having the same settling velocity as that of a sand particle. Therefore, the SES of a grit particle is smaller than its actual size. In the current design practice, most high efficiency grit removal systems are designed for particle sizes lower than 0.15 mm, typically with a sand equivalent size (SES) of 0.075–0.15 mm (Metcalf & Eddy/AECOM 2014).

Grit removal from wastewater is mainly achieved by settling of grit particles from wastewater, but there are other types of degritting equipment also e.g., those utilizing

Fig. 5.1 Mechanically cleaned coarse screens used in wastewater treatment. *Photo* Courtesy of Headworks International, Houston, TX

Table 5.3 Types of grit removal treatment systems

Types	Description
Aerated grit chambers	This type of grit removal system involves the wastewater flowing in a spiral pattern by means of air fed in the grit chamber along one side which causes a perpendicular spiral velocity pattern to flow through the tank
Vortex—type grit removal systems	The vortex type grit chamber consists of a cylindrical tank, the flow enters the tank tangentially thereby creating a vortex type of flow pattern Grit present in the wastewater settles under gravity into the bottom of the tank hopper. The wastewater leaves from the top of the tank. The grit settling at the tank hopper is removed by a grit pump
Detritus tanks	A square degritter tank which is also known as detritus is based on a constant level with a shorter retention settling tank. This type of grit removal system employs a grit washing system to remove organic matter and an optional grit auger and a rake arrangement that removes and classifies grit from the grit sump
Horizontal flow grit chambers	This is the oldest type of grit removal system which operates on maintaining an upstream velocity of 0.3 m/s. Velocity in the chamber is controlled by a rectangular control section (Parshall flumes). Heavy grit particles in the wastewater settle to the bottom of the chamber and the lighter organic particles remain suspended and exit the channel along with the wastewater
Hydrocyclones	Hydrocyclones use differences in specific gravities and high centrifugal forces to separate solids from water or oil. Due to the high centrifugal force, heavier grit particles are collected on the sides and drop to the bottom of the cyclone and the lighter particles are removed from the top of the cyclone

the principles of vortex, centrifugal force, etc., that are commercially available, as listed in Table 5.3 above.

5.3 Primary Treatment

Primary treatment is a group of unit operations which utilizes gravity and floatation to remove solids and floating matters present in the wastewater. In conventional wastewater treatment systems handling domestic wastewater the role of primary treatment is to remove a large fraction (50 to 70%) of suspended solids associated with the influent. However, in a petroleum refinery wastewater treatment system, the primary treatment steps consist of a significant amount of oil and grease removal in addition to solids removal.

Physical and chemical characteristics of crudes may pose unique challenges to the refinery wastewater treatment operation. These are more pronounced when refineries process opportunity crudes—the types of crude oil that refiners can purchase at steep discounts against typical crude oil due to their low API gravity and high sulfur content.

Some of these crudes cannot be effectively handled by units in the refinery such as desalters and sour water strippers. Emulsification of oil in the desalters can cause inadvertent discharges of oil from the desalter. Effective Oil/Water separation is a critical primary treatment step of refinery wastewater because getting the oil content down to acceptable levels is imperative for effective biological treatment. Oil and grease (O&G) above the range 100 to 300 mg/L are harmful to the microorganisms in biological treatment. Additionally, excessive oil in the bioreactor can result in foaming and frothing which could lead to deterioration of the bioreactor performance. Details of these adverse impacts of excessive oil on biological treatment process are discussed later in this chapter and in the literature (Jenkins et al. 2003). This problem is not limited to traditional activated sludge process only. MBR and MBBR, which are the recent developments of biological treatment, are also not immune. The effect is more pronounced because the presence of more than 10 mg/L of oil will hinder mass transfer in the MBBR media or foul the MBR membranes. Therefore, it is imperative that the primary treatment equipment is properly maintained and operated to ensure optimal performance of the biological treatment system. Table 5.4 shows the expected concentrations at the inlet to the oil/water separation section of the treatment plant.

There are three types of oil present in the feed to the wastewater treatment plant.

- Free Phase Oil—this consists of oil droplets that typically rise to the surface because their specific gravity is less than water. With enough residence time, free oil can be separated from the water by gravity.
- Emulsified Oil—this consists of oil droplets that are small in diameter (generally less than 25 μm) and form a stable emulsion with water and are therefore difficult to separate by gravity alone.
- Dissolved Oil—this fraction of oil is dissolved in the water up to the solubility limit of the various components in the oil. This fraction is removed from water using biological treatment downstream of the oil/water separation section of the wastewater treatment plant.

All the above types of oil combined together is Total Oil. A detailed discussion of the types of chemical compounds falling under the definition of Oil & Grease, and the analytical methods, have been presented in Chap. 1.

Figure 5.2 presents the typical block flow diagram of a refinery wastewater treatment plant. Refinery wastewater treatment consists of two stages of oil/water separation—primary and secondary, in order to get oil content down to suitable levels for biological treatment. The equalization step provides surge capacity as well as handles variations in flow to the treatment plant. Biological treatment, clarification and tertiary treatment are downstream of Equalization. Placement and location of the equalization step is specific to a treatment plant depending on the variations of flow, and the frequencies of flow and Oil & Grease concentration fluctuations. This is further discussed later in this chapter.

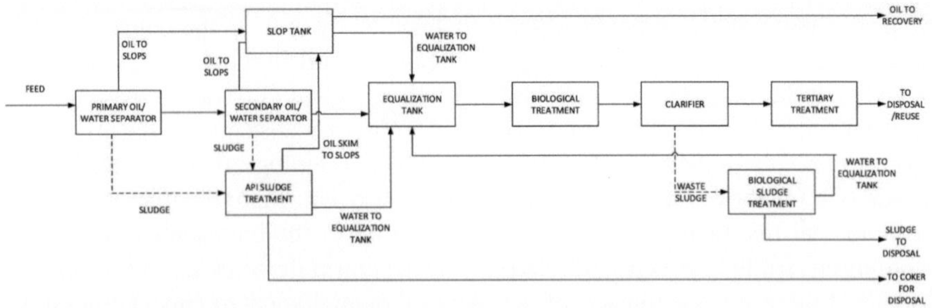

Fig. 5.2 Refinery Wastewater Treatment System Block Flow Diagram

5.3.1 Primary Oil/Water Separation

Bulk removal of oil and suspended solids, including a major part of the free phase oil and some emulsified oil, is conducted in this step. The equipment utilized are:

- Primary Clarifier or Sedimentation Tank
- American Petroleum Institute (API) Separator
- Corrugated Plate Interceptor (CPI) Separator

5.3.1.1 Primary Clarifier

The main function of primary clarifiers is to remove suspended solids. It also removes some amount of oil and grease. In refinery applications it is a less commonly used equipment for primary treatment as oil and grease form a major part of the wastewater. However, primary clarifiers are sometimes used for streams containing solids but low concentrations of oil and grease. One such example is segregated stormwater streams from non-process areas that can be collected, treated, and recycled within the refinery complex for reuse. However, the main application of primary clarifiers in a refinery is the make-up water (instead of wastewater) treatment.

5.3.1.2 American Petroleum Institute (API) Separators

The most commonly used equipment for primary oil/water separation is the API Separator. They perform the first stage of oil-water separation and conducts the bulk removal of free phase oil droplets (~150 μm or greater size) and some emulsified oil. Depending upon the source of wastewater the API separator influent oil concentration can be as high as 60,000 mg/L as shown in Table 5.4. A survey of refinery wastewaters (API 1985) demonstrated the influent oil concentration ranging from 200 to 80,000 mg/L, with the majority of data points lying between 1,000 and 8,000 mg/L. The spread of effluent concentration data is relatively small, in the range of 40 to 200 mg/L.

Table 5.4 Oil/water separation inlet concentration

Component	Units	Concentration
Oil & Grease (O&G)	mg/L	Up to 60,000
Total Suspended Solids (TSS)	mg/L	Up to 30,000

API separators are used in petroleum refinery wastewater treatment systems where the wastewater contains oil and oily sludge. These separators are tanks or surface impoundments that provide sufficient retention time for solids to settle and oil droplets to coalesce and float up. The operating principle of API separators is based on the difference in specific gravity that allows the heavier matter to settle below lighter liquids. Organics/oil that floats on the top surface is skimmed off and the sludge settles to the bottom of the separator and is removed continuously.

Typically, the API Separator is located below grade to facilitate gravity flow of the wastewater to the treatment plant. At some refineries the separator is located above grade at an elevation of 1.5 to 2 m to allow for proper sludge removal from the bottom of the separator. These above ground separators are fed from lift stations that collect wastewater streams in the refinery and pump the wastewater to the equipment.

Figures 5.3a, b and 5.4 present the plan and elevation views, and schematic representation of API Separators. Figure 5.5 a and b depict the installation of an API separator at a refinery and its internal components.

Typically, API separators are rectangular in shape and are up to 5 m in width and can be more than 20 m long. They consist of multiple bays to accommodate the flow and allow for periodic cleaning of accumulated sludge from the separator. As the wastewater enters the separator it is intercepted by a trash trap that consists of inclined parallel bars to trap any large solids that might be present in the wastewater. The wastewater then passes under an oil retention baffle and through a flow distributor into the main chamber of the separator. The main chamber is designed to provide quiescent conditions to allow oil droplets to rise to the surface. In the main chamber free oil rises to the surface and is removed by a slotted pipe located at the far end of the separator. The slotted pipe extends the entire width of the separator and is equipped with a long handle so that it can be manually adjusted to skim oil as the situation needs. Oil skimmed by the slotted pipe is drained into a sump and then pumped to a slop tank. This tank is heated and the residence time in this tank allows separation of oil and water layers by gravity. Oil collected on the top surface of the slop tank is recovered and sent to crude tanks for blending and processing. Water layer from to slop tank is directed to the wastewater treatment plant.

The water from the chamber flows under the oil retention baffle located adjacent to the slotted pipe, and over the adjustable overflow weir into the outlet chamber from where it flows onward. The bottom of the chamber is equipped with a moving scraper that pushes the settled sludge towards the sludge sump in the separator. The collected sludge is removed from the sludge sump.

The rise rate of the oil droplets (vertical velocity) is a function of the surface loading rate in the separator. The rise rate is the velocity at which oil particles move toward the separator surface as a result of the differential density of the oil phase

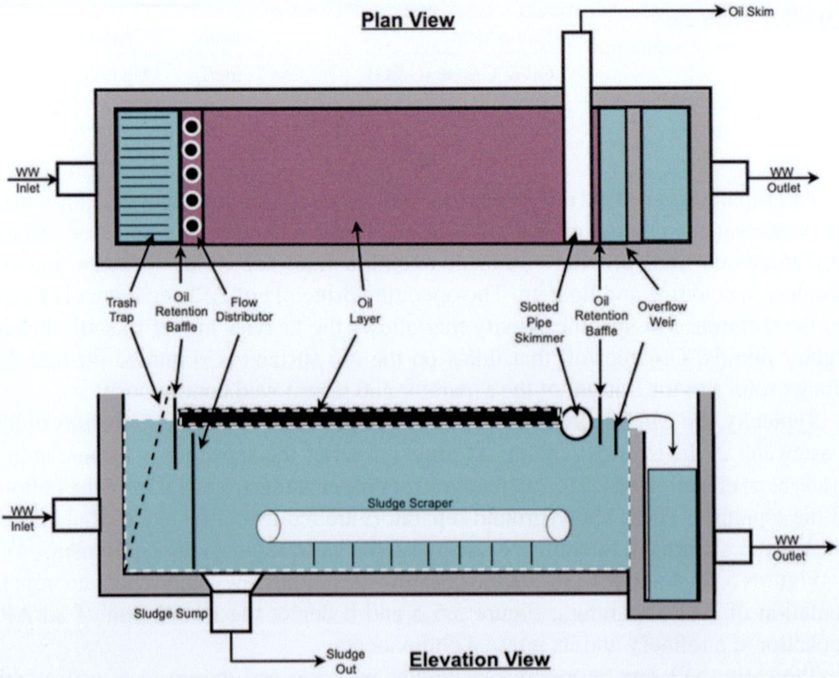

Fig. 5.3 **a** and **b** Plan and Elevation Views of an API Separator (Adapted from website: https://www.monroeenvironmental.com/water-and-wastewater-treatment/api-separators)

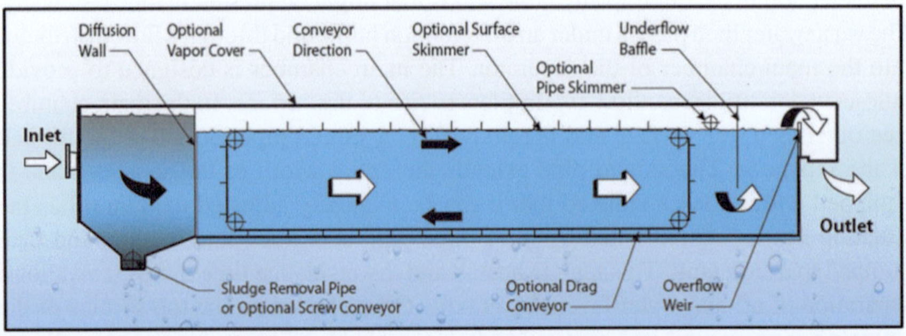

Fig. 5.4 Schematic Representation of an API Separator (Adapted from website: https://www.monroeenvironmental.com/water-and-wastewater-treatment/api-separators)

and the aqueous phase of the wastewater. The surface loading rate is the flow rate of wastewater in the separator divided by the horizontal surface area of the separator. In a typical separation scenario, any oil droplet with a rise rate greater than or equal to the surface loading rate will reach the separator surface and be removed. Typical hydraulic loading rate of gravity oil separators, including API Separators are in the range of 12 to 30 m^3/m^2.day (300 to 800 gpd/ft^2) (Ford 2012).

Fig. 5.5 **a** Typical API Separator Installation and 5.5 **b** Internal Mechanisms. *Photo* Courtesy of Evoqua (Xylem) Water Technologies, Waukesha, WI

The design of the separator is based on Stoke's Law (see equation below) which applies to laminar flow conditions.

$$V_t = \frac{g * (\rho w - \rho o) * D^2}{18\mu} \quad (5.1)$$

where,

V_t is the terminal (rise velocity) of the design oil globule, in centimeters per second.

g is the acceleration due to gravity (981 cm/sec^2).

ρ_w is the density of water at the design temperature (gms/cm^3).

ρ_o is the density of oil at the design temperature (gms/cm^3).

μ is the absolute viscosity of wastewater at the design temperature, in (gm/cm—sec).

D is the diameter of the oil globule to be removed, in centimeters.

In strict terms Stoke's law applies when the flow is in laminar flow conditions. Flow regimes are defined as follows:

Laminar flow regime $N_{Re} < 2,100$.

Transition flow regime $N_{Re} > 2,100 < 4,000$.

Turbulent flow regime $N_{Re} > 4,000$.

N_{Re} is defined by the following equation.

$$N_{Re} = \frac{\rho u D}{\mu} \tag{5.2}$$

where,

N_{Re} is the Reynolds number (dimensionless).

ρ is the fluid density (gm/cm^3).

u is the fluid velocity (cm/sec).

D is the diameter (cm).

μ is the viscosity (gm/cm—sec).

It must be recognized that the flow regime in an API Separator is frequently in the transition regime and the design assumes that Stoke's law also applies under these flow conditions.

While Stoke's Law forms the theoretical basis historically the detailed design of API separators has developed from the results of a study conducted by the American Petroleum Institute (API) in 1948. Since then, numerous oil/water separators have been designed and installed based on design criteria developed by API. These methods are also followed by the many equipment vendors who supply this equipment.

In 1985 API conducted a survey of oil/water separators operating at refineries and published the results in API publication 421 (API 1990). Additional data on the performance of API Separators is provided by Thompson (1973). Both these sets of data are combined and summarized in the Fig. 5.6.

This chart shows the relationship between the inlet loading—surface area of the separator (m^2) divided by the flow to the separator (m^3/min), and the outlet oil concentration in the wastewater (mg/L). The expected outlet oil and grease (O&G) concentration from properly designed and operated API Separators should be less than 100 mg/L.

API Separators have a relatively large open area containing oil and can therefore be a significant source of emissions/odors in the wastewater treatment plant in a refinery. Some refineries have chosen to cover their API Separators as an odor abatement measure. Figure 5.7 shows a diagram of a covered API Separator.

The separator is usually covered with a domed roof cover (usually aluminum). The vapor space is blanketed with nitrogen in order to prevent the formation of explosive mixtures and vented by pressure control to a control device such as a dedicated flare.

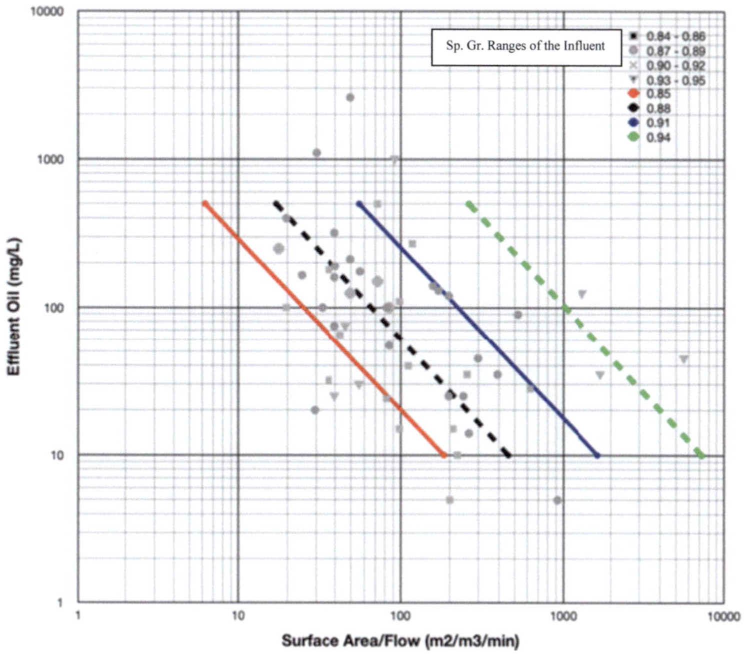

Fig. 5.6 API performance data (Source: Monograph on Refinery Environmental Control: Management of Water Discharges—Design and Operation of Oil/Water Separators, API Publication 421, 1st Edition, February 1990)

Carbon canisters are not used for this application because they would be quickly exhausted due to the potential high loading of volatile organic compounds (VOCs). The vent line from the separator cannot be connected to the refinery flare due to the presence of nitrogen in the discharge gas. If connecting to the refinery flare is to be considered, refinery fuel gas will need to be used as the blanketing gas. The capital cost of running the fuel gas line from the main header in the refinery to the API Separator as well as running the vent line to the nearest flare header will need to be taken into account.

Another option for the primary oil/water separation unit is to use a specially designed floating roof tank in place of an API Separator. The floating roof eliminates the need for vapor control since there will be no emissions from a tank with this type of roof. This tank can also serve a dual purpose of being the oil/water/sludge separator as well as the equalization tank in the plant. Since the API Separator and Equalization tank represent a significant portion of the capital costs in a refinery wastewater treatment plant, using a tank for dual purpose results in substantial savings in capital costs when compared to an API separator plus and an Equalization tank. This is a viable option for new wastewater treatment plants as well as retrofitting existing plants at refineries. Figure 5.8 is a diagram of a floating roof API separator tank.

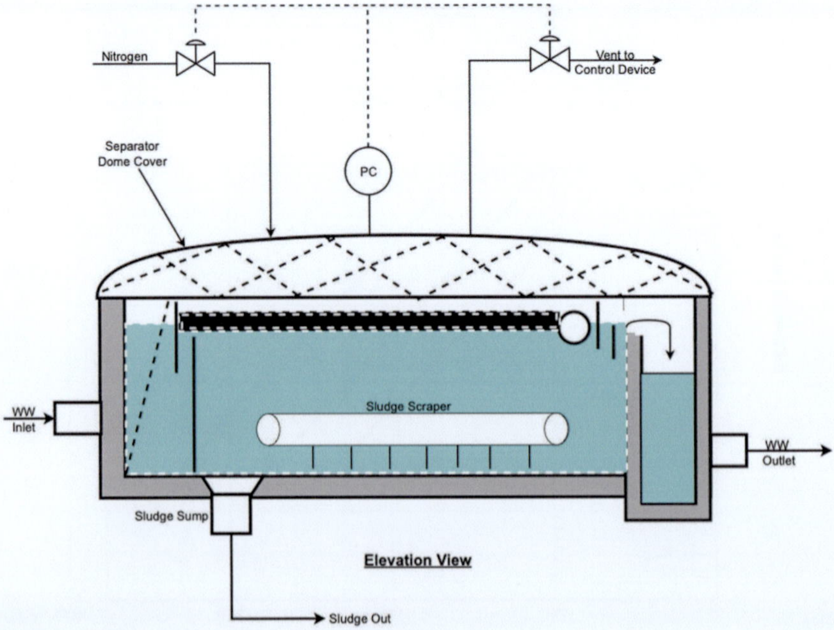

Fig. 5.7 Covered API separator

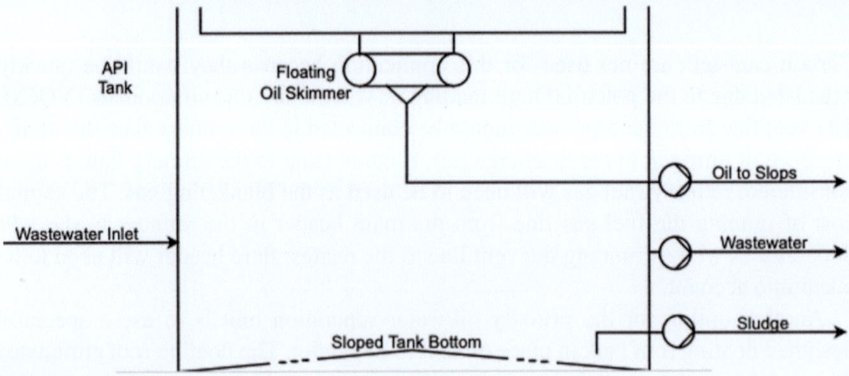

Fig. 5.8 Floating roof API tank

This tank is designed with a maximum height to diameter ratio of 1.5 and a maximum diameter of about 15 m. If larger capacities are required multiple tanks in parallel can be used. The bottom of the tank is sloped towards the walls of the tank to allow the sludge to collect towards the wall as much as possible. The wastewater inlet is located about two meters above the bottom of the tank in order to ensure that it is well above the height of any accumulated sludge. The sludge is removed from the bottom of the tank using diaphragm pumps. If necessary, multiple sludge outlets around the circumference of the tank can be installed to facilitate the removal of sludge as much as possible. It must be understood that not all the sludge that accumulates in the tank will be completely removed from the tank by the sludge pumps. The tanks will be required to be cleaned periodically (similar to the cleaning schedule of crude tanks in the refinery) to remove the residual sludge. The wastewater outlet is located 180° across from the inlet nozzle to allow for maximum residence time to facilitate effective oil/water separation. The oil outlet is located towards the bottom of the tank (below the height of the roof stops) and is connected to the floating oil skimmer that is attached to the floating roof of the tank. The refineries that have installed this type of tanks have reported that they consistently achieve an outlet O&G concentration of less than 80 mg/L.

The lift stations that are used to pump the wastewater to API separators can be a source of emissions/odors in the treatment plants. Unfortunately, it is not possible to cover, and vapor control these lift stations because it would interfere with proper gravity draining into the sump. The only emissions minimization strategy that can be implemented in this case is to minimize the cross—sectional area of the sump that is exposed to the atmosphere and if necessary, route the discharge of the vent to an elevated safe location.

5.3.1.3 Corrugated Plate Interceptor (CPI) Separators

CPI Separators also perform bulk removal of free phase oil from influent wastewater following a principle similar to that of API separators. Hydraulic loading rate is an important parameter that determines the oil removal performance of these separators. Increase in surface area results in the decrease of hydraulc loading rate and helps improve the separation efficiency. However, for API Separators, this also requires a corresponding increase in the footprint area and the tank size, and hence the cost. CPI separators overcome this limitation significantly as the surface area is increased severalfold by installing a pack of parallel plates intercepting influent flow. Typically, the plates are inclined between 45 and 60° and spaced in the range of 2 to 2.5 cm between each other. This arrangement enhances the total surface area, and thus decreases the surface loading rate, resulting in an increase in separation efficiency. The effective surface area of each plate A_{eff} is the horizontal projected area A_p of each plate such that:

$$A_{\text{eff}} = A_p * \text{Cosine}\,\alpha, \text{ where } \alpha \text{ is the angle of incline of each plate} \tag{5.3}$$

A CPI separator influent is directly intercepted by the inclined surface of the plates and distributed among the parallel plates. Solid particles flow down along the surfaces of the plates and collect in a hopper. Oil globules coalesce together and float up to form a layer on the top surface which is skimmed off by a rotating skimmer.

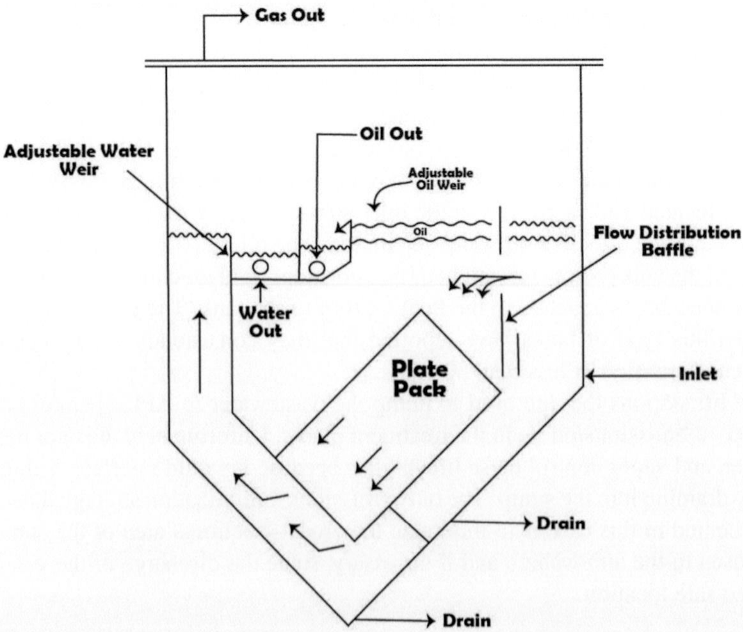

Fig. 5.9 Corrugated plate interceptor (CPI) (Adapted from Ken Arnold 1999)

It may appear that by simply increasing the number of plates, and hence the total surface area, the hydraulic loading rates can be reduced, and oil-water separation efficiency can be increased infinitely. However, to increase the number of plates within a given volume, the space between the plates will need to be reduced which will create fouling of the space between the plates, eventually leading to total blockage of the separator tank. Even while operating under normal loads with the number of plates that meets the specified spacing between plates, fouling of CPI separators cannot be avoided. Therefore, routine maintenance and cleaning of CPI separator plates are necessary to ensure smooth operation continuously. At some refineries the API and CPI separators are located above grade at an elevation of 1.5 to 2 m to allow for proper sludge removal from the bottom of the separator. Figure 5.9 presents a schematic view of a CPI Separator.

5.3.2 Secondary Oil-Water Separation

Primary oil-water separators remove bulk of free phase oil by utilizing the difference in specific gravities of water and oil. Secondary separation, however, is a more difficult task as it separates dispersed oil present in water in the form of emulsion. Most refineries use flotation units for secondary oil/water separation. Droplets of emulsified oil in water exist in microparticulate form in the size range of 5 to 20 μm.

They remain in suspension for infinite time and are not amenable to separation only by the difference of specific gravity. Stability of the oil droplets in water is due to their surface negative charge, which is contributed by several factors—high pH, high concentrations of surface active agents like naphthenic acids, turbulent mixing, presence of fine particulate solids, surfactants that are typically added as dispersants in cooling water systems, brine stream from desalters, etc. The goal is to consistently achieve an O&G concentration of less than 30 mg/L prior to biological treatment.

For effective oil-water separation from emulsions it is first necessary to destabilize the fine oil droplets by neutralization of surface charge so they can come close to each other and coalesce together to form larger drops. This is done by adding coagulant metal salts alum or ferric chloride. However, large concentrations (10 to 100 mg/L) of these salts are necessary to neutralize the oil droplets, which produce large quantities of aluminum and ferric hydroxide sludge that require proper disposal. Moreover, both of these salts potentially depress pH and require addition of alkali to neutralize. Dosage of high molecular weight cationic polymers with high charge density significantly reduces the dosing rates, and hence the sludge production rates. Combinations of polymers as coagulant and flocculant are utilized to optimize the polymer dosing rates.

The next step is to attach the larger drops of oil onto the surfaces of fine air bubbles so they can form agglomerates of air bubbles with oil drops on the surfaces and rise to the water surface as aggregates. Oil and suspended solids concentrate on to the water surface as froth and are continuously removed by skimmers. This is the basis of secondary oil—water separation aided by air (gas) floatation. Heavy solids settle at the bottom of the separators in a hopper and are removed continuously.

There are two main types of secondary oil—water separators that rely on air (gas) floatation. These are Dissolved Air/Gas Floatation (DAF/DGF) and Induced Air/Gas Floatation (IAF/IGF) as discussed below.

5.3.2.1 Dissolved Air and Gas Flotation (DAF and DGF)

The function of a DAF system is to remove fine particulate matter and oil droplets from influent wastewater. This is the second oil-water separation step after the first step by API or CPI separator which removes mainly the free phase oil and some emulsified oil (5 to 20 μm). Primarily a DAF separator consists of a flash mixing tank, a flocculation tank, and a flotation unit as shown in Fig. 5.11.

In a typical DAF operation, the pH of the incoming stream is adjusted by dosing acid or alkali (H_2SO_4 or NaOH) solution as required. Polyelectrolyte is dosed in the flocculation tank to aid in the formation of flocs. Typically, a paddle type mixer is employed for this purpose. Subsequently the effluent from the flocculation tank enters the flotation tank. A part of the effluent from the flotation tank is recycled back after being saturated with air at a pressure of about 4 to 6 bar in the Saturation tank. As the pressure of the air-water solution is dropped in a let-down valve and the flow is introduced in the DAF tank via a sparger near the bottom of the tank, microbubbles of air (10 to 100 μm size) are released at the bottom of the tank. The air bubble sizes gradually increase, and oil droplets attach on to their surfaces as they rise to the water surface in the DAF tank.

Efficiency of oil-water separation depends on the air bubble size distribution, and from this viewpoint DAF separators perform at a superior level due to the size of microbubbles generated by the release of air dissolved in water under high pressure. From their experimental studies Han, et al. (2007) observed bubbles ranging in size between 10 and 110 μm under the saturation pressures between 2 and 5 bars, but the distribution curves were observed to have shifted left (Fig. 5.10) with the rise in pressure. This implies that rise in air saturation pressure favors the formation of smaller bubbles. The authors concluded that, as the pressure increases the bubble sizes decrease which helps increase in floc size, resulting in enhanced solid and Oil/Grease removal capability by a DAF unit. Similar results have been reported in the literature about the relation between air saturation pressure and bubble size distribution in the DAF tanks and its effect on oil-water separation.

The scum generated at the top of the DAF tank is collected and removed by a skimmer and routed to the Oil Sump. Properly designed and operated DAF units can consistently achieve an outlet O&G concentration of <25 mg/L. Any solids present are collected at the bottom and is routed to the Sump. A schematic view of a DAF separator is presented in Fig. 5.11. Typical design parameters for DAF units are as tabulated in Table 5.5.

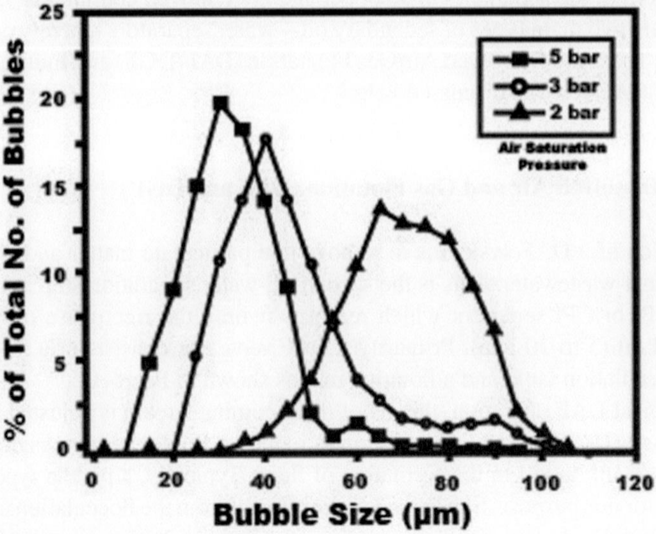

Fig. 5.10 Bubble size distribution in a DAF separator as a function of air saturation pressure (Adpted from Han et al. 2007)

There are some variations in DAF design. These can be rectangular or cylindrical in shape depending primarily upon the footprint area and any other preferences the site may have. Another variation is in the mode of addition of polyelectrolyte as flocculant. The traditional method is by dosing and mixing flocculant with the incoming wastewater in a mixing tank as shown in Fig. 5.11. The mechanical details associated with a DAF tank are presented as Fig. 5.12. However, many DAF system manufacturers replace the mixing tank and the mixer combination by a long serpentine tube at the entrance of the DAF tank. The wastewater flows through the tube and the flocculant is dosed into the influent line. This allows adequate time for mixing of the flocculant with the DAF influent and aging the polymer to become effective. This replacement eliminates the need for a mixing tank and mixer and the associated cost.

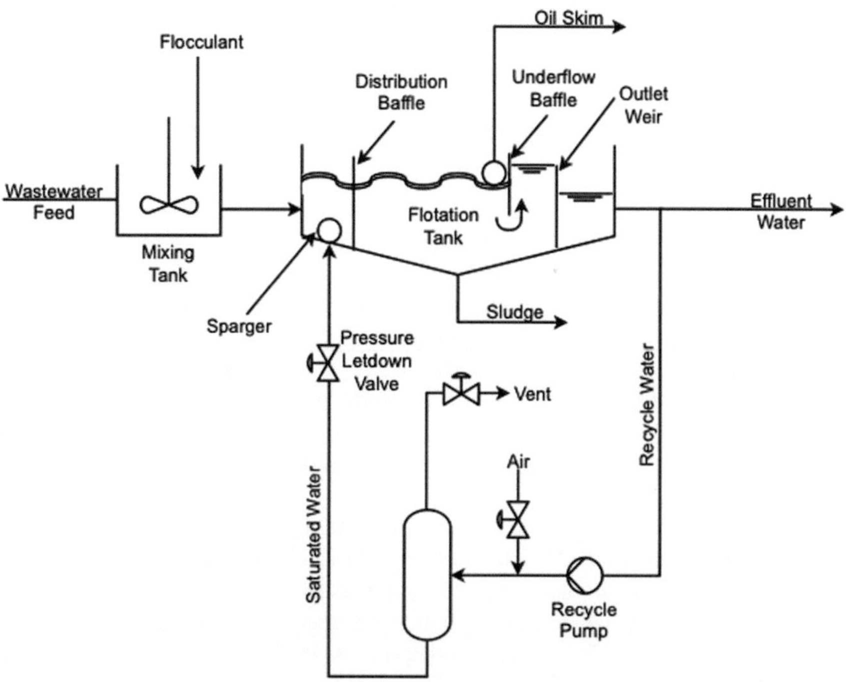

Fig. 5.11 Schematic view of a dissolved air flotation unit

Table 5.5 Typical range of design parameters for DAF systems

Treatment	Air/solid ratio	Solids loading rate (kg/m^2-d)	Hydraulic loading rate (m^3/m^2-d)	Solids capture (%)
Secondary oil/ water separation	0.04–0.07	90–200	90–250	80–95

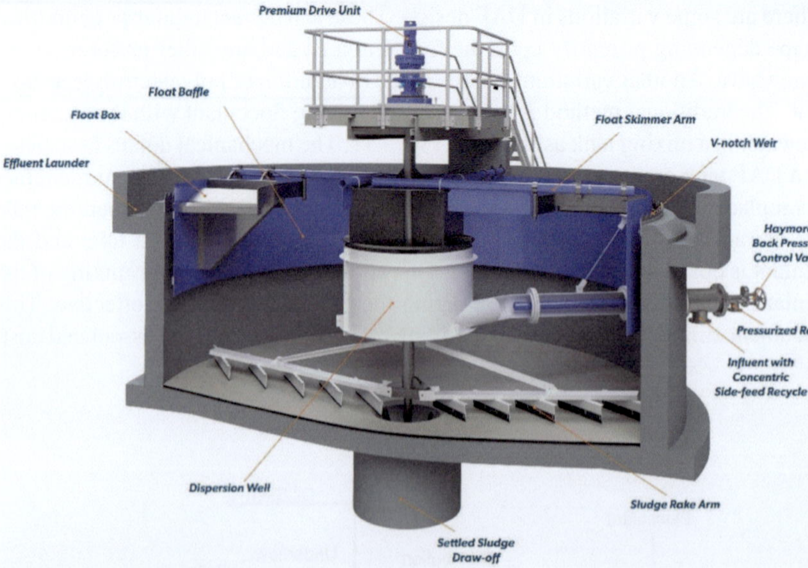

Fig. 5.12 Typical DAF installation at a refinery wastewater treatment plant. *Photo* Courtesy of WesTech Engineering Co., Salt Lake City, UT, USA

Another variation of the traditional design is the air saturation system. Several DAF system suppliers offer air introduction directly into the recycle pump which are especially designed to aspirate air from the atmosphere and dissolve air into water under high pressure. The replacement of a separate air saturation tank, supply of pressurized air by an air compressor, and all other associated equipment also result in significant cost saving. Among several system vendors offering pumps capable of taking in air from the atmosphere and saturating it with recycle water under high pressure are Nikuni Pumps (Nikuniamerica.com/DAF) and Brise Pumps (siemens-energy.com/water).

The rising air bubbles in a DAF tank pass through the water surface and are released to the atmosphere along with emissions of light organic compounds. Thus, DAF units can be a significant source of emissions and odors due to their relatively large open cross-sectional area. An option to alleviate the several disadvantages of a DAF unit, including volatile organic compounds (VOC) and odor emissions, is to use a Dissolved Gas Flotation (DGF) unit. Figure 5.13 presents a schematic view of a DGF unit and Fig. 5.14 presents the view of a DGF tank with mechanical details.

When nitrogen is used in place of air as a floatation medium in a DGF it is called a dissolved nitrogen floatation (DNF) unit. The flotation tank as well as the mixing tank in DNF are blanketed with nitrogen and covered with vapor tight covers. The vent gas is sent by pressure control to a vapor control device such as a carbon canister. The feed to the unit is sent to a covered mixing tank where a flocculent is added. The wastewater from the mixing tank is sent to the covered DNF tank. A part of the effluent from this tank is recycled and sent to the pressurizing drum where nitrogen is

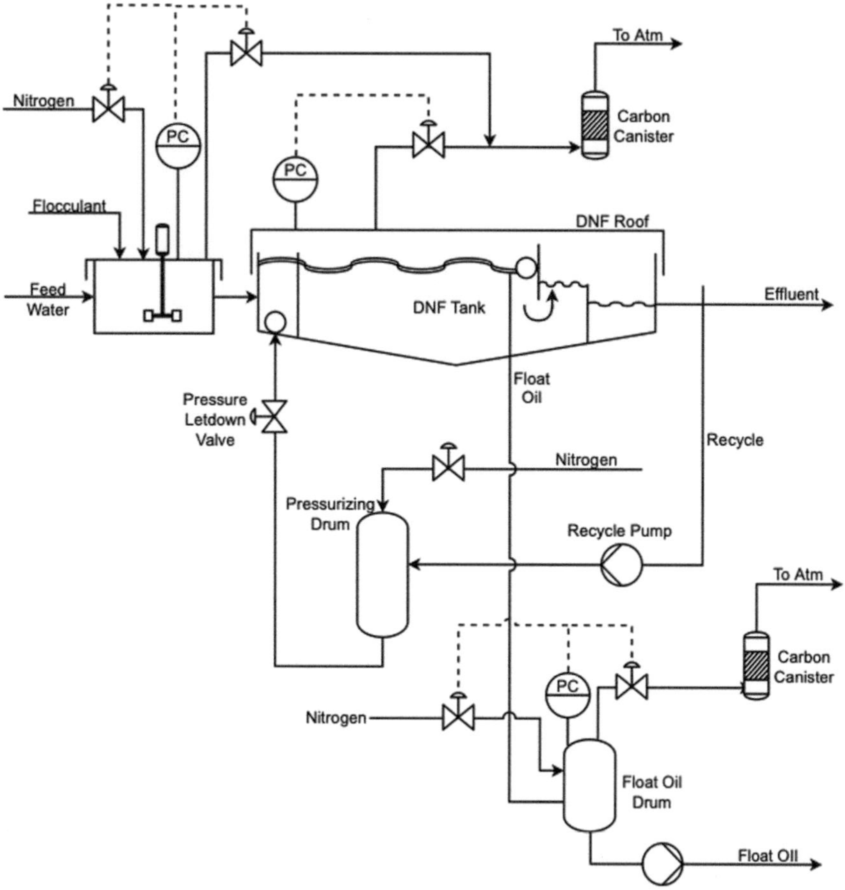

Fig. 5.13 Schematic view of a dissolved nitrogen flotation (DNF) unit

added. Nitrogen gas is dissolved in the water at a pressure of about 4 to 6 bar. Water from this tank is routed through a letdown valve to the sparger near the bottom of the DNF tank where the bubbles are released. The bubbles help float the oil droplets on the water surface. The float oil is removed and sent to the Float Oil drum and then sent to disposal or recycle. The vent from the Float drum is sent to a granulated activated carbon (GAC) canister. This arrangement of fully enclosed DNF units offers several benefits over traditional DAF units:

- It prevents air emission of volatile organic compounds (VOC) and odors
- Nitrogen being inert prevents potential explosions that could be caused by light hydrocarbons in presence of oxygen
- Nitrogen also helps prevent corrosion potential of metallic components in a DGF/DNF unit.

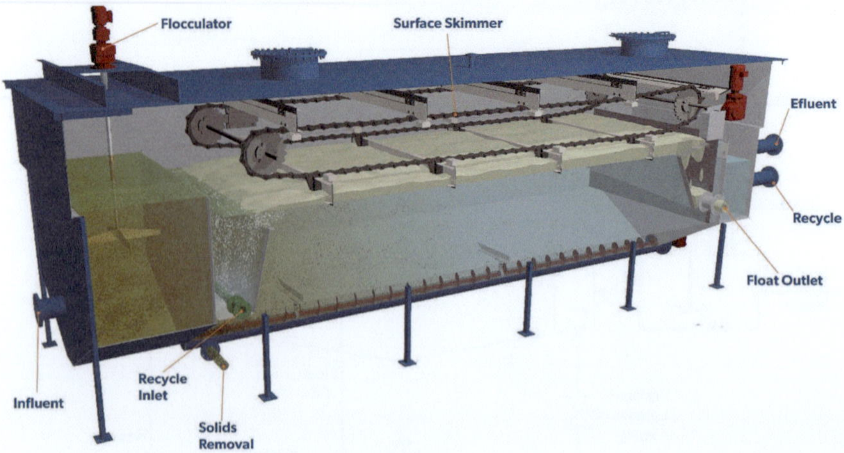

Fig. 5.14 Mechanical details of dissolved gas flotation (DGF) Unit. *Photo* Courtesy of WesTech Engineering Co., Salt Lake City, UT, USA

5.3.2.2 Induced Air Flotation (IAF) and Induced Gas Flotation (IGF)

The other type of secondary oil—water separation system used in the petroleum industry is an Induced Air Flotation (IAF) System. These are mainly used in upstream operations (Exploration and Production). Advantages of the IAF systems are their compact size and lower capital cost. The effluent oil content from IAF is about 15–40 mg/L.

The main difference between IAF and DAF is in the way air is introduced into the wastewater to create microbubbles. In the case of DAF, air is conveyed into the separator tank by dissolving it into a part of the treated water under high pressure and returning it into the DAF tank. In IAFs air is introduced by two alternate means—(a) hydraulic, or (b) mechanical, that do not involve dissolution of air in water.

Mechanical induction of air in an IAF utilizes a rotor-disperser device. The wastewater is introduced into the inlet well of the float tank. It is equipped with an agitator that has multiple impellers which creates vacuum that pulls air from the headspace, mixes the air with water, and disperses it in the tank in the form of microbubbles. As the bubbles rise, they attach oil droplets on to their surface and push them to the top surface where they are skimmed off via the oil skim pipe.

In the case of the hydraulic air introduction system, a split stream of treated effluent is returned to the IAF tank at high velocity which creates a vacuum that pulls air from the headspace. The air-water mixture is distributed through eductors creating microbubbles of air that are released at the bottom of the separator tank. As the bubbles rise, they grow in size and attach droplets of oil creating a froth on the top surface. The oil layer is removed by a continuously rotating skimmer as described earlier.

IAF systems are housed in horizontal cylindrical tanks that are partitioned into three to four cells in series. The liquid moves from one cell to the next before leaving the unit and the float passes over the weir on one or both sides of the unit. The

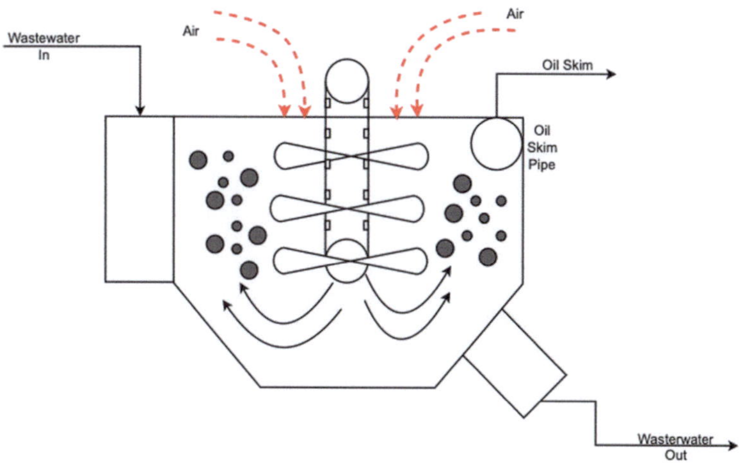

Fig. 5.15 Schematic view of a single cell of a mechanical IAF system

influent is treated progressively as it flows from one cell to the next and in the last cell treated water is degassed before it is discharged. The schematic view of a single cell of a mechanical IAF system is presented in Fig. 5.15 and a multicell hydraulic IGF system is presented in Fig. 5.16.

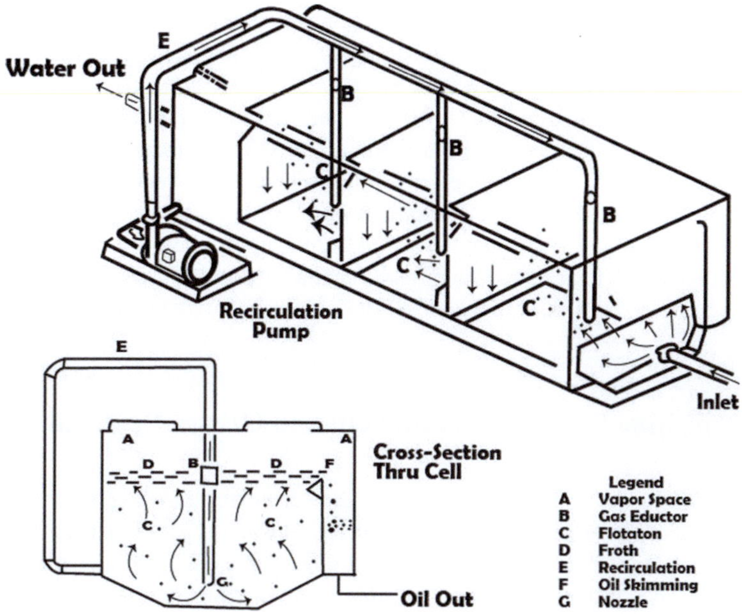

Fig. 5.16 Schematic view of hydraulic IAF unit (Meyer, 2012)

Table 5.6 Comparison of main operational characteristics, and specific advantages and disadvantages of DAF/DGF and IAF/IGF

Operating parameter and performance	DAF/DGF	IAF/IGF
Hydraulic loading rate, m/hr (gpm/ft^2)	4–11 (2–5)	12–24 (5–10)
Hydraulic residence time, mins	10–20	1–2/cell, 4–8 for four cells
Bubble size, μm	10–100	500–1000
Float production flow (% of Influent)	<1	2–5
Effluent oil & grease concentration under stable and normal operation (mg/L)	5–20	15–40
Footprint requirement	High	Low
Capital cost	High	Low
Maintenance requirement	High	Low

As in the case of dissolved air floatation (DAF) units, some applications use covered IGF instead of IAF to prevent emissions, corrosion, and explosion potential of hydrocarbon vapors in presence of oxygen. In an IGF, when nitrogen is used as flotation gas and it is referred to as INF.

Based on pilot and field scale operations it has been reported that the mechanical IGF performs better than hydraulically operated IGF (Movafaghian et al. 2003). One of the reasons is that hydraulic units require closer attention as they are more prone to scaling and fouling of the eductor nozzles that requires complete shutdown for maintenance of the eductor systems.

Microbubbles formed by both hydraulic and mechanical means are in the size range of ~500 to 1,000 μm. The drawback of the IGF compared to DGF is that it does not create enough fine bubbles, and this limits its effectiveness in removing oil from the wastewater. IGFs are not commonly used in refinery wastewater services and are more prevalent in the upstream sector of the oil supply chain where they are used in produced water treatment. Operational characteristics, and specific advantages and disadvantages of DAF/DGF and IAF/IGF are summarized in Table 5.6.

5.4 Polishing Step for Residual Oil and Grease Removal

Some refineries use walnut shell filters as the secondary oil/water separation unit and a few others use these units after secondary oil/water separation as the final step to further remove emulsified and dissolved oil in both upstream and downstream operations. The system consists of multiple tanks in which the wastewater is sent to one tank where the oil is adsorbed onto the surfaces of walnut shells. These filters utilize walnuts shells of ~3 mm size and specific gravity 1.4 as the filtration media with bed depths ranging between 1.2 and 1.8 m. Typical loading rates for such filters are in the range of 470 to 880 $m^3/m^2/d$ (8 to 15 gpm/ft^2). The influent oil concentration to these filters should not exceed 100 mg/L. Under these operating conditions the treated effluent oil concentrations are expected to be <10 mg/L (Meyer 2012). However, at higher loadings and influent oil concentrations these filters result in reduced run time and require excessive and frequent backwash. When the shells become saturated, the feed is switched to a second tank where the adsorption of oil continues. In the meantime, the shells from the first tank are removed and discarded. The tank is then refilled with new walnut shell media.

Various other filter media have been tried for this type of application, but the superior surface characteristics for coalescence and filtration make the walnut shell filters most suited for removing emulsified and dissolved oil from water and wastewater originating from petroleum production and processing.

Typically, the primary and secondary oil/water separation steps followed by most refineries are capable of meeting the oil removal goals from wastewater prior to biological treatment and walnut shell filtration is not commonly used in modern refineries.

5.5 Equalization

The purpose of an equalization unit in the refinery wastewater treatment plant is to provide surge capacity to handle variations in flow. It is an important step to ensure a relatively constant flow to the downstream treatment equipment. The equalization can be provided either by a tank or a pond. In case of excessive carryover of oil due to a malfunction of the oil-water separator equipment the oil layer is skimmed from the top surface of equalization tank.

One of the considerations in the design of the equalization unit is the location of the equalization tank/pond in the treatment train.

Figure 5.17 presents two options for the location of equalization tank in a refinery wastewater treatment system. Each of the above options has its own pros and cons. The main advantages of Option 1 are—(a) it provides a surge capacity to the biological treatment reactor thereby avoiding a large volume to accommodate the peaks and valleys of hydraulic and chemical loads, and the associated capital cost, and (b) minimizes the VOC emissions potential from the equalization tank/pond, potentially avoiding the need to cover it. However, its disadvantage is that it does not protect the secondary Oil/Water separation DAF/DGF or IAF/IGF from large swings in hydraulic and chemical loads leading to the need for large sizes of these equipment and the associated higher cost, and potential loss of performance resulting from such variations in loads. Also, the secondary Oil/Water separation equipment needs to

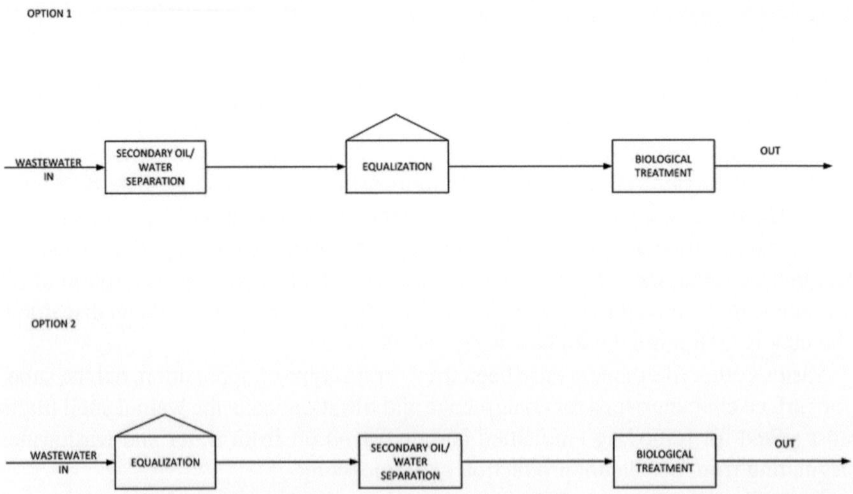

Fig. 5.17 Location to equalization unit

be covered and nitrogen blanketed for air emissions control and potential explosion hazard prevention as described earlier.

In case of Option 2 the equalization tank is located between the primary oil-water separation and secondary oil—water separation steps. Refinery wastewater streams vary widely in their flows and chemical characteristics. While the primary oil-water separators conduct a bulk removal of oil and solids from the wastewater, which may be somewhat tolerant to such fluctuations, performance of the equipment in downstream treatment processes are very sensitive to such fluctuations. Placement of equalization tank prior to the secondary Oil/Water separator dampens the effects of the peaks and valleys in the hydraulic loads entering the secondary oil/water separators. This function is very important as the treated effluent quality from secondary Oil/Water separators, e.g., DAF/DGF and IAF/IGF are adversely affected by fluctuations in their influent parameters, which also seriously impact the biological treatment process downstream. However, in this case the Equalization tank will cause VOC emissions and the resultant odors.

The relative advantages and disadvantages of the two options are summarized in Table 5.7.

The function of an equalization tank can be summarized as follows:

- Receives wastewater from the secondary Oil/Water separator (in Option 1) or primary Oil/Water separator (in Option 2), and other streams that may bypass the separators.
- Provides a large retention volume and mixing of all the incoming streams to feed the downstream wastewater treatment equipment uniform flows to ensure their effective performance.
- Serves as a reservoir to feed the downstream treatment systems on a continuous basis.

Table 5.7 Advantages and disadvantages of option 1 and 2

Option	Advantage	Disadvantage
1	Provides required surge capacity to the Biological unit thereby facilitating smooth operation of the unit	The secondary oil/water separation unit could be subject to wide variations in flow thereby making it subject to poor performance of the unit due to flow variations
	The biological treatment unit can be designed without providing for extra capacity to handle the variations in flow	The secondary oil/water separation unit would have to be designed for a larger capacity to handle the potential flow variations thereby resulting in higher capital cost
	Since all the oil has been removed, there would be no emissions or odors from the equalization unit and therefore this unit can be an open tank/pond	
2	Provides surge capacity to both the secondary oil/water separation unit as well as the biological treatment unit	
	Both the secondary oil/water separation unit and the biological unit can be designed without allowing for extra capacity which would result in smaller units and lower capital costs	The potential emissions and odors from the equalization unit will require the tank to be covered with a floating roof. Open ponds cannot be used for this service if it is located upstream of the secondary oil/water separation unit

- An equalization tank serves as the first line of defense to protect the downstream biological treatment process against toxic and shock loads by diverting flows during such incidents to a standby off-specification tank temporarily. After the incident(s) are over, and the normal flow from the equalization tank to the downstream treatment train is resumed, the liquid stored in the off-spec tank is slowly blended with the mainstream flow and released for treatment.

Equalization tanks are sized based on hydraulic retention time (HRT), which is a function of hourly variation of flow during a day. A procedure for sizing equalization tanks is presented in the literature (Eckenfelder et al. 2009). Many refineries provide a surge capacity of up to 24 h in the equalization unit. Another approach is to base it on the hydraulic loading rate, which is typically ~3.3 $m^3/d/m^2$, or 80 gpd/ft^2 (Ford 2012).

Whether option 1 or option 2 is chosen, the problem of the API separator being subject to potential variations in flow still exists. Since the design of the API separator is based on having relatively constant flow, the quiescent zone of the separator can be upset by flow variation thereby affecting performance, and as a worst case resulting in washout of the oil and sludge from the separator. While the separator can be overdesigned to compensate for the variations in flow to the separator this will result in a larger separator and higher capital costs. Installing an API Tank (as discussed in earlier section) will alleviate this problem and satisfy the need of providing adequate surge capacity as well as the required oil/water separation needs.

5.6 Secondary Treatment

Secondary treatment of refinery wastewater is conducted after hydraulic and chemical equalization. It is a two-step process. The first step consists of biological treatment to remove large concentrations of dissolved and fine particulate contaminants primarily in the forms of complex organic and nitrogenous compounds. The net growth of biomass is separated and wasted in the second step consisting of solid/liquid separation. The discussion included in the following section covers biological treatment and solid/liquid separation systems used in the downstream sector of the oil supply chain.

5.6.1 Biological Treatment

5.6.1.1 Background and Historical Perspective

The purpose of biological treatment is to remove soluble organic matter and suspended particles not removed by primary treatment. Conventional biological treatment can remove around 80 and 90% of BOD_5 and TSS respectively.

Over a hundred years have passed since the development of the activated sludge process in 1914, which was a major breakthrough in the science and technology of water pollution control. It was the work of groups of chemists, microbiologists and sanitary engineers under the Royal Commission on Sewage Disposal that led to the invention of activated sludge process in England in 1914. Around the same time, it was also developed at the Lawrence Experiment Station in Massachusetts arising out of a series of experiments aimed at purification of sewage by oxidation with air. It was a milestone in human history considering its impact on public health and sanitation.

Suspended growth biological processes have been widely used for a long time for biological wastewater treatment. It still continues to be the leading technology for wastewater treatment throughout the world and the work horse of the industry. In its basic form, it utilizes a microbial population that grows and thrives on organic pollutants in the wastewater as food. Air is supplied externally to the microbes as the source of oxygen for their respiration, and also to provide mixing energy for the biomass to remain suspended in water. Treated water is freed from the biomass in a settling basin and discharged. The activated sludge process falls under the broad category of suspended growth wastewater treatment processes.

Over the course of time, however, other types of biological treatment processes also emerged and found wide acceptance. In a parallel course with suspended growth processes, the 'fixed film treatment processes' were also under development in both England and the US. Its roots can be traced back to the 1870's with the advent of filtration columns as devices to support microbial population on the media surfaces. All biological wastewater treatment processes are broadly categorized into suspended growth, fixed film (or attached growth) and hybrid processes. For high load conditions, traditional suspended growth treatment systems require a large footprint and energy. Attached growth and hybrid growth systems, on the other hand, can handle higher loads and are often more efficient requiring less footprint area.

It is emphasized that the success of all types of biological treatment processes—both suspended and attached growth systems, depends heavily on the effectiveness of oil and grease removal step in the primary treatment. The normally acceptable O&G:MLVSS ratio in suspended growth biological treatment processes is in the range 0.04 to 0.1 (Ford 2012).

Based upon the metabolic functions and energy requirements for their growth and sustenance microbial populations carry out wastewater treatment processes under aerobic, anaerobic, and anoxic conditions. Accordingly, another means of classification of biological treatment processes is:

- Aerobic Processes
- Anoxic Processes
- Anaerobic Processes

The difference between the three types of treatment processes based on their reaction mechanism has been discussed in detail in Chap. 1. Each of the above processes can be conducted in both attached and suspended growth reactors.

Growth of biomass in a reactor depends on many factors such as: temperature, loading rates, availability of substrate (food), and availability of energy source. In all the biological systems (attached, suspended or hybrid) biomass grows through the following stages.

1: Lag Phase: The bacteria assimilate to their environment.

2: Log—growth Phase: The bacteria population rises rapidly.

3: Stationary Phase: The bacteria reach a stable population, where the growth rate is almost equal to the death rate.

4: Endogenous Phase: The death rate of bacteria is higher than the growth rate of bacteria causing a decline in population of bacteria (biomass).

The growth phases of microorganisms have been discussed in detail in Chap. 1.

During the log-growth phase, the microorganisms grow at their maximum rate by consuming substrate at the maximum rate. For aerobic suspended growth processes, substrate consumption rate in g of substrate/(m^3*d) during this phase is expressed by the equation below:

$$\left(-\frac{dS}{dt}\right) = \frac{k * S}{K_s + S} * X \tag{5.4}$$

where k = Maximum specific substrate consumption rate, g of substrate/(g of biomass*d)

X = Biomass concentration, g of biomass/m^3

S = Substrate concentration, g of substrate/m^3
K_s = Substrate concentration at half of the maximum substrate consumption rate, g of substance/m^3

If all the substrate is converted to biomass, then the rate of substrate consumption will be related to the biomass production rate by the following mathematical equation:

$$\frac{dX}{dt} = \left(-\frac{dS}{dt}\right) * Y \tag{5.5}$$

where, Y = Growth yield coefficient of biomass, g of biomass growth/g of substrate consumed
Accordingly, the maximum specific biomass growth rate is related to the maximum specific substrate consumption rate by the equation:

$$\mu_m = Y^* k$$

where μ_m = maximum specific growth rate of bacteria, g of biomass growth/(g of biomass*day)

$$\text{Therefore, } k = \mu_m / Y \tag{5.6}$$

Combining Eqs. 5.4 and 5.6, the substrate consumption rate can be expressed in terms of the maximum specific growth rate as the classical Monod's kinetic equation as follows:

$$\left(-\frac{dS}{dt}\right) = \frac{\mu_m * S * X}{Y(K_s + S)} \tag{5.7}$$

A death rate term K_d*X is added to Eq. 5.7 to account for the decay of biomass during the endogenous phase to give a complete mathematical expression of the observed phases of life of microbial population. Equation 5.8 is obtained after combining Eqs. 5.5, 5.6 and 5.7, and adding the death rate term as follows:

$$\frac{dX}{dt} = \frac{k * S}{K_s + S} * \mathrm{Y} * X - K_d * X \tag{5.8}$$

where K_d = specific death rate constant of biomass (g of biomass decay/g*day)
Equation 5.8 provides a relationship between biological growth and substrate utilization that has been commonly used for aerobic biological treatment processes.

5.6.1.2 Nutrient Requirements

For bacterial growth it is necessary that all the macro and micronutrients be present in the wastewater. Nitrogen, and phosphorus are major macronutrients which constitute approximately 12 and 2% of dry weight of biomass, respectively. In refinery wastewater, an adequate concentration of nitrogen is typically present but phosphorous addition might be required to ensure the bacteria can grow. Many oil refineries use phosphorous compounds in cooling water treatment to control calcium carbonate scale formation. In those cases, phosphorus is present in cooling tower blowdown. Since cooling tower blowdown is typically sent to the wastewater treatment plant it may partially or totally satisfy the nutrient phosphorus needs for biological treatment. In some instances, additional phosphorus is added by dosing phosphoric acid into the system. Micronutrients are also important constituents for the growth of bacteria cells. Typically, refinery wastewaters contain these elements which are derived from crude petroleum.

In addition to ensuring nutrients are above the minimum required for biological growth, it is also important to ensure that the concentrations of nitrogen and phosphorous in effluent from refinery wastewater treatment plants are not too high, as above a certain level, they can cause eutrophication and have adverse effects on the surrounding environment and the quality of receiving water bodies. The treatment methods used for removal of nitrogen has been described in detail in the Activated Sludge Treatment system later in this section. Removal of phosphorus from wastewater is not typically required in treatment systems in the oil processing industry.

A brief overview of some of the most commonly used treatment processes, their categorization, and guidelines for design under each category has been discussed in the following sections of this chapter. Primarily the aerobic treatment processes have been the emphasis of these discussions, with some references to anoxic and anaerobic processes.

5.6.1.3 Aerobic Biological Treatment Systems

Suspended Growth Systems

In an aerobic suspended growth treatment, the microorganisms are well mixed and remain in suspension under the aerobic condition. The microorganisms use the organic constituents as food for their growth and congregate together to form the active biomass. Most common aerobic suspended growth systems in refinery wastewater treatment use the following reactor configurations.

- Aerated Lagoons
- Activated Sludge Process
- Sequencing Batch Reactor
- Membrane Bioreactor (MBR).

Aerated Lagoons

In this system wastewater is treated in earthen in ground ponds/basins. Aerated lagoons are large, suspended growth, flow-through biological reactors. The basin is equipped with several surface aerators. These aerators float on the surface of the

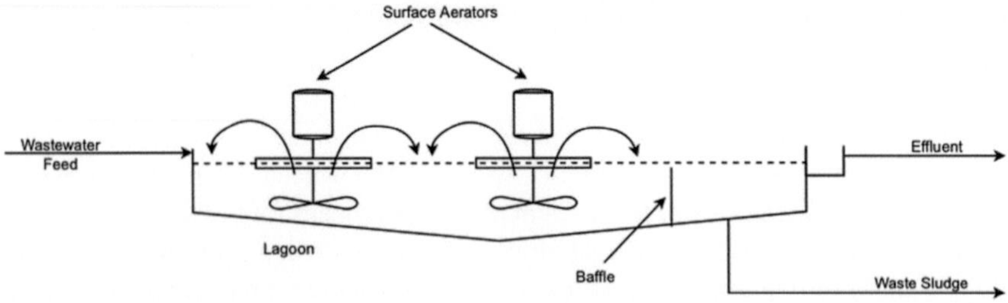

Fig. 5.18 Aerated Lagoon wastewater treatment system

Table 5.8 Typical design guidelines for Aerated Lagoons

Parameter	Value
HRT (days)	3–10
SRT (days)	3–10
Liquid depth (m)	2–6
Oxygen requirement ($kgO_2/kgBOD_5$)	0.7–1.6
Energy requirement (kW/1000 m^3)	~2

water and are equipped with large, submerged impellers. The impellers rotate fast enough to throw the water about one meter above the surface of water. This action results in oxygen from the air being transferred into the wastewater and used by the microorganisms to conduct biochemical oxidation reactions. The lagoons cover a large area and are typically equipped with multiple aerators in order to cover the entire surface area. In some instances, the lagoons are installed with submerged diffused aerators instead of surface aerators. Some lagoons have an area that is baffled off and maintained as a quiescent zone in order to allow the sludge to settle. The sludge is removed periodically (once in several years) from this section. Figure 5.18 presents a schematic representation of an aerated lagoon based treatment system.

Aerated lagoons are relatively inexpensive to build and are simple to operate but require a large area of land and volume of reactor. They are mostly used in petroleum refineries where minimal secondary treatment requirements may be acceptable, and land is available at low cost. Although this type of treatment system can be effective, given the relatively high concentrations of contaminants in refinery wastewater, the treatment pond/lagoon could cover an area of several hectares. This is a major drawback since most refineries do not have the required real estate available to install such a system.

The design parameters for aerated lagoons are as shown tabulated in Table 5.8.

Activated Sludge Process

Activated sludge is a continuous suspension of aerobic biological growths in wastewater and contains entrapped suspended colloidal, dissolved organic and inorganic material. The microorganisms use the organic matter as the carbon source and energy for microbial growth and convert the food into cell tissue and oxidized products such as carbon dioxide.

The activated sludge process is a very robust and reliable treatment process and is used extensively in both domestic and industrial wastewater treatment applications including petroleum refineries around the world. In its basic form it is used to remove organic pollutants (BOD and COD) from wastewater where microorganisms are mixed thoroughly with the particulate and dissolved organic matters in wastewater that they consume to fulfill their growth and energy requirements. As in the case of aerated lagoons, oxygen is supplied in the form of air by diffused aeration or mechanical aeration devices.

There are two important parameters that affect the proper functioning of biological treatment by suspended growth processes. These are Food to Microorganism Ratio (F:M) and Sludge Age. The terms Sludge Age, Solids Retention Time (SRT) and Mean Cell Residence Time (MCRT) are used interchangeably and defined later in this section. Activated sludge is the term that refers to the active mass of microorganisms that carry out aerobic wastewater treatment reactions in suspension. The suspension of biomass in wastewater in the bioreactor is called mixed liquor (ML) and the concentration of biomass in mixed liquor is termed mixed liquor suspended solids (MLSS), expressed in terms of mg/L (or, g/m^3). Conceptually, MLSS is similar to TSS where the suspended solids are comprised of biomass. MLVSS refers to concentration of volatile solids primarily comprising of organic matter in biomass. Typically, MLVSS comprises about 65 to 80% of MLSS in mixed liquor. The terms aerobic bioreactor, aeration tank and aeration basin have been used interchangeably in this text to define the tank that carries out wastewater treatment biochemical reactions.

The growth of new cells produced within a bioreactor depends on the oxidation of the organic matter (BOD consumed), the sludge age, and efficiency of the microbial population. The amount of the organic matter oxidized increases with sludge age and the amount of sludge produced decreases with increase in sludge age.

Mixed liquor flows from the aeration tank to the secondary clarifier where the activated sludge is settled for solid–liquid separation and thickening of solids at the bottom. The majority of the settled sludge is returned to the aeration basin as Return Activated Sludge (RAS) that provides the population necessary to carry out the oxidation of BOD associated with the wastewater influent. In order to maintain a steady state MLSS in aeration basin a part of the settled solid stream from secondary clarifier, which equals the net growth of biomass, is wasted as Waste Activated Sludge (WAS).

F:M ratio is defined as the rate of food (BOD) influent to a bioreactor (g/day) per unit mass (g) of biomass as MLVSS present in the reactor. It is expressed as:

$$\frac{F}{M} = \frac{\text{Bioreactor Influent Rate of } BOD\ (g/day)}{Mass\ of\ Biomass\ Present\ in\ Bioreactor\ as\ MLVSS\ (g)} \tag{5.9}$$

The unit of F:M ratio is day^{-1}. The numerator is determined by multiplying influent BOD (g/m^3) by influent flow rate, Q (m^3/day) and the denominator is obtained by multiplying the MLVSS (g/m^3) by the volume, V (m^3) of wastewater in the bioreactor. F:M ratio of 0.2 to 0.5 is typical for a healthy aerobic activated sludge treatment process removing carbonaceous BOD. A very high or low F:M ratio will produce dispersed flocs which will be difficult to settle. However, refinery wastewater treatment operators typically maintain a low ratio to ensure effective destruction of difficult to degrade organic components and also to minimize sludge growth and disposal requirement.

The net Growth rate of microorganisms can be calculated by using Eq. 5.8. It includes the net yield (Y) and death or decay rate of microorganisms (K_d*X). This provides the amount of biomass present in the reactor at any given time. It is important to understand this as it helps in designing the treatment system.

The sludge age, Solids Retention Time (SRT) or Mean Cell residence Time (MCRT) is the time solids spend within the bioreactor system. It is determined by dividing the mass of solids in the bioreactor by the mass of biomass wasted in the waste sludge or in the effluent.

$$\mathrm{SRT} = \frac{VX}{(Q - Qw)Xe + Q_w * X_r} \tag{5.10}$$

where,

V = reactor volume (m^3).
X = Biomass concentration (mg VSS/ L).
Q = Wastewater influent flowrate (m^3/d).
Qw = Waste sludge flowrate (m^3/d).
Xe = Biomass concentration in effluent (mg VSS/L).
Xr = Biomass concentration in recycle stream (RAS) (mg VSS/L).

Hydraulic Retention Time (HRT) is defined as the amount of time wastewater remains within the bioreactor. It is calculated as follows:

$$\mathrm{HRT} = \frac{Volume\ of\ the\ bioreactor}{Flow\ rate\ of\ waste\ water} \tag{5.11}$$

The unit of both SRT and HRT is days. Solids Retention time (SRT) is the average time solids are within the reactor whereas HRT is defined as the average time wastewater is within the reactor. In the activated sludge process SRT is much larger than HRT. Only in case of once through processes e.g., aerated lagoons, solids exit the reactor as a part of the treated wastewater stream and SRT and HRT are the same.

F:M Ratio and SRT are among the two most important process control parameters of activated sludge process. They work in opposite directions to each other. When F:M ratio increases the SRT decreases as in the case of a carbonaceous BOD removal process. On the other hand, when F:M ratio decreases SRT increases as in the case of extended aeration, refractory organics removal and nitrification processes. It is also important to maintain required temperature and dissolved oxygen concentration in the system for efficient and healthy growth of the microbial population.

Microorganisms found in wastewater treatment plants are predominantly mesophilic in nature. The optimal operating temperature should be between 20–35 °C, though they can operate at temperatures below 10 °C.

It is important to maintain adequate dissolved oxygen in aerobic wastewater systems to maintain optimal substrate utilization rates and biomass production rates. For consistent BOD removal a minimum dissolved oxygen concentration of 1 mg/L is typically used in activated sludge systems.

Activated Sludge Processes for Nitrification

Nitrification:

Nitrification involves biochemical oxidation of ammonia/ammonium ions to nitrate. It takes place in two steps as follows:

$$\text{Nitritation}: \ 2NH_4^+ + 3O_2 \rightarrow 2NO_2^- + 4H^+ + 2H_2O \tag{5.12}$$

$$\text{Nitratation}: \ 2NO_2^- + O_2 \rightarrow 2NO_3^- \tag{5.13}$$

These conversions are done by two different species of bacteria. The most common bacteria are a combination of Ammonia oxidizing bacteria (AOB's) such as Nitrosomonas and nitrite oxidizing bacteria (NOB's) such as Nitrosporia, and Nitrobacter. The first step, Nitritation, is carried out by AOB and the next step, Nitration, is completed by NOB.

For a completely mixed activated sludge system operated at steady state, the net specific growth rate is inversely proportional to the solids residence time (SRT).

$$\mu_{net} = \frac{1}{SRT} \tag{5.14}$$

Traditionally, nitrogen removal from treated refinery wastewaters has not been required in the past. Recently, however, regulations are becoming increasingly stringent and requirement for nitrification of ammonia to nitrate is becoming more common. Discharge of uncontrolled ammonia with treated effluent to the receiving body of water causes multiple adverse impacts including a drop in dissolved oxygen (DO) at the point of discharge, toxicity to aquatic life, and eutrophication. Although eutrophication cannot be overcome by conducting nitrification alone without subsequent denitrification of nitrate to nitrogen gas, the other two effects of drop in DO, and toxicity can be effectively removed by nitrification.

Some of the important parameters to ensure successful nitrification include:

- **Temperature**

The optimum nitrification temperature range is 15 to 35 °C. The reaction rates slow down significantly when the temperature is below 10 °C, and the activity drops when the temperature is above 38 °C.

- **Dissolved oxygen (DO)**

It is necessary to maintain a DO level at or above 2 mg/L for successful nitrification. Nitrification imposes a large oxygen demand. Every gram of ammonia—nitrogen (i.e., every gram of nitrogen contained in ammonia) requires 4.57 g of oxygen, which is supplied by aeration blowers to the reactors through diffusers.

- **pH**

The optimum pH range for nitrification is 7.5–8, and a pH of less than 6.5 can be inhibitory. Nitrification produces hydrogen ions, creating an acidic condition that needs to be neutralized by adding alkalinity to maintain a stable pH. Alkalinity consumption for nitrification is 7.14 g $CaCO_3$ per gram of ammonium—nitrogen. If the wastewater does not have an adequate buffering capacity, alkalinity has to be added from an external source.

- **F:M Ratio and SRT**

Nitrifiers are autotrophic organisms and grow at a much slower rate, typically by an order of magnitude, than BOD oxidizing heterotrophic organisms. Therefore, in order to maintain a healthy population in the bioreactor it is essential to operate the process at high SRT and low F:M ratio. Typically, SRT over 5 days and F:M ratio of 0.08 to 0.12 are necessary to maintain nitrification. However, since growth rate is a function of temperature, nitrification can be achieved successfully at high temperature even at lower SRT values and higher F:M ratios than the typical recommended ranges. Some refinery wastewater treatment activated sludge reactors operate at very high SRT, approaching 40 days, in order to minimize sludge production and disposal requirements.

Various Types of Activated Sludge Processes for BOD Oxidation and Nitrification

The various types of aerobic activated sludge processes for BOD oxidation and nitrification, and a few of their major design parameters are summarized in Table 5.9.

They vary over wide ranges. The optimum ranges depend upon actual operating experience over lengths of time under variable operating conditions.

Conventional Activated Sludge (CAS) and Extended Aeration are the two types of activated sludge processes most widely used in refinery wastewater treatment plants. Both involve design and operational variation of the basic activated sludge process. The reactor configuration and operating experience of both types are discussed in the following section.

Wastewater treatment scheme by conventional activated sludge process is presented in Fig. 5.19. Influent to biological treatment is fed to the aeration tank which is typically equipped with submerged aerators. Either coarse bubble or fine

Table 5.9 Typical design guidelines for various types of activated sludge processes for BOD oxidation and nitrification 2014 (Adapted from Metcalf & Eddy/AECOM)

Process	Process description	Sludge retention time (d)	F:M ratio (d^{-1}) based on kg/d of BOD/ kg of MLVSS	MLSS (mg/L)	Recirculation Ratio
Conventional activated sludge (CAS)	In this process the wastewater enters a complete mix or plug flow reactor and flows into a secondary clarifier. A stream from the secondary clarifier is reintroduced into the reactor containing suspended solids. For completely mix CAS reactors substrate concentration and F:M ratio are constant throughout the reactor volume. For plug flow CAS reactors F:M ratio is high at the entrance and gradually decreases along the length as food is depleted. Accordingly, oxygen demand also changes from high to low	3–15	0.2–0.4	1500–3000	0.25–0.5
Extended aeration	This process is a modification conventional activated sludge (CAS) process which is designed with high SRT (20–40 days) and high HRT of around 24 h. with the intent of maximizing biosolids stabilization in the aeration basin to reduce sludge disposal requirement. In refinery wastewater treatment applications this also helps breakdown difficult to degrade organics	20–30	0.05–0.15	3000–6000	0.5–2.0

(continued)

Table 5.9 (continued)

Process	Process description	Sludge retention time (d)	F:M ratio (d^{-1}) based on kg/d of BOD/ kg of MLVSS	MLSS (mg/L)	Recirculation Ratio
Tapered aeration	A type of activated sludge processes with a plug flow reactor in which the aeration air is not applied at a uniform rate. It is high around 40% near the reactor inlet and around 20% at the end of the reactor	3–15	0.2–0.4	1500–3000	0.25–0.5
Step feed aeration	This process is a modification of conventional activated sludge process in plug flow configuration. Wastewater enters the reactor at 3 to 4 points along the reactor to equalize the F:M and reduce peak oxygen demand. This system provides higher SRT than conventional Plug flow process	3–15	0.2–0.4	2000–3500	0.25–0.75

(continued)

Table 5.9 (continued)

Process	Process description	Sludge retention time (d)	F:M ratio (d^{-1}) based on kg/d of BOD/ kg of MLVSS	MLSS (mg/L)	Recirculation Ratio
Contact stabilization	This process is similar to CAS system but uses two tanks instead of one aeration reactor. Tank one is the stabilization tank which is for stabilization of return activated sludge (RAS) and the other, the contact tank is for treatment of wastewater. The contact tank has a residence time of 30–60 min and MLSS concentration in this tank is lower than stabilization tank. Residence time in stabilization tank is 1 to 2 h	5–15	0.2–0.6	1000–3000 in contact tank/ 3000–10,000 in stabilization tank	0.5–1.0
High purity oxygen	This process usually is conducted using 3 or 4 reactor tanks. In these the influent wastewater, RAS and high purity oxygen are added in the first tank through eductor jets close to the bottom of the reactor. The amount of oxygen introduced decreases over the reactors from a high of 40–60% in first reactor to a low of 20% in last reactor	1–4	0.25–1.0	2000–5000	0.25–0.5

bubble diffused aerators are used most frequently, and the aeration system is designed to both provide the required oxygen to the system as well as keep the tank well mixed and keep the biomass in suspension.

Figure 5.19 shows addition of nutrients nitrogen and phosphorus in the aeration tank. However, as discussed earlier, typically nitrogen is present in sufficient concentration in refinery wastewater and may not need further supplementation. Phosphorus dosing will be necessary in most applications. Typically, a Mixed liquor

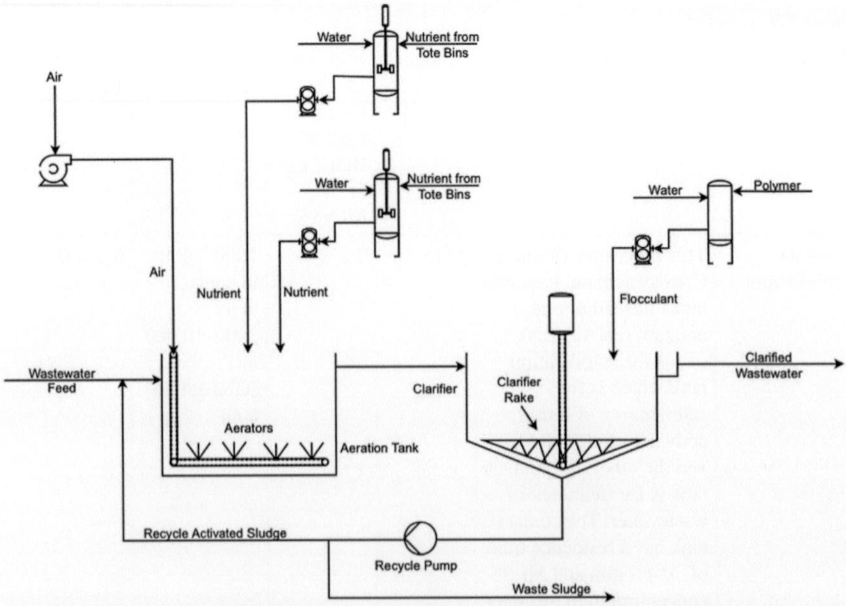

Fig. 5.19 Refinery wastewater treatment by conventional activated sludge (CAS) process

suspended solids (MLSS) concentration of 1,500 to 3,000 mg/L is maintained in the aeration tank. The mixed liquor from the aeration tank flows to the clarifier where the sludge is allowed to settle. Polymer may be added to aid in the flocculation of biomass and promote solid–liquid separation. A part of the sludge is recycled back to the aeration tank as return activated sludge (RAS) while the remainder is removed as waste activated sludge (WAS).

The activated sludge system can remove up to 85 to 90% of the COD contained in the wastewater feed and the clarifier can remove up to 99%of the MLSS in the feed to the clarifier. Typical characteristics of conventional activated sludge process influent and effluent characteristics are listed in Table 5.10.

Typically, the activated sludge reactors in refineries are operated in extended aeration mode. The design parameters for extended aeration type suspended growth rectors are listed in Table 5.9. It must be noted that these are designed to operate under very high sludge retention time (SRT) and low food to microorganism (F:M) ratios. There are two main reasons for refineries to operate their activated sludge reactors in extended aeration mode:

- Break down some of the hard to degrade organics including phenols and long chain molecules to minimize the risk of non-compliance of COD or TOC effluent limits.
- Minimize the sludge production, and thereby save on sludge handling and disposal cost, by oxidizing a large part of net growth of sludge in situ in the reactor.

Table 5.10 Typical influent and effluent characteristics of refinery conventional activated sludge systems

Parameter	Influent characteristics	Maximum influent concentrations under upset conditions of upstream process units	Bioreactor effluent characteristics with BOD removal + nitrification	Clarifier effluent characteristics with BOD removal + nitrification
Temperature, °C	20–35	40	20–35	20–35
pH, su	6.5–9.0	8.5	6.5–8.0	6.5–8.0
BOD, mg/L	120–150	–	10 (soluble)	10 (soluble)
COD, mg/L	200–400	≤1,000	50–60 (soluble)	50–60 (soluble)
Ammonia-N, mg/L	25–50	≤100	1–5	1–5
TN, mg/L	35–55	–	2–6	2–6
TP, mg/L	8–15	–	5–8	5–8
Oil & grease, mg/L	20–50	–	1–5	1–5
Aromatics	25–35	–	2–5	2–5
Phenols	5-20	≤50	≤1	≤1
H_2S and other reduced sulfur compounds	1–5	≤100	0–1	0–1

In addition to these, high SRT also helps achieve nitrification, if needed.

The downside of operating at high SRT is poor settling of solids in the secondary clarifiers. This is because high SRTs adversely impact the production of extracellular polymeric substances (EPS) by microbial cells which help them flocculate together and settle in the clarifier in large flocs. This situation can be mitigated by adding polymeric flocculant aides in the secondary clarifier as shown in Fig. 5.19.

In an attempt to achieve near complete elimination of sludge production and disposal, a refinery on the US Gulf coast started operating the activated sludge reactor at SRT above 40 days by minimizing sludge wasting rate from the secondary clarifier and returning about 100% of the sludge to the reactor as return activated sludge (RAS). This condition created the growth of filamentous microorganisms, causing foaming and bulking of the sludge in the reactor basin. The situation got out of control as the foam spilled over the reactor basin covering the surrounding areas with a layer of slippery foam, and ultimately leading to shutdown of the reactor.

Filaments represent a wide variety of bacterial genera. The conditions for their development and growth also are different from each other and are specific to the individual types. In each case, the type of filamentous organism must be identified microscopically to determine its identity, the cause(s) of its appearance and growth,

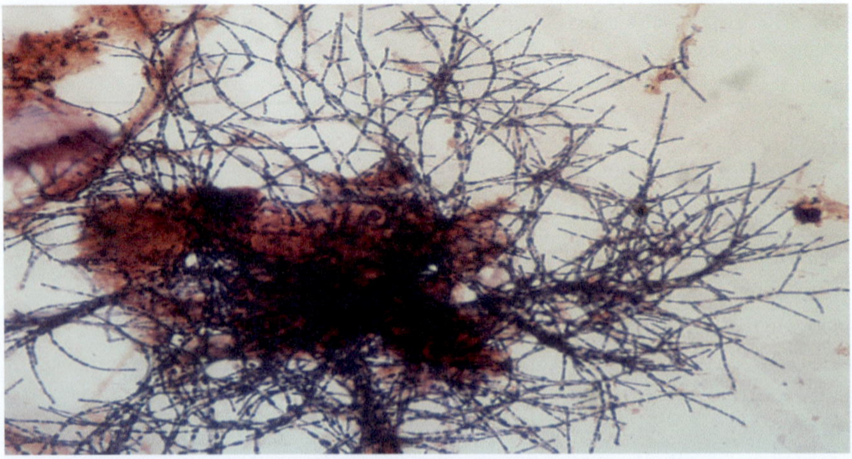

Fig. 5.20 Filamentous microorganism in refinery secondary clarifier foam sample (Visible under 1000 × Direct illumination gram staining, Jenkins 2013; adapted from Hydrocarbon Processing, Basu 2018)

and identify the remedial measures to mitigate the situation. In the example described above, the filament was identified as *Nocardia spp* due to excessive oil in bioreactor feed, coupled with the very high SRT. The high level of oil is indicative of inadequate primary treatment performance leading to carry over of oil into the reactor basins. Figure 5.20 shows the magnified image of filamentous structure of these organisms obtained from a foam sample taken from the site during this incident at the refinery. Identification of microorganisms in general, including filamentous growth has been introduced in Chap. 1 and a detailed discussion of this subject has been provided by Jenkins et al. (2003).

Activated Sludge Processes for Total Nitrogen (TN) Control

For total nitrogen removal systems nitrification is combined with denitrification in a two-step treatment process:

1. Nitrification—Conversion of Ammonia—nitrogen to nitrates (discussed earlier in this section)
2. Denitrification—Conversion of Nitrates to Nitrogen gas which is released into atmosphere

Denitrification

Denitrification is an anoxic biochemical process involving reduction of nitrate to nitrogen gas. Denitrifiers are heterotrophic organisms, and they fulfill their metabolic oxygen requirement by extracting oxygen atoms from nitrate ions. Although total nitrogen (TN) control in refinery wastewater treatment is not commonly mandated by the regulatory bodies several refineries are currently adding a denitrification step in the treatment train above and beyond nitrification. This is mainly driven by the urge to save operating cost as denitrification produces 3.57 g alkalinity as $CaCO_3$ per gram of nitrate—nitrogen that is denitrified, which reduces the amount of alkalinity consumed for nitrification by 50% and returns over 50% of oxygen that is consumed for nitrification.

In denitrification, nitrates are reduced to nitrogen gas by using carbon as food source (electron donor). The biochemical reaction is represented by the following equation.

$$6NO_3^- + 5\,CH_3OH \rightarrow 3N_2 + 5CO_2 + 7H_2O + 6OH^- \quad (5.15)$$

In the above reaction methanol (CH_3OH), which is widely used, is shown as the external carbon source for denitrification reaction. However, other readily biodegradable organic substrates e.g., acetic acid, glycerin, etc. and their derived commercial products are also used depending upon their availability and price. Literature reports (Metcalf and Eddy/AECOM 2014) a typical ratio of BOD to nitrate—nitrogen of 4:1 to maintain or increase denitrification rates.

Factors affecting denitrification process are:

- **Temperature**

The optimum temperature range for denitrification is 15 to 35 °C.

- **Dissolved oxygen (DO)**

The DO concentration is a critical factor in denitrification as an increase in DO decreases denitrification rate. Therefore, it is essential to incorporate processes, structural designs, and techniques to reduce and maintain DO concentration below 0.5 mg/L in anoxic stages.

- **pH**

The ideal pH for denitrification is 7–9. A pH outside of this range will adversely affect denitrification rates.

Integrated BOD Oxidation and Total Nitrogen (TN) Control

For Total Nitrogen Removal one of the most used treatment technologies is the modified Ludzack Ettinger (MLE) process (Fig. 5.21). In this process the influent enters the pre-anoxic stage where it mixes with the internal recycle (IR) and return activated sludge (RAS) streams. Denitrification occurs in this pre-anoxic reactor which also removes some of the incoming BOD which serves the purpose of a carbon source. Then the wastewater flows into the aerobic basins where further BOD oxidation and nitrification take place. A large part of MLSS that contains nitrates is returned to denitrification basin as internal recirculation (IR) stream from the aeration basin along with return activated sludge stream (RAS) from the clarifier. Typically, an IR rate of 2–4 times the influent is applied to achieve maximum denitrification with the carbon available in the raw influent. This can help attain effluent TN levels as low as about 6 mg/L.

At some refineries, the extended aeration reactors are of special configuration that are offered by technology providers as shown in the Figs. 5.22 and 5.23 below. Each of these systems consists of both the reactor and the clarifier that are packaged together within the same assembly. Figure 5.22 represents a reactor—clarifier assembly in two concentric cylindrical tanks. The inner tank acts as the secondary clarifier. The annular space has two sections separated by a partition baffle—the larger section is the aerobic treatment reactor equipped with fine bubble diffusers for extended aeration and the smaller section acts as an aerobic digester of the waste sludge. The aerobic reactor carries out both COD oxidation and nitrification. An additional section in the annular space can also be included to carry out pre-anoxic denitrification if total nitrogen removal is required. This type of combined biological reactor—clarifier system is utilized for secondary treatment of wastewater in a refinery in Washington state, USA.

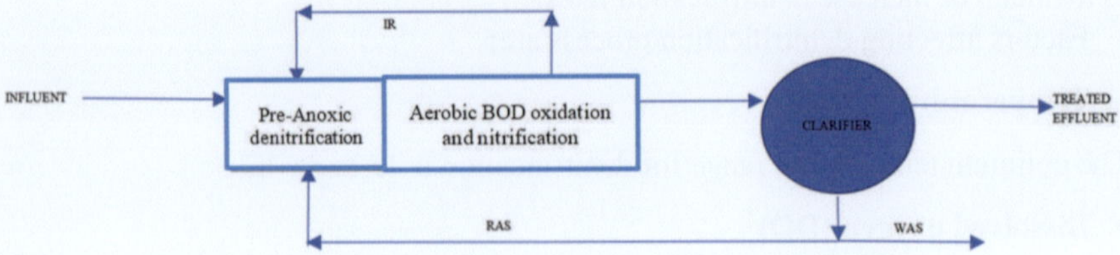

Fig. 5.21 Modified Ludzack Ettinger (MLE) process for total nitrogen removal

Fig. 5.22 Extended aeration reactor and clarifier in concentric tanks. *Photo* Courtesy of Evoqua (Xylem) Water Technologies, Waukesha, WI

Fig. 5.23 Orbal extended aeration package system. *Photo* Courtesy of Evoqua Water Technologies (Xylem), Waukesha, WI

Figure 5.23 depicts an Orbal extended aeration treatment system with the ability to carry out COD removal, nitrification and also denitrification, if required. The influent wastewater flows through the channels configured in the form of a racetrack. As mentioned earlier in this section, a refinery in California carries out its secondary wastewater treatment using this type of reactor. Orbal is marketed by Evoqua (currently Xylem). Details of the working principle are available at their website: https://www.evoqua.com/en/evoqua/products-services/aerobic-wastewater-treatment/aerobic-systems/the-orbal-system/.

Effective BOD removal and TN control requires continuous adjustment of sludge wasting, return and internal recirculation rates in real time involving rigorous and time—consuming calculations. Various programs are available commercially that run dynamic simulations over multiple days to provide the user insight into the operation. These simulations can be used to find the required sludge wasting and return rates, internal recycle rate, aeration requirement and calibration of process design with operating data. Further details regarding the workings of process simulators used to model the design and operation of wastewater treatment plants can be found in Chap. 6.

Treatment of Reduced Sulfur Compounds

Primarily hydrogen sulfide and organic sulfur compounds, exist in sour crudes. This is a significant issue with crudes from many geographical locations. Globally, there is a push to use low sulfur fuels in order to reduce sulfur oxide (SO_x) emission from the combustion off gases. Removal of reduced sulfur compounds from petroleum products is a two-step process—(a) hydrogenation and hydrogesulfurization of all organic sulfur compounds to H_2S, and (b) removal of H_2S by steam stripping to free fuels from sulfur, thus minimizing SO_x in the combustion product gases. The source water stripper bottoms streams contain H_2S, hydrocarbons and other contaminants that make up parts of the effluent sent to the wastewater treatment plant.

The presence of hydrogen sulfide in wastewater poses several operating and safety problems:

- H_2S is malodorous and creates health and safety risk to operators and others in nearby areas;
- It is corrosive to equipment, pipes, and vessels; and
- H_2S interferes with biological treatment process.

H_2S can be removed from wastewater by biological oxidation in activated sludge aeration basins. However, high sulfide concentration promotes the proliferation of certain types of filamentous microorganisms of *Thiothrix* genus. It is a group of sulfur oxidizing bacteria which create bulking sludge that does not settle in secondary clarifiers and interferes with solid–liquid separation. *Thiothrix* bacteria conducts partial oxidation of H_2S to elemental sulfur that deposit within their cell structure as shown in the figure below Fig. 5.24.

Due to these issues with H_2S it is desirable to remove it prior to biological treatment.

In water H_2S can exist in solution as hydrogen sulfide (H_2S), bisulfide (HS^-) ions and sulfide (S^{2-}) ions. Their relative distribution is a function of pH as can be observed from the distribution diagram that is discussed in Chap. 1.

At some refineries, chemical oxidation is used for removing H_2S from refinery wastewater prior to biotreatment. Oxidation reactions can elevate sulfur from –2 oxidation state to 0 (elemental sulfur), +2 (thiosulfate), +4 (sulfite) and +6 (sulfate) states. These are represented by the following reactions:

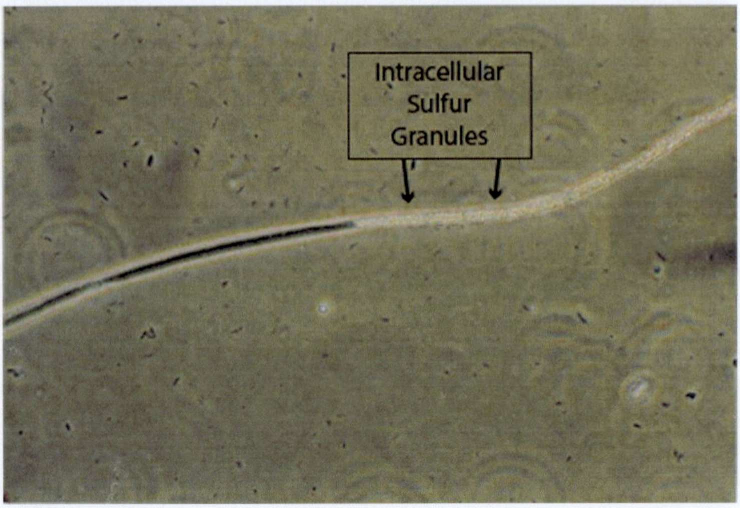

Fig. 5.24 Aerobic *Thiothrix sp* bacteria with deposit of intracellular sulfur granules (Basu et al. 2001) (Magnification 1000 X)

$$HS^- + 0.5\,O_2 \rightarrow S_0(\text{elemental sulfur}) + OH^- \tag{5.16}$$

$$HS^- + O_2 \rightarrow 0.5\,S_2O_3^-(\text{thiosulfate}) + 0.5\,H_2O \tag{5.17}$$

$$HS^- + 1.5\,O_2 \rightarrow SO_3^{2-}(\text{sulfite}) + H^+ \tag{5.18}$$

$$HS^- + 2\,O_2 \rightarrow SO_4^{2-}(\text{sulfate}) + H^+ \tag{5.19}$$

H_2S oxidation is kinetically limited and can produce partially oxidized sulfur species. Production of elemental sulfur is undesirable as it creates solid particles that can deposit on solid surfaces in tanks and equipment. Addition of a catalyst enhances the rate of reaction and produces ionic species containing higher oxidation states of sulfur. Metal ions e.g., Ferrous (Fe^{2+}) and Ferric (Fe^{3+}) ions are good catalysts that promote oxidation of H_2S to thiosulfate, sulfite and sulfate salts that remain dissolved in water typically do not cause operational problems. This reaction is conducted in an aerated pre-oxidation tank where air serves as the source of oxygen and also supports mixing. Ferrous sulfate solution is dosed as the catalyst.

The overall reaction for complete oxidation of sulfide to sulfate ions is:

$$H_2S + 2\,O_2 + 2\,NaOH \rightarrow Na_2SO_4 + 2\,H_2O \tag{5.20}$$

It is desirable that the complete oxidation reaction takes place in the pre-oxidation tank. However, some amounts of partially oxidized products—thiosulfate and sulfite ions are carried over from this tank and oxidized biologically to sulfate ions in the aeration basin.

Powdered Activated Carbon Treatment® (PACT) System

PACT is an extension of the activated sludge system. It is very similar to the conventional activated sludge system but is enhanced with the use of activated carbon which is added to the aeration tank. Figure 5.25 shows a typical PACT system. Both biochemical oxidation and carbon adsorption occur in the tank simultaneously thus enhancing the removal of contaminants in the wastewater. This is especially useful in instances where there are components recalcitrant (refractory) to biological treatment present in the wastewater. The subject of recalcitrance of certain chemicals in refinery wastewaters to biological treatment has been discussed later in this chapter. In this system the waste sludge from the clarifier is sent to a thickener. The decant from the thickener is returned to the aeration tank and the waste sludge is sent for disposal. Fresh carbon is added to the aeration tank as required. PACT systems have been reported to have removed more than 90% of the COD contained in the wastewater feed to the reactor.

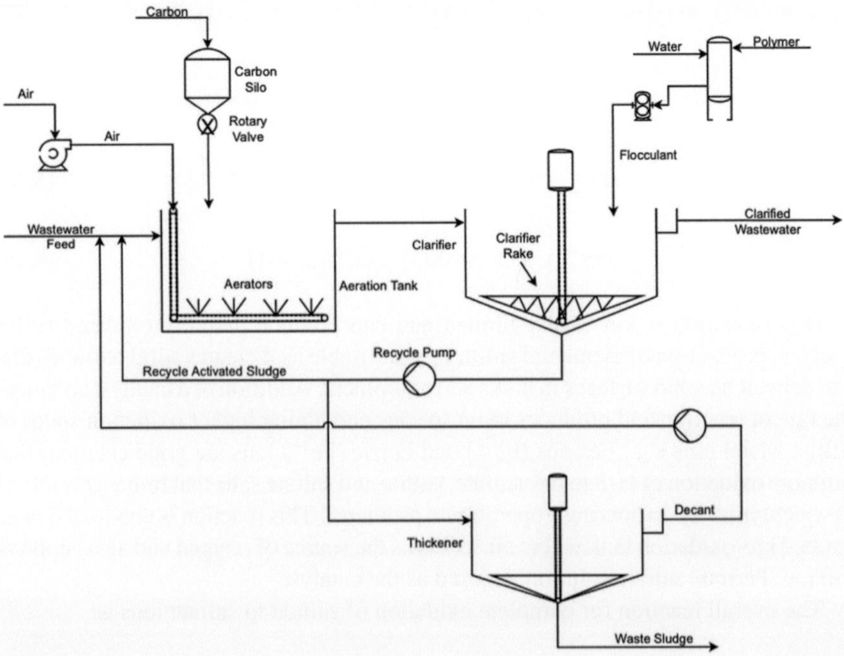

Fig. 5.25 PACT unit

5.6.2 Secondary Clarification

Secondary clarification is an integral part of the activated sludge process. The main purpose of a secondary clarifier is to conduct solid–liquid separation of the mixed liquor from the aeration basin by settling sedimentation in order to produce good quality effluent that is low in suspended solids and BOD. The settled solids are concentrated in the collection hopper as a sludge blanket at the bottom of the clarifier. Under steady state operating conditions, a well-functioning secondary clarifier produces effluent of TSS in the range of ~10 to 20 mg/L and settled solids concentration of ~0.8 to 1%. Several factors such as flow variations, density currents, solids concentration, solids loading, surface area, retention time and overflow rate influence the design and performance of clarifiers. WEF Manual of Practice (MOP) FD 8 (WEF 2005) on clarifier design is among several standards that provide the guidelines.

Clarifiers are designed for average and peak flow periods. For the latter, the retention time gets reduced due to increase in the overflow rate and consequently overloading the clarifier during these periods. The ratio of recycle and waste activated sludge (RAS and WAS) is calculated depending on the wastewater characteristic to provide ideal growth environment for the microorganism. Important design criteria that need to be considered for designing of the secondary clarifiers are:

- Surface Overflow Rate
- Solids Loading Rate
- Weir Loading
- Depth and Detention Time

Surface Overflow Rate

Surface Overflow Rate represents the hydraulic loading per unit surface area of a tank per unit time and is expressed as $m^3/d\text{-}m^2$. This parameter for both the average flow and peak flow need to be considered to achieve optimum performance of secondary clarifiers.

Solids Loading Rate

The solid loading rate is an important variable for the design of secondary clarifiers. It is expressed as kg MLSS/d-m^2. The solid loading rate is a parameter to understand the thickening limitations for a clarifier.

Weir Loading

Weir loading influences the removal of solids particularly in rectangular secondary clarifiers. The loading should however ensure uniform withdrawal over the entire periphery of the tank to avoid short—circuiting or dead pockets. Performance of some clarifiers for SS removal can be improved by increasing their weir length.

Depth and Detention Time

The side water depths (SWD) of secondary clarifiers is the depth of the water surface from the bottom of the clarifier tank. It varies from 2.4 to 4.2 m and detention times vary from 2 to 4 h.

Sludge Volume Index (SVI)

At most of the wastewater treatment plants (WWTP), the thickening characteristics of wastewater sludge is monitored regularly by using simple sludge volume index (SVI) tests. This test monitors any changes in the mixed liquor (MLSS) settling properties as a function of operational, seasonal changes or change in influent wastewater streams. Lower SVI values have been associated with an efficient clarifier performance and high sludge settleability.

$$\text{SVI, mL/g} : \frac{settled\ volume\ of\ sludge\left(\frac{mL}{L}\right) * 1000\, mg/g}{suspended\ solids\left(\frac{mg}{L}\right)} \tag{5.21}$$

The SVI is calculated by placing a mixed liquor sample in a 1 L container and measuring settled volume after 30 min and corresponding MLSS concentration as expressed in Eq. 5.21.

Design guidelines for secondary clarifiers are given in Table 5.11.

Clarifiers can be constructed from concrete. However, in refinery applications clarifiers are typically made of epoxy coated carbon steel tanks or glass lined steel tanks.

Table 5.11 Typical design guidelines for secondary Clarifier (Adapted from Metcalf & Eddy/AECOM, 2014)

Type of settling	Overflow rate (m^3/m^2-d)		Solids loading (kg/m^2-h)		Side water depth (m)	Weir loading (m^3/m-d)
	Average	Peak	Average	Peak		
Secondary settling for activated sludge	16–28	36–56	0.8–1.2	2.0	>3.0–3.5	185
Secondary settling for extended aeration	8–16	24–32	0.2–1.0	1.6	>3.0–4.0	185

Dissolved air floatation (DAF) units are also frequently used for refinery secondary solid–liquid separation and can be used instead of clarifiers. These are more common for moving bed biological reactors (MBBR) than suspended growth reactors. This has been discussed further later in this chapter.

5.6.3 *Special Types of Aerobic Suspended Growth Processes*

Sequencing Batch Reactor (SBR) and Membrane Biological Reactor (MBR) belong to a special category of suspended growth biological treatment reactors that do not utilize a secondary clarifier for solid–liquid separation of the mixed liquor. Working principles of these are discussed below. Although this discussion is primarily in reference to BOD removal only but both technologies are routinely utilized to achieve additional treatment goals of nitrification and denitrification if ammonia removal and Total Nitrogen removal are required.

5.6.3.1 Sequencing Batch Reactor (SBR)

SBR is a fill-and-draw type of activated sludge system for wastewater treatment (Fig. 5.26). In this system, wastewater is fed to a single reactor, treated to remove BOD, ammonia-N, COD/TOC, if applicable and then discharged. There is no clarifier or DAF to separate the mixed liquor, instead the SBR reactor includes a settling phase. Typically, two or more batch reactor tanks are used in parallel in a predetermined sequence of operations to enable flow to be continually fed to the system. Each reactor tank operates in typically 6-to-8-h cycles, consisting of 3 to 4 cycles every day. A 6-h cycle, for example, typically consists of the following steps:

- Anoxic/Anaerobic Fill for 1 h
- Aerobic React for 4 h
- Quiescent Settle for 30 min
- Decant and Settled Sludge Wasting for 30 min

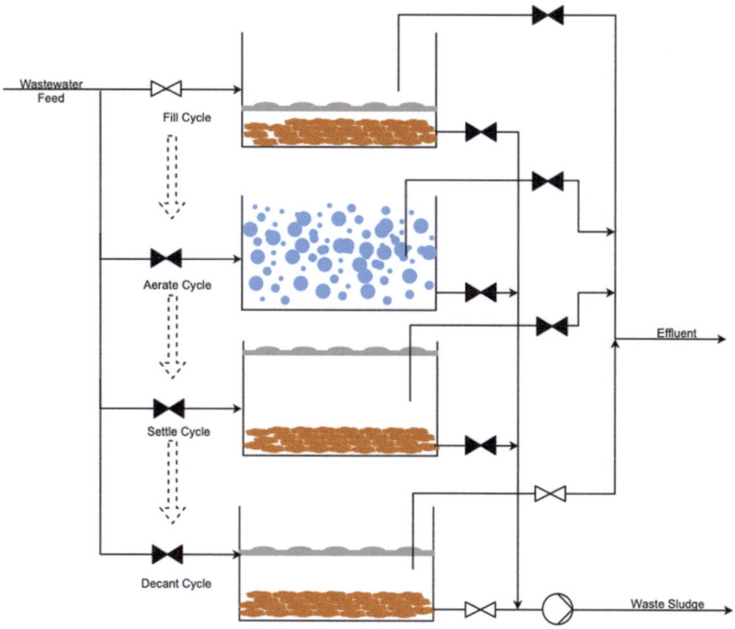

Fig. 5.26 Typical SBR treatment system

The first step of anoxic/anaerobic fill utilizes the influent BOD to denitrify the nitrate formed during the aerobic react step from the preceding cycle. This step proceeds in absence of aeration. Aeration is resumed during the aerobic react step to conduct BOD oxidation and nitrification of the wastewater. A mixer agitates the contents of the reactor tank to mix it thoroughly and keep the biomass in suspension. Upon completion of the react step the aeration and mixer are stopped, and the solids are allowed to settle and thicken under quiescent conditions. The SBR tank during this step performs the role of a secondary clarifier. Finally, the net growth of biomass is wasted from the bottom of the tank to maintain the desired MLSS in the tank. This concludes one complete cycle, and the tank is then ready to start the next cycle to receive a new batch of influent and resume a new anoxic/anaerobic fill step.

Note that the above cyclic steps of anoxic fill and aerobic react are included assuming a total nitrogen (TN) removal process. However, if only BOD oxidation with or without nitrification is required then the fill step also can be aerobic without the need for anoxic step. This will result in shorter cycles and allow the system to handle higher flow rates. The duration of each step in a cycle and the total number of cycles in a day are determined based on the characteristics of the wastewater and the treated effluent quality goal.

The biological treatment process of SBR and conventional activated sludge (CAS) are essentially the same. The difference between the two technologies is that the SBR does not use a separate secondary clarifier as it performs biological treatment and

Table 5.12 Typical design guidelines for sequential batch reactor

Process	Flow regime	Sludge retention time (d)	F:M ratio (d^{-1})	MLSS (mg/L)
Sequential batch reactor	Batch—fed, complete mix	5–15	0.05–0.30	1500–5000

clarification in a single tank using a timed controlled sequence. An equalization tank precedes SBR treatment system to collect the influent and feed it to a reactor tank based on its sequence of operation as controlled by a programable logic controller (PLC). Although the same tanks carry out the functions of biochemical reactions and solids settling but typically SBR requires a pre-equalization tank to distribute feed into the reactor tanks in sequence. It may also need a post equalization tank to uniformly discharge the treated effluent at a constant rate to the receiving water bodies instead of large hydraulic surges in short times.

Design guidelines for a SBR process is summarized above in Table 5.12.

The SRT and F:M ratios depend upon the operational duty of the SBR. For a BOD oxidation only, operation is conducted at low SRT and high F:M ratios. However, for nitrification and total nitrogen (TN) removal high SRT and low F:M ratios are necessary. As with other types of suspended growth processes SBRs for refinery wastewater treatment are also operated under high aerobic SRT and low F:M ratio to maximize breakdown of complex organic molecules and minimize sludge production and sludge wasting. An SBR system is used at a refinery in Kansas for COD oxidation and nitrification.

5.6.3.2 Membrane Bioreactor (MBR)

The aeration basin of an MBR system is typical of an activated sludge process, but it operates at a high MLSS of typically 6,000 to 8,000 mg/L. The mixed liquor from the tank is sent to a second tank which contains microfiltration (MF) or ultrafiltration (UF) membranes for solids—liquid separation in place of settling sedimentation by a secondary clarifier. The membranes are used to filter solids in the mixed liquor. The oil and grease content of the feed needs to be tightly controlled to less than 10 mg/L to prevent fouling of the membranes and maintain its performance. The membrane tank is subjected to a low-pressure vacuum that pulls water through the membranes and pumps the filtered water to the next process step while retaining solids in the reactor. Compressed air is injected into the system to scour the exterior surfaces of the membranes and keep it free from accumulation of solids.

A typical schematic of a MBR system is shown in Fig. 5.27. The permeate stream is the treated effluent of high quality, typically BOD and TSS of <2 mg/L each. The separated solids stream concentration is typically 8,000 to 10,000 mg/L. A part of this, that accounts for the net growth, is wasted as WAS and the remaining part is recycled as RAS.

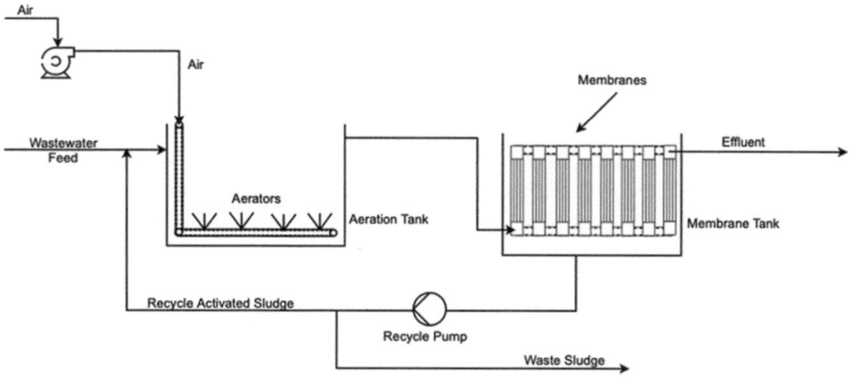

Fig. 5.27 Typical MBR treatment system

MBR systems are still not used extensively in the oil refining industry due to high capital cost compared to conventional activated sludge. However, since activated sludge systems often require tertiary filtration, MBR becomes more cost competitive, as it is equivalent to having a clarifier and effluent filter combined in the MBR treatment system.

There are pros and cons associated with MBR systems when compared with conventional activated sludge (CAS) treatment systems. The main advantage of an MBR system is the effluent quality with very low TSS/Turbidity and BOD making the treated effluent suitable for reuse. In applications where further treatment such as reverse osmosis is used, MBR can be more cost effective than the alternative option of using media filtration and MF or UF after biological treatment by a conventional activated sludge system consisting of aeration tank and secondary clarifier. The other major advantage is its high operating MLSS which allows the MBR reactors achieve high solids retention time (SRT) and low F:M ratio. This significantly reduces the footprint area, reactor volume and the associated capital cost. On the other hand, MBR systems are relatively capital intensive due to the cost of membranes and the need for their replacement after completion of their operating life. High operation and maintenance cost results from the need to carry out chemical clean in place (CIP) and routine maintenance of membranes requiring occasional downtimes. The membranes used in MBRs are typically polymeric, but some system suppliers use ceramic membranes which are more resistant to chemical contaminants.

Design guidelines for a MBR process is as summarized below in Table 5.13.

Table 5.13 Typical design guidelines for membrane bioreactor

Process	Sludge retention time (d)	F:M ratio (d^{-1})	MLSS (mg/L)	Recirculation ratio	Membrane flux (L/ m^2–H)
Membrane bioreactor	10–25	0.1–0.3	6000–10,000	2–4	12–30

5.6.4 Attached Growth Systems

Attached growth biological systems employ inert media on which microorganisms grow and form a biofilm the mechanism of formation of biofilms on solid surfaces and their role in attached growth wastewater treament processes are discussed in detail in Appendix C. The organic matter is removed from the wastewater as it passes over the biofilm attached to the media. These media can be rocks, redwood, or a wide variety of synthetic polymeric media. Attached growth systems are also employed for nitrification applications where the biofilm is comprised of nitrifiers that feed on ammonia—N in wastewater to conduct their metabolic activities and grow in population.

In the attached growth processes, the thickness of the biofilm ranges from 0.1 to 10 mm. A stationary liquid layer, also called the diffusion layer, separates the biofilm from the bulk liquid. Oxygen transfers through the stagnant liquid and the biofilm, while the products of metabolic biochemical reactions diffuse in the opposite direction.

The most common types of attached growth system are trickling filter, rotating biological contactor (RBC) and moving bed biofilm reactor (MBBR). The media in trickling filters remains fixed but in an MBBR the media moves around in the reactor, and in a rotating biological contactor the media rotates in the tank around a fixed shaft.

Since the advent of attached growth processes trickling filters and RBC have been the two types used in wastewater treatment plants. Both are used in old petroleum refineries—trickling filters as roughing reactors and RBCs for total biological treatment. However, various modifications of suspended growth processes have dominated the full-scale installations in both municipal and industrial applications. Recent development of MBBR has led to the resurgence of interest in attached growth processes, primarily driven by their advantages over traditional suspended growth processes in terms of footprint requirements, ease of conversion and related economic benefits, as detailed in the following sections. An in-depth discussion of attached growth processes is provided in Appendix C for ease of understanding.

5.6.4.1 Trickling Filter

Trickling filters can be used as a pretreatment step prior to treatment in a suspended growth unit in an oil refinery. Typically, these are designed as roughing reactors to shave off up to 30% of the influent BOD load. It consists of a tank that is filled with random plastic packings, or in some cases different sized rocks with a high surface area to volume ratio in a shallow bed. Wastewater is fed to the top of the filter by rotating distributor arms with spray nozzles so that the feed is evenly distributed over the cross section of the tank. Organic materials present in the wastewater are removed by bacteria growing in the biofilm that develops on the media surfaces. The wastewater trickles down the filter, into the underdrain system. This under drained

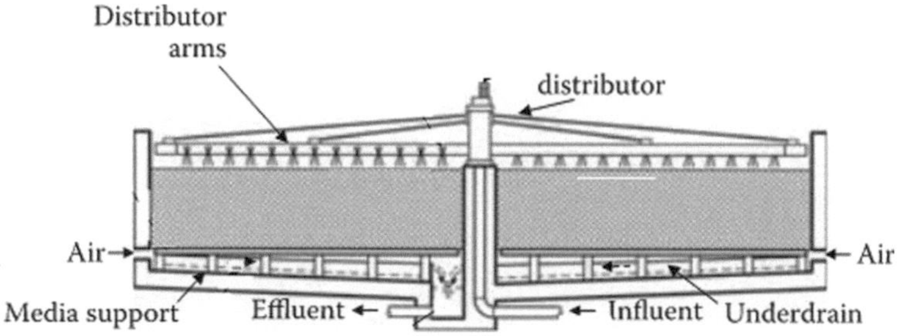

Fig. 5.28 Schematic of Trickling Filter

effluent also contains a portion of the biological film that is detached (or sloughed off) from the media. Air in the trickling filter rises from the bottom through the void spaces in the media supplied by natural or forced draft.

The effluent is then fed to a clarifier to remove the solids particles carried over along with the treated effluent. In some systems, a portion of treated effluent is recycled which helps to prevent formation of water pools (called ponding) on the filter, maintain uniform head loss, dilutes the influent, and reduces the nuisance from odors. A typical schematic of trickling filters is shown in Fig. 5.28 above. Four main components of the trickling filter are the media, air distribution system, water distribution system and the underdrain system.

Trickling filters can be classified based on the loading rate applied to them as follows:

- Low-rate filter
- Intermediate rate filter
- High-rate filter
- Roughing filter

If ammonia removal is desired, it is usually conducted in a separate nitrification filter following BOD removal. A few important operating parameters of trickling filters are as follows.

BOD loading (Organic Loading)—In trickling filters organic loading is defined as the amount of BOD load (kg/d) applied to the volume of the reactor (m^3).

$$\text{Organic loading} = \frac{\text{BOD}\left(\text{kg/m}^3\right) \times \text{Flow}\left(\text{m}^3/\text{d}\right)}{\text{Volume of the filter}\left(\text{m}^3\right)} \tag{5.22}$$

Hydraulic loading-Hydraulic loading is defined by the total flow (m^3/d) applied across the surface area of the filter.

$$\text{Hydraulic loading} = \frac{\text{Flow}\left(\text{m}^3/\text{d}\right)}{\text{Surface area of the Filter}\left(m^2\right)} \tag{5.23}$$

Recycle ratio is defined as the return of filter effluent back to the influent flow of filter.

$$\text{Recycle ratio} = \frac{\text{Filter effluent recycle flow}\left(\text{m}^3/\text{d}\right)}{\text{Filter Influent Flow}\left(\text{m}^3/d\right)} \tag{5.24}$$

Various design criteria for these filters are summarized below in Table 5.14.

The main advantages of trickling filters are—they are simple to operate, have very low power requirements, and they achieve long solids retention time as the biomass is immobilized on to the media surfaces and are not removed from the system along with the flow. Some refineries use trickling filters to serve as roughing reactors ahead of activated sludge reactors. At certain locations it also serves to remove heat from the influent wastewater to lower the temperature below 35 °C, the optimum range for biological treatment. This is particularly important for refineries with wastewater temperatures approaching 40 °C or higher. Typically, a drop of 5 °C has been reported.

Two refineries, one in the Texas Gulf Coast and the other in California use Trickling Filters as part of their treatment train. In the California refinery, the trickling filter is followed by suspended growth biological reactors (Orbal[1] extended aeration process), and at the Texas refinery two parallel trains of activated sludge reactors operating in extended aeration mode follow the tricking filter.

Trickling filters, however, have some major disadvantages. Since most of these installations are old and the mechanical parts, especially the moving arms become aged resulting in uneven distribution of the influent. Very often trickling filters are subject to plugging by solids in the influent, leading to channeling. This limits the total surface area of the media that is available for treatment and results in an increased load to the suspended growth reactors downstream of the trickling filters. Trickling filters are sources of odor and VOC emissions. They also become infested by higher life forms, e.g. snails, on their top surface, which prey on the biomass and reduce the effectiveness of treatment. There is specifically a concern in maintaining the health of nitrifying trickling filters as nitrifiers are preferentially subject to attack by snails.

Table 5.14 Typical design parameters of trickling filters

Parameter	Low-rate filter	Intermediate to high-rate filter	Roughing filter
BOD_5 loading (kg/m^3-d)	0.08–0.32	0.24–2.4	1.6–6
Hydraulic loading (m^3/m^2-d)	1–4	4–40	40–100
Recirculation ratio	0–1	0.1–2	0–2
Depth (m)	1–2.5	2–2.5	1–10
BOD removal (%)	80–90	80–90 single-stage 90–95 two-stage	40–70

[1] *Marketed by Evoqua (Xylem) Technologies.*

5.6.4.2 Rotating Biological Contactor (RBC)

Rotating biological contactors (RBC) consist of a series of parallel circular disks which are mounted on a shaft that rotates slowly. 30 to 50% of the disk area of RBC is always submerged in a tank through which the wastewater flows. The biological growth (biofilm) develops over the surfaces on both sides of each disk. As the disks rotate, the biofilm is exposed alternately to wastewater and air. In wastewater, the biofilms absorb and break down organic matter and grow. The biofilm thickness ranges from 1 to 5 mm with solids content of about 5–10% Fig. 5.29.

The main components of RBC are disks, shaft, drive unit, housing, and settling basin. The disks are circular, closely spaced, and parallel to each other, and made up of high-density polyethylene, or PVC. These can be 2 to 4 m in diameter and 10 mm thick. These are mounted on a shaft 3 to 4 m long which passes through the center of the disks. There could be 3 to 4 shafts located in a tank which rotate at 2 to 6 revolutions per minute. Larger treatment units could also consist of several tanks that operate in parallel or can also be operated in series. Fiberglass reinforced plastic (FRP) covers are normally provided over each segment of the RBC to protect the assembly against cold and damaging weather, to protect from UV light, to prevent algae growth and avoid emission of nuisance odor. Two examples of the use of RBCs at refineries are a refinery in Salt Lake City, Utah and another refinery in Kenai, Alaska.

In an RBC system, oxygen is transferred by two mechanisms—directly from the atmosphere to the non-submerged biofilm and from the DO in the bulk liquid of wastewater to the submerged biofilm. Hydraulic Loading and organic loading are the main process parameters of RBC. These are defined below. Typical process design guidelines for RBC systems are as shown in Table 5.15.

Hydraulic loading—Hydraulic loading is defined by the total flow (m^3/d) applied across the total surface area of the discs within each RBC drum.

$$\text{Hydraulic loading} = \frac{\text{Flow}\left(m^3/d\right)}{\text{Surface area of the media}\left(m^2\right)} \tag{5.25}$$

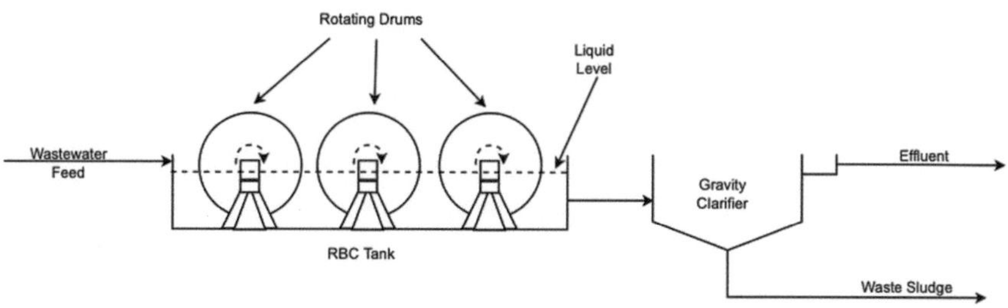

Fig. 5.29 Wastewater treatment process flow schematic of an RBC

Table 5.15 Typical design guidelines for RBC system

Parameter	BOD removal	BOD removal and nitrification
Hydraulic loading (m^3/m^2-d)	0.08–0.16	0.03–0.08
Surface organic loading ($gBOD_5/m^2$-d) Overall	8–20	5–16
Maximum in first stage	24–30	24–30
Hydraulic detention time (h)	0.75–1.5	1.5–4
No of Stages (Stage)	1–2	4

BOD loading (Organic Loading)—Organic loading is defined as the amount of BOD load (g/d) applied to the total surface area of the discs present within each RBC drum (m^2).

$$\text{Organic loading} = \frac{\text{BOD}\left(\text{g/m}^3\right) * \text{Flow}\left(\text{m}^3/\text{d}\right)}{\text{Surface area of the media}\left(\text{m}^2\right)} \tag{5.26}$$

The main advantages of RBCs are simple construction and easy operation. Another major advantage is that RBCs do not require separate aeration as the biomass on the attached films receives oxygen from the ambient air when the disc surfaces become exposed to air during their continuous rotation.

Disadvantages of RBC are similar to those of trickling filters. While channeling is not a major problem, challenges include limitation in oxygen diffusion to the depth of the films, which leads to septicity, incomplete treatment, and odor and VOC emissions. Another major disadvantage is the incidents of shaft failure. With time the discs accumulate biofilm, which gradually become thick and heavy. This has led to breakdown and failure of shaft on which the discs are mounted.

Since new regulations are increasingly requiring compliance with stringent limits on ammonia over and above the COD limits in effluent discharges, refineries are faced with upgrading the wastewater treatment plants to include nitrification. This requires significant expansion of RBC treatment systems. Under such circumstances many refineries are choosing to either totally replace the existing RBCs with new technology or add supplemental reactors by utilizing other treatment technologies downstream of the existing RBCs. In effect, the use of RBCs as a biological wastewater treatment process in refineries is gradually diminishing.

5.6.4.3 Moving Bed Biofilm Reactor (MBBR)

The MBBR process is a dynamic attached growth process wherein the microorganisms grow and remain attached on the media surfaces as biofilms, as in the case of trickling filter and rotating biological contactors (RBC). However, as opposed to the other attached growth processes, the reactor tank in a MBBR system is partially filled with high-density polyethylene (HDPE) media which are continuously in motion within the reactor. The media are kept suspended by a diffused air aeration system for an aerobic process or by a mechanical or hydraulic mixing system for an anoxic or anaerobic process. A sieve is installed at the exit of the tank to keep the carrier media from escaping the tank with the effluent.

MBBR media are especially designed for their shape, size and configuration with the goal of maximizing surface area for attachment of biofilms. As shown in Fig. 5.30 the media consist of an outer and an inner, also called protected, surface area. It is the protected surface that supports the growth and attachment of biofilms. Biofilms are attached onto the media surfaces by slimy, glue like exocellular polymeric substances (EPS). This biofilm formation happens very quickly in aerobic wastewater treatment systems. Although the fixed film is visible after a few days, a microscopic layer develops within minutes to hours of start-up.

Fixed film on media surfaces accommodates large population of biomass in limited volume. The biomass population attached on to the media surface is equivalent to 1,000 to 5,000 mg/L as suspended solids, but its effectiveness in BOD removal and nitrification is far greater than that of a suspended growth reactor of equal volume with comparable MLSS concentrations (Rusten et al. 1995). As a result of this special feature MBBR process offers many advantages over the other attached suspended and attached growth processes, as follows:

- MBBR system requires a lower hydraulics retention time (HRT) compared to competing systems, leading to smaller footprint area, reactor volume, and associated capital cost.
- Biofilm on MBBR media increases or decreases as required to handle the variations in organic loading.
- MBBR systems are resilient to peak flows and to toxic and shock loads.
- Biomass developed on MBBR media surfaces is less vulnerable to bulking and foaming resulting from the growth of filamentous bacteria.

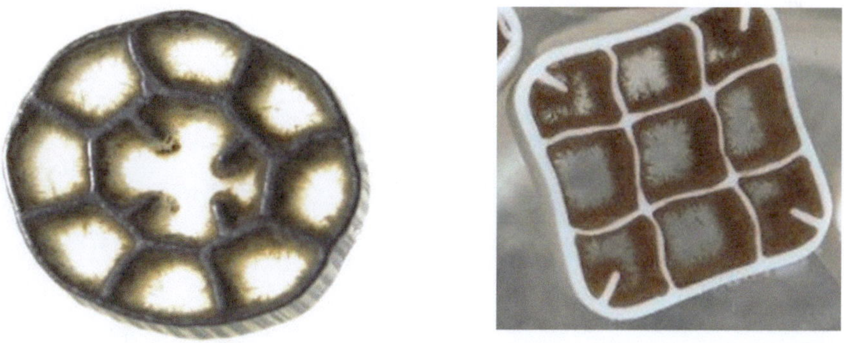

Fig. 5.30 Growth of attached BOD oxidizing biofilm on MBBR Media AC 450 (Left) and Nitrification Biofilm on MBBR Media AC 920 (Right) protected surfaces. *Photo* Courtesy of Headworks International, Houston, TX

MBBR is a once through process without the need for any sludge recycle. From that viewpoint it is easy to operate as the operator does not need to decide the sludge wasting rate to maintain MLSS and SRT in the reactor as in the case of a continuous activated sludge (CAS) reactor. Secondary clarification is required after the MBBR tank to separate sloughed off solids and reduce the TSS of the treated effluent. Due to the poor settling property of the solids with SVI ~120 or more a DAF separator is better suited for solid–liquid separation of MBBR effluent. A typical schematic of a MBBR system is shown in Fig. 5.31.

Some of the disadvantages of MBBR compared to activated sludge systems are higher energy requirement to facilitate the need for higher dissolved oxygen concentration, in the bulk liquid the need to use proprietary media, the need to remove media for diffuser maintenance and the need to improve pretreatment by use of improved screening technologies. Improved pretreatment is required to avoid accumulation of inert particles, dust, sand, and plastic bags. This accumulation can affect the efficiency of the aeration system installed within the reactor and also diffusion of substrates and products in and out of the biofilm.

The MBBR treatment system can be modified as required to meet the regulations. The capacity of MBBR system can be increased by increasing the media total surface area, which is proportional to the media fill within the reactor up to 60–65%. The amount of media in the reactor is quantified as a percentage of the active volume of reactor tank and is referred to as media fill fraction. Increase in media fill allows for greater organic loadings. Media surface area is expressed in terms of media volume and specific protected surface area according to Eqs. 5.26 and 5.27.

Total surface area consists of the combined internal and external surface areas of the media and protected surface area counts only the internal surface area. It is the protected surface area which provides active surface area for attachment of biofilms.

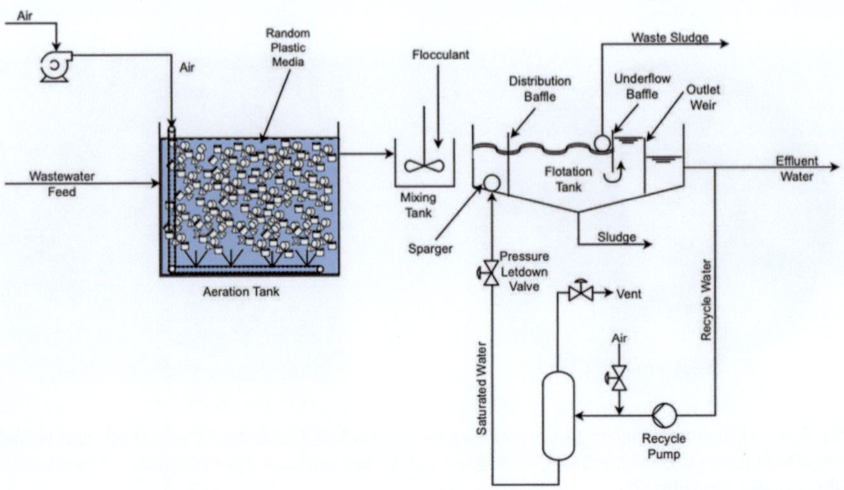

Fig. 5.31 Typical MBBR system

The external surface area is not counted for calculating the volume of media required for a given treatment duty due to the lack of attachment of biofilm resulting from continuous shearing under turbulent conditions in the reactor tank.

The media fill in the reactor provides the amount of media which is used to calculate overall media surface area available within the reactor. The organic loading of a MBBR system is defined as the BOD load (g/d) over the media surface area of the reactor.

$$\text{Organic loading} = \frac{\text{BOD}\left(\frac{\text{g}}{\text{m}^3}\right) * \text{Flow}\left(\text{m}^3/\text{d}\right)}{\text{Surface area of the media}\left(\text{m}^2\right)} \tag{5.27}$$

$$\text{Volumetric Organic loading} = \frac{\text{BOD}\left(\frac{\text{g}}{\text{m}^3}\right) * \text{Flow}\left(\text{m}^3/\text{d}\right)}{\text{Volume of the media}\left(\text{m}^3\right)} \tag{5.28}$$

where,

Volume of the media (m^3) = % Fill of Media in MBBR Reactor*Volume of the Reactor Tank (m^3).

Surface area of the media = Specific protected (inner) surface area of media (m^2/m^3)*Volume of media (m^3).

Design guidelines for a MBBR process is as summarized below in Table 5.16.

Table 5.16 Typical design guidelines for moving bed biofilm reactor

Parameter	Value
Surface loading rate, g/m^2-d	
· BOD removal	5–20
· Nitrification	0.4–1.4
Volumetric loading, kg/($m^3$of media volume)*d	
· BOD removal	1.7–8 (Note)
· Nitrification	0.1–0.4
Hydraulic retention time (HRT), hour (for BOD Removal only. For Nitrification it is higher)	1.5–6
Media internal protected specific surface area, m^2/m^3	400–800
Media % fill	30–60

Note The upper limit is based on media specific surface area 800 m^2/m^3 and average media surface loading rate of 10 g/m^2/day

5.6.5 Hybrid Systems—Integrated Fixed—Film Activated Sludge (IFAS)

IFAS also is a media based process, very similar to MBBR. Its main difference with MBBR is that it is accompanied by a return activated sludge (RAS) stream in addition to the waste activated sludge (WAS) from the mixed liquor solid–liquid separator (gravity clarifier or DAF). Therefore, the mixed liquor in an IFAS reactor is a combination of biomass in suspended growth and the sloughed off cells from attached growth on the media. As a result, due to the combined effect of both attached and suspended growth, the treatment effectiveness is significantly enhanced compared to a conventional activated sludge reactor or an MBBR alone. Adding media to grow fixed film and increasing the biomass population in an activated sludge reactor can effectively increase the MLSS of the reactor by 150 to 200% (Metcalf & Eddy/AECOM 2014). An IFAS system can achieve the same level of treatment of a given wastewater in a smaller reactor volume and a smaller footprint area than those required individually by conventional activated sludge (CAS) and MBBR processes. Correspondingly, it enhances the solids retention time (SRT) in the reactor without adding any volume.

The mixed liquor from IFAS has better settling property than the effluent from MBBR. Therefore, if an activated sludge plant with existing gravity clarifier is upgraded to an IFAS the clarifier can continue to perform solid–liquid separation unless there is any hydraulic limitation. In addition, the overall solids loading rate of the clarifier may be reduced by retaining more biomass on the media and reducing the mixed liquor suspended solids (MLSS).

IFAS is a relatively new treatment technology which was introduced commercially in the mid-1990s. Its application to the refinery wastewater treatment is limited. However, as many existing refineries with activated sludge plants are ageing IFAS has the potential to expand its application in new installations as well as upgrade of existing treatment systems within the same footprint.

A typical schematic of an IFAS system for BOD and total nitrogen (TN) control is shown in Fig. 5.32. The first compartment of the reactor is anoxic with a mechanical mixer to blend the influent with the internal recycle stream while keeping the media suspended and thoroughly mixed. Typical design guidelines for IFAS systems are as presented in Table 5.17.

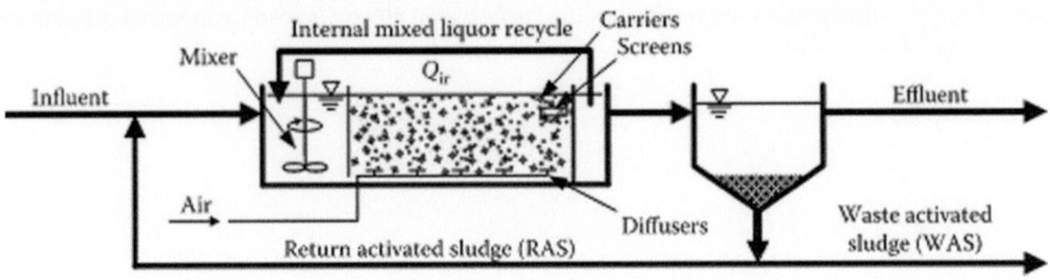

Fig. 5.32 Process flow schematic of IFAS wastewater treatment system

Table 5.17 Typical design guidelines for IFAS

Parameter	Media type			
	Sponge	Plastic (HDPE) wheels	Plastic (HDPE) chips	Fabric
Organic loading surface area removal rate, g BOD_5/m^2-d	2–10	2–10	2–10	2–10
Volumetric organic loading (VOL), kg BOD_5/m^3-d	1.5–3	1.5–3	1.5–3	1.5–3
Hydraulic retention time (HRT), h	3.5–5	3.5–5	3.5–5	4–6
MLSS, g TSS/L	2.5	2–3	2–3	3
Media specific surface area, m^2/m^3	150–250	150–400	300–600	50–100
Media % fill	15–20	20–50	20–50	45–55

The nitrification process in IFAS system, specifically the proportion of nitrification happening in biofilm relative to the suspended growth, is highly dependent on the suspended growth SRT in the IFAS system. If the SRT in the system is kept below the minimum required SRT for nitrification in suspended growth (typically >5 days depending on the reactor temperature), entire nitrification will happen in the biofilm, provided the loading rates on the media are in the recommended range. If the SRT is more than the minimum required SRT for nitrification in suspended growth, partial nitrification will occur first in the suspended phase and then in the biofilm proportional to their respective conditions.

5.6.6 *Current Trends in New Wastewater Treatment System Installations*

Wastewater treatment plants at many refineries have aged. Many treatment plants need to accommodate increases in flows and loads based on the expansion of the refineries. Also, process equipment sizes need to be increased to respond to the regulatory requirement for ammonia and nitrogen control. Among the various treatment technologies, the most utilized ones in the recent past have been the relatively new technologies MBR and MBBR, and to a lesser extent IFAS. This is primarily driven by the main advantage of small footprint requirement by these over the traditional technologies. These technologies are also superior to conventional activated sludge process in maintaining long effective SRT in the reactor basin for retention of biomass against potential washout by shock ad toxic loads without the need for adjusting sludge return (RAS) and wasting rates (WAS) as explained in the earlier section. A few case studies of these three technologies at recent installations are presented below.

5.6.6.1 Case Studies for MBR

Co-Op Refinery Complex-Regina, Saskatchewan, Canada

This refinery used to fulfill its water needs of ~500 m^3/hr (2,200 gpm) by withdrawing a part of it from groundwater wells and the balance by purchasing from the city. About 180 m^3/hr (800 gpm) was being lost due to evaporation from cooling towers and other losses, and 320 m^3/hr (1400 gpm) was being contaminated by hydrocarbons in the refinery as part of normal operations and was being discharged to the city wastewater treatment plant. Due to its location in a water stressed region the refinery implemented a wastewater improvement project in 2016 to treat the wastewater on site by adding a dissolved gas floatation (DGF) unit and MBR based biological treatment. Data collected during the period from early 2016 through mid-2018 showed the DGF removed ~85% TSS and Oil and Grease. The DGF effluent is fed to two parallel aeration basins for COD removal and nitrification. Total nitrogen (TN) control is carried out by denitrification in a pre-anoxic basin. Solid–Liquid separation of the mixed liquor occurs in three parallel MBR tanks which produce high quality effluent with average COD, BOD and total nitrogen in the final effluent of <80, 2 and 10 mg/L, respectively (Hilts et al. 2018).

According to the authors of the referenced article, the upgrade of the wastewater treatment system to an MBR enabled the Co-op Refinery complex to achieve almost 100% recycle and reuse of treated wastewater by processing the MBR effluent using RO and made it suitable for use as the boiler feed make up water.

Petrobras—Henrique Lage Refinery—Sao Paolo, Brazil

This refinery (Cerqueira 2013) upgraded their existing activated sludge based treatment system to MBR to handle high flows and loads resulting from expansion of the refinery. Since this refinery is located in a heavily populated area sourcing the additional water required would be a challenge. Petrobras decided to install MBR system to allow reuse of treated effluent. Prior to the upgrade the treatment train the plant consisted of API separators (2 parallel), equalization tank (1), DAF separators (2 parallel), activated sludge reactors (4 parallel) and gravity clarifiers (2 parallel). Of all the wastewater streams the desalter brine is the most difficult due to its high salinity, large fraction of non-biodegradable COD and high ammonia concentration. The upgrade consisted of segregation of desalter brine from the low salinity wastewater and treating it separately. A part of the low salinity stream (300 m^3/h) continues to be treated by the existing treatment train followed by an Actiflo ballast type separator to achieve high level of TSS removal by enhancing flocculation. The remaining part of the low salinity wastewater is blended with the desalter brine in the ratio of 1:2 and fed to the new MBR treatment train. This new train consists of an API separator, DAF separator, walnut shell filters (3 in parallel), equalization tanks (2 in parallel), two parallel trains of biological reactors, each consisting of a pre-anoxic denitrification tank, followed by an aeration tank in series for COD oxidation and nitrification. As this wastewater lacks in required alkalinity to conduct nitrification caustic soda (NaOH) solution is added to supplement alkalinity. The motivation behind implementing denitrification is to recover 50% of alkalinity and dissolved oxygen to achieve saving in operating costs associated with chemicals and energy. The mixed liquor from each train is introduced into 2 parallel tanks containing ultra-filtration (UF) membrane system to function as MBR. The bioreactor is operated with an MLSS concentration of 10,600 mg/L (± 300 mg/L), VSS:TSS ratio of 0.75 and an SRT of 80 to 100 days.

Blending of desalter brine with low salinity wastewater helps maintain MBR influent oil and grease, and temperature within or close to the design values. Performance of the MBR treatment system was monitored during July 2012 to August 2013 and compared with the design values of some of the important parameters. The results are listed in Table 5.18 below:

Marathon Petroleum Corporation—Detroit Michigan Refinery—This refinery has a capacity to handle 140,000 barrels per day of crude, which includes a 80,000 barrels per day of Canadian heavy sour crude. Wastewater produced at this facility originally used to be treated by an Induced Gas Flotation (IGF) unit, followed by multimedia filters prior to discharge to the municipal sewer system for further treatment and disposal. Along with the increase in flow and load, the refinery was faced with a new discharge permit that required a strict limit on emission of volatile organic carbon (VOC), particularly benzene. As a result, the refinery installed a new 15 MLD (~4 MGD, or 3,300 gpm) MBR based wastewater treatment system comprising of three parallel trains of biological reactor systems following the IGF, for COD oxidation, benzene removal and total nitrogen (TN) control. Each train consists of a pre-anoxic

Table 5.18 MBR system parameters downstream of walnut shell filter—design versus performance

	Design influent	Operating influent	Design permeate	Operating permeate
Influent Rate, m^3/hr	225	125–175	–	–
BOD, mg/L	230	125–330 (Estimated, Note 2)	≤5	<5
COD, mg/L	800	300–800	≤60	50–90
TSS, mg/L	<5	–	N/A	<2
TKN, mg/L	90	N/R	N/A	N/R
TN, mg/L	N/A	N/A	≤20	N/R
NH_3-N	80	60–140	≤1	<0.5
Oil & Grease	<5	10–70	≤1	N/R
Turbidity, NTU	N/A	20–80	≤1	N/R
Temperature, °C	25–32	25–37	25–35	27–37

Notes

1. N/A = Not Applicable, and N/R = Not Recorded
2. Influent BOD is estimated from observed average COD:BOD ratio of 2.4

Table 5.19 Marathon detroit refinery MBR performance

	Influent	Effluent
COD, mg/L	550	70
Ammonia-N, mg/L	40	<5
Benzene, μg/L	135	ND
Oil & Grease, mg/L	10	<5
Turbidity, NTU	10	<0.5

Note Numbers Have Been Rounded Off to Nearest Multiple of 5, ND denotes not detected

reactor for denitrification (20% of total volume), aerobic reactor for COD oxidation and nitrification (72% of total volume) and UF membrane tank (8% of total volume).

The design of full-scale installation was based on the results of a pilot test program. The influent and effluent characteristics are reported in the table above (Arntsen et al. 2017) (Table 5.19).

Operating data collected over two years 2016 and 2017 demonstrated MBR was a good choice for producing high quality effluent along with removing benzene. On an average, COD removal was observed at levels higher than 85%. Removal of most of the contaminants, including benzene and ammonia, was observed to be near complete although the fraction of benzene removed by biodegradation versus air stripping has not been quantified. The treated effluent is suitable for reuse as cooling water make-up. During the reported period of operation, the operation was observed to be smooth without any deterioration of the membranes due to fouling.

The above case studies on MBR processes have been reported by Suez Water Technologies (currently Veolia Water) from their installations at various sites. In all cases ZeeWeed 500 D UF membrane cassettes were used, and the bioreactors were operated with mixed liquor concentrations (MLSS) of ~10,000 mg/L. Refinery wastewater characteristics, especially COD, Oil & Grease and TKN, vary frequently over wide ranges. These variations result from the changes in crude slate, desalter performance and efficiency of API Separators and DGF. However, according to the reported case studies, MBR process at each installation very effectively fulfills or exceeds the treatment goals.

5.6.6.2 Case Studies for MBBR

MBBR has been a well-established treatment technology for domestic as well as many types of industrial wastewaters. However, its adoption in refinery wastewater treatment has been slow due to the chemical characteristics of the wastewater. Significant research on both bench and pilot scale is being carried out globally to demonstrate the efficacy of MBBR in refinery wastewater treatment. Qaderi, et al. (2018) conducted bench scale experimental studies using synthetic wastewater closely resembling the characteristics of biological treatment influent (downstream of primary separation of TSS and Oil & Grease) of the Tehran refinery in Iran. The wastewater test samples were produced by following the composition of a combination of diesel oil (C 16 to C 20) and crude (C 8 to C 37) in the ratio of 1:2. The authors investigated the effects of three parameters - wastewater strength, hydraulic retention time (HRT) and media fill ratio on treatment performance. The experimental program consisted of examining the effect of varying each of these parameters, one at a time, on the total petroleum hydrocarbon (TPH) removal efficiency while keeping the other two constant. Wastewater strength was varied in the range of 165 to 585 mg/L of TPH, HRT was tested in range of 11 to 35 h, and media fill ratio was varied in the range of 28 to 62%. Removal efficiencies improved with increase of HRT and fill ratio but increase in feed TPH concentration adversely impacted the removal efficiency. The best results were obtained when TPH removal efficiency reached 97% under the conditions of 23 h of HRT, 45% media fill ratio and MBBR influent TPH concentration of ~165 mg/L. This, and other bench scale studies have demonstrated the effectiveness of MBBR for wastewater treatment at this refinery.

Jolibois et al. (2018) reported that the Valero Jean-Gaulin Refinery in Levis, Quebec, Canada, as part of the upgrade of their Wastewater Treatment Plant, conducted pilot scale studies on an MBBR system. The pilot tests using a 2 stage MBBR system were successful for the removal of over 70% of influent COD and 5% influent TKN under average operating conditions. These results were utilized in the design of the full-scale system that has been operational on site since 2019.

Dale et al. (2017) reported case studies of wastewater treatment at two refineries A and B in the US and Sweden, respectively. At both locations MBBR was installed with the goal of COD oxidation, benzene removal and phenol removal. Some of the design and operating parameters of the two reactors are listed in the Table 5.20.

The goal of the study at these two full scale plants was to examine the effectiveness of MBBR in removing phenol and benzene to a target concentration of $\leq 10\ \mu g/L$ under normal operating conditions. Phenol concentrations were monitored for two categories—Pure Phenol and Total Recoverable Phenols. The recoverable phenols included several substituted phenolic compounds in addition to pure phenol.

Table 5.20 MBBR reactor setup at two plants

	Reactor placement	Reactor volume, m^3	% Fill	Media type	Specific surface area, m^2/m^3	Surface area loading rate (SALR) g $TCOD/m^2$.d	Process	# of stages
Plant A	Downstream of deoiling tank	4554	26	Veolia K1	500	1.5–9.8	COD oxidation + nitrification	1
Plant B	Downstream gravity oil separation + filtration	95	55	Veolia K3	500	18–22	COD oxidation	1

Note
TCOD = Total COD mg/L
sCOD = Soluble COD, mg/L

The results for Plant A demonstrated that during the five months of monitoring the treated effluent:

- COD concentration was reduced from average TCOD of 558 mg/L (214 to 1179 mg/L range) and average sCOD 369 mg/L (168 to 569 mg/L range) to average sCOD of 56 mg/L (4 to 66 mg/L range)
- Influent benzene concentration was reduced from a range of 4,700 to 14,000 μg/L to below detection limit of 4.4 μg/L
- Influent pure phenol was reduced from a range 540 to 1500 μg/L to below 10 μg/L
- Influent total recoverable phenol was reduced from a range 540 to 4,900 μg to 10 to 30 μg/L.

In summary, the Plant A demonstrated that up to an SALR (Surface Area Loading Rate) of 5,000 mg sCOD/m^2/d the plant consistently achieves effluent benzene and phenol concentration ≤10 μg/L under normal operating conditions.

In case of Plant B:

- The influent sCOD was reduced from a range of 114 to 177 mg/L down to 50 to 70 mg/L
- Influent benzene was reduced to non-detect level except for only one occasion when the effluent concentration was reported to be 24 μg/L
- Total recoverable phenol in the treated effluent was reduced from a range of 1,300 to 2,000 μg/L down to 40 to 48 μg/L.

Along with the study of the full-scale treatment plants the authors also conducted pilot scale studies using synthetic wastewaters containing benzene and phenol, and an actual wastewater sample from another refinery in Sweden (not refinery B). The results of the pilot tests demonstrated that up to an SALR of 20,000 mg sCOD/m^2/d an MBBR based treatment system operating at ~30^{0}C could sustain the pure phenol and benzene concentrations of ≤10 μg/L in the effluent.

Based on the above observations in bench, pilot and full-scale operation it was concluded that MBBR is a robust technology for the treatment of refinery wastewater.

5.6.6.3 Case Study for IFAS

Frontier refinery in Cheyenne, Wyoming upgraded their existing activated sludge—based wastewater treatment plant to an IFAS in 2007 to fulfill to the new regulation requiring 2.18 mg/L NH_3-N in the treated effluent (Casey et al. 2008). This was accomplished by adding Entex BioPortz media of specific surface area ~580 m^2/m^3 in existing activated sludge reactor basins at a fill ratio of 13%.

The influent wastewater undergoes primary treatment by an IGF separator to remove TSS, and emulsified and free phase oil. Two parallel IFAS basins carry out biological treatment for COD oxidation and nitrification. Effluent from each basin flows into a gravity clarifier for solid–liquid separation. Treated effluent is discharged trough the outfall. Conversion to IFAS enabled the refinery to avoid increasing the volume and footprint of the bioreactors thereby avoiding additional capital expenditure while complying with the newly imposed ammonia-N limit in the treated effluent.

5.6.6.4 Granular Activated Sludge (GAS)

This technology refers to a dynamic attached growth treatment system containing flocs of large size ($\geqq$ 200 microns) without attrition under aerobic treatment conditions. In this technology the reaction environment is controlled in such a way that aerobic types of microorganisms are developed on the outer surfaces and exposed to DO concentration in the bulk liquid phase, and the anoxic populations develop internally to the floc particles under limited DO condition. This unique characteristic allows them to perform complete nitrogen and organic removal (Nitrification, denitrification, BOD removal). The sludge granules produced are more robust than conventional activated sludge flocs and contain different types of microorganisms carrying out specific biochemical reactions. The dense sludge granules provide mixed liquor concentration of ~8 to 10 g/L and <60 mL/g SVI compared to 2,500 to 4,500 mg/L MLSS and >100 mL/g of SVI for conventional activated sludge process. This results in very good settling properties for granular sludge and a potential reduction in clarifier size. Figure 5.33 presents a conceptual depiction of the structure, microbiological populations and their respective functions within a sludge granule.

Granular sludge can be produced by providing a suitable reaction environment similar to conditions in an SBR. Currently this technology is new and not known to be used in petroleum refinery wastewater treatment. An example application is presented in Chap. 3.

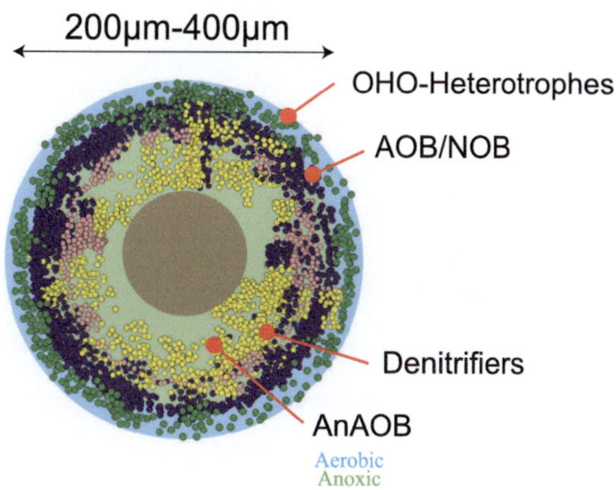

Fig. 5.33 Aerobic granular sludge floc

5.6.7 *Aeration Systems for Aerobic Biological Treatment Processes*

An aeration system introduces oxygen in the form of air to wastewater for the aerobic bacteria to oxidize the organic matter and ammonia (in case of nitrification), produce energy and yield the growth of new cells. It also provides the mixing energy for suspended growth processes, MBBR and IFAS reactors. Providing air to the biomass in a reactor is one of the most important and energy consuming steps in an aerobic biological wastewater treatment process.

The two major types of aeration system used in biological treatment are:

- Diffused aeration
- Mechanical aeration

5.6.7.1 Diffused Aeration System

In diffused aeration, air is supplied through porous diffusers or through air nozzles near the bottom of the tank. Diffused aeration system includes diffuser/nozzle, piping, and blower/compressor. The key parameters affecting the oxygen transfer are bubble size, diffuser air rate, diffuser placement and tank depth. Operation of the diffused air aeration system in an aerobic treatment plant is presented in Fig. 5.34a. Components of a diffused aeration system is depicted in Fig. 5.34b.

Air diffusers used are of various types. A few of the commonly used types and their important attributes are listed in Table 5.21.

Fig. 5.34 **a** Diffused air system operating in an aeration tank, **b** Components of an diffused air aeration system

Table 5.21 Types of Aeration System

Air diffuser type	Description	Oxygen transfer efficiency	Oxygen transfer rate (kgO_2/kW-h)
Bubble diffuser • Coarse • Fine • Medium	The Bubble or Porous diffusers are available mostly in disk, dome, and membrane shapes for production of fine bubbles. For coarse bubbles pipes/diffusers with nozzles are used	• 4 to 20% • 10 to 30% • 6 to 15%	• 0.6 to 1.2 • 1.2 to 2.0 • 1.0 to 1.6
Tubular diffuser	They are tubular type diffusers	7 to 10%	1.2 to 1.6
Jet diffuser	Jet type diffusers	10 to 25%	1.2 to 2.4
Aspirator device	Aspirator Jet diffusers	20 to 30%	1.5 to 2.5

5.6.7.2 Mechanical Aeration System

Mechanical aerators consist of fully or partly submerged impellers that are attached to motors, mounted on a float or on fixed structures. Mechanical aerators can be classified into the following types:

- Aerators with vertical axis
- Aerators with horizontal axis.

These can further be sub classified as surface aerators and submerged aerators.

Surface Aerators with Vertical Axis

Surface aerators have their impellers partly or fully submerged and have an advantage of flexibility in tank shape and size. These aerators entrain air form the atmosphere and distribute it into the reactors. Floating surface aerators can be used for ponds or lagoons where a rigid support will be impractical or where the water level fluctuates. The transfer rate of surface aerators range between 1.2 to 2.7 kgO_2/kW-h.

These aerators can also be classified based on the type of their impellers used—centrifugal, radial—axial, or axial; and based on their speed—high or low. Usually, centrifugal impeller types of aerators are low speed aerators and axial types are high speed.

Submerged Aerators with Vertical Axis

Most surface aerators are dependent on atmospheric air entrapment for efficiency and agitation of the water surface whereas submerged/turbine type aerators have their impellers submerged and piped air or oxygen is delivered to a point below the impeller. The impeller disperses air into fine bubbles and mixes the content of the tank.

Oxygen transfer efficiency of submerged aerators ranges from 15 to 35% with transfer rate of 1.2 to 2.0 kgO_2/kW-h. These types of aerators cause mixing of air and oxygen transfer by violent agitation of the water surface.

Some models of these type of aerators have a draft tube which is used to control the flow and mixing pattern either up flow or downflow. There are some designs with hyperboloid mixer/aerator which have a separate air system and consist of downflow impellers that provide higher efficiencies than self-induced aeration systems. Figure 5.35 below shows an example of surface aerators.

Surface Aerators with Horizontal Axis

These aerators are mounted on a horizontal shaft and resemble a brush. They have typical oxygen transfer rate of 1.5–2.1 kgO_2/kW.h. The aerator bars are made up of plastic or steel. These have been used mainly for oxidation ditch type biological treatment system.

Submerged Aerators with Horizontal Axis

The submerged aerators with horizontal axis are also blades, bars or discs connected to a submerged rotating shaft. They have an oxygen transfer rate of 1.5–2.1 kg O_2/kW.h. The spacing between discs is used as a mechanism to control the mixing and air flow of the system.

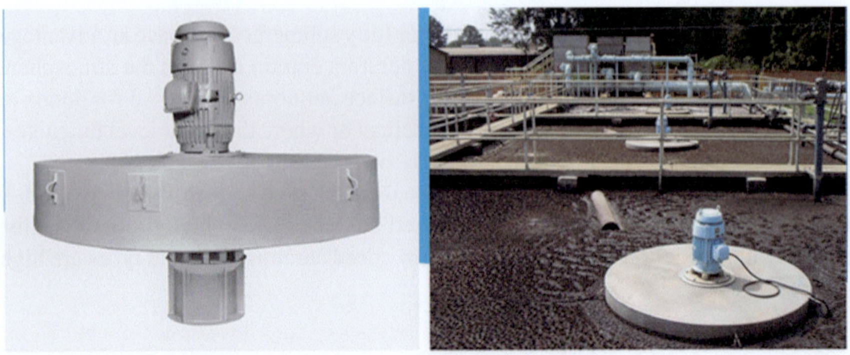

Fig. 5.35 Mechanical Aerators with Vertical Axis

5.6.7.3 Aeration System Calculations

The amount of air required in an aeration basin for oxidation of BOD and nitrification of ammonia—N depends on various factors including flow, influent concentrations, effluent concentrations, temperature, and several other factors. Air is provided to aeration basins by surface or submerged aerators when mechanical aeration is used. This requires the aerators, supports and appurtenances. In the case of diffused aeration, the system consists of aeration blowers, air supply piping and joints, and the diffusers. Estimation of air requirement is important for design of the aeration system, and sizing and selection of the equipment. Majority of new refinery wastewater treatment plants use diffused aeration for feeding oxygen into the reactor basin. Calculation of air requirement is a three-step process—(a) calculation of mass flow rate (kg/hr) of Actual Oxygen Requirement (AOR) under actual operating conditions; (b) calculation of mass flow rate (kg/hr) of Standard Oxygen Requirement (SOR) under standard conditions of pressure of 1 bar, temperature of 20 °C and dissolved oxygen concentration of 0 (zero) mg/L; and (c) calculation of volumetric flow rate of air corresponding to the SOR as determined in Step (b) above. The step by step calculation procedure is presented below.

(a) Actual Oxygen Requirement (AOR) for BOD Removal and Nitrification under Field Operating Conditions

$$AOR = Flow\left(\frac{\mathrm{m}^3}{d}\right)\left[\begin{array}{c}\{O_2\, ratio\, for\, BOD\, removal * BOD\, removed\left(\frac{mg}{l}\right)\}\\ +\{O_2\, ratio\, for\, nitrification * NH3 - N\, nitrified\left(\frac{mg}{l}\right)\}\end{array}\right]/[24 * 1000] \tag{5.29}$$

where:

O_2 ratio for BOD removal = 1.0 g of O_2 required per g of BOD_5, and
O_2 ratio for nitrification = 4.56 g of O_2 required per g of NH_3–N

For plants that operate to remove BOD only, the term for oxygen requirement for nitrification needs to be removed.

(b) Application of Correction Factors for Calculating Standard Oxygen Requirement (SOR)

$$AOR = SOR\left(\frac{\Omega * \tau * \beta * C *_{\infty,20} - C}{C*_{\infty,20}}\right)\left(1.024^{T-20}\right)(\alpha)(F) \tag{5.30}$$

where:

AOR = actual oxygen transfer rate under field conditions, (kg O_2/h)
SOR = standard oxygen transfer rate in tap water at 20 °C and zero dissolved oxygen (kg O_2/h)
β = salinity—surface tension correction factor (typically 0.95–0.98)
$C*_{\infty,20}$ = oxygen saturation concentration at sea level and at standard temperature 20 C (mg/L)

$C*_{\infty,20} = C * s, 20\,(1 + de(\frac{Df}{Ps}))$
Ω = Pressure correction factor = $\frac{Pb}{Ps}$
P_b = pressure at test site (kPa)
P atm, s = standard atmospheric pressure (kPa)
de, mid depth = mid depth correction factor (typically 0.25–0.45)
D_f = depth of diffuser in basin, m
C = average dissolved oxygen concentration (mg/L)
C*s,20 = dissolved oxygen surface saturation concentration at 20 °C and 1 atm (mg/L)
T = operating temperature (°C)
α = oxygen transfer correction factor for waste
F = fouling factor (typically 0.65–0.9)
τ = Temperature correction factor = $(\frac{C*_{\infty,20}}{C*s,20})$

(c) Calculation for Volumetric Air Flow Requirement Corresponding to SOR

$$Air\ Flow\ Requirement = SOR/\left[\rho_{air} * \%O_2 * Depth * \left[\frac{T.E}{100}\right]\right] \tag{5.31}$$

where:
Air Requirement (m^3/hr.)
SOR = Standard oxygen requirement kg O_2/hour
Atmospheric Pressure = 33.8 ft of Water Column = 101,325 Pa
Submergence depth of aeration grid = 0.15 m Depth
T.E. = oxygen transfer efficiency = 6% per meter of diffuser submergence
ρ_{air} = Density of air at 25 °C = 1.184 kg/m^3

Example Aeration Calculation for BOD Oxidation:
Assuming:
Inlet BOD = 300 mg/L
Outlet BOD = 30 mg/L (90% removal of inlet BOD)
No nitrification required
Flow = 1000 m^3/d

$$AOR = \left[1,000\left(\frac{m^3}{d}\right)\left[1 * \left(300\frac{mg}{1} - 30\frac{mg}{1}\right)\right]/[24 * 1000]\right]$$

AOR = 11.25 kg/hr

$$AOR = SOR\left(\frac{1 * 0.9 * 0.95 * 9.64 - 3}{9.64}\right)\left(1.024^{25-20}\right)(0.8)(0.9)$$

$\alpha = 0.8$
$\beta = 0.95$
$\tau = \frac{C*_{st}}{C*_{s20}} == 8.263/9.09 = 0.9$

$C*_{\infty,20} = C*_{s20} * \left[1 + de * \left(\frac{D_f}{P_s}\right)\right] = 9.64\,\mathrm{mg/L}$
$C*_{st} = 8.263\,\mathrm{mg/L}$ @ 25 C and sea level
$C*_{s20} = 9.09\,\mathrm{mg/L}$
de = 0.4
$D_f = 0.15$ m (Typically 0.15 m from bottom of the tank)
$P_s = 1$ atm = 101.325 kPa
T = 25 °C
F = 0.9
$\Omega = Pb/Ps = 1$ at sea level
Therefore,
SOR = 26 kg/hr
Air density, at 25 °C = 1.184 kg/m^3
O_2 in air = 23.2% = 0.232
OTE—6%/m of diffuser submergence.
Submergence depth = SWD—diffuser height = 5–0.15 m

$$AirRequirement = 25.5/\left[1.184 * 0.232 * 4.85 * \left[\frac{6}{100}\right]\right]$$

Therefore, Total aeration requirement for the system ~**320 m³/hr**.

An example calculation for aeration requirement for BOD oxidation and nitrification with the aid of process modeling has been presented in Chap. 6.

5.6.8 Inhibition of Wastewater Treatment by Reactor Environmental Conditions and Chemical Matrices

Toxic and/or shock loads can potentially wash out biomass population from the reactor which may require fresh seeding to restart the process. Autotrophs are much more sensitive to shock and toxic loads compared to the heterotrophic population, which are robust. For these reasons maintenance of a high SRT with the help of abundant inventory of biomass is essential to ensure the success of nitrification. SRT can be effectively controlled by controlling the sludge wasting rate. In the case of different variations of conventional activated sludge processes high SRT can be achieved by reducing the WAS rate and vice versa. However, the other alternatives are MBR among suspended growth processes and MBBR among the attached growth processes.

The difference between inhibition and toxicity is that inhibition is the impairment of biological function whereas toxicity is an adverse effect on biological metabolism of microorganisms. This makes inhibition a reversible process whereas toxicity is normally irreversible. Inhibition results in—(a) reduced levels of treatment in terms of pollutant removal, (b) recalcitrance (non-responsiveness) of certain constituents to biological treatment, (c) alteration of microbiological population diversity, (d) poor solid–liquid separation by settling due to change in bioflocculation property, etc. However where inhibition may not wash out the biomass population from a reactor, escape of inhibitory constituents through biological wastewater treatment systems can lead to toxicity of the discharge to the receiving water bodies. The discussion about inhibition in this section is limited to BOD oxidation, and nitrification, by certain constituents of refinery wastewaters. These are in addition to the environmental effects of reaction conditions on nitrification, which include DO, pH and temperatures outside their optimum ranges, as discussed earlier.

Certain chemicals are toxic to nitrifiers and are included in Nutrient Control Design Manual (USEPA 2010). These can be broadly classified as organic and inorganic chemicals. Two very common organic contaminants in refinery wastewater are phenol and naphthenic acids. Phenol used to be considered hard to degrade biologically above 25 mg/L in the past, but research and practice have shown that phenol can act as a food source and be broken down by aerobic microorganisms to concentrations as high as 500 mg/L (Sawyer et al. 2003).

Naphthenic acids (NA) are a complex group of alkyl-substituted acyclic, monocyclic and polycyclic carboxylic acids present in crude oil, oil sands process water and tailings ponds, as well as in refinery wastewater. NA potentially inhibit nitrification and denitrification in aerobic heterotrophic and nitrifying microbial cultures, and fermentation/methanogenesis in anaerobic cultures. The nitrification inhibition threshold of NA is 80 mg/L in a heterotrophic/nitrifying mixed culture in suspended growth. However, that limit is 400 mg/L in the presence of an enriched culture of nitrifying population alone (Misti et al. 2013). Chemicals such as naphthenic acids may pose toxic effect to aquatic life based on the toxicity monitoring assays. In such cases, the secondary effluent BOD values are misleading about its real quality as the nonbiodegradable contaminants pass through unaffected and the true characteristics are indicated by the COD and TOC values. This may warrant regulators to impose COD and/or TOC as discharge compliance parameters.

Among the inorganic constituents, cyanides are widely known to cause inhibition to biological treatment of wastewater. The threshold concentration of cyanide to BOD oxidizing heterotrophic bacteria is 100 mg/L. However, cyanide is not as forgiving to the nitrifiers. The threshold is very close to zero (Eckenfelder et al. 2009). Typically, cyanide bearing wastewaters are pretreated by chemical oxidation treatment prior to feeding it into a biological reactor and discussed in detail later in chapter.

Total dissolved solids (TDS), including chloride ions representing salinity, reduces the metabolic functions of microorganisms and the removal rates of substrates (Hong 2007). Figure 5.36 presents the gradual decrease in the ammonia—N removal efficiency and nitrification rate as a function of increasing salt concentration. This applies to produced water treatment also in upstream water management operations and has been discussed in Chap. 3.

Heavy metals and metalloids could be inhibitory to microbes, although several microorganisms have evolved to develop detoxification mechanisms to counter the toxic effects of these inorganic metals (Igiri et al. 2018). Especially nickel, copper and chromium have been reported inhibitory to nitrification above the threshold

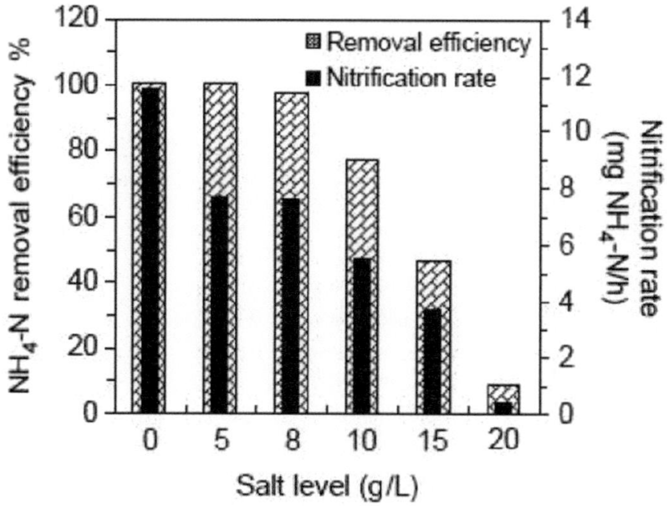

Fig. 5.36 Effect of Salt Concentration on Ammonia—N Removal Efficiency and Nitrification Rate (Cui et al. 2006, Adapted from Water Science and Technology, vol. 53., No. 6, pp 115–122)

concentrations ranges of 0.25–0.50, 0.05–0.50, and total 0.25–1.9 mg/L, respectively. Inhibitory effects of some of the commonly encountered inorganic and organic species in industrial wastewaters, and their thresholds concentrations are listed in Table 5.22.

For all of the inhibitory and toxic compounds discussed in this section, the concentrations at which they impact the process detrimentally can be increased if the biomass is fed with a constant, low concentration of the potential toxin. This technique, known as acclimation, breeds a population that is resistant to the toxins, and typically requires a much longer sludge age than a system without toxins. For a specific constituent—the concentration corresponding to 50% of its inhibitory effect is expressed as IC_{50}.

5.6.9 Microbial Community Studies in Refinery Wastewater

The composition of refinery wastewater plays a pivotal role in shaping the activity of specific bacterial groups. Fluctuations in loads can rapidly alter microbial activity and diversity, ultimately leading to a decline in effluent quality. This deterioration persists until the root cause of the issue is addressed or until sufficient active microbial mass is established or introduced into the biological system (Kim et al. 2020). Consequently, dynamic changes in influent composition causes seasonal shifts in the abundance and diversity of bacteria, involved in nitrogen, sulfur, and organic compounds degradation (Soares-Castro et al. 2019).

While extensive research has been conducted on microbial populations in municipal wastewater, there remains a substantial knowledge gap regarding industrial wastewaters like refinery effluents due to insufficient research in this area. This

Table 5.22 Threshold inhibitory concentrations of inorganic and organic contaminants in wastewater from petroleum industries (USEPA 1987 and 2004) to biological treatment

Type of contaminant	Contaminant	Source of contaminant	Threshold concentration for BOD oxidation, mg/L	Threshold concentration for nitrification, mg/L
Inorganic	Ammonia	Amines and other nitrogenous compounds in petroleum processing	480	Function of pH and temperature (as Described above)
	Arsenic	Crude petroleum	0.1	1.5
	Boron	Production chemical carried over with crude Petroleum	0.05–10	N/A
	Cadmium	Crude petroleum	1–10	5.2
	Copper	Cooling tower blowdown	0.10–1.0	0.05–0.5
	Chromium (Total)	Cooling tower blowdown	1–100	0.25–1.9
	Cyanide	Produced in petroleum cracking	0.05–20	0.05–0.50
	Iron	Cooling tower blowdown and in crude	5–500	N/A
	Lead	Equipment material of construction	1–5	0.5
	Mercury	Crude petroleum	0.10–5.0	2.0–12.5
	Nickel	Crude petroleum	1–2.5	0.25–0.5
	Vanadium	Crude petroleum	20	N/A
	Zinc	Crude petroleum	0.30–5.0	0.08–0.50
Organic	Phenols	Produced in cracking and coking	90–1,000	1–10
	Naphthenic acids	Crude petroleum	N/A	80

N/A = Not Available

section delves into specific studies exploring microbial communities in refinery wastewater and discusses the application of amplicon sequencing for analyzing and comprehending microbial composition within the wastewater matrix.

Soares-Castro et al. (2019) conducted a study on the microbial community composition in oil wastewater using 16S rRNA sequencing. Their research provided detailed insights into the intricate dynamics of microbial communities within an oil refinery wastewater treatment plant. The study examined seasonal fluctuations in microbial populations and their impacts on treatment processes. By employing 16S-metabarcoding, the study analyzed bacterial communities across various treatment plant compartments during winter and monsoon seasons identifying operational taxonomic units (OTUs) and their relative abundances. Significant seasonal variations in microbial populations, particularly key functional groups involved in nitrogen removal such as ammonia-oxidizing bacteria (AOB) and nitrite-oxidizing bacteria (NOB) were observed.

During the monsoon season, AOB and NOB populations notably decreased, leading to reduced nitrification efficiency. Notably, AOB abundance was absent in the clarifier and decreased in the guard pond, while NOB abundance decreased in both compartments. The study also identified putative denitrifying autotrophs and heterotrophic denitrifiers, indicating a complex network of microbial populations involved in nitrogen removal. However, overall denitrifier abundance decreased during the monsoon season compared to the winter season. The increase in ammonia content during the monsoon season negatively impacted nitrification, potentially inhibiting AOB and NOB activities. The study also noted dissimilatory nitrate reduction to ammonia (DNRA) during the monsoon season, contributing to increased ammonia levels in treated wastewater. These findings underscore the importance of considering seasonal variations in microbial dynamics when optimizing wastewater treatment processes. Shifts in microbial populations, influenced by factors such as temperature and organic load, can significantly impact nitrogen removal efficiency. Continuous monitoring and adaptive management of treatment processes are essential for maintaining effectiveness across seasons.

Figuerola and Erijman (2010) investigated bacterial populations demonstrating unstable nitrification in a full-scale activated sludge plant receiving petroleum refinery wastewater. They employed real-time PCR targeting betaproteobacterial AOB (CTO and amoA), Nitrobacter-like NOB (FGPS), and Nitrospira-like NOB (NRS) with specific primers. Key findings revealed that the inhibition of ammonia oxidation was correlated with phenol concentration. Interestingly, while the abundance of ammonia—oxidizing bacteria (AOB) did not correlate with nitrification performance, the total count of nitrite—oxidizing bacteria (NOB) notably decreased during periods of nitrite accumulation or absence of nitrification.

In a comprehensive study of a refinery wastewater treatment plant, Steeves et al. (2022) aimed to identify the causes behind the loss of complete nitrification. Using advanced biomass characterization techniques, particularly 16S rRNA sequencing, the study monitored the plant and the microbial community dynamics. Throughout the study, significant operational and engineering enhancements were made, which not only improved process reliability and treatment efficacy but also appeared to stabilize the desired microbial community—a factor deemed crucial for consistent complete nitrification. The findings revealed a high abundance of denitrifying organisms; however, complete nitrification was notably achieved when the abundance of simultaneous nitrifying and denitrifying (SND) organisms was low, coupled with a stable, moderate presence of ammonia oxidizing (AOB) and nitrite oxidizing (NOB) organisms. The resulting data portrayed three nitrification states: complete, incomplete, and absent. This correlated with the concentrations of ammonia and nitrite in the effluent. Post—September 2019 marked a period of consistent low levels of these compounds, coinciding with fewer process upsets, which was attributed to both the refinement of the primary treatment process, especially the Dissolved Air Flotation (DAF) unit, and the strategic use of powdered activated carbon (PAC) in the bioreactors. These collective findings not only elucidate the vital role of microbial community composition in nitrification efficacy but also underscore the importance of adaptive management and continuous monitoring for optimizing wastewater treatment processes across varying conditions (Steeves et al. 2022).

5.6.9.1 Aquatic Toxicity and Its Measure

As opposed to inhibition of biochemical reactions in wastewater treatment, aquatic toxicity refers to the impact of chemical species in the treated effluent of wastewater treatment plants discharging to the receiving water bodies. Aquatic toxicity is the potential or capacity of a test material to cause adverse effects on living organisms. It is a result of dose or exposure concentration and exposure time, modified by such variables as temperature, chemical form, and availability. An aquatic toxicity test is a procedure in which the responses of aquatic organisms are used to detect or measure the presence or effect of one or more substances, wastes, or environmental factors, alone or in combination.

EPA promulgated a Whole Effluent Toxicity (WET) testing protocol to regulate the quality and characteristics of effluents discharged into a receiving body of water in the US. It describes the toxic effect of an aqueous discharge on the test organisms based on their response in terms of lethality, impaired growth or reproductive ability. WET test is a very important criterion set by the regulators to grant the discharge permit to a wastewater treatment plant operator for the fulfilment of the National Pollutant Discharge Elimination System (NPDES) according to the US Clean Water Act (CWA).

LC_{50} is a measure of acute toxicity. It is the median lethal concentration of a toxicant that kills 50% of exposed organisms at a specific time of observation (e.g., 96-h LC_{50}).

EC_{50} is the toxicant concentration estimated to cause a specified effect in a designated proportion of 50% of the test organisms. The effect is usually sublethal. EC_{50} values for toxic substances can be determined by conducting Microtox test on the substance under consideration. Microtox test instrument and reagents are marketed by Microbics Corporation, Carlsbad, CA. The test procedure is detailed in the literature published by Microbics Corporation (1995).

Chronic toxicity effects are typically determined from the lethal or sublethal effects endpoints for test duration of 7 to 9 days. It is the No Observed Effect Concentration (NOEC) or Lowest Observed Effect Concentration (LOEC) of the discharge from a treatment plant that impacts on the growth or reproduction of test organisms.

Further discussion of aquatic toxicity, its impacts on environment and aquatic life, biomonitoring assays, and mitigation measures are included in detail in Chap. 7.

5.6.10 *Tertiary Treatment*

Tertiary treatment processes are used to remove suspended solids, refractory compounds, inhibitory inorganic and organic compounds, and nitrogen compounds from the secondary effluents. There are various processes such as filtration, nitrogen control, carbon adsorption, chemical oxidation, disinfection, microfiltration, ultrafiltration, ion exchange and reverse osmosis, which have been used for tertiary treatment of wastewaters for different applications. Working principles of these processes are very similar in both water and wastewater treatment applications.

Two main drivers to add a tertiary treatment step for refinery wastewater are:

- Compliance with regulatory requirements, and
- Reuse of secondary effluent

5.6.10.1 Tertiary Treatment to Meet Regulatory Requirements

Most common regulatory requirements for refinery wastewater treatment discharge are BOD and TOC and TSS. However, regulations are gradually becoming increasingly stringent about limiting COD, ammonia-N, total nitrogen (TN), toxic organic compounds, selected heavy metals and metalloids. Tertiary treatment processes are added to address these needs.

TSS Control

Effluent filtration is the tertiary treatment process to ensure compliance with the TSS discharge limit. Typically, the refinery wastewater treatment effluent TSS is <30 mg/L, but it may contain as high as 80 to 100 mg/L of suspended solids (TSS) under certain upset conditions of the secondary clarifier. Also, at many locations refinery wastewater discharge permits limit the TSS in the discharge to as low as 15 mg/L. In such cases, in order to comply with the regulatory limits consistently, the secondary effluent will require further improvement by tertiary treatment to lower the TSS in the effluent and the most commonly used unit operation for this purpose is Pressure Sand Filtration. A schematic view of a pressure sand filtration unit is shown in Fig. 5.37.

A pressure sand filtration unit typically consists of multiple drums containing dual media. These dual media filters comprise a layer of anthracite over sand. Normally one of the drums is maintained in standby mode. The larger particles contained in the wastewater are trapped by the anthracite and the finer particles are trapped by the sand. The pressure drop across each bed is monitored continuous and when it reaches a certain level that vessel is taken out of service and the standby unit is brought into service. The out-of-service drum goes into backwash mode where the flow through the drum is reversed to remove the trapped solids.

Further details of pressure sand filtration units are presented in Chap. 2.

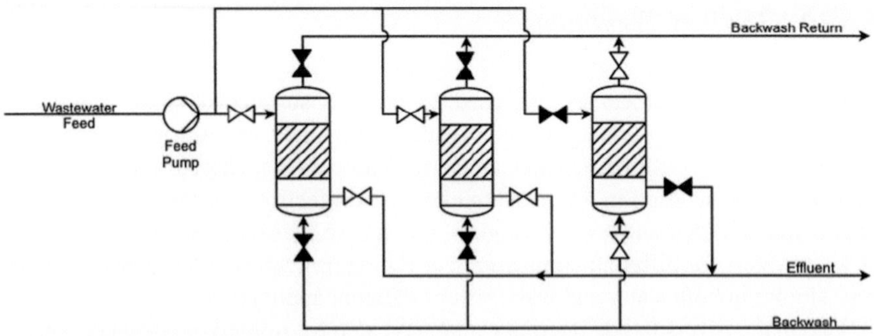

Fig. 5.37 Pressure sand filtration unit

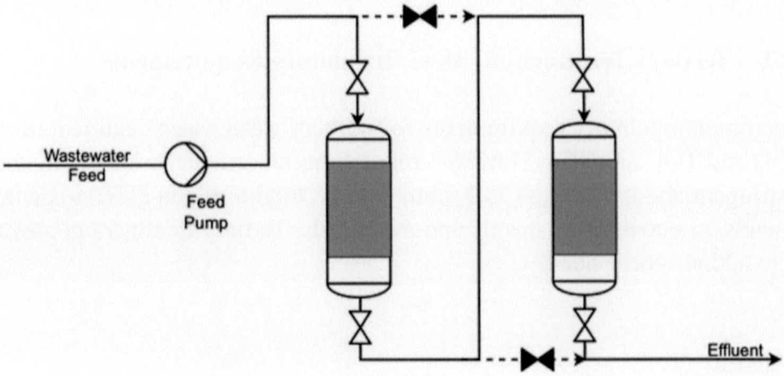

Fig. 5.38 Granular activated carbon unit

Control of Refractory Organics

If the wastewater generated in the refinery contains refractory organic compounds, they will pass through the biological treatment system untreated that might need to be removed. In such cases the secondary effluent may have low BOD, but the COD and TOC may be high. Such plants may exceed limits if the regulation requires control of COD and/or TOC. In these instances, treatment with activated carbon could be used as a polishing step. Figure 5.38 shows a Granular Activated Carbon (GAC) unit for removal of residual organic compounds.

In this process the wastewater is passed through a drum containing a bed of granular activated carbon (GAC) into which the organic compounds contained in the wastewater are adsorbed. A second drum containing GAC is maintained in standby mode. The operation of the operating bed is constantly monitored and as soon as break-through occurs in a bed, that bed is taken out of service and the standby bed is brought into service. The spent carbon bed is removed for regeneration and replaced by a fresh batch.

Before a GAC system is installed bench scale tests are conducted, to test the efficacy of contaminant removal. If the GAC does not exhibit an acceptable removal efficiency, then alternative treatment method needs to be considered. One such alternative is chemical oxidation as described below.

Chemical Oxidation

Chemical oxidation is a process to remove refractory organic chemicals by breaking down to various oxidized products. A part is completely oxidized to CO_2 and H_2O, and another part is partially oxidized to a mixture of organic acids and other breakdown products. A diagram of a chemical oxidation unit is presented as Fig. 5.39.

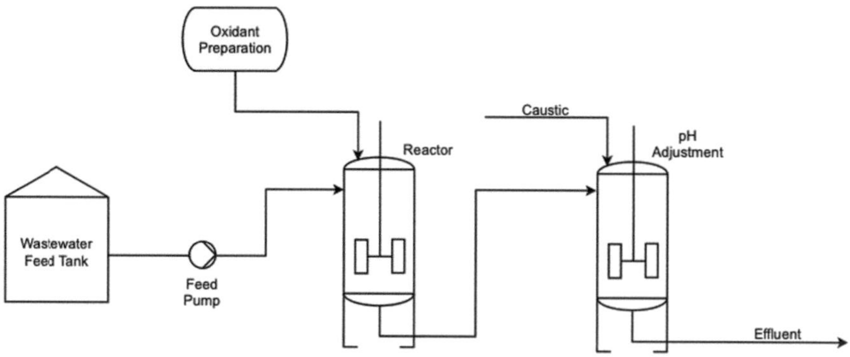

Fig. 5.39 Chemical Oxidation Unit

The following oxidation reagents are generally used in a chemical oxidation system:

- hydrogen peroxide
- chlorine dioxide
- ozone

Chemical oxidation by H_2O_2 or ozone can be enhanced in some cases to an advanced oxidation process (AOP) by the use of UV light to produce hydroxyl radicals as a catalyst, but this needs to be evaluated on a case-by-case basis.

In a chemical oxidation unit, the wastewater is sent to the oxidation reactor from a feed tank which provides any surge capacity that is required. Chemical oxidant (Hydrogen peroxide, ozone or chlorine dioxide) is prepared fresh to enhance the reactivity of the oxidant and fed to the reactor which is equipped with an agitator. The effluent from the reactor is sent to a second vessel for pH adjustment as necessary. Due to high operating costs, chemical oxidation may not be used on the entire wastewater stream. Instead, it is applied to a slip stream of the effluent so that when its effluent is mixed with the total effluent it can meet the required effluent discharge limits.

GAC adsorption and chemical oxidation are the tertiary treatment steps to mitigate situations involving treatment of wastewaters containing recalcitrant contaminants. Working principles of GAC adsorption and chemical oxidation have been discussed in detail in Chap. 2.

Removal of Inorganic Contaminants

Inorganic contaminants in refinery wastewaters consist of ammonia, cyanide, heavy metals and metalloids. If ammonia and Total Nitrogen (TN) control are required, it is integrated with secondary treatment. For older plants that are constrained by footprint area, a relatively easy solution is conversion of existing activated sludge reactor to MBR, or MBBR, or IFAS process without significantly impacting the infrastructure. This has been discussed in detail in the earlier section and not repeated here.

The threshold concentration of cyanide for BOD oxidizing heterotrophic bacteria is 0.05 to 20 mg/L and that for nitrifiers is much lower at 0.05 to 0.50 mg/L. Therefore, wastewaters containing high concentrations of cyanide need to be treated prior to their entry into the bioreactor to mitigate the toxic effect and prevent the biomass from washing out. Most commonly cyanide is broken down by a multistep oxidation by chlorine solution as follows:

$$Cl_2 + H_2O \rightarrow HCl + H^+ + OCl^- \quad (5.32)$$

$$CN^- + OCl^- + H_2O \rightarrow CNCl + 2OH^- \rightarrow CNO^- + Cl^- + H_2O \text{ (at pH 9 to 10)} \quad (5.33)$$

$$2CNO^- + 3HOCl \rightarrow 2HCO_3^- + N_2 + 3Cl^- + H^+ \quad (5.34)$$

Chemical oxidation, however, requires high operating cost due to the consumption of large amount of chlorine. Alvarado-Lopez, et al. (2023) isolated a consortium of bacterial population from gold mine tailings containing metal cyanide complexes that could break down 5 to 120 mg/L of cyanide in synthetic wastewater by ~98% at pH > 9 in 15 days. Luque-Almagro et al. (2016) identified cyanotrophic microorganisms that utilize cyanide and its derivatives as a nitrogen source for their growth and metabolic activities in industrial wastewater treatment applications. Petrozzi and Dunn (1994) demonstrated >99% degradation of cyanide bound with almond seed extract by adapted microorganism which used cyanide as a nitrogen source and supplemental COD as carbon source for their growth.

Heavy metals typically present in refinery wastewater are copper, iron, mercury, zinc, nickel and chromium. All of these metals, except chromium, are removed by hydroxide precipitation at the pH of their respective minimum solubility. Figure 5.40 presents the solubility of heavy metals as a function of pH.

Figure 5.41 presents a schematic view of metals removal by hydroxide precipitation process. Water with dissolved metals is introduced into a reactor tank. The

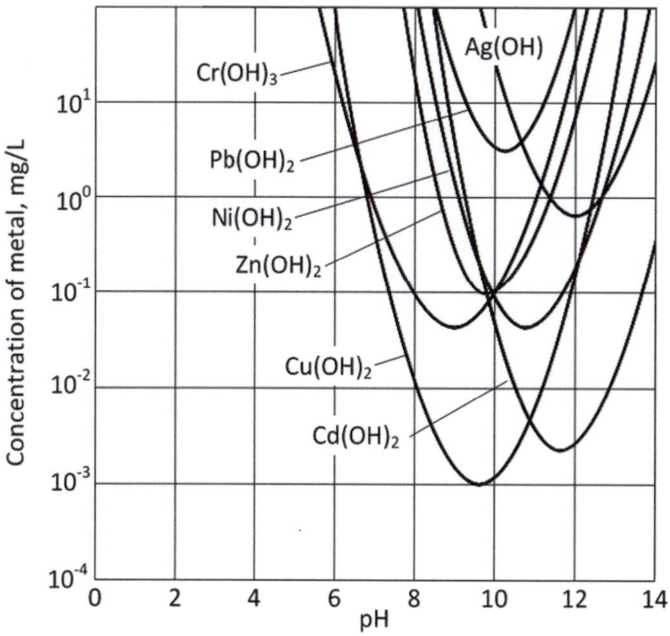

Fig. 5.40 Solubility of heavy metals as function of pH. (Source: Metcalf and Eddy/AECOM, 2014)

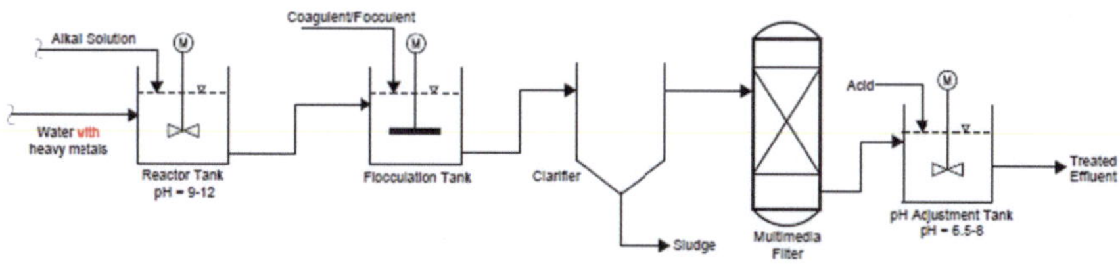

Fig. 5.41 Schematic view of metal removal process from wastewater by hydroxide precipitation

pH is raised to 9 to 10.5, which is the range of minimum solubility for a majority of the metals associated with refinery wastewater, by adding an alkaline solution e.g., a caustic soda (NaOH) solution. Fine particles of metals hydroxides are formed in this reactor. The particulate matter then forms large flocs in a flocculation tank by addition of a coagulant/flocculant chemical. Solids are then separated from water by settling sedimentation in a clarifier. Water overflows from the clarifier into a multimedia filter for further removal of solids. The pH of filter effluent is then adjusted to 6.5 to 8 in a well-mixed tank by adding an acid. Final effluent leaves the pH adjustment tank for disposal or reuse. Solids settled in the clarifier is collected and disposed.

Chromium removal is a multi-step process. It exists in water in trivalent (Cr^{3+}) and hexavalent (Cr^{6+}) forms. Hexavalent chromium is first reduced in a reactor tank to the trivalent form by reacting it with ferrous ions in an acidic medium. After this step all the chromium ions are dissolved in water in Cr^{3+} form. In the next step the pH of water is raised to 9 in a second reactor, which corresponds to the minimum solubility of Cr^{3+} ions. From this point on the treatment process follows the steps presented in Fig. 5.41 for removal of chromium by hydroxide precipitation as $Cr(OH)_3$. Chemical reactions involved in the removal of chromium from wastewater are presented in Sect. 1.3 of Chap. 1

Selenium is a micronutrient for human health as it protects the body from reproductive malfunction and infections. It also is essential for thyroid hormone metabolism and DNA synthesis in human cells. But overdose of selenium has adverse health effects. EPA and various state agencies have set up exposure limit of selenium to the human body.

Food and drinking water are the primary routes of selenium to the human body it originates from soil. Certain parts of the world, including the western states of the US are rich in selenium in their geological structure which adds to its excessive intake into the human body. Also, selenium finds its way into natural bodies of water through industrial discharges, including petroleum production and processing water and wastewater, notably:

- Desalters and sour water stripper effluents of petroleum refineries
- Produced and flowback water from oil and gas exploration activities
- Flue gas desulfurization blowdown water from fossil fuel fired power plants

Selenium is present in the environment in four different forms—as selenite (SeO_3^-) ions dissolved in water in +4 oxidation state, selenate (SeO_4^-) ions dissolved in water in +6 oxidation state, as elemental selenium solids in 0 oxidation state and as hydrogen or metal selenide in –2 oxidation state. Selenite and selenate are the predominant forms of selenium present in water and wastewater. To protect human health and well-being EPA established an enforceable selenium limit of 50 μg/L (ppb) in drinking water. In the western states of the US, especially in California, selenium in refinery wastewater is regulated. Among several methods to comply with this stringent limit, co-precipitation with ferric chloride as presented in Fig. 5.42, has been demonstrated to be reliable and cost effective. Electrocoagulation at optimum acidic pH range of 4.5 to 6.5 using iron electrodes improves the performance of traditional coagulation process.

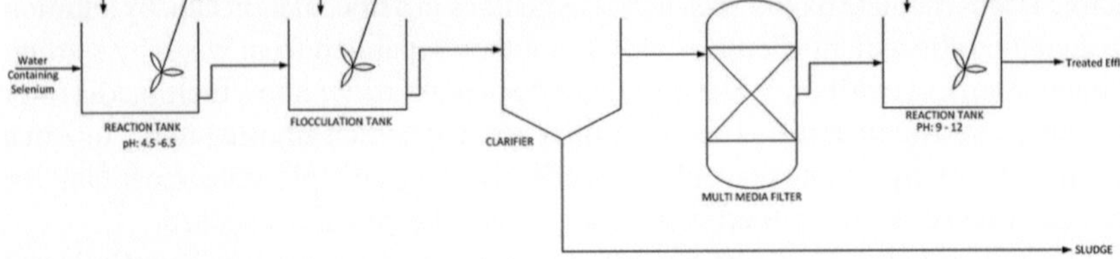

Fig. 5.42 Schematic View of Selenium Removal

5.6.10.2 Tertiary Treatment for Reuse of Refinery Wastewater Secondary Effluent

Microfiltration, ultrafiltration, ion exchange and reverse osmosis are treatment processes to produce high quality water that can be reused for cooling water make up and boiler feedwater makeup. Working principles of these processes are discussed in detail in Chap. 2 and are not repeated here and case studies of water reuse are presented in Chap. 8.

5.6.10.3 Impact of Processing Opportunity Crudes on Refinery Wastewater Treatment Systems

Opportunity crudes are those produced from unconventional resources. These include crudes originating from (a) thermal processing and/or open pit mining of Canadian oil sands, (b) synthetic crudes from partially upgrading Canadian heavy crudes and Venezuelan high sulfur crudes, and (c) tight oils produced by hydrofracturing of rocks from shale formations. Refiners buy these crudes due to their availability at lower prices compared to those obtained from conventional resources. However, blending opportunity crudes with conventional crudes present several challenges in the operation of refinery process equipment which are designed to handle conventional crudes only. Petroleum Environment Research Forum (PERF) sponsored a comprehensive study to investigate the impact of processing opportunity crudes on the operation of wastewater treatment plants. This included a survey of eleven refineries owned and operated by the PERF member companies. The study considered the impact of processing the West Texas Intermediate (WTI) crude in addition to that of processing the three types of opportunity crudes on the operation of their wastewater treatment plants. The reason for including WTI crudes in this study is the fact that a large part of this crude is comprised of the production from hydrofracturing operation at the Permian Basin. The Canadian heavy crudes and synthetic crudes are characterized by low API Gravity in the range of 15 to 33, high sulfur content in the range of 0.12 to 3.3% (w/w), and high TAN in the range of 0 to 3.0. Tight oils from shale formations and WTI crudes are much lighter with API gravities in the ranges of 32 to 45 and 40 to 45 respectively, TAN in the ranges of 0 to 0.36 and 0.05 respectively, and sulfur content in the ranges of 0.10 to 2.0% and 0.02 to 0.49% (w/w) respectively. A summary of the results of this study is reported (O'Donnell 2022) below.

Nine out of the eleven (11) refineries in the survey reported wastewater flow rates in the range of 11,000 cu.m/d (2,000 gpm) to 95,000 cu.m/d (17,400 gpm) with an average flow of 30,500 cu.m/d (5,600 gpm). This was an end to end survey of the major components of the wastewater treatment plants starting from the desalter to the secondary clarifiers.

Seven respondents reported effect of processing unconventional crude slates on their desalter performance. A total of 25 desalters at these seven facilities were studied. The results are:

- Two (2) facilities reported incompatibility of blending heavy sour crudes with light crudes from shale formations, especially at elevated temperatures creating the possibility of emulsions formation.
- Ten (10) out of the 25 desalters had > 1000 mg/L of oil in their effluent when processing heavy Canadian crudes. Poor performance was attributed to a combination of factors e.g., use of a single, as opposed to multiple stage desalters, and high frequency of mudwashing resulting in turbulence and emulsion formation.
- It was concluded that high dosage of deimulsifier, with the right deimulsifier chemistry can mitigate the poor desalter performance in handling the heavy Canadian crudes.

The primary oil-water separation step consists of mainly API separators in most refineries. One of the sites only CPI separator and three use a combination of API and CPI separators. It was observed that CPI separators face operating challenge due to clogging of spaces between parallel plates by large amounts of solids associated with opportunity crudes. In general, primary separators are subject to poor oil-water separation due to emulsion formation. The separated oil also is tightly attached to oil droplets in slop tanks. A few remedial measures have been identified as—(a) increased wash water flow in the desalter, which also has a potential negative effect of increased brine temperature that may adversely impact biological treatment downstream, (b) using a dedicated separator for desalter brine before blending it with the rest of the wastewater streams feeding the primary separator, (c) using a coagulant/flocculant chemical upstream API separator.

The equalization tank is subject to frequent accumulation of solids resulting from settling of large amounts of suspended solids associated with heavy Canadian crudes in the form of sand and clay. This leads to gradual reduction in active volume, and hydraulic retention time, and hence the effective level of treatment. A mitigation step is the use of a continuously operating mechanical mixer which keeps the solids in suspension and prevents settling inside the tank.

The secondary oil—water separation step at these refineries consists of DNF (6 refineries), DAF (3 refineries) and INF separators (2 refineries). The DNF and DAF separators demonstrated better performance than INF. Their effluent oil and grease concentrations are in the ranges of 5 to 20, 10 to 27 and 30 to 50 mg/L, respectively. Nine (9) of the eleven (11) refineries use deimulsifying chemical to improve secondary oil—water separation performance.

Several impacts of processing opportunity crudes on biological treatment of wastewater have been reported:

- Increased COD load due to high oil concentration escaping upstream oil-water separation steps into biotreatment.
- High amine content in wastewater due to their addition as a masking agent to control high H_2S concentrations thereby increasing COD and ammonia loads and DO demand.
- Creating inhibition to nitrification.
- Passage of refractory organics into biological treatment reactors leading to their escape into receiving body of water and creating toxicity to the aquatic life

- Foaming and frothing of bioreactor and secondary clarifier due to formation of filamentous microorganisms.

Poor settling of biosolids in secondary clarifiers is a common occurrence in all the refineries surveyed and all of them, except for one, add coagulant to facilitate settling in secondary clarifiers. Even with coagulant addition all experience poor settling and foaming and frothing. It was identified that this is more common with the use of Canadian heavy crudes. Several of the eleven refineries has some form of existing tertiary treatment steps such as multimedia filters, or aerated settling ponds, or lagoons for removal of carried over solids from secondary clarifiers to prevent potential permit violations. None of the refineries participating in this survey had to build any additional tertiary treatment unit to manage wastewater treatment operation resulting from the use of opportunity crudes. One refinery reported effluent toxicity resulting from the use of high TAN crude. However, they discontinued the use of that crude and resolved that situation. Another facility reported high ammonia effluent toxicity resulting from the presence of large concentration of amines used for suppressing H_2S in the crude. They upgraded their biotreatment process to nitrification to prevent escape of large ammonia concentration in the discharge.

The common observation was that most of the adverse impact on wastewater treatment were from using Canadian heavy crudes and synthetic crude of Venezuelan origin. A few refineries reported that discontinuation of synthetic crudes helped them overcome many of the operating problems. However, no such operating problem was reported with the use of shale oil produced by hydrofracturing and WTI crudes.

For upgrading existing wastewater treatment facilities and construction of new facilities MBBR or IFAS reactors, followed by DAF secondary clarifiers, have been recommended by the study report in order to increase the COD load and incorporate nitrification within the same footprint area as a conventional activated sludge (CAS) process.

5.7 Handling and Treatment of Wastewater Residuals

In a refinery wastewater treatment plant, sludge is typically produced from the following unit operations:

- API separator—bottom sludge;
- Dissolved air/gas flotation (DGF) and induced air/gas flotation (IGF) systems—float and bottom sludge.
- Refinery wastewater biological treatment sludge
- Sanitary wastewater biological treatment sludge

Characteristics of each type of sludge is different from one another and is handled and managed separately, as follows.

5.7.1 API Separator Sludge

The need for treatment as well as the level of treatment required for the sludge from the API separators is dictated by the configuration of process units in the refinery as well as local environmental regulations. This sludge after dewatering and/or deoiling can be sent either to offsite disposal or to the delayed coker process unit in the refinery (provided the refinery has a coker unit). Figure 5.43 shows a typical API sludge treatment unit.

The sludge is first sent to a Decanter tank (floating roof tank to minimize emissions) where oil is skimmed with a floating skimmer and sent to refinery slops. The separated water is pumped back to the refinery wastewater treatment plant. The sludge phase is sent to a centrifuge that separates oil (as centrate) and is sent to slops. The sludge from the centrifuge is sent to disposal.

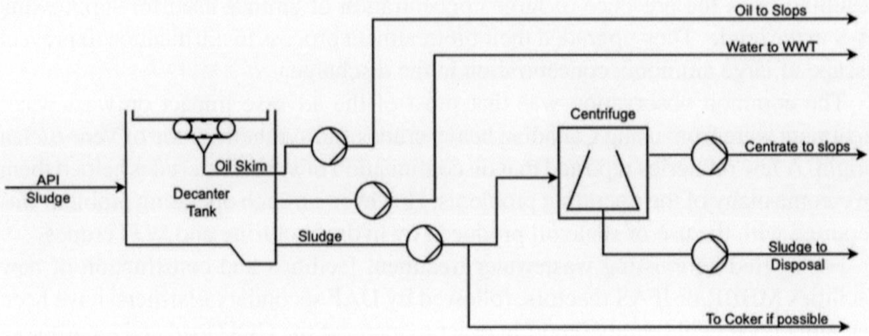

Fig. 5.43 API sludge treatment unit

5.7.2 IAF/DAF Float

The float stream from an IAF (or, IGF)/DAF (or, DGF) typically contains emulsions and chemicals (flocculants that were added) and a significant amount of water, and therefore makes it unsuitable for adding it to the refinery slop system. In order to reduce disposal costs, it is beneficial to remove the water contained in this stream. A DAF float treatment system is shown as Fig. 5.44.

The unit consists of a tank to which the float stream is added. Emulsion breaker chemicals are also added to this tank. The contents of the tank are circulated and heated up using a heat exchanger in order to break up the emulsions. The water that separates out is sent back to the refinery wastewater treatment plant. The remaining float from the tank is sent to disposal.

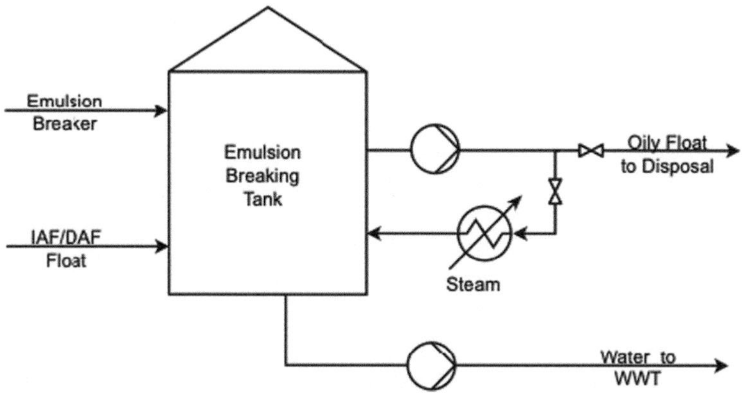

Fig. 5.44 IAF (or, IGF)/DAF (or, DGF) float treatment

5.7.3 *Waste Biological Sludge*

Treatment of the sludge generated in the wastewater treatment plant typically includes a sequence of thickening, stabilization, and dewatering processes. These steps are described in the following sections. A typical sludge treatment system is shown in Fig. 5.45.

Thickening

Thickening is usually the first step in sludge treatment and is typically carried out in a gravity thickener tank, or gravity belt thickener. A thickener can reduce the total volume of sludge to less than half of the original volume. A sludge removal and thickening system is selected depending on sludge characteristics of the solids produced from biological treament processes. The biological solids generated from a secondary clarifier contains about 0.8 to 1% solids and can be thickened up to 4% solids. However, if a DAF is used as a secondary clarifier the solids concentration

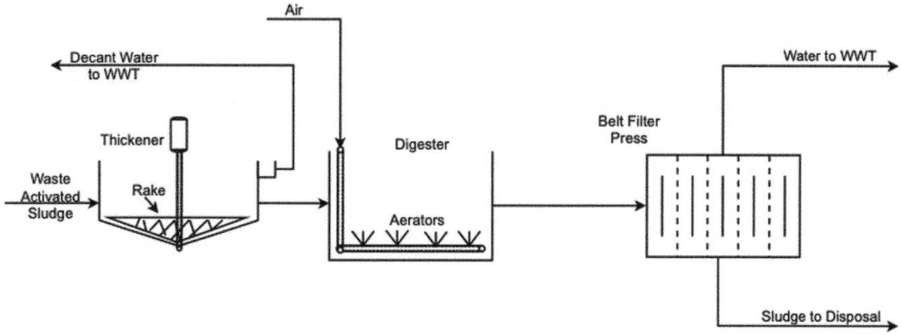

Fig. 5.45 Biological Sludge Treatment

of the float is ~4% and may not require any additional thickening step prior to stabilization.

Sludge Stabilization

Sludge stabilization is a process in which organic components of the solids (volatile solids) are decomposed into stable solids. Stabilization process helps to reduce the total mass of sludge, destroys undesirable microorganisms, eliminates bad odor and makes it easier to dewater and/or dry the sludge.

In petroleum refinery operation, aerobic digester is the most used system for sludge stabilization.

Aerobic digestion is similar to the conventional activated sludge process. Sludge is stabilized using aeration for mixing the solids and reducing the volatile organic solids by oxidation in an open top reactor. This process is sensitive to temperature, tank geometry and concentration of feed solids. The solids produced through aerobic digestion may satisfy the regulatory requirements for disposal by land application. The solids digested aerobically have been observed to have poor mechanical dewatering characteristics.

Important factors to consider while designing conventional aerobic digestion system are as follows:

1. Temperature: As the reactor is open to air, the biological reaction rate will vary with change in outside temperature
2. Feed solids concentration: It is difficult to provide aeration and mixing to feed solids content of more than 4%
3. Tank volume: It will depend on the solid's residence time necessary, and the detention time required to achieve desired volatile solids (VS) reduction.

Also, the amount of oxygen required, mixing energy requirements and required volatile solids reduction play an important role in designing the system. Table 5.23 below shows the important design criteria range.

Other methods for stabilization used are alkaline stabilization and composting. However, these are less common and not discussed here. More details for these methods can be found in literature (WEF 2018).

Dewatering

Digested sludge is dewatered before disposal. Dewatered sludge still contains a significant amount of water—often as much as 85%—but, even with that much

Table 5.23 Design parameters for aerobic digestion

Parameter	Value
SRT at (15–20° C) (days)	60–40
Volatile solids loading (kg/m^3-d)	1.6–4.8
Oxygen requirements (kgO_2/kg VSS destroyed)	2.3
Reduction volatile suspended solids (%)	38–50

moisture, sludge no longer behaves as a liquid and can be handled as a semi-solid material. Sludge drying beds are the simplest method of dewatering. Drying takes place by a combination of evaporation and gravity drainage through the sand. After about six weeks of drying, the sludge cake, as it is called, may have a solids content of about 40%.

Alternatives to sludge drying beds include the rotary press, centrifuge, and belt filter press. These mechanical systems require less space than sludge drying beds, and they offer a greater degree of operational control. However, they usually have to be preceded by a sludge conditioning step, in which polymer is added to the liquid sludge to flocculate the solids and improve dewaterability (Table 5.24).

When a refinery waste activated sludge is combined with primary sludge from API Separator bottoms it contains a mixture of oil and waste sludge. For dewatering this type of sludge, a three—phase centrifuge is employed. It is one of the most common dewatering equipment used in the refining industry. In this, the wastewater is separated into oil, dried solids, and water. The dried solid generated from this type of combined sludge is not suitable for land application and is disposed of to a hazardous waste landfill. The recovered oil is sent back to the slop tank. Water is returned to the inlet of the DAF separator for treatment to remove BOD/COD and TSS.

Disposal

The destination of treated sludge usually is the land. As described above, dewatered waste activated sludge may be land applied if it fulfills the regulatory requirements but if the sludge originates from a combination with API separator bottom stream, then it needs to be buried underground in a hazardous waste landfill. Where a suitable site for sludge disposal is not available sludge may be incinerated. Incineration completely evaporates the moisture and converts the organic solids into inert ash. Certain refineries blend dewatered sludge with the feed to delayed coker to produce petroleum coke, a valuable product.

Table 5.24 Types of sludge dewatering Systems

Types	Type of feed	Polymer required (g/kg dry TS)	Cake solids (% TS)	Solids capture, %
Rotary press	Aerobically digested WAS, 1–3% solids	8.5–17.5	20–35	90–95
Solid bowl centrifuge	Aerobically digested WAS, 1–3% solids	10–15	18–25	95+
Belt filter press	Aerobically digested WAS, 1–3% solids	6–10	12–22	85–95
Screw press	Aerobically digested WAS, 1–3% solids	8.5–17.5	15–20	88–95

5.8 Marketing Terminals

This section is a discussion of wastewater treatment in the Marketing sector (Distribution Terminals) of the oil supply chain. This includes the treatment systems typically encountered in this sector.

The rainwater that is collected from the truck loading stations and the tank drains could be contaminated with hydrocarbons due to accidental spills and leaks and therefore will need to be treated prior to discharge. The sources of this wastewater were discussed in Chap. 4. Figure 5.46 shows a typical Terminal wastewater pre-treatment system.

Wastewater from tank drains and the truck loading stations are sent to a surge tank where oil is skimmed off using a floating skimmer and sent to disposal. In the diagram the surge tank is shown with a cone roof, but it could be a tank with no cover depending on local regulations and requirements. The water from the tank is sent to a CPI Separator for separation of oil, sand and grit. The separator is equipped with inclined parallel corrugated plates that aid in the separation of oil and solids. The oil is skimmed off using a collection pipe located at the top of the separator. Any sludge that collects at the bottom of the separator is periodically pumped put. The wastewater is sent for further treatment if required.

Table 5.25 shows the expected contaminants in the effluent from CPI separators in marketing terminals.

At many locations discharge of wastewater from a CPI separator could be prohibited and be required to be treated further to remove dissolved hydrocarbons. The expected variations in flow as well as the prevailing effluent discharge limits will dictate the type of treatment that will be required. One option would be to install a small packaged biological treatment system. However, most terminal operations are sparsely staffed and therefore prefer a system that requires minimal operator

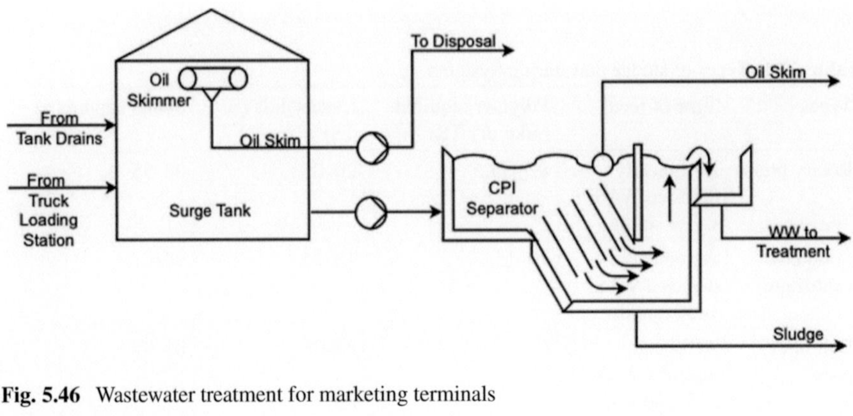

Fig. 5.46 Wastewater treatment for marketing terminals

Table 5.25 Expected contaminants and their concentrations of marketing terminal CPI separator effluent

Contaminant	mg/L
Chemical oxygen demand	Up to 3,000
Biological oxygen demand	Up to 2,200
Total suspended solids	Up to 50
Free hydrocarbons	Up to 50

attention. One such option would be to install a constructed wetland[2] provided the required real estate is available.

Wetlands are transitional areas between land and water and could include a range of environments including wet meadows, swamps, and marshes. In a wetland the flow is very slow which allows sedimentation of substrates. Constructed wetlands, also referred to as Reed beds, are a wastewater treatment system that rely upon natural soil, plant, and hydrologic processes to improve water quality. Reed beds provide effective treatment of carbonaceous contaminants in wastewaters such as wastewaters containing petroleum hydrocarbons because microbial populations naturally occurring in wetland sediments and decomposing plant matter will assimilate carbon in the wastewater and transform the compounds to carbon dioxide and water through bacterial metabolic function. Plants provide a naturally regenerating source of organic substrates for biological treatment as well as a natural hydrologic "pump" to bring water and contaminants into the biologically active sediment layer and root zone. Plants also shade the water and prevent excessive algal growth.

Reed beds are frequently built as a surface impoundment vegetated with plants adapted to year round inundation in shallow water (<30 cm). Alternatively, reed beds may be constructed as horizontal subsurface flow wetlands, where plants are installed in a gravel or porous media, and water flows through the biologically active root zone below the media surface. Wastewater typically is applied at one end of the system to flow by gravity to an outlet on the opposite side.

Developments in wetland treatment technology over the past decade have introduced the concept of wetland "intensification", where air is introduced mechanically. One form of wetland intensification has been the addition of diffusors into the wetland subsurface to distribute air directly into the flow stream and the biologically active media. This approach requires the addition of air blowers, ductwork and diffusors and a power supply into the design. Figure 5.47 below is the diagram of a typical Reed Bed system. A conceptual diagram of an enhanced wetland system is presented as Fig. 5.48.

As the use of constructed wetlands for treatment has progressed, research and field experience have identified multiple factors that influence performance. These include:

- wetland area,
- flow rate to be treated,
- target effluent concentrations,

[2] A Handbook of Constructed Wetlands-epa.gov.

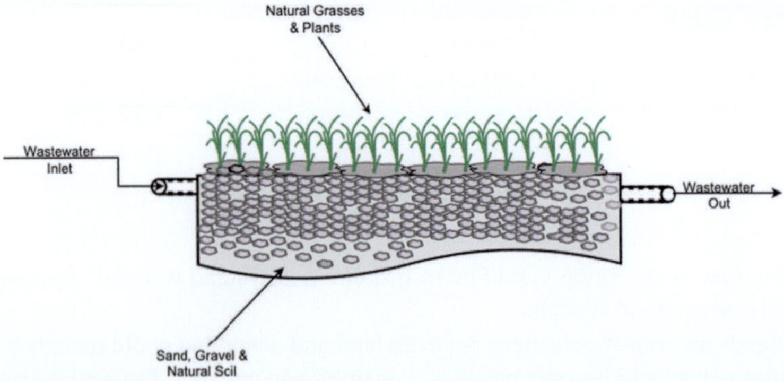

Fig. 5.47 Reed bed system

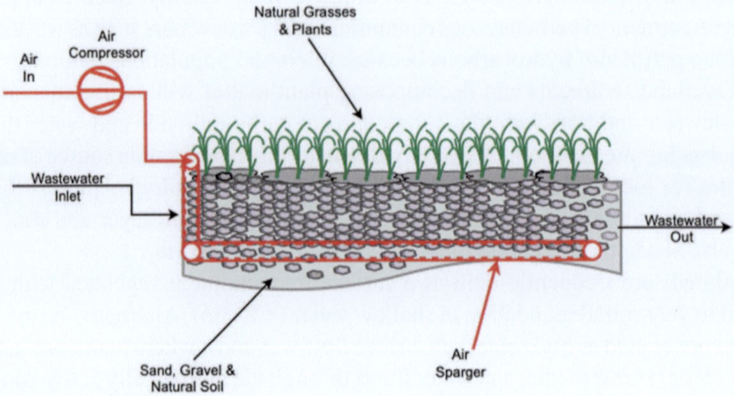

Fig. 5.48 Enhanced reed bed system

- type of system,
- wastewater type and strength,
- pollutant removal characteristics,
- system hydraulics, and
- regional climate.

Available area is the most important constraint to the performance of a wetland treatment system. The treatment process is typically characterized as area—based and is dependent upon the interaction between wetland surfaces and the water being treated. Also, higher concentrations require greater area for treatment, and low target effluent concentrations will also increase area requirements.

As a rule of thumb, the area required for a reed bed treatment system would be up to 300 m^2 per m^3/hr of flow.

Abbreviations and Glossary of Terms

16S rRNA	Metagenomic Analysis Technique using 16S ribosomal RNA Genes
AnAOB	Anaerobic Ammonia Oxidizing Bacteria (also called Anammox bacteria)
AOB	Ammonia Oxidizing Bacteria
AOR	Actual Oxygen Requirement
API Separator	American Petroleum Institute Separator
BOD_5	5-day Biochemical Oxygen Demand (BOD_5 and BOD have been used interchangeably in the book both mean 5 day BOD)
CAS	Conventional Activated Sludge
CPI Separator	Corrugated Plate Interceptor Separator
DAF	Dissolved Air Floatation
DGF	Dissolved Gas Floatation
DNF	Dissolved Nitrogen Floatation
DNRA	Dissimilatory Nitrate Reduction to Ammonia
DO	Dissolved Oxygen
EC_{50}	Toxicant Concentration to Cause Half Maximum Response by Microorganisms
EPS	Extracellular Polymeric Substances
F:M	Food: Microorganism Ratio
FRP	Fiberglass Reinforced Plastic
GAC	Granular Activated Carbon
GAS	Granular Activated Sludge
HDPE	High Density Polyethylene
HRT	Hydraulic Retention Time
IAF	Induced Air Floatation
IC_{50}	Toxicant Concentration Inhibitory to 50% of Organismal Population
IFAS	Integrated Fixed Film Activated Sludge
IGF	Induced Gas Floatation
IR	Internal Recycle
kW.h	Kilowatt Hour
LC_{50}	Toxicant Concentration Lethal to 50% of Organismal Population
MBBR	Moving Bed Biological Reactor
MBR	Membrane Bioreactor
MCRT	Mean Cell Residence Time
MF	Microfiltration
MIC	Microbiologically Induced Corrosion
ML	Mixed Liquor
MLE	Modified Ludzack Ettinger Process
MLVSS	Mixed Liquor Volatile Suspended Solids
MOP	Manual of Practice

NA	Naphthenic Acids
NOB	Nitrite Oxidizing Bacteria
O&G	Oil and Grease
OHO	Ordinary Heterotrophic Organisms
PACT	Powdered Activate Carbon Treatment
PERF	Petroleum Environment Research
PLC	Programmable Logic Controller
PVC	Polyvinyl Chloride
RAS	Return Activated Sludge
RBC	Rotating Biological Contactor
SAA	Surface Active Agents
SBR	Sequencing Batch Reactor
sCOD	Soluble Chemical Oxygen Demand
SES	Sand Equivalent Size
SND	Simultaneous Nitrifying Denitrifying Bacteria
SOR	Standard Oxygen Requirement
SRT	Solids Retention Time
SVI	Sludge Volume Index
SWD	Side Water Depth
TDS	Total Dissolved Solids
TKN	Total Kjeldahl Nitrogen
TN	Total Nitrogen
TOC	Total Organic Carbon
TP	Total Phosphorus
TPH	Total Petroleum Hydrocarbon
TSS	Total Suspended Solids
UF	Ultrafiltration
UV	Ultraviolet
VOC	Volatile Organic Carbon
VSS	Volatile Suspended Solids
WAS	Waste Activated Sludge
WET	Whole Effluent Toxicity
WTI	West Texas Intermediate
Y	Net Yield Coefficient

References

American Petroleum Institute (1990) Monograph on Refinery Environmental Control - Management of Water Discharges - Design and Operation of Oil-Water Separators, API Publication 421, First Edition, February 1990

Alvarado-Lopez MJ, Garrido-Hoyos SE, Raynal-Gutierrez ME, El-Kassis EG, Luqui-Almagro VM, Rosano-Ortega G (2023) Cyanide biodegradation by a native bacterial consortium and its potential for gold mine tailing biotreatment, Water, 15:1595

American Petroleum Institute Survey (1985) API Refinery Survey
A Handbook of Constructed Wetlands-epa.gov\
Arntsen B, Bigham J, Kosic A, Penny J, Katz S (2017) Implementing membrane bioreactor technology to upgrade marathon petroleum's detroit refinery wastewater treatment. Proceedings of the 2017 Annual Meeting of AFPM, San Antonio, TX
Basu S (2018) Impact of opportunity crudes on refinery desalter and wastewater treatment performance—part 2, *Hydrocarbon Processing*
Basu S, McAuley J, White N, Lamb S, Pitt P (2001) Microbiologically induced corrosion in a wastewater treatment plant, *Material Performance*, July 2001, pp 2-7
Casey M, Grillo R, Kalluri R, Hubbel S, Fluornoy W (2008) Integrated fixed film/ activated sludge (IFAS) for refinery and petrochemical wastewater treatment plant upgrades. Proceedings of the 69th Annual Industrial Water Conference, October 26–30, San Antonio, TX
Cerqueira AC, Lopes T, Santiago V, Vallero M, Trovati J, Arntsen B, Syed W (2013) Design and performance of the first full scale membrane bioreactor plant treating oil refinery effluent in Brazil, *Proceeding of the WEFTEC 2012*, New Orleans, LA
Cui YW, Peng YZ, Peng CY, Gan XQ, Ye L (2006) Achieving biological nitrogen removal via nitrite by salt inhibition. Water Sci Techn 53(6):115–122
Dale C, Wenta R, Eckenberg M, Jacobsson S (2017) Phenol and benzene removal in moving bed biofilm reactors—a review of operating data from two full scale refineries
Eckenfelder WW, Ford DL, Englande AJ (2009) Industrial Water Quality, 4th Edition, McGraw Hill
Figuerola ELM, Erijman L (2010) Diversity of nitrifying bacteria in a full—scale petroleum refinery wastewater treatment plant experiencing unstable nitrification. J Hazard Materials 181(1):281–288
Ford D (2012) Pretreatment/primary oil removal, back to basics: wastewater treatment fundamentals for petroleum refineries and petrochemical plants. Workshop 102, WEFTEC 2012, New Orleans, LA
Han M, Kim T, Kim J (2007) Effects of floc and bubble size on the efficiency of the dissolved air floatation process. Water Sci Technol IWA Publish 56(10):109–115
Hilts J, Eastuke B, Robertson E, Knight M, Edwards D, Elliott J, Arntsen B (2018) A review of the co-op refinery complex's wastewater improvement project. *Proceedings of the WEFTEC 2018*, New Orleans, LA
Igiri BE, Okoduwa SIR, Idoko GO, Akabuogu EP, Adeyi AO, Ejiogu IK (2018) Toxicity and bioremediation of heavy metals contaminated ecosystem from tannery wastewater: a review. J Toxicol, September 27, 2018:2568038
Jenkins D, Richard MG, Daigger GT (2003) Manual on the causes and control of activated sludge bulking, foaming, and other solids separation problems, 3rd edn. CRC Press
Jenkins D (2013) Personal communications
Jolibois Francois et al. (2018) Refiner wastewater treatment using two-stage attached growth bioreactors for organics removal and nitrification, WEFTEC 2018 Proceedings
Ken Arnold MS (1999) Surface production operations Volume 1, 2nd Edition. Woburn, MA: Butterworth—Heinemann
Kim E, Yulisa A, Kim S, Hwang S (2020) Monitoring microbial community structure and variations in a full-scale petroleum refinery wastewater treatment plant. Bioresou Techn 306:123178
Luque-Almagro VM, Moreno-Vivian C, Roldan MD (2016) Biodegradation of cyanide wastes from mining and jewelry industries. Current Opinions in Biotechnology, 38:9–13
Microbics Corporation (1995) Microtox acute toxicity basic test procedure, Carlsbad, CA
Metcalf & Eddy/AECOM (2014) Wastewater engineering, treatment and reuse, 5th edition. McGraw—Hill, New York
Meyer C (2012) Secondary oil/water/solids separation, back to basics: wastewater treatment fundamentals for petroleum refineries and petrochemical plants, Workshop 102, WEFTEC 2012, New Orleans, LA

Misiti T, Tandukar M, Tezel U, Pavlostathis SG (2013) Inhibition and biotransformation potential of naphthenic acids under different electron accepting conditions. Water Res 47(1):406–418. https://doi.org/10.1016/j.watres.2012.10.019. Epub 2012 Oct 17. Erratum in: Water Res. 2014 Jan 1;48:645. PMID: 23134740

Movafaghian S, Chen J, Wheeler SS, Guidry RW (2003) Introduction of a new generation of induced gas flotation equipment, presented at the society of petroleum engineers (SPE) Latin American and Caribbean Conference, Port of Spain, Trinidad

O'Donnell J (2022) Impact of unconventional crudes processing on WWTP operations, WEFTEC Proceedings 2022

Petrozzi S, Dunn IJ (1994) Bioprocess Engineering, 11:29–38

Qaderi F, Sayahzadeh AH, Azizi M (2018) Efficiency optimization of petroleum wastewater treatment by using serial moving bed biofilm reactors. J Clean Prod jclepro 2018.04.257

Rusten B, Hem L, Odegaard H (1995) Nitrification of municipal wastewater in novel moving bed biofilm reactors. Water Env Res 67:75–86

Sawyer CN, McCarty PL, Parkin GF (2003) Chemistry for environmental engineering, 5th edition. McGraw Hill, New York

Soares-Castro P, Yadav TC, Viggor S, Kivisaar M, Kapley A, Santos PM (2019) Seasonal bacterial community dynamics in a crude oil refinery wastewater treatment plant. Appl Microb Biotechn 103(21):9131–9141

Steeves H, Boswell C, Mason D, Norman P, Schmidt J and Watson L. (2022) Understanding loss of nitrification at a refinery wastewater treatment plant using 16S rRNA Sequencing, Proceedings of WEFTEC 2022

Thompson SJ (1973) Data improves separator design. Hydrocarbon Processing: 81–83, October 1973

USEPA (1987) Guidance manual on the development and implementation of local discharge limitations under the pretreatment program, December 1987, pp 3-44 to 3-49, Washington, DC

USEPA (2004) Local limits development guidance appendices, Washington, DC

USEPA (2010) Nutrient Control Design Manual, EPA/600/R-10/100

Water Environment Federation (2018) Design of water resource recovery facilities, MOP 8, 6th Edition, WEF and McGraw

Chapter 6
Water and Wastewater Treatment Process Modeling for Petroleum Industrial Operations

Andrew R. Shaw, Leslie Miller, Kristen Jenkins, A. J. Gerbino, Rajeev Goel, Nicholas Piccolo, and Spencer Snowling

Abstract Treatment of water and wastewater streams in petroleum production and processing comprises of many unit operations and processes to produce a treated effluent that is safe and suitable for discharge or reuse. There is very little control over influent characteristics, which can vary widely, but the effluent quality must fulfill the regulatory requirements for discharge, or the specifications for reuse. This requires a robust design that can meet the treatment goals under all conditions. Computer models are tools that aid in the design of individual unit operations and processes, and the treatment train as a whole. Models are also very important for tracking the performance of an operating plant with respect to its intended design and taking necessary corrective actions during upsets. This chapter outlines the principles behind

A. R. Shaw (✉)
Black & Veatch, Houston, TX, USA
e-mail: ShawAR@bv.com

L. Miller
OLI Systems, Parsippany-Troy Hills, NJ, USA
e-mail: leslie.miller@olisystems.com

K. Jenkins
Kiewit Engineering Group Inc., Lenexa, KS, USA
e-mail: Kristen.Jenkins@kiewit.com

A. J. Gerbino
AQSim, Glen Ridge, NJ, USA
e-mail: aj.gerbino@aqsim.com

R. Goel · S. Snowling
Hatch, Ltd., Hamilton, ON, Canada
e-mail: rajeev.goel@hatch.com

S. Snowling
e-mail: spencer.snowling@hatch.com

N. Piccolo
McMaster University, Hamilton, ON, Canada
e-mail: npiccolo9@gmail.com

S. Basu et al. (eds.), *Water Management in Petroleum Industries*, Water Resources Development and Management, https://doi.org/10.1007/978-981-19-3159-8_6

the development of models for both the upstream production as well as downstream processing sectors.

Keywords Activated sludge model · Activity · Arrhenius equation · Biological treatment process modeling · Electrolyte modeling · Exploration and production modeling · Fugacity · Inhibition · Ionic Activity Product (IAP) · Modeling for fouling and corrosion · Non-ideal equilibrium · Scale index · Sour water stripper modeling · Speciation calculations · Stripping · VOC emissions

6.1 Introduction

Process models support the design of water and wastewater treatment systems associated with petroleum industries which require the solution of multiple mathematical equations simultaneously with the help of computers. Models are also a great tool in the interpretation of operating data supporting critical decision making in real time. Experience gained from full scale operating plants greatly helps in the calibration of models, making the results accurate and useful. Procedures using widely used process simulators for the modeling of several important processes in upstream and downstream operation of petroleum industries are presented in the following sections.

6.2 Principle of Electrolyte Modeling

Water is an electrolyte. Water and wastewater streams associated with oil and gas operations contain many components and operate over broad ranges of temperatures, pressures, and concentrations. Many chemicals react in water to form one or more new species according to chemical equilibrium reactions, which is called speciation. Speciation, combined with physical equilibrium of separate phases, can produce unexpected results like scaling or corrosion.

Water chemistry in upstream exploration and production and downstream refining is sufficiently complex to require an advanced electrolyte modeling tool. These tools can aid operators in identifying scaling, solids formation and corrosion issues prior to their occurrence. The section on water stability in Chap. 2 introduced the basic concepts of corrosion and scaling potential in terms of various $CaCO_3$ saturation indices in water. Computer modeling is an extension of this exercise as it examines the scaling potential of other metal salts in addition to $CaCO_3$.

This section presents an overview of the theory behind electrolyte modeling related to oil and gas industrial operations, followed by details on commercially available electrolyte models. An overview of electrolyte modeling is presented to provide the background information to examine the corrosion and/or scaling potential of produced water in upstream operations and to support the design of sour water strippers in downstream operations.

6.2.1 Overview

Electrolyte solution models include those that:

- Treat electrolytes on an undissociated basis;
- Assume complete dissociation of dissolved solids into constituent ions; or
- Speciation-based models, which explicitly treat the solution chemistry.

Speciation calculations become necessary when solution chemistry is complex or is very high in ionic strength (total dissolved solids concentration). Speciation variations also have a significant effect on phase equilibria such as salt solubility, especially in multi-salt, mixed-solvent systems. Speciation chemistry is the inherent reason for nonideal properties of an electrolyte solution and therefore needs to be accounted for in any calculation. Water from oil and gas production or refining operations is complex enough to require such speciation-based modeling. A detailed discussion on the principle and procedure for speciation is available in the literature (Wang et al. 2002). This section is focused on application of speciation-based modeling for non-ideal systems in upstream and downstream water and wastewater treatment processes.

An example of speciation complexity is shown in the following four-component system: water, ammonia, carbon dioxide and hydrogen sulfide. The list of species and reactions include:

Components: H_2O, NH_3, CO_2, H_2S
Species: H_2O, H^+, OH^-, CO_2, CO_3^{-2}, HCO_3^-, NH_3, NH_4^+, $NH_2CO_2^-$, H_2S, HS^-, S^{-2}
Speciation reactions:

$$H_2O \leftrightarrow H^+ + OH^-$$

$$CO_2 + H_2O \leftrightarrow H^+ + HCO_3^-$$

$$HCO_3^- \leftrightarrow H^+ + CO_3^{-2}$$

$$NH_3 + H_2O \leftrightarrow NH_4^+ + OH^-$$

$$NH_2CO_2^- + H_2O \leftrightarrow NH_4^+ + CO_3^{-2}$$

$$H_2S \leftrightarrow H^+ + HS^-$$

$$HS^- \leftrightarrow H^+ + S^{-2}$$

Figures 6.1 and 6.2 are parity plots showing the calculated and experimental solubilities for CO_2, H_2S, and NH_3 at 20 and 60 °C. When Henry's constants alone are used to predict dissolved species concentrations (Fig. 6.1) the error is up to 10,000× compared to measured values.

When aqueous speciation is included, the error decreases significantly (Fig. 6.2).

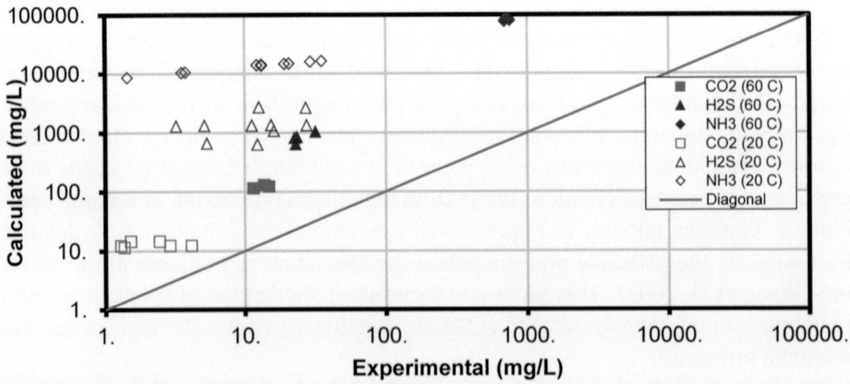

Fig. 6.1 Parity plot for CO_2, H_2S, and NH_3 vapor pressures when only the Henry's constant is used to predict solubility (with no aqueous speciation)

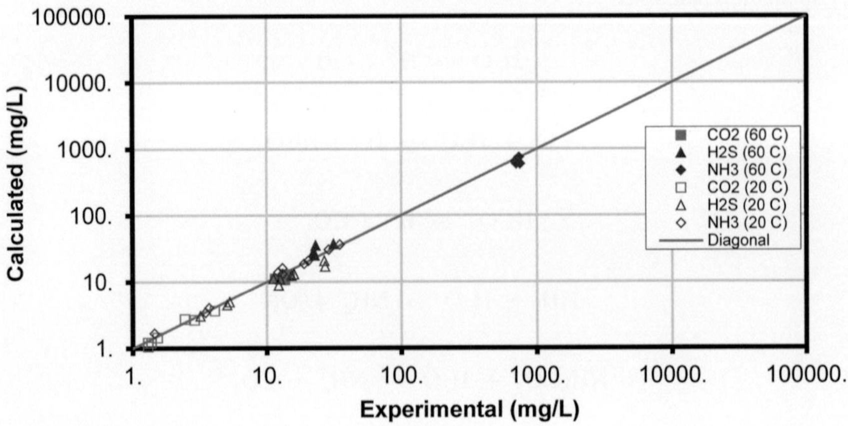

Fig. 6.2 Parity plot for CO_2, H_2S and NH_3 solubility in water when a full speciation model is used to predict solubility

6.2.2 Thermodynamic Framework

The thermodynamic framework of a simulation tool should be reviewed to determine its applicability to the modeling problem at hand. In electrolyte solutions there are two critical frameworks, the Equation of State, and the activity coefficient.

Equation of State

An Equation of State is used to compute the equilibrium constants for chemical reactions as a function of temperature and pressure. The Helgeson-Kirkham-Flowers-Tanger (HKFT) equation of state (Wang et al. 2002) is a robust model used by advanced water-based simulators.

The five principal properties used in the HKFT equation of state include, the Gibbs Free Energy, enthalpy, entropy, heat capacity and volume. These thermodynamic properties are sourced from peer-reviewed journals, thermophysical data books, conference proceedings, company reports, theses, and electronic data repositories (e.g., NIST's TRC databank, which provides comprehensive thermodynamic thermophysical properties data, models and standards).

Equilibrium under Non-Ideal Conditions, Activity and Fugacity

An activity coefficient model is used to calculate the effects of species concentration on the overall reaction equilibrium. The activity coefficient model works with the equation of state to produce a more accurate equilibrium prediction. An equation of state is used to predict equilibrium at ideal conditions. Under ideal conditions the effects of electric fields generated by ions in solution are ignored. This is the case when a solution is very dilute. The discussions on liquid–liquid, vapor–liquid and solid–liquid equilibria in Chap. 1 assume ideal solutions and need to be corrected when applied to electrolytic solutions at high concentrations. The electric fields created by the dissolved ionic species have a significant effect on the overall equilibria and must therefore be quantified. Under non-ideal conditions the concentration term, 'C_i' of each ion in an equilibrium equation is replaced by the term activity 'a_i'. Molar concentration and activity of an ionic species in an electrolyte solution are related to each other by a quantity referred to as the activity coefficient, Υ_i such that $a_i = \Upsilon_i * C_i$.

The Debye-Hückel equation and its many extensions were developed to calculate the value of the activity coefficient caused by these electric fields, or electrostatic interactions. Detailed discussions on the calculation procedure of activity coefficients by the application of Debye- Hückel equation and its simplified version by Güntelberg approximation are available in literature (Snoeyink and Jenkins 1991). More advanced activity models like the Mixed Solvent Electrolyte framework (MSE) extends these models to what is known as the fused salt limit, where water is no longer present. Such systems include 100% H_2SO_4, salts dissolved in alcohols, or non-aqueous ionic liquids.

As in the case of solid–liquid equilibria, the vapor–liquid equilibria are also affected by non-ideality. Gases under high pressure depart significantly from ideal and the pressure term in the Henry's Law (Chap. 2) is replaced by fugacity such that the concentration of a substance (solute) dissolved in the liquid phase (in this case water) in equilibrium with a gas at pressure P is expressed as:

$$C_{aq} = K_H * f \tag{6.1}$$

where,

C_{aq} Concentration of the solute in aqueous phase
K_H Henry's Constant at a given temperature (25 °C is the standard temperature)
f Fugacity
P Pressure in the gas phase.

Fugacity f and Pressure P are related to fugacity coefficient Φ by the following equation.

$$\Phi = f/P \tag{6.2}$$

For a gas at low pressure, which behaves as an ideal gas, $\Phi = 1$. The higher the pressure of the gas, the more it behaves increasingly non-ideally, and greater is the departure of the value of Φ from 1.

There is a large body of published work in the field of electrolyte modeling for the equations, activity and fugacity. To gain an in-depth understanding of the subject the reader is encouraged to review the materials presented in Smith et al. (2021), Anderson (2005) and Jemal (1999).

6.2.3 Important Calculations for Industrial Processes Using Electrolyte Models

Electrolyte models are used in every industry to predict fluid properties in various systems. Common examples include the scrubbing of combustion gases to remove harmful components like Hg, SO_2, SO_3, and HCl. More recently, combustion gas scrubbing is used to remove CO_2 as the first step in carbon capture. Another example is the gas stripping of volatile components from water so that the water can be reused. A third example is predicting when a free acid phase will condense from a gas stream. A particularly pernicious example of this is at the top of the atmospheric crude distillation unit in an oil refinery. Vapor from this unit is sent off for further processing. This vapor contains naphtha, light hydrocarbon gases, NH_3, H_2S and HCl. As the vapor cools, NH_3 and HCl combine to form an NH_4Cl salt. This salt condenses from the vapor, and depending on the temperature, can form a solid molten salt, or reacts with water in the vapor to create a free acid phase. These acids form as the H_2O present in the vapor condenses, thereby creating an acidic liquid that dissolves metal (Wang et al. 2002).

Table 6.1 Example applications where electrolyte thermodynamic models are useful for predicting the performance of a process

Oil and gas, geothermal production	Refining	Chemical processing	Environmental/water treatment
Mineral scaling	Amine-HCl condensation	Reactive distillation	Demineralization using ion exchange
Corrosion	Sour water stripping	Evaporation and crystallization	Water softening
Glycol regeneration	Amine sweetening	Alcohol washing	Zero liquid discharge
Sour gas sweetening	Sulfur recovery	Dehydration	Carbon adsorption
Produced water treatment and disposal	Desalting		Reverse osmosis and membrane separation

Table 6.1 lists some common processes that use electrolyte simulation in its design or optimization.

6.3 Modeling Applications in Upstream Water Treatment in Exploration and Production Operations

Produced water is a complicating factor during oil and gas production. It contributes to flow assurance problems like mineral scaling, corrosion, and gas-hydrate formation. Operational remedies, such as glycol injection to prevent gas hydrate formation, cause secondary challenges like scale buildup in the glycol regeneration units.

6.3.1 Scale Prediction Studies for Oil or Gas Assets

Scale prediction modeling is a common flow assurance study in oil and gas production. The objective of scale prediction is to determine if solids will form and accumulate within the production line and threaten productivity. The most common type of scale study is to test whether scale will form on the production piping that brings the oil, gas, and produced water from the reservoir to the surface.

This section provides a general description of a scale risk analysis. An example case file is provided at the end of the section. It includes some of the challenges faced when performing a study. The example was created using OLI Studio: ScaleChem V11.0 (OLI Systems, Inc.).[1]

[1] OLI Systems, Inc. is a commercial developer of electrolyte thermodynamic software. The software is used to calculate the properties of water streams or to simulate the mass and energy balance of water-dominated processes.

6.3.2 Scale Prediction Study for a Given Well

A scale prediction study can be divided into several steps: gathering pertinent data; creating a mass balance of the production system; and reviewing the fluid properties at bottom hole conditions and at the sampling points.

The work starts by gathering the pertinent data. This includes the water, gas, and oil analyses, a general block flow diagram of the process, the production location where samples were taken, the temperature and pressure profile of the well, and the daily flow rates. The measured data is used to simulate the producing well. The stepwise procedure is as follows:

Step 1 Combine the water, gas, and oil analyses and create a multi-phase (gas, water, oil) fluid composition;
Step 2 Compute the fluid properties at reservoir conditions and at an equilibrium state;
Step 3 Calculate the changes in fluid properties as the fluid flows to the surface.

The above processes provide an estimate of mass flow rate, chemistry of constituents, and phase equilibrium relationships of the fluid exiting the reservoir, flowing up the well, and exiting the wellhead at the surface.

To ensure that the predictions are consistent with field observations, the calculated phase compositions and flow rates are compared to measured data taken at known conditions. A good simulation will predict the proper flow of each phase (gas, water, and oil) and an accurate distribution of species that partition to multiple phases. If the predicted compositions for example, do not match observed values, then the original water, gas, and oil analyses are reviewed to see if something was not measured in one of the phases.

The final step described above is where the user computes the potential for solids to precipitate from the production fluid. The Saturation (S) and Excess Solute (ES) of each mineral phase is computed at each point along the flow path. The term saturation is the ratio between the concentration in solution and the concentration at equilibrium and is used to determine if a solid is undersaturated (will not form) or supersaturated (will eventually form).

$$Saturation\,(S) = \frac{\boldsymbol{IAP}}{\boldsymbol{K_{sp}}} = \frac{Ion\,Activity\,Product}{Solubility\,Product\,Constant} \propto \frac{C}{C_0} \tag{6.3}$$

The variables C and C_o are the measured concentration and concentration at saturation, respectively. The oilfield frequently uses the names Scale Tendency (ST) and Scale Ratio (SR) interchangeably for the above saturation term.

$$Scale\,Tendency\,(ST) = Scale\,Ratio\,(SR) = Saturation\,(S) \tag{6.4}$$

The excess solute is the mineral mass that precipitates from solution. It is usually reported in units of mass rate (g/day or lb/day) or in concentration (mg/L). The terms

Ionic Activity Product (IAP), Scale Tendency (ST), Sale Ratio (SR), Saturation (S) and Scale Index (SI) are explained in greater detail below.

Referring to Chap. 1, the solubility product K_{sp} of an ionic compound in water in solid–liquid equilibrium $aA_{(s)} \leftrightarrow bB_{(aq)} + cC_{(aq)}$ is given by:

$$K_{sp} = [B]^b * [C]^c \tag{6.5}$$

where [B] and [C] are molar concentrations of ions B and C. Under non-ideal conditions the concentration terms are replaced by activity terms such that:

$$K_{sp} = a_B^b * a_C^c \tag{6.6}$$

where a_B and a_C are activities of B and C.

For non-ideal solutions under non-equilibrium conditions, the terms Ionic Activity Product (IAP) is defined as:

$$IAP = a_B^b * a_C^c \tag{6.7}$$

As explained above, the Scaling Tendency (ST), which are also alternatively called Scale Ratio (SR), or Saturation (S), is the ratio of IAP and K_{sp}:

$$S = IAP/K_{sp} \tag{6.8}$$

The term Scale Index (SI) is the logarithmic form of the saturation equation and is expressed as:

$$Scale\ Index\ (SI) = Log_{10}(S) \tag{6.9}$$

When $S < 1$ and $SI < 0$, the solution is sub saturated and there is not scaling possibility.

When $S = 1$ and $SI = 0$, the solution is in solid–liquid equilibrium and saturated.

When $S > 1$ and $SI > 0$ the solution is super saturated, and it will scale the piping and equipment.

Since production conditions change over time, scale risk analyses are performed many times during the life of the well. Production changes observed include decreasing reservoir pressure, increasing volumes of water, and use of enhanced recovery techniques like gas lift or water injection. A more detailed description of steps taken during a scale study or scale assessment is presented below.

6.3.2.1 Gathering Data

Good experimental data is critical to any scale study. The most important information is the produced water (PW) analysis. This analysis measures the concentration of salinity ions (Na^+ and Cl^-), scaling ions (Ca^{2+}, Ba^{2+}, Fe^{3+}, and SO_4^{2-}), alkalinity

(HCO_3^- and other weak acids), and other key measurements (total sulfide and total organic acids) (Note: complete speciation and common ion charges are not shown so that the user reads this as the total amount of each dissolved metal or anion). A scaling study without a good analysis of produced water is not practical because there is no effective way to recreate a water composition without certain essential measurements.

Also important are the conditions under which the water sample is taken. Volatile components like CO_2, H_2S, and organic acids partition across the water-hydrocarbon (HC) boundary as temperature and pressure vary. Thus, knowing the sampling location, and the temperature and pressure conditions where the produced water was last in contact with hydrocarbon is important. A common sampling location is at or near the gas–liquid separator (see Fig. 6.7). A produced water sample taken here, where temperature and pressure are moderate, will have a different composition than that of a sample taken upstream, like at downhole (DH) conditions.

Conditions where one measures fluid properties like pH and conductivity also affect the values. The pH of a produced water sample will drift higher when left out at ambient temperature and pressure. This is because CO_2 degases and solids precipitate. For this reason, produced water samples containing dissolved gas like CO_2 and H_2S should be kept in a pressurized vessel so their concentrations can be measured accurately. Otherwise, they will degas once the sample is depressurized, and the characteristics of the water will change.

Adding a hydrocarbon (HC) analysis to the scaling study is important when the water content is low. The HC phases are sources and sinks for CO_2, H_2S, H_2O, and organic acids. For example, in fluids with a small amount of water relative to oil and gas, a relatively larger fraction of these components will be in the HC phase. As temperature and pressure change, these components may partition from the hydrocarbon into the water phase, or vice versa. This partitioning will change the produced water properties and therefore the scaling risk. For example, in low water-content wells, where the temperature is constant but the pressure decreases, H_2O will evaporate from the water into the gas phase and cause solids to become supersaturated (Nieuwland and Collins 2004).

Also important to an effective scale study is accurate temperature (T) and pressure (P) versus depth data. Since T and P do not change linearly with well depth, obtaining these conditions at multiple locations makes the predictions more accurate and relevant to specific locations in the process.

Accurate flow rate data is also important. Flow rates are used to calculate overall mass balance and the mass of scale that may form. Hydrocarbon flow data is reported in standard units Sm^3/day, Std bbl/day or Std ft^3/day. Standard conditions for the oil and gas are 15.6 °C and 1.013 bar pressure (ASTM 2019) and 15.0 °C and 1.013 bar pressure, respectively (ISO 1996). Reporting flow at standard volumes makes it straightforward to convert the hydrocarbon into mole and mass flows.

Lastly, information on the reservoir mineralogy can be useful. The solid–liquid equilibrium between the reservoir minerals and the multiphase fluids help to establish the concentration limits of dissolved ions like calcium, iron, carbonate, and sulfate.

There is a certain amount of rigorous work involved with validating data including confirmation that there are no missing species, cross-referencing data with offset well data (when available), reviewing the flow diagram, and confirming the process

Table 6.2 Essential data needed to develop a scaling threat assessment in a produced water

Cations	Anions	Gas	Other properties	Conditions and flows
Na^+ Ca^{2+} Mg^{2+} Ba^{2+} Fe^{3+}	Cl^- $SO_4{}^{2-}$ $HCO_3{}^-$ HS^- (or total sulfides)	p_{H2S} p_{CO2}	TDS (measured only) pH Density Conductivity	**T and P** at sampling conditions **T and P** at reservoir conditions Gas, Water, Oil flow rates

Table 6.3 Secondary analyses that can be useful when assessing the scale risk in some oilfield waters

Sr^{2+}	If $SO_4{}^{2-}$ is high (possible $SrSO_4$ scale)
Pb^{2+} and Zn^{2+}	If Sulfide scales are a risk and the reservoir brine is known to contain these metals
Organic acids	If alkalinity is high and $CaCO_3$(s) is calculated to be highly supersaturated
PVT or crude assay (Note 1)	If crude oil is produced
Condensate analysis	If gas condensate is produced
Br^-	To review the Cl:Br ratio and estimate if the brine is derived from a halite reservoir (Note 2)
Formation minerals	If the whole fluid is to be saturated with the minerals in the formation

Notes
1. PVT assay is the ratio of crude oil phase volume downhole under reservoir pressure and temperature conditions to that on the surface
2. Halite reservoirs are made of sedimentary rocks that are formed by evaporation of seawater leaving sodium chloride (salt) and other minerals

conditions and mixing scenarios information. Completing this part of a fluid study takes a sizable portion of the entire scale assessment. Table 6.2 contains the essential data needed to perform a scaling study.

Table 6.3 contains additional measurable data that will improve the accuracy of the scale study but is not necessarily essential to completing a basic study.

6.3.2.2 Alkalinity Measurement and Its Importance

Alkalinity of water is its capacity to buffer itself from change in pH when mixed with an acid or base. A more detailed definition of alkalinity has been presented in Chap. 1 and also can be found in aqueous chemistry textbooks (Stumm and Morgan 1996). In the case of groundwater and surface water, buffering is primarily provided by bicarbonate ions (as discussed in Chap. 1). However, produced water contains a large variety of minerals dissolved in high concentrations. Its total alkalinity is contributed by a wide range of species dissolved in water including bicarbonate, carbonate, bisulfide, borate, acetate, propanoate, and others proton-accepting species.

An alkalinity measurement is performed by titrating the water sample with a strong acid (usually a standard solution of H_2SO_4 or HCl) until the pH decreases to about 4.5. Any species in water that accepts/complexes with a H^+ ion when the acid is added contributes to the total alkalinity value. This action, accepting/complexing with a H^+ ion, is referred to as acid buffering. Buffers are species that limit the change in pH of a water when an acid or base is added to it. Buffers are essential to natural and biological systems, and the reader is encouraged to read further about this topic.

Since each species has different affinities for an H^+ ion, its contribution to a water's alkalinity depends on the water's starting pH. Figure 6.3 shows the concentration of several buffering species as a function of pH. The shaded area is the typical pH range over which an alkalinity titration occurs. The right boundary is the upper pH recorded for most types of produced water. The left boundary is the alkalinity titration endpoint, pH = 4.5 Methyl Orange end point, see titration curve, Fig. 1.5, chap. 1. Species that contribute to alkalinity decrease in concentration within this region. Bicarbonate and bisulfide contribute almost 100% of their concentration to alkalinity and are therefore excellent buffers in this pH region. Most of the acetate ion is consumed by the end of the titration, but not all. To consume all the acetate buffer, the endpoint pH would need to be about 2.5. Thus, only a portion of the acetate contributes to a produced water alkalinity. Borate and carbonate, which react with acid at high pH, contribute little to the alkalinity, because their concentrations are already low before the titration experiment is even started.

The starting concentration for each species in Fig. 6.3 is 1 mM (1 mm/L) and so the contributions of HCO_3^- and HS^- to the total alkalinity in Fig. 6.3 is the same—each will consume ~1 mmol of acid down to a pH of 4.5. The concentration of bicarbonate in actual produced water, however, is much higher than bisulfide,

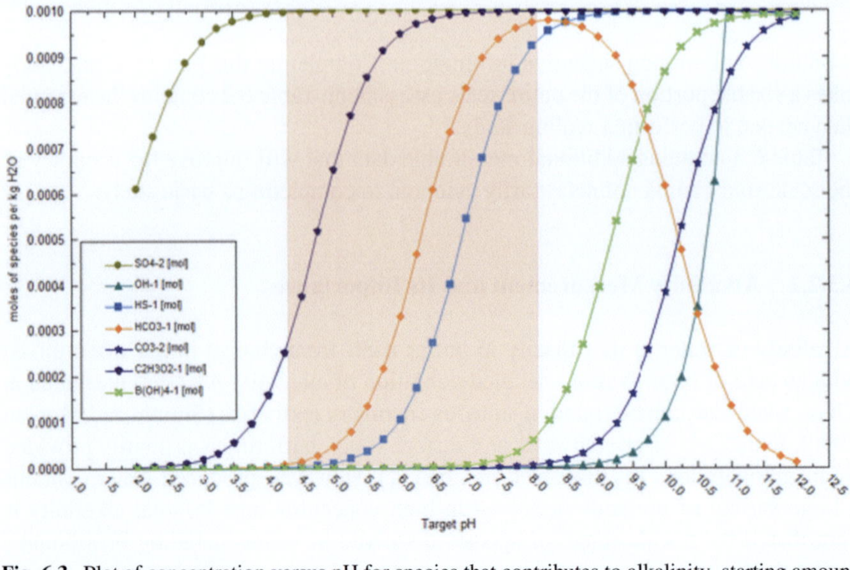

Fig. 6.3 Plot of concentration versus pH for species that contributes to alkalinity. starting amount for each species is 1 mM

and in most cases, higher than organic acids. Therefore, the total alkalinity value is regularly converted to a bicarbonate concentration. This conversion is fine, so long as the concentrations of bisulfide (HS^-) and acetate ions (and other organic acid anions) are low. If their concentrations are high, then a separate organic acid measurement or a separate inorganic carbon measurement is needed. There are methods published in literature that describe how to separate bicarbonate alkalinity from interfering ions (Tomson et al. 2005).

6.3.2.3 Choosing a Credible Thermodynamic Tool

The above analysis and production data can be combined into a scale prediction tool. This tool should be capable of

- Calculating phase equilibrium when four phases are present: gas, liquid (brine or produced water), oil, and solids
- Predict properties across a broad temperature and pressure range
- Predict the activity coefficients of species at varying salinities
- Predict the proper speciation of ions at different compositions,
- Predict a gas-phase fugacity as high pressures.

Figure 6.4 shows barite ($BaSO_4$) solubility as a function of temperature from 25 to 250 °C and NaCl concentration from 0.01 to 5 M. Barite's solubility can differ by up to 30×, depending on differences in water salinity and temperature. This plot illustrates

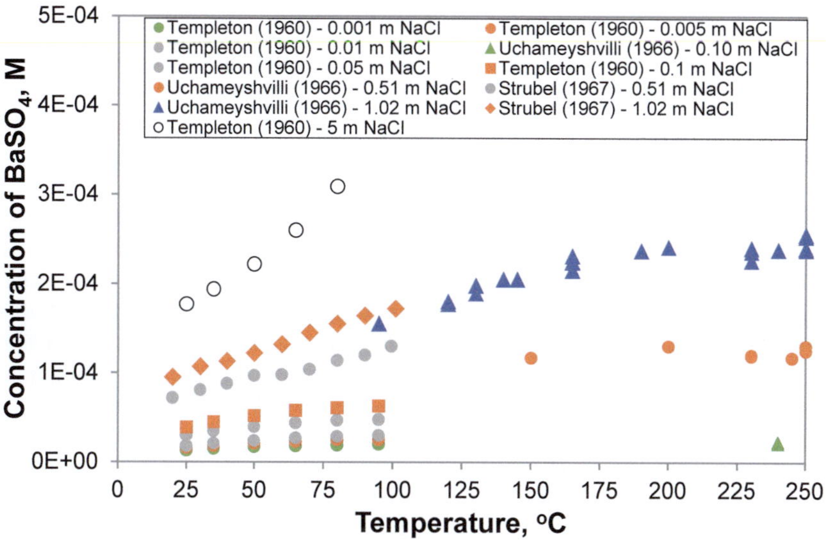

Fig. 6.4 Barite ($BaSO_4$) solubility as a function of temperature and salinity. Experimental data retrieved from multiple authors

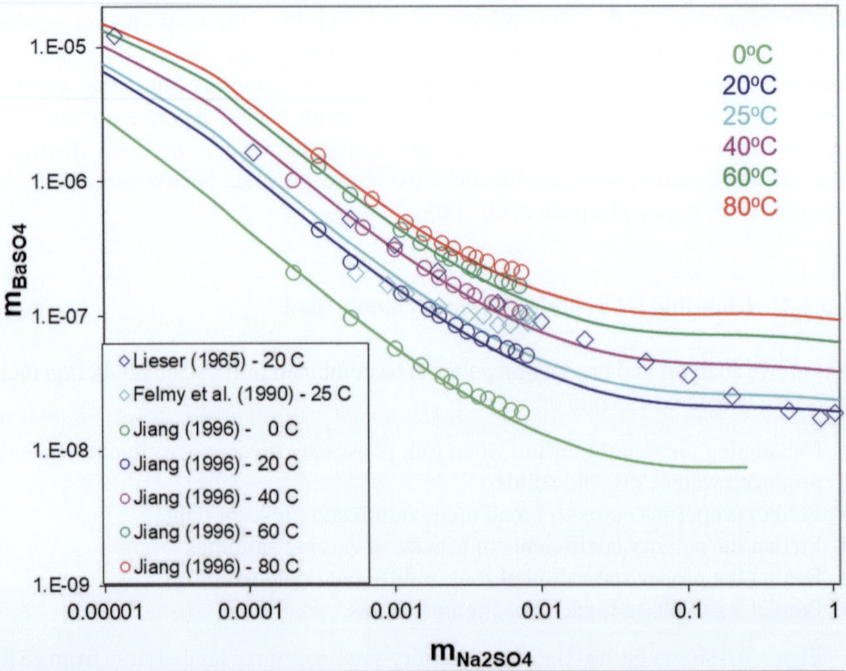

Fig. 6.5 Barite solubility at different temperatures as the Ba^{2+}:$SO_4{}^{2-}$ Molar ratios change.[2] Symbols are experimental data. Lines are model predictions

why it is important to measure the salinity of produced water and the need for a modeling software to predict solubility at a given range of temperature and pressure.

Figure 6.5 is a plot of barite solubility as the Ba:SO_4 molar ratio in solution varies. As expected, the $BaSO_4$ solubility decreases as SO_4 is added, but the decrease is not linear. This is because dissolved species interact with each other and change the properties in an unexpected way. This illustrates why the modeling tool that can predict such activity and speciation behavior is important. If a simple scaling model were used, the $BaSO_4$ solubility curve would decrease linearly as Na_2SO_4 increases, and the model would over-predict the amount of scale that forms in a produced water.

Figure 6.6 is the measured CO_2 solubility in water vs. pressure. The dashed line is CO_2 solubility predicted by the Henry's equation and the symbols are experimental data. As pressure increases, CO_2 solubility increases, but its solubility becomes non-linear above 100 bar. The reason is that gas molecules interact at high pressure and this interaction changes their properties. A modeling tool like a Henry's equation does not capture such high-pressure effects and if used, will over-predict CO_2 solubility. This in turn will cause a model to under-predict $CaCO_3$ and $FeCO_3$ scaling tendencies. It is for this reason that a model that can predict the *fugacity* of a gas is important for scale risk studies.

[2] Image provided by OLI Systems. reference cited based on their investigation.

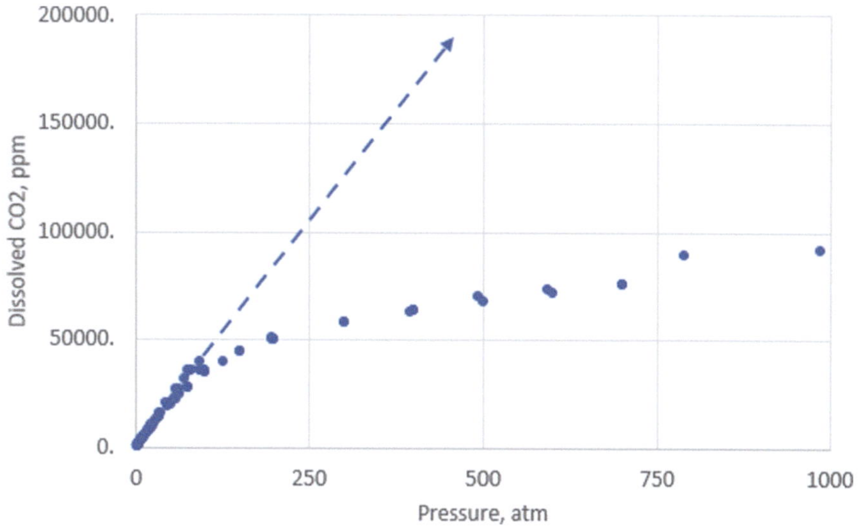

Fig. 6.6 Carbon dioxide solubility (symbols) in water at the bubble point pressure. The line is the solubility predicted by Henry's Law.[3] (*Note* Bubble point pressure is the pressure in the reservoir at which the first gas bubbles come out of the crude)

6.3.2.4 Stepwise Procedure for Creating a Representative Fluid

Step 1: Reconciling a Produced Water Analysis

There are several modeling adjustments that can be made to a produced water analysis after it is entered into a scale prediction model. The first step is balancing charges. Produced water analyses are observed to be charge imbalanced because not every species is measured, and because there is a margin of error with analytical methods. Balancing charge is usually a matter of increasing or decreasing the Na^+ or Cl^- concentrations. This is reasonable because most produced waters are derived from seawater. The computed conductivity can be compared to the measured conductivity to determine if Na^+ or Cl^- needs to be increased or decreased.

A second adjustment is to bring into alignment the calculated and measured properties, like pH, alkalinity, total dissolve solids, density, and conductivity. The pH is often an unreliable measurement because it can drift over time due to various reasons including degassing of CO_2, H_2S and other volatile components, as well as precipitation of solids. The alkalinity is sometimes insufficient because there may be measurable quantities of organic acids in the produced water, and this interferes with the experiment's purpose to measure the dissolved carbonate concentration (see above discussion). Table 6.4 contains three adjustment approaches based on what data is available.

---

[3] Image provided by OLI Systems. No data reference shown.

Table 6.4 Reconciling a water analysis when different sets of data are available

Scenario	Use these measurements	Adjust these concentrations	Calculate these concentrations
No TFA and pH is reliable	Alkalinity + pH	CO_2, H^+	$HCO3^-$ + CO_2
TFA is present and pH is reliable	Alkalinity + TFA + pH or Alkalinity + TIC + pH	CO_2, H^+ or Acetic acid, H^+	$HCO3^-$ + CO_2 TFA^- + HCO_3^- + CO_2^0
TFA present and pH is unreliable	Alkalinity + TFA + P_{CO2} or Alkalinity + TIC + P_{CO2}	HCO_3^- or Acetic acid	HCO_3^- + CO_2 + pH TFA^-, TFA + CO_2 + pH

Notes

TFA = Total Fatty Acids, a term that refers to the organic acids found in oilfield brines, e.g., formic, acetic, propionic, butyric, etc.

TIC = Total Inorganic Carbon, a measurement used to distinguish inorganic carbonate and bicarbonate from dissolved organic carbon. This is an important measurement when determining total carbonate alkalinity in a brine that contains appreciable amounts of organic acid (organic acids interfere with the carbonate alkalinity titration)

Inaccurate HCO_3^- concentration leads to an incorrect calcite ($CaCO_3$) and siderite ($FeCO_3$) scaling risk. Thus, efforts to compute a more accurate HCO_3^- concentration reduces prediction uncertainty. Also, since the pH measurement is heavily dependent on conditions (temperature, pressure, contact vapor composition), relying on pH alone for computing a representative water may lead to inaccurate results.

Step 2: Entering the Hydrocarbon Analyses

Hydrocarbon analyses are important inputs. They are easy to enter. They are comprised of pure components (C1 to Cn species), pseudo-components, distillation curves, inorganic gases (CO_2, H_2S, N_2, H_2), and reactive organics (mercaptans, organic acids, etc.). Pseudo-components are a number of similar components grouped together as a single component. They are fractions of the crude oil with its own set of properties. Each pseudo-component has a defined molecular weight, boiling point, specific gravity, and critical properties. Thermodynamic software uses pseudo-components to simulate the oil properties. This is because crude oils contain hundreds of compounds, and it is impractical to measure each of them.

Hydrocarbon analyses are often reported as individual gas and oil analysis or as a multiphase composition. Each of these reports can be used, depending on the capability of the software tool. In most cases, dissolved H_2O in gas and oil is not reported. This missing mass directly affect the scaling study if the water content is low.

Step 3: Creating a Representative Fluid

The end goal is to create a representative, multiphase fluid in liquid–gas-solid equilibrium, and with the correct mass flow rate. Figure 6.7 is an image of a production well.

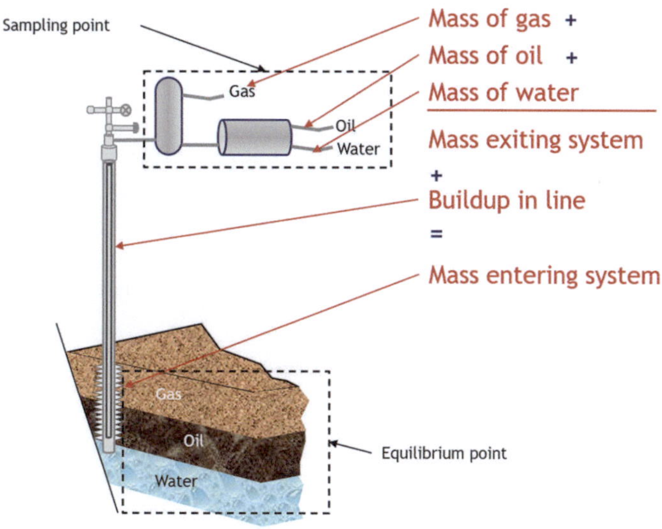

Fig. 6.7 Mass balance of fluid as it enters the well from the producing zone and when it exits the system at the surface

The well has a mass inlet (reservoir) and a mass outlet (separator). Phase volume rates are measured at the surface and are converted to phase mass using density measurements or predictions. The material lost from the reservoir to the surface is the accumulation in the well. Computing the deposition rate of the fluid across the production profile helps determine this accumulation.

Creating a representative fluid starts by combining the water and hydrocarbon analyses at their reported flow rates. The production rates are measured downstream of the separation unit. Separation is at a given temperature and pressure, and these conditions are important when reconstructing the fluid. These volumetric flows are converted to mass flow rates using calculated densities at standard conditions.

Two key parameters are the reservoir temperature and pressure. A common assumption made in scale prediction studies is that the fluid is in equilibrium with the reservoir minerals when it exits the formation. This assumption is somewhat asymmetric, in that it is not useful for sub-saturated minerals, but it serves as a boundary condition when dissolved ion concentrations exceed mineral solubility. This assumption has flaws as it assumes a homogenous, isotropic rock matrix and phase equilibrium in the vertical column of the liquids However, it does provide a reasonable bounding for minerals like NaCl because it can be used to back calculate the missing H_2O from the gas and oil analyses. For instance, if H_2O is not added to the hydrocarbon analyses, then there is a deficiency in the fluid's mass balance. Consequently, H_2O will partition from the aqueous to the hydrocarbon phase when the calculation is run. This partitioning erroneously increases the ion concentrations. Since NaCl is unlikely to be supersaturated because it precipitates rapidly, its solubility envelope can be used as an indication that H_2O is missing.

Table 6.5 Produced Water analysis for example case study

Cations (mg/L)		Anions (mg/L)		Neutral constituents (mg/L)		Measured properties	
Na^+	15,260	Cl^-	24,500	$B(OH)_3$	107	Temperature	25 °C
K^+	180	SO_4^{2-}	160	SiO_2	60	Pressure	1 atm
Ca^{2+}	850	HCO_3^-	See Alkalinity	–	–	pH	6.7 su
Mg^{2+}	135	Acetate$^-$	386	–	–	Alkalinity	673 mg/L
Sr^{2+}	65	–	–	–	–	Density	1.026 kg/L
Ba^{2+}	1.3	–	–	–	–	Conductivity (mho)	–
Fe^{2+}	2.1	–	–	–	–	TDS (mg/L)	–

6.3.3 *Case Study: Predicting Scaling Risk in a Typical Oil and Gas Well*

6.3.3.1 Background

An on-shore well is producing a light oil and some associated gas. The reservoir is a calcite-cemented sandstone at moderate temperature and pressure. The produced water contains about 45,000 mg/L TDS.

Table 6.5 presents an example of a produced water analysis. It lists the produced water composition. It was sampled at the separator and measured in the laboratory at ambient conditions.

6.3.3.2 Water Analysis

The water analysis is entered into a scale prediction software package (Fig. 6.8). The electroneutrality is balanced automatically by adding 625.2 mg/L Cl^- to the input value (25,125.2 vs. 24,500 mg/L). Note that the electroneutrality is determined by adding the concentration of individual cations and anions by their respective equivalent weights and making sure that the total cationic charges equal the total anionic charge.

For example, in this case, the charge for Na^+ ions = concentration of Na^+ ions/eq. wt. of Na^+

= (15,260 mg/L)/(23 mg/meq), i.e. 663.48 meq/L.

The charge of SO_4^{2-} ions = concentration of SO_4^{2-} ions/eq. wt. of SO_4^{2-} ions.

= (160 mg/L)/(96/2 mg/meq) i.e. 3.33 meq/L.

The simulator calculates all the negative and positive charges contributed by the anions and cations as shown above and, to achieve the electroneutrality, automatically predicts the concentration of chloride if there is a deficit in the total anionic charge, and sodium if there is deficit in the total cationic charge. In this example, the deficit is in the total anionic charge which has to be compensated for by adding 625.2 mg/L of additional chloride ions to the reported concentration of 24,500 mg/L.

The HCO_3^- concentration was not measured directly. Rather, the lab measured the total alkalinity. The total alkalinity value is the contribution of HCO_3^- and acetate ($C_2H_3O_2^-$). The user can estimate the HCO_3^- concentration from the total alkalinity and acetate (−1) values or, depending on the software package, have the software calculate it automatically. In this example, we will do both, estimate the HCO_3^- concentration and then have the software correct the value. Figure 6.8 shows that 300 mg/L HCO_3^- is added as an initial guess. The guess is based on subtracting the value 386 mg/L (acetate concentration) from the 673 mg/L value (alkalinity measurement). Although this is not an equal subtraction, since only a fraction of acetate is neutralized in the alkalinity titration, it is a reasonable starting point. We want a value entered because it also affects the charge balance.

There is a systematic approach to studying scaling risk in a well. Scaling software programs can examine the potential for scaling risks from produced water for a large number of scenarios. An example presented in Fig. 6.9 demonstrates a typical set of scaling studies that can be conducted by the OLI software. However, there are several instances where the data is uncertain or where assumptions need to be made. As a result, scale prediction studies have a degree of uncertainty. Good thermodynamic modeling software can mitigate the impact of these uncertainties.

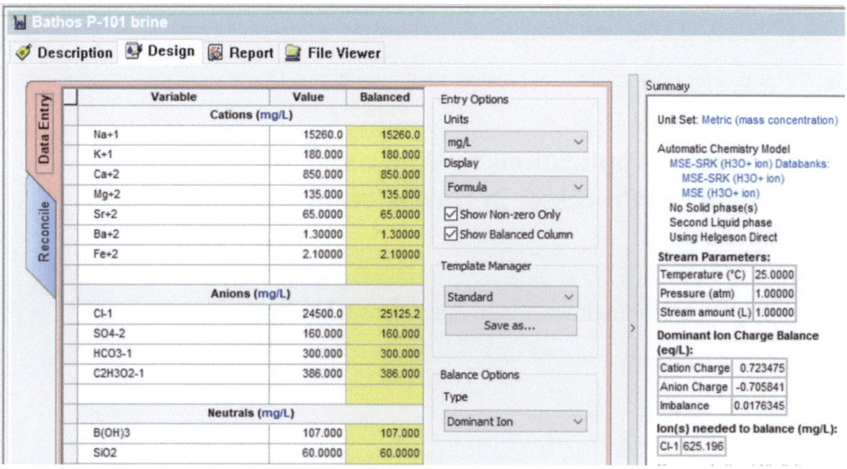

Fig. 6.8 Produced water dissolved ionic composition

Basic scale study for a single well	Scale prediction is from the reservoir to the surface using the current temperature, pressure and production profile. The saturation ratio and excess solute for all relevant solids is reported by the software.
Early to Late-life study on a single well	This scale study is for different brine ratios (formation water to oil ratios in the reservoirs) at the temperature and pressure profiles. A single gas and oil analysis is used for all water compositions. Each scale scenario has a different temperature and pressure profile.
Simulation study with gas lift	Gas injection is done as a means of Enhanced Oil Recovery (EOR) to boost production. Scale prediction is done from the reservoir to the gas injection point. Then the scale prediction with the fluid + gas mass rate is calculated from the injection point to the surface
Injection water breakthrough	An injection water stream is equilibrated with reservoir minerals at the reservoir temperature and pressure. This equilibrated brine is added to the production fluid at varying ratios (e.g., 10 to 90% of injection water). Scale prediction from the bottom of the well to the surface using the temperature and pressure profile is then run.
Multiple fluid scale predictions	Complex projects generally include the following: 1) Multiple wells gathering a specific location, 2) Many waters (produced or industrial) with variable flow rates mixing at one or more surface ahead of an injection or water treatment plant, 3) A water injection project with multiple injection and producing wells, 4) A water treatment plant including filtration, evaporation, reverse osmosis, ion exchange, and hardness removal.
Shale Production in a field	A single oil analysis is used to simulate production in a hydrofractured field. The brine is specific to the well and the gas is specific to the well. Rod pumps are used in some wells, and these pumps are part of the scale and corrosion scenario.
Production over time	A single gas, brine, and oil analysis are used to set the original conditions in the well. As the well is produced, the water cut changes and the fluids are re-saturated. A new brine is computed from this re-saturation and the scaling studies are done on the new equilibrated fluid. Lift gas is added as the water cut increases and as the reservoir pressure decreases. This gas addition is part of the scale study that is performed using the Facilities calculation tool.
Multi-well production from a single field	Separator water analyses from two wells are entered into the software. A single gas and oil analysis are used for both wells. Fluid saturation is performed for two dates. Early (low water cut) and mid-life (moderate water cut). Facilities and contour diagrams are used to determine the scaling risk
Single well production with abundant data	Separate oil and gas analyses are provided for multiple gas-oil-ratios (GOR). Each has a different composition. The gas, oil, and brine are combined in the reservoir to produce a whole fluid. Each reservoir calculation represents a different gas-oil-ratio and a different water cut. A scale tendency calculation is run on each combination.
Halite scale in evaporative production	A stock-tank oil and gas are provided also a single-phase oil and gas analysis. Stock tank oil and gas is the oil after it is stabilized in stock tanks upon vaporization of gas in storage temperature and pressure conditions. The brine and hydrocarbon fluids are saturated to create a whole fluid. Then additional saturator blocks are used to evaporate the brine to the point where halite precipitates
Multi-field study	Water analyses from multiple reservoirs are entered. Whole oil analyses are created for several separate samples, some from the same reservoir but different wells. Gas analyses are provided for a few wells. Reservoir saturation calculations are done for specific wells. Scale studies for each well is done over a five-year period and each scale study is done with a different water cut (composition).

Fig. 6.9 Scenarios of produced water scaling studies conducted by upstream process simulators

6.4 Modeling Applications in Water and Wastewater Treatment Processes in Downstream Refining Operations

Wastewater treatment in downstream operations include:

Preliminary treatment

- Influent Screens
- Desalters
- Sour Water Strippers

Primary Oil–Water separation

- API Separators
- CPI Separators

Secondary Oil–water separation

- Dissolved Air Floatation (DAF) Separator
- Dissolved Gas Floatation (DGF) Separator
- Induced Air Floatation (IAF) Separator
- Induced Gas Floatation (IGF) Separators

Biological treatment

- Suspended growth processes
  - Conventional Activated Sludge
  - Extended Aeration
  - Sequencing Batch Reactor (SBR)
  - Membrane Bioreactor (MBR)
- Attached growth processes
  - Trickling Filters
  - Rotating Biological Contactors (RBC)
  - Moving Bed Biological Reactors (MBBR)
- Hybrid process—Integrated Fixed Film Activated Sludge (IFAS) process

Many of the process equipment in the downstream sector are designed based on proprietary technology and in-house knowledge of the equipment manufacturers. For a few others e.g., API and DAF/DGF separators, preliminary sizing is done using standard equations available in the literature or engineering standards. A procedure for API separator design based on API Publication 421 is presented in Chap. 5. These utilize simple spreadsheet-based calculations for preliminary design. However, there are a few that follow significant process design steps before they are sized and installed. Most notably these include the sour water strippers and biological treatment reactors. Process design of these systems require solving complex and iterative mathematical equations with the help of computer aided process simulations. A procedure for these designs is discussed in the following sections.

6.4.1 Simulation Overview for Sour Water Stripping

The function and working principle of sour water strippers have been introduced in Chap. 5. It is discussed here in further detail to set the stage for the step-by-step procedure for the design of these strippers.

A Sour Water Stripper (SWS) is a trayed or a packed column that uses live steam or heat from a reboiler to remove H_2S, NH_3, and CO_2 from wastewater. The SWS is critical to the operation of the refinery as it ensures that the water quality is acceptable for water reuse, wastewater treatment plant operations, and sulfur recovery units. For example, H_2S concentrations in wastewater that are too high can significantly increase oxygen demand in the biological wastewater treatment plants, causing upsets in the plant and personnel safety concerns. Similarly, too high NH_3 concentrations in wastewater discharges will impact aquatic life in the receiving water bodies due to toxic effect and high oxygen demand. However, some nitrogen is required as it provides nutrients to the biological system. In summary, high concentration of H_2S and NH_3 drive both the capital and operating costs high to meet the requirements for their removal by oxidation in biological wastewater treatment plants. These have been discussed in detail in Chap. 5.

Modelling the SWS is critical for the stripper design, operation, and troubleshooting. Many challenges exist when modeling this aqueous-based tower as the chemistry is complex; the system contains both H_2S, a weak acid, and NH_3, a weak base, creating a scenario where simple changes in water quality and pH have large impacts on removal efficiencies. Additionally, because of the complex interactions between water quality and removal efficiencies, the modelling of the SWS becomes even more essential. For the following discussion the sour water stripper is considered as a column with trays as opposed to a packed column. Historically using stage equilibrium modeling methods, SWS tray efficiencies vary with estimates between 15 and 45%, indicating that the system rarely approaches equilibrium. To model the SWS more effectively, one should use an electrolyte-based modelling software with a mass transfer rate-based approach. Typical mass transfer models use the Two-Film Theory, which accounts for the diffusion of gases through the vapor and liquid phase at steady state and does not assume equilibrium. This improves the efficiency predictions as gas/liquid concentrations are calculated on each tray and are not assumed to be the same across the whole tower. An example of modeling a SWS with mass transfer is included in the 6.4.1.5. Process Modeling of Refinery Sour Water Stripper using Electrolyte Chemistry Model later in this chapter.

6.4.1.1 Chemistry of Sour Water and Its Effect on Stripping Process

Unlike most fractionation towers in the oil refinery which are largely hydrocarbon-based, the SWS is an aqueous system driven by acid–base chemistry and mass transfer rates. Sour water contains both ammonia (NH_3) and acid gases (H_2S and CO_2). Ammonia forms ammonium ion in aqueous solutions having a pKa value of 9.25 (at 25 °C):

$$NH_3 + H^+ \leftrightarrow NH_4^+$$

In contrast, H_2S and CO_2 act as weak acids in aqueous solutions. H_2S ionizes to HS^- with a pKa = 7.1 and CO_2 ionizes to $HCO_3{}^-$ with pKa = 6.35 (at 25 °C):

$$H_2S \leftrightarrow H^+ + HS^-$$

$$CO_2 + H_2O \leftrightarrow H^+ + HCO_3^-$$

Hydrogen sulfide removal at acidic pH, as found in sour water, is efficient and usually occurs on the first several trays of the SWS. In contrast, the protonated ammonium ions, as found at the same pH range, is non-volatile and will largely remain in the liquid phase, therefore it is quite common to inject a small amount of a strong base (NaOH) to shift the pH into a range where ammonium ion, NH_4^+, shifts back to the volatile NH_3. pH adjustments are usually made onto a tray far enough down the column where most of the H_2S has already been stripped out and ammonia is the main remaining component (Weiland and Hatcher 2012).

6.4.1.2 Process and Single Point Source Modeling

Modelling the pH-dependent reactions found in a SWS requires that one considers acid–base chemistry, vapor–liquid equilibrium, and mass transfer rates. The thermodynamic framework used to model these systems should be closely considered as chemical speciation and phase equilibria play key roles in the stripping performance. Both process modeling and single point analysis are necessary to evaluate operations, scaling tendencies (fouling), and corrosion risk.

6.4.1.3 Process Flow and Chemistry Modelling

Process modeling is a valuable tool in ensuring energy efficient operations while meeting downstream process unit requirements. Flowsheet process modeling software using complex feed water chemistry, operating conditions, and tray internals specifications to calculate phase equilibria, pH, chemical speciation, and mass transfer should be used for these systems. The process model includes the sour water stripper column, overhead systems, reboiler or steam injection, heat integration, and other ancillary equipment. Once built, the process model is used to:

- Design, analyze, and optimize the SWS tray number, required heat duty via the reboiler or steam injection, and overhead systems;
- Evaluate overhead and bottoms streams for compatibility in downstream process units (i.e., corrosion and fouling risks);
- Determine if caustic injection is required and the proper injection location/dosage;
- Troubleshoot operations and optimize existing units.

6.4.1.4 Modelling for Fouling and Corrosion Risk

The SWS is highly susceptible to fouling, which is why often trays are used instead of packing. A detailed comparison of packed versus tray columns for sour water stripping has been presented in Chap. 4. The operating pH has the unintended consequence of solid precipitation in complex sour water streams. Thermodynamic analysis by electrolyte modeling (using a modeling software) of these streams at varying water qualities and operating conditions (temperature, pressure, pH, and composition) can be used to evaluate scaling (fouling) potential. These insights allow the engineer to adjust operating conditions to avoid potential costly tower and heat exchanger fouling.

Similarly, H_2S-containing sour water comes with an increased corrosion risk. With changing vapor/liquid composition and temperatures throughout the trayed tower, corrosion risk will also change. Individual streams can be analyzed at various operating conditions, to aid in the selection of materials for trays, heat exchangers, and overhead systems. This can further optimize the tower design and establish operating envelopes that avoid highly corrosive conditions.

6.4.1.5 Process Modeling of Refinery Sour Water Stripper Using Electrolyte Chemistry Model

There are multiple references to corrosion occurring in equipment processing sour water. The purpose of this case study is to conduct a thorough study of the chemical composition of the vapor and liquid phases flowing through the unit. The flowsheet software used in this case is OLI Systems Flowsheet ESP.[4] The generated model is used to create the stripping process and to compute the composition of each stream and on each stage of the column. The corrosion engineer or material scientist can then use the computed results, plus their knowledge of corrosion, to determine if the existing process is at risk.

Figure 6.10 is the simplified process flow diagram used in this study. It represents a generic stripping column in which heated sour water is introduced at the top of a twenty-tray tower. The tower overhead is cooled, and the liquid reflux is returned to the top of the tower. The remaining acid gas exits the process, and the stripped water leaves the bottom of the tower.

In approaching how to model this system, the following six steps are used,

Step 1 Define the Chemistry Model
Step 2 Build the Process Flowsheet
Step 3 Enter Stream Conditions
Step 4 Enter Process Conditions
Step 5 Run and View Results
Step 6 Review Calculation Results.

Step 1: Define the Chemistry Model

The chemistry model is one of two components required for modeling the system. In defining the chemistry model, the composition, and the desired thermodynamic model must be defined. The second component is the process model described in Step 2. Table 6.6 defines the sour water composition used in this example.

[4] OLI Systems, Inc. is a developer of electrolyte thermodynamic software. The software is used to calculate the properties of water streams or to simulate a water-dominated process.

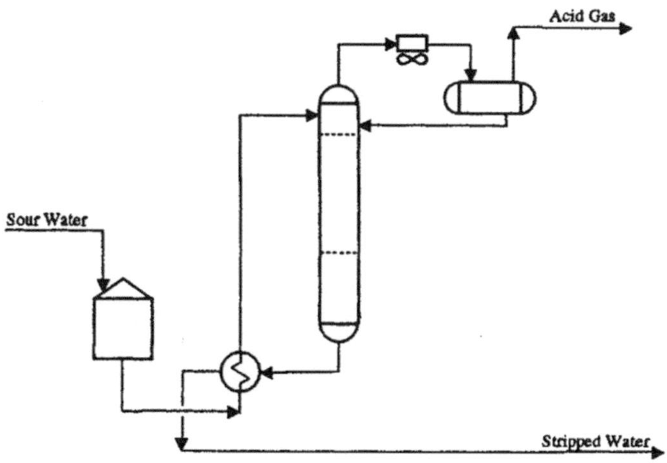

Fig. 6.10 Simplified process flow diagram of a generic sour water stripper

Table 6.6 Sour water influent characteristics

Influent chemistry	Concentration (mg/L)
H_2O	Solvent
H_2S	8000
NH_3	6000
CO_2	3000
SO_2	1500
NaCl	100
$NaHCO_3$	100
$CaCl_2$	100
$MgCl_2$	100
KCl	100

In modeling a Sour Water Stripper, the appropriate electrolyte modeling method, in this case the Mixed Solvent Electrolyte (MSE) framework, must be selected. The MSE thermodynamic model is the state-of-the-art calculation method used in electrolyte chemistry simulation. The thermodynamic frameworks available for electrolyte modeling, like MSE and E-NRTL (electrolyte non-random two-liquid), depend on the process modeling software used, thus care should be taken in selecting a thermodynamic framework for electrolyte simulation if the user is not familiar with the frameworks available in their software.

Once the appropriate thermodynamic model is selected, the chemical species used in the model can be defined (see Fig. 6.11). Here, the chemical species found in the influent sour water are defined that can be used in defining influent streams.

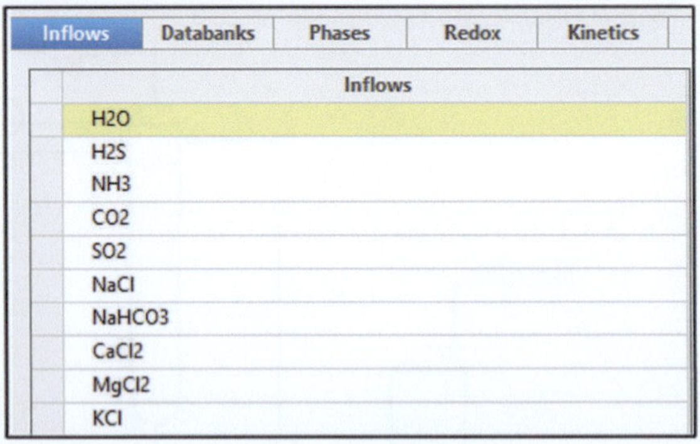

Fig. 6.11 Chemical species defined in the chemistry model

After the chemical species are defined and the thermodynamic model is selected, the software can generate the complete chemistry model: aqueous species, gases, oils, and solid phases. As mentioned in the sections above, chemical speciation is critical to accurate electrolyte simulation predictions as defining the chemistry model may be the most important step in building an electrolyte process simulation like this.

Step 2: Build the Process Flowsheet

The second step in modeling the system is building the process flowsheet. This is done by selecting the appropriate process units (heat exchangers, separators, and column) and connecting them with process streams. Figure 6.12 is a screenshot of the OLI Flowsheet ESP model used in this example.

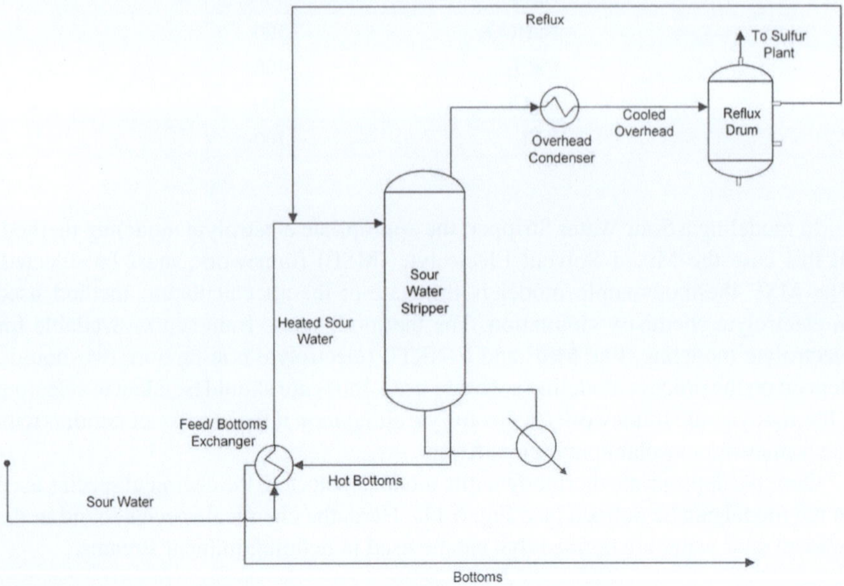

Fig. 6.12 OLI systems flowsheet esp used to model the electrolyte chemistry in a sour water stripper

Step 3: Enter Stream Conditions

After building the simulation, the inlet stream conditions are input. In this example, the only influent stream is the sour water, all other streams will be calculated when the simulation is run. To define the sour water stream, temperature, pressure, flowrate, and contaminant concentrations are required. This may be provided by the refinery if modeling is done for an existing tower, or as part of the design basis if this is a new design. Figure 6.13 is a depiction of the properties window used to define the influent sour water stream.

Parameters	
Temperature (°C)	50.0
Pressure (atm)	1.1
Total Flow (kg/hr)	25000.0
Inflows (ppm (mass))	
H_2O	9.81e5
H_2S	8000.0
NH_3	6000.0
CO_2	3000.0
SO_2	1500.0
NaCl	100.0
$NaHCO_3$	100.0
$CaCl_2$	100.0
$MgCl_2$	100.0
KCl	100.0
NaOH	0.0
Total	**1.0e6**

Fig. 6.13 The properties panel used to define the sour water influent stream

Step 4: Enter Process Conditions

After defining the influent stream(s), the operating conditions for each process unit are defined. The conditions for heat exchangers, either a discharge temperature or a heat duty, is added, in addition to pressure specifications if applicable. Additionally,

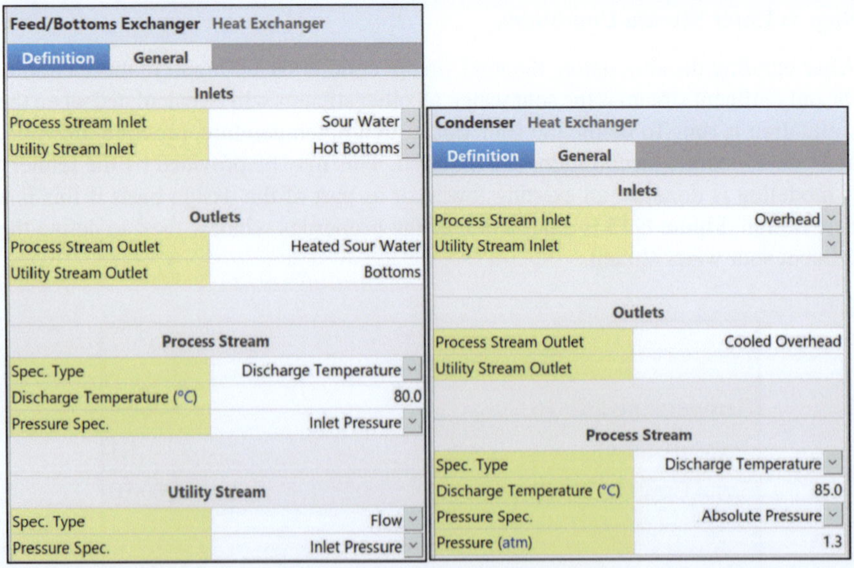

Fig. 6.14 Properties panels used to define the two heat exchangers in the sour water stripper process

a heat exchanger can use a utility stream, here the tower bottoms are used as the utility stream for the feed/bottoms heat exchanger. In other words, the heat from the bottoms stream is used to heat the feed stream. Figure 6.14 shows the properties window for both heat exchangers—Feed/Bottoms Exchanger and Condenser.

Specifying the sour water stripper column is more complex and requires that a calculation method, inlet/outlet stream locations, tower design features, and tower specifications/control schemes are appropriately defined.

In choosing the calculation method, either equilibrium or mass transfer limited, can be selected. As described in the previous section, using a mass transfer limited calculation method is more accurate than the equilibrium method for estimating stripping efficiency in aqueous systems as these systems rarely approach equilibrium.

Figure 6.15 lists the required input fields when selecting a mass transfer rate calculation. This calculation method requires certain tower internals information: a

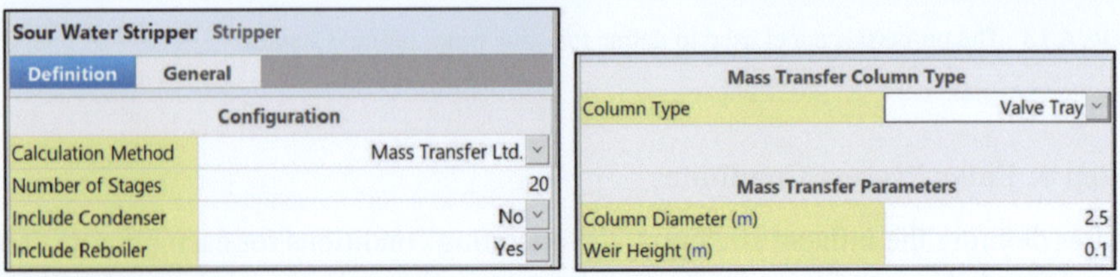

Fig. 6.15 The specifications required in defining the sour water stripper calculation method and tower design features for the mass transfer calculation

column type (tray, packing, etc.), column diameter, and tray weir height (if using trays). The tower internals are only used to estimate a liquid volume on each tray and is not intended to conduct a design evaluation. The simulation engineer should perform a deeper investigation into tower design to determine if the specifications are optimal.

Further specifications required in defining the column are the inlet and outlet locations (Fig. 6.16). In this example, the sour water enters at the top (Tray 20) and the second inlet, Reflux, also enters at Tray 20. The outlet streams include the tower

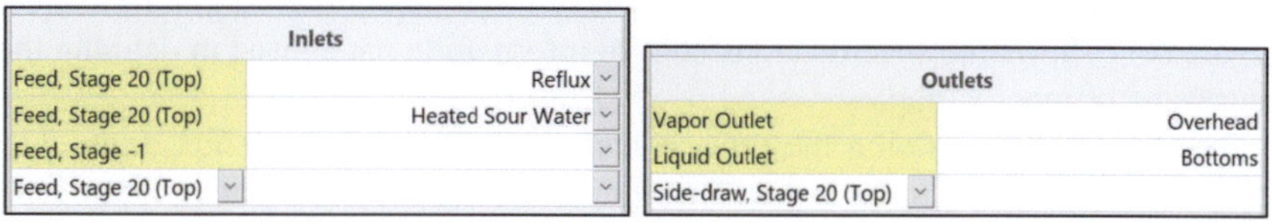

Fig. 6.16 All inlets and outlets from the column are defined in the properties tab for the sour water stripper unit

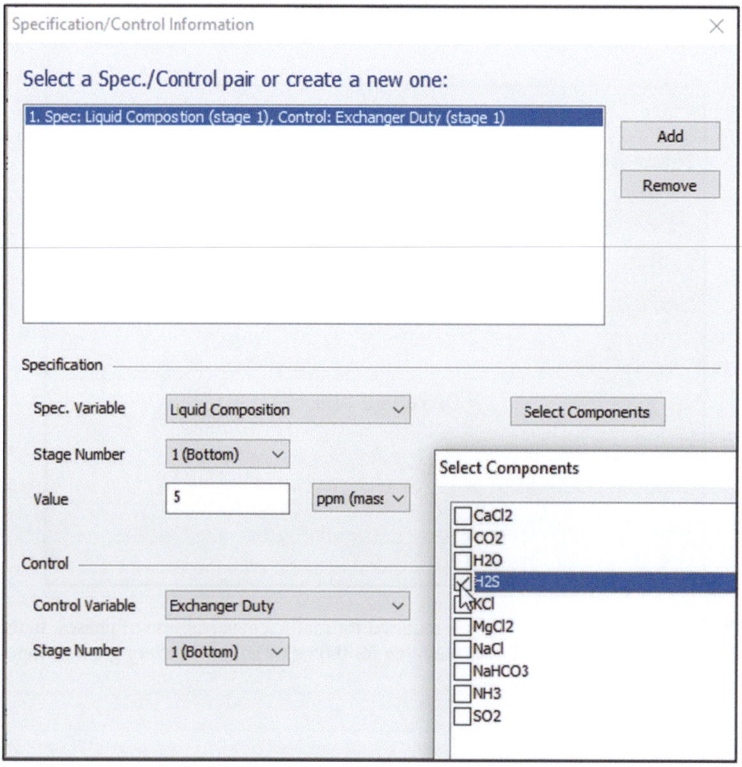

Fig. 6.17 The specifications for control panel for the sour water stripper defining the bottom water composition (5 mg/L H_2S) and the control variable—the exchanger duty

overhead and the tower bottoms, which are the vapor outlet from the tower (top) and liquid outlet from the tower (bottom), respectively.

Finally, specification for the control information can be added to the tower. This allows the user to define control variables allowing the software to calculate the required operating conditions. In this example, the process goal is to achieve <5 ppm H_2S in the tower bottoms. The control variable used to meet this is the reboiler duty. Other variables can be selected like feed or side-draw flowrates. Additionally, more than one specification can be added to a single column, variables like vapor composition or tray temperature can also be added. Care should be taken when adding specifications as unrealistic scenarios may result in a model that cannot be solved. Figure 6.17 shows the specifications/control information panel used in defining the sour water stripper variables.

The last step in defining the process units is in defining the reflux drum. The reflux drum is a phase separator and at equilibrium will separate vapor, liquid, and solids at 100% efficiency (i.e., no liquid is carried over in the vapor phase). In actual operations, phase separations are never 100% efficient, therefore the software allows the user to define entrainment to account for inefficient separations. For this example, a small amount of liquid (5%) is carried out of the reflux drum with the vapor and is defined as Entrained Liquid as seen in Fig. 6.18.

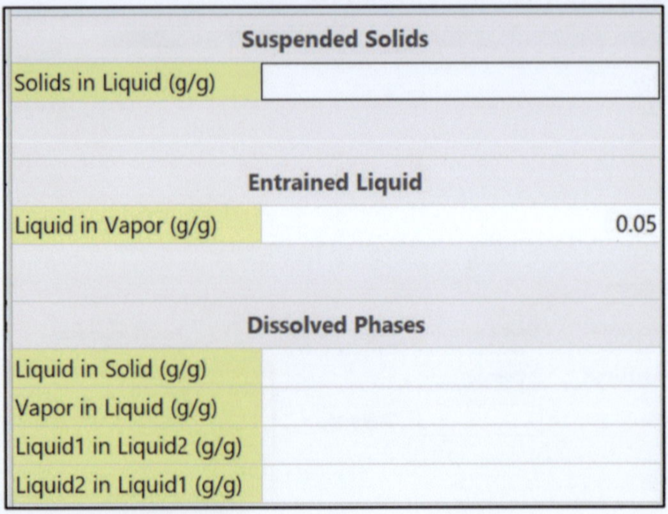

Fig. 6.18 Defining entrainment ratios is required for inefficient separation of phases. In the panel above, the vapor phase is defined by accounting for 0.05 g of liquid for every gram of vapor in the stream

Step: 5. Run and View the Case

Once the chemistry model, streams, and process units are defined, the case can be run. When the case is running, chemical specification and phase equilibrium is calculated for all streams and trays considering the operating conditions.

Step 6: Reviewing the Calculation Results

Reviewing the calculations is essential in determining that the case is representative of the process. Here we analyze the bottoms water composition that show we are meeting the 5 ppm H_2S concentration specified in the simulation (Table 6.7). However, depending on the refinery wastewater treatment plant requirements, the NH_3 concentration could be too high. The engineer may choose to modify the simulation to further reduce NH_3, likely by sodium hydroxide injection in the lower half of the tower would be required.

Once a simulation is analyzed and determined to be representative. The stream conditions and compositions can be used in assessing the risk for corrosion in the overhead system. According to the calculations, Table 6.8 lists the concentration of sour components in the reflux stream. Use this data to evaluate whether corrosion is possible at these conditions and make appropriate design or process changes to mitigate this risk.

Table 6.9 is the calculated liquid and vapor loads on each tray with pH and vapor/liquid temperature. This information is used to determine design anomalies and corrosion risks throughout the tower. Simulation and/or design improvements can be made to mitigate identified risks, upgrading materials of construction, adjusting tower operating temperatures, or changing feed locations, are all options that can be tested to improve the design based on the results presented in this table. Table 6.10 presents the SWS model heat exchanger duties calculated by the model.

Table 6.7 Calculated bottoms water composition

Bottoms	
T (°C)	119.566
pH	7.35577
CO_2 Liq1, apparent (mg/L)	0.0154348
H_2S Liq1, apparent (mg/L)	4.72467
NH_3 Liq1, apparent (mg/L)	1291.28

Table 6.8 Calculated reflux stream composition used to analyze the corrosion risk in the tower overhead system

Reflux	
T (°C)	85.0
pH	8.18022
CO_2 Liq1, apparent (mg/L)	18,361.7
H_2S Liq1, apparent (mg/L)	29,272.3
NH_3 Liq1, apparent (mg/L)	57,759.9

Table 6.9 Calculated sour water stripper parameters per tray

Tray	pH	Liquid, T (°C)	Vapor, T (°C)	Total flowrate (liquid, kg/hr)	Total flowrate (vapor, kg/hr)	NH_3 Vapor (kg/hr)	H_2S Vapor (kg/hr)	CO_2 Vapor (kg/hr)
20 (Top)	7.71	98	110	26,857	1663	181	233	95.9
19	7.76	107	113	27,299	2452	252	199	69.0
18	7.82	111	114	27,482	2894	283	165	49.3
17	7.89	113	115	27,557	3077	288	133	34.1
16	7.94	114	115	27,594	3153	281	107	23.4
15	7.98	114	116	27,616	3189	268	86	16.2
14	8.00	115	116	27,634	3212	251	70	11.1
13	8.02	115	117	27,650	3229	232	57	7.66
12	8.03	116	117	27,666	3246	212	47	5.25
11	8.03	116	117	27,681	3261	191	38	3.56
10	8.03	117	118	27,696	3276	171	30	2.39
9	8.01	117	118	27,710	3291	151	24	1.58
8	7.99	117 ara>	118	27,724	3305	132	19	1.03
7	7.96	118	119	27,737	3319	114	14	0.66
6	7.93	118	119	27,750	3333	96.6	10.5	0.41
5	7.88	118	119	27,762	3346	80.7	7.61	0.25
4	7.83	119	119	27,774	3358	66.1	5.33	0.15
3	7.77	119	119	27,785	3370	52.7	3.61	0.09
2	7.69	119	119	27,796	3381	40.4	2.34	0.05
1 (Bottom)	7.36	120	120	24,405	3391	29.3	1.45	0.028

(continued)

Table 6.9 (continued)

Tray	NH_3 Liquid (kg/hr)	NH_4^+ Liquid (kg/hr)	H_2S Liquid (kg/hr)	HS^- Liquid (kg/hr)	S^{2-} Liquid (kg/hr)	CO_2 Liquid (kg/hr)	HCO_3^- Liquid (kg/hr)	CO_3^{2-} Liquid (kg/hr)
20 (Top)	46.8	149	7.65	186	0.014	1.43	73.0	1.86
19	67.6	124	5.06	155	0.018	0.84	49.6	1.39
18	78.5	101	3.39	126	0.019	0.49	33.5	1.01
17	82.7	84.0	2.36	101	0.018	0.29	22.8	0.72
16	83.1	70.6	1.71	82.3	0.016	0.18	15.7	0.51
15	80.9	60.2	1.28	67.0	0.014	0.12	10.9	0.36
14	77.1	52.0	0.99	54.7	0.012	0.077	7.54	0.25
13	72.3	45.4	0.78	44.5	0.010	0.052	5.22	0.17
12	67.0	40.1	0.62	36.0	0.008	0.035	3.59	0.12
11	61.3	35.6	0.50	28.9	0.007	0.024	2.44	0.076
10	55.5	31.9	0.40	22.9	0.005	0.017	1.64	0.049
9	49.7	28.8	0.32	17.8	0.004	0.012	1.09	0.030
8	44.0	26.2	0.26	13.6	0.003	0.008	0.71	0.018
7	38.6	24.1	0.21	10.1	0.002	0.005	0.45	0.011
6	33.5	22.3	0.16	7.34	0.001	0.004	0.28	0.006
5	28.7	20.8	0.13	5.17	0.001	0.003	0.17	0.003
4	24.2	19.6	0.10	3.52	0.000	0.002	0.10	0.002
3	20.1	18.5	0.08	2.32	0.000	0.001	0.06	0.001
2	16.3	17.6	0.06	1.47	0.000	0.001	0.04	0.000
1 (Bottom)	6.56	14.8	0.009	0.109	0.000	0.000	0.001	0.000

Table 6.10 Calculated heat SWS model heat exchanger duties

Heat exchanger	Duty (kJ/hr)
Feed/bottoms exchanger	3.1896E + 06
Reboiler	7.5444E + 06
Condenser	−2.5242E + 06

6.4.1.6 Summary and Conclusions of Sour Water Stripper Case Study

The purpose of this case study is to obtain the steady-state composition and process conditions at various points in a sour water stripper unit. A chemical corrosion assessment is then performed using these results and other relevant factors. Based on the evaluation, mitigation steps can be taken either through selecting appropriate materials of construction or changing process conditions.

Unlike most fractionation towers in the oil refinery which are largely hydrocarbon-based, the SWS is an aqueous system driven by acid–base chemistry and mass transfer rates. As discussed earlier, aqueous systems are highly complex and require specialized thermodynamic frameworks for accurate vapor/liquid equilibrium predictions. With these models, SWS designs can be improved ultimately saving costs in design, construction, and operations.

Process design of sour water strippers using OLI as the modeling tool has been described above. Other such commercial simulators e.g., RADFRAC module of Aspen Plus are also available in the market.

6.4.2 Simulation of Biological Treatment Processes

Wastewater produced from petroleum refineries is a complex blend of organic and inorganic contaminants. Refinery wastewater characteristics depend on the makeup of the incoming crude, refining processes (cracking, coking, hydrotreating, alkylation, polymerization, isomerization and catalytic reforming), in-plant control measures and other operational practices. Typically, the wastewater from petroleum refining facilities contains mixtures of oil, grease and various other hydrocarbons among organic compounds; inorganic and organic compounds of sulfur and nitrogen; and compounds containing trace metals. The petroleum refinery wastewaters are high strength wastewaters with many toxic contaminants and need to be treated through sequential treatment units consisting of physical, chemical, and biological processes. Biological treatment processes play an important role in removing the dissolved organic and inorganic contaminants contained in petroleum refinery wastewater. Design, optimization, and control of biological processes is key to achieving an effluent quality that meets the discharge standards. The discussion on modeling of biological treatment in this section is limited to activated sludge processes only, which account for about 90% of the refinery wastewater treatment systems in operation currently, although other types of processes, especially MBR and MBBR are increasingly being designed and installed at both new plants and to upgrade existing facilities.

Process modeling is an established and proven approach for planning, designing, and optimizing municipal and industrial wastewater treatment plants (WEF 2009). Since the mid-1980s, there has been significant advancement in modelling biological activated sludge and attached growth processes. Biological wastewater treatment modelling became popular after the publication of Activated Sludge Model no. 1 (ASM1) for COD Removal, Nitrification and Denitrification by the IWA Task group on modelling in 1986 (Henze et al. 1987) and implementation of this model in several commercial simulators. Several subsequent advancements in modelling solid settling, gas–liquid transfer and chemical precipitation allowed process modelling to become an integral and indispensable part of wastewater treatment plant design and optimization studies.

Biological process models incorporate a set of non-linear ordinary differential equations incorporating the stoichiometry and kinetics of biological reactions performed by different groups of microorganisms. The framework of biological models allows for both steady state and dynamic analysis. The dynamic analysis feature of the process models makes these powerful predictive tools for operational control and optimization. The process models are useful to track mass flows, determine maximum and safe operating capacities, and to predict key performance indicators like effluent quality, sludge production, oxygen demand, and chemical demand for different operational conditions.

The activated sludge models originally developed for municipal wastewater treatment plant have been successfully used in modelling industrial wastewater treatment facilities treating food waste, meat processing, breweries, and wineries with little or no change in the underlying model structure (Bury et al. 2002; Goodfellow et al. 2004; Schraa et al. 2004; Stricker and Racault 2005; Takács et al. 1998; Eremektar et al. 2002; Ky et al. 2001). There are also application examples where activated sludge models are used in refinery wastewater treatment plant modelling successfully (Marinetti et al. 2009, Maqueda et al. 2006, Trojanowicz and Wojcik 2014).

The scope of activated sludge models is often expanded to answer more specific questions related to petroleum refinery wastewater treatment including fate of specific contaminants, Volatile Organic Carbon (VOC) emissions, inhibitions etc. Unlike the IWA model frameworks, there is no standard approach to modelling refinery wastewater treatment plants, however some commercial simulator companies have developed proprietary models for petroleum refinery wastewater treatment. This chapter will explore the advances and application of process modelling in simulating petroleum refinery wastewater treatment plant.

6.4.2.1 Key Aspects of Refinery Activated Sludge Models

Even though the biological treatment technologies (conventional activated sludge, tricking filters etc.) for treating petroleum refinery wastewater are similar to those which are used for treating municipal wastewater, additional attention has been paid to include following aspects in developing process models for refinery wastewater:

- Volatilization and Stripping
- Sorption
- Inhibition
- Temperature Effects.

In addition, modeling the fate of contaminants in petroleum refinery wastewater requires a detailed characterization of the organic compounds with respect to the physical, chemical and biological characteristics. Additional biological conversion pathways for sulfur compounds may be necessary to model the oxygen demand in sulfide oxidation, H_2S volatilization and chemical requirements for pH adjustment. A brief description on modelling the above listed processes is provided below.

Volatilization and Stripping

Volatile hydrocarbon contaminants in refinery wastewater may be removed by volatilization or stripping, resulting in fugitive emissions from the plant. Volatilization is a process in which a contaminant is transferred from the water surface to air under quiescent or wind driven condition. Stripping is defined as contaminant loss from wastewater which is turbulent through agitation by mechanical means or through head loss (Melcer et al. 1994). Both volatilization and/or stripping can cause significant loss of a contaminant during wastewater conveyance and treatment. During wastewater treatment, biological reactors with mechanical or diffused aeration system can result in measurable stripping of contaminant. The VOC emissions from refineries in United States are regulated under the National Emission Standards for Hazardous Air Pollutants (NESHAPs) promulgated after the 1990 Clean Air Act Amendments are found in 40 CFR Part 63 (see discussion in Chap. 7). These standards provide guidelines to meet the compliance with wastewater rules that regulate emissions of semi-volatile and volatile organic compounds (VOC). Several U.S. EPA regulations define VOCs as those organic compounds with a vapor pressure of greater than 0.01 kPa (Pincince 1995). The EPA regulations and their requirements have been discussed in detail in Chap. 7. The extent of volatilization of a contaminant is a function of vapor pressure of the contaminant, typically determined by Henry's constant and open surface area and wind velocities above the surface. A procedure to estimate the mass transfer constant for volatilization above open surface reactors is found in Mackey and Yeun (1983). Contaminant stripping generally depends on the mechanical power of aerators in mechanically aerated activated sludge reactor. For diffused aeration, the air flow rates, and diffuser type (fine, versus coarse bubble) affect the extent of contaminant stripping. Depending on the chemical property of a compound, the loss of contaminant from biological reactors due to volatilization and stripping could be significant and needs to be properly accounted. Quantification of contaminant loss is important not only to manage the regulatory VOC emission compliance, but also to maintain safe operating conditions at treatment facilities. The other process design implications of volatilization/stripping include reduced oxygen demand and sludge production from biological reactors.

Volatilization/stripping from biological reactors is modelled using the equation below.

$$r_v = K_L^{voc} a.\left(C - C^*\right).V \tag{6.10}$$

where,

r_v rate of volatilization, mg/h
$K_L{}^{VOC}$ overall mass transfer coefficient, m/h
a specific interfacial area for mass transfer, m^2/m^3
C concentration of volatile compound in the water, mg/m^3
C* equilibrium concentration of compound in water, mg/m^3
V volume of process vessel, m^3.

For low contaminant concentrations equilibrium between the liquid and gas phases can be represented by Henry's law:

$$C^* = \frac{C_g}{K_H} \tag{6.11}$$

where,

K_H Henry's law coefficient, m^3 liquid/m^3 gas.

For open reactors, the C_g value in air above the reactor is negligible and the volatilization/stripping equation reduces to following expression.

$$r_v = K_L^{voc} a.C.V \tag{6.12}$$

Considering two film resistance theory (Metcalf & Eddy/AECOM, 2014), the overall mass transfer rate is related to the mass transfer rate through liquid film (k_L^{voc}) and gas film (k_g^{voc}) by following expression:

$$K_L^{voc} = \frac{k_L^{voc} k_g^{voc} K_H}{\left(k_L^{voc} + k_g^{voc} K_H\right)} \tag{6.13}$$

where,

$K_L{}^{VOC}$ overall mass transfer coefficient based on liquid film resistance, m/h
$k_L{}^{VOC}$ liquid phase mass transfer coefficient, m/h
$k_G{}^{VOC}$ gas phase mass transfer coefficient, m/h.

When, K_H value is high, $k_g^{voc} K_H \gg k_L^{voc}$ resulting in situation where the mass transfer rate is limited by liquid film resistance. In such situation, the overall mass transfer rate can be approximated by:

$$K_L^{voc} \cong k_L^{voc} \tag{6.14}$$

However, for low volatility compounds $\left(k_g^{voc} K_H \ll k_L^{voc}\right)$, overall mass transfer rate is controlled by gas film resistance. For semi volatile compound, both the liquid film and gas film resistance should be considered.

The value of overall mass transfer rate (K_L^{voc}) in volatilization/stripping equation depends on the aeration system i.e., mechanical, fine bubble, coarse bubble and can be estimated from the oxygen transfer coefficient K_L^{oxy} using the following correlation proposed by Thibodeaux (1979).

$$k_L^{voc} a = k_L^{oxy} a. \left(\frac{D_{voc}}{D_{oxy}}\right)^n \cong K_L^{oxy} a. \left(\frac{D_{voc}}{D_{oxy}}\right)^n \tag{6.15}$$

where,

$K_L{}^{oxy}$ oxygen transfer coefficient, m/h
D_{VOC} diffusion coefficient of compound in water, cm^2/s
D_{oxy} diffusion coefficient of oxygen in water, cm^2/s
n proportionality parameter between 0.5 and 1.0.

The k_L^{oxy} is approximated to K_L^{oxy} as the K_H value of oxygen is 40 and the overall mass transfer rate are controlled by liquid film thickness. Now, the overall $K_L^{voc} a$ can be estimated using the following expression:

$$K_L^{voc} a = k_L^{voc} a \frac{K_H}{\left(\frac{k_L^{voc}}{k_g^{voc}} + K_H\right)} \tag{6.16}$$

The above expression requires a knowledge of $\frac{k_L^{voc}}{k_g^{voc}}$ for the aeration system. Hsieh et al. (1993) observed a ratio ranging from 2.2 to 3.6 for diffused aeration systems. Munz and Roberts (1989) suggested a value of 40 for mechanical aerated systems.

Example 1 Calculate the Rate of Volatilization of Benzene and Toluene in an Open Reactor

An industrial wastewater treatment facility treats petroleum refinery wastewater that contains benzene and toluene. The plant uses an open completely mixed reactor with diffused aeration. Given the following information, calculate the rate of volatilization of benzene and toluene.

1. Wastewater temperature, T = 25 °C
2. The open reactor has a surface area of 2000 m^2 and depth of 5.0 m

3. Reactor effluent benzene concentration, $C_{ben} = 1.0$ mg/m^3
   Reactor effluent toluene concentration, $C_{tol} = 1.0$ mg/m^3
4. Oxygen transfer coefficient, $K_L{}^{oxy} = 0.03$ m/h
5. Henry's law coefficient for benzene @25 °C, $K_{Hben} = 9.41$ m^3 liquid/m^3 gas
   Henry's law coefficient for toluene @25 °C, $K_{Htol} = 11.15$ m^3 liquid/m^3 gas
6. Diffusion coefficient of oxygen in water @25 °C, $D_{oxy} = 2.42 \times 10^{-5}$ cm^2/s
   Diffusion coefficient of benzene in water @25 °C, $D_{ben} = 1.14 \times 10^{-5}$ cm^2/s
   Diffusion coefficient of toluene in water @25 °C, $D_{tol} = 2.58 \times 10^{-6}$ cm^2/s
7. Assume $\frac{k_L^{voc}}{k_g^{voc}}$ for both benzene and toluene are 3.0
8. Assume proportionality parameter, n = 0.5.

Solution

1. Determine the liquid phase mass transfer coefficients.

   Based on Eq. 6.15,

$$k_L^{\text{ben}} \cong K_L^{\text{oxy}}\left(\frac{D_{\text{ben}}}{D_{\text{oxy}}}\right)^n = (0.03\,\text{m/h})\left(\frac{1.14\times10^{-5}\ \text{cm}^2/\text{s}}{2.42\times10^{-5}\ \text{cm}^2/\text{s}}\right)^{0.5}$$
$$=2.06\times10^{-2}\ \text{m/h}$$

$$k_L^{\text{tol}} \cong K_L^{\text{oxy}}\left(\frac{D_{\text{tol}}}{D_{\text{oxy}}}\right)^n = (0.03\,\text{m/h})\left(\frac{2.58\times10^{-6}\ \text{cm}^2/\text{s}}{2.42\times10^{-5}\ \text{cm}^2/\text{s}}\right)^{0.5}$$
$$=9.79\times10^{-3}\ \text{m/h}$$

2. Determine the overall mass transfer coefficients.

   Based on Eq. 6.16,

$$K_L^{\text{ben}} = k_L^{\text{ben}}\frac{K_H}{\left(\frac{k_L^{\text{ben}}}{k_g^{\text{ben}}}+K_H\right)} = \left(2.06\times10^{-2}\ \text{m/h}\right)\frac{9.41\text{m}^3\ \text{liquid/m}^3\ \text{gas}}{(3.0+9.41\text{m}^3\ \text{liquid/m}^3\ \text{gas})} = 1.56\times10^{-2}\ \text{m/h}$$

$$K_L^{\text{tol}} = k_L^{\text{tol}}\frac{K_H}{\left(\frac{k_L^{\text{tol}}}{k_g^{\text{tol}}}+K_H\right)} = \left(9.79\times10^{-3}\ \text{m/h}\right)\frac{11.15\text{m}^3\ \text{liquid/m}^3\ \text{gas}}{(3.0+11.15\text{m}^3\ \text{liquid/m}^3\ \text{gas})} = 7.71\times10^{-3}\ \text{m/h}$$

3. Determine the rate of volatilization.

   Volume of the open reactor, $V = \left(2000\,\text{m}^2\right)(5.0\,\text{m}) = 10000\,\text{m}^3$.

   Specific interfacial area for mass transfer, $a = \frac{2000\text{m}^2}{10{,}000\text{m}^3} = 0.2\,\text{m}^2/\text{m}^3$

   For open reactors, the equilibrium concentration of compound in air is close to zero.

   Therefore, based on Eq. 6.12,

$$r_v^{\text{ben}} = K_L^{\text{ben}}a.C.V = \left(1.56\times10^{-2}\ \text{m/h}\right)\left(0.2\,\text{m}^2/\text{m}^3\right)\left(1.0\,\text{mg/m}^3\right)\left(10000\,\text{m}^3\right) = 31.2\,\text{mg/h}$$
$$r_v^{\text{tol}} = K_L^{\text{tol}}a.C.V = \left(7.71\times10^{-3}\ \text{m/h}\right)\left(0.2\,\text{m}^2/\text{m}^3\right)\left(1.0\,\text{mg/m}^3\right)\left(10000\,\text{m}^3\right) = 15.4\,\text{mg/h}$$

Sorption

Some of the contaminants in petroleum refinery wastewater can be transported from the liquid phase onto biomass or other solids by the sorption process. Petroleum refinery wastewater containing polynuclear aromatic hydrocarbons (PAHs) are hydrophobic compounds and sorption to biological solids may be a significant mechanism for the removal of PAHs from refinery wastewater (Stringfellow and Alvarez-Cohen 1999). Fatone et al. 2011, studied the fate of VOC compounds in municipal wastewater treatment plants and observed significant partitioning of these compounds to activated sludge. The adsorption of VOC compounds to activated sludge results in generation of sludge contaminated with hydrocarbon and therefore require proper disposal to meet the appropriate disposal regulations. The adsorbed VOC compounds can also increase the apparent yield of the solids.

Sorption and desorption of contaminants by wastewater treatment plant solids has been observed to be a relatively rapid process (Dobbs et al. 1989; Weber and Jones 1986). Therefore, the sorption process is either modelled through an instantaneous equilibrium between the liquid/solid contaminant concentration or by using a kinetic equation with high sorption rate coefficient. For the instantaneous equilibrium model, the liquid–solid equilibrium is described as below.

$$C_{sorb,solid} = K_p.C \tag{6.17}$$

where,

$C_{sorb,solid}$ concentration of contaminant in solid phase, mg/kg (ug/g)
K_p sorption partition coefficient, m^3/kg (L/g)
C concentration of contaminant in liquid phase, mg/m^3 (ug/L).

Dobbs et al. (1989) studied the sorption properties of different wastewater sludges (primary, secondary and digested sludge) and found that the sorption is related to volatile fraction of the solids. Therefore, in sorption equation, the VSS concentration is used to express the amount of sorbent in the biological reactors. Considering this, the sorbed mass with respect to liquid volume can be expressed as below.

$$C_{sorb,liquid} = C_{sorb,solid}.X_{VSS} = K_p.C.X_{VSS} \tag{6.18}$$

where,

X_{VSS} volatile suspended solids (VSS) concentration, kg/m^3.

The sorption equilibrium equation is generally valid at low contaminant concentrations. The sorption partition coefficient for a given compound can be related to the octanol/water partition coefficient of that compound. The following correlation reported by Dobbs et al. (1989) is used to estimate the sorption partition coefficients where experimentally determined coefficients are not available.

$$K_p = 10^{(0.58 log K_{OW} + 1.14)}.10^{-3} = 10^{(0.58 log K_{OW} - 1.86)} \tag{6.19}$$

where,

K_{OW} octanol/water partition coefficient.

The factor 10^{-3} is used to convert the L/kg unit for K_p to m^3/kg.

The dynamic modelling of the sorption process requires estimation of rate of sorption and simple expressions as below have been used.

$$r_{sorb} = k_{sorb}.\left(K_p.C.X_{VSS} - C_{sorbed,liquid}\right) \tag{6.20}$$

where,

r_{sorb} sorption process rate, mg/kg/d
k_{sorb} sorption rate constant, 1/d.

The above kinetic expression includes both the sorption and desorption processes and at steady state the net rate of sorption $r_{sorb} = 0.0$ as the sorbed concentration becomes equal to the equilibrium concentration estimated based on the liquid/solid partition coefficient $\left(K_p\right)$. The time for sorption equilibrium is reported to be from a few seconds to several hours. The data on rate of sorption (k_{sorb}) is scarce. However, for many practical purposes, a high sorption rate that leads to a sorption equilibrium in several minutes may be sufficient considering the typical long retention times in bioreactors.

The reversible nature of sorption has several environmental implications as the sorbed compounds may find their way into water, soil or air environment if the biosolids are not disposed securely. Sorbed compounds in sludge can impact the sludge treatment and disposal options for a wastewater treatment plant.

Example 2—Calculate the Concentration of Sorbed Benzene and Toluene in a Reactor

Consider the same plant given in Example 1. Use the additional information given below to calculate the concentration of sorbed benzene and toluene with respect to liquid volume.

1. Log octanol/water partition coefficient of benzene, $\log\left(K_{OW}^{\text{ben}}\right) = 2.13$

   Log octanol/water partition coefficient of toluene, $\log\left(K_{OW}^{\text{tol}}\right) = 2.69$.
2. Mixed liquor concentration, MLSS = 3000 mg/L.

   VSS to TSS ratio is 75%.

Solution

1. Determine the sorption partition coefficients.

   Based on Eq. 6.19,

$$K_p^{\text{ben}} = 10^{\left(0.58\log(K_{OW}^{\text{ben}})-1.86\right)} = 10^{(0.58\times 2.13-1.86)} = 0.237\text{m}^3/\text{kg}$$
$$K_p^{\text{tol}} = 10^{\left(0.58\log(K_{OW}^{\text{tol}})-1.86\right)} = 10^{(0.58\times 2.69-1.86)} = 0.501\text{m}^3/\text{kg}$$

2. Determine the volatile suspended solids (VSS) concentration.

$$X_{VSS} = (3000\text{mg/L})(75\%)(\text{g}/1000\text{ mg}) = 2.25\text{g/L}\ (\text{kg/m}^3)$$

3. Determine the concentration of sorbed benzene and toluene.

   Based on Eq. 6.18,

$$C^{\text{benzene}}_{\text{sorb, liquid}} = K_p^{\text{ben}}.C_{\text{ben}}.X_{VSS} = \left(0.237\text{m}^3/\text{kg}\right)\left(1.0\text{ mg/m}^3\right)\left(2.25\text{kg/m}^3\right) = 0.533\text{mg/m}^3$$

$$C^{\text{toluene}}_{sorb,\text{liquid}} = K_p^{\text{tol}}.C_{\text{tol}}.X_{VSS} = \left(0.501\text{m}^3/\text{kg}\right)\left(1.0\text{ mg/m}^3\right)\left(2.25\text{kg/m}^3\right) = 1.13\text{mg/m}^3$$

Inhibition

The biological oxidation of petroleum wastewater compounds is typically modelled using Monod kinetics. A detailed description of Monod kinetics is presented in Chap. 5. While standard Monod kinetics work well for most organic compounds, the petroleum refinery wastewater may contain compounds which are inhibitory to microorganisms requiring different kinetic models for degradation. A material is considered inhibitory if it inhibits microbial growth (reversible inhibition) or leads to death/decay of microorganism (irreversible inhibition). The typical compounds of concern are phenolic compounds and naphthenic acids, for which nitrifiers are usually the most sensitive populations. Growth equations can include terms for inhibition on one or more components (e.g., Haldane equation).

Schroder et al. (1997) lists the inhibition kinetic models used to model phenol inhibition. Some of the inhibition models applied in practice are as listed in Table 6.11. Haldane substrate inhibition model is a case of competitive inhibition, where at high concentration too many substrate molecules crowd the enzyme functional groups leading to no product formation. It is the preferred choice in modelling substrate inhibition for refinery wastewater.

Hill and Robinson (1975) studied the substrate inhibition kinetics of phenol degradation by pure culture bacteria at 30 °C and pH 6.5. The inhibition coefficient (K_{IH}) in the Haldane inhibition model was determined to be 470 mg/L. Blum and Speece (1991) determined the IC_{50} concentration of different organic compounds on heterotrophic, Nitrosomonas and methanogenic bacteria. Hockenbury and Grady (1977) provided a literature review of the inhibitory concentration of various organic compounds to nitrification process. They also extended the list by their own experimental study. Carrera et al. (2004) evaluated different inhibition models for inhibition of ammonia and nitrite oxidation.

Modelling substrate inhibition is however not without challenges. While there is substantial experimental information about inhibitory concentration of compounds found in refinery wastewater, the large variability in numerical values is often the reflection of the differences in experimental methods and microbial characteristics employed in these studies. The other challenge in including substrate inhibition

Table 6.11 Biological inhibition models used in practice

Inhibition model	Expression
Haldane	$r = r_{max} \cdot \frac{S}{K_s + S + \frac{S^2}{K_{IH}}}$
Aiba	$r = r_{max} \cdot \frac{S}{K_s + S}\left[exp\left(-\frac{S}{K_{IA}}\right)\right]$
Luong	$r = r_{max} \cdot \frac{S}{K_s + S}\left[1 - \left(\frac{S}{S_m}\right)\right]^n$
Teissier (Edwards 1970)	$r = r_{max} \cdot \left[exp\left(-\frac{S}{K_{IE}}\right) - exp\left(-\frac{S}{K_s}\right)\right]$
Laidler	$r = r_{max} \cdot \frac{S}{K_s + S} \cdot \frac{K_{IL}}{K_{IL} + S}$
Webb (Edwards 1970)	$r = r_{max} \cdot \frac{S.\left(1 + \frac{S}{K_{2W}}\right)}{K_s + S + \frac{S^2}{K_{IW}}}$
Han and Levenspiel (1988)	$r = r_{max} \cdot \frac{S}{K_s\left[1 - \left(\frac{S}{S_m}\right)\right]^m + S.}\left[1 - \left(\frac{S}{S_m}\right)\right]^n$
Yano and Koga (1969)	$r = r_{max} \cdot \frac{S}{K_s + S + \frac{S^2}{K_{1YK}} + \frac{S^3}{K_{2YK}}}$

kinetics expression in biological growth is in achieving steady state solutions as including inhibition expressions can lead to multiple solutions.

Temperature Effects

The temperature of municipal wastewater is typically below 35 °C. However, depending on upstream refining processes and operations, the petroleum refinery wastewater temperature may be higher and may necessitate cooling for optimum biological treatment. The temperature dependence of kinetic parameters in Activated Sludge Models (ASMs) is described by the Arrhenius equation.

$$k_T = k_{20} \cdot \theta^{(T-20)}$$

where,

k_T kinetic parameter value at temperature T (units depend on parameter).
k_{20} kinetic parameter value at (units depend on parameter).
θ Arrhenius temperature coefficient (unitless).
T Wastewater Temperature, C

This equation, however, is typically used in the temperature range of 20 °C ± 15 °C. As the liquid temperature in the biological reactor increases above 35 °C, activity of certain microorganisms start to get reduced. Nitrifying bacteria (Nitrosomonas and Nitrobacter) appear to be more sensitive to increased temperature. An optimum temperature range of 30–35 °C is recommended (Neufeld et al. 1986,

Shammas 1986, Antoniou et al. 1990, van Hulle et al. 2007). To account for the change in bacterial activity with temperature a bell-shaped curve for growth rate is typically used. Stark (1996) provided a comparison of commonly available temperature growth rate models. Goel and Schraa (2013) while modelling an industrial wastewater treatment plant developed a double Arrhenius model to describe the temperature-growth rate relationship. The advantage of this over other multi-parameter models is its compatibility with the presently used Arrhenius model for low temperature and more importantly, the two additional model parameters (T_{opt}—optimum temperature and θ_2 temperature coefficient at high temperature) are easily determinable. Appropriate attention is required to optimize the process with respect to temperature as, on one hand it affects the performance of biological treatment, on the other the capital and operational cost of cooling facilities could be exorbitant depending on the plant capacity and the difference between the wastewater temperature and maximum permissible temperature in biological treatment.

6.4.2.2 Commercial Simulators

Wastewater Treatment Modelling Platforms

Several wastewater modelling platforms have been developed during the past three decades to provide consultants, managers, and operators with tools to analyze and optimize their treatment facilities. While the underlying model structures for the original dynamic-mechanistic models (such as ASM1, as discussed above) was developed for municipal wastewater, this modelling approach has also been successfully applied to industrial wastewater systems through the development of refinery-specific equation matrices and unit process models. These models are designed to provide "out of the box" modelling experiences for users through the development of industry-specific models that have been calibrated for full-scale industrial applications.

These platforms conduct mass balance for COD, nitrogen, phosphorus and sulfur across each unit process in the system, and therefore across the entire treatment facility as a whole. By entering influent loading, unit sizes and operation, the models can predict the level of treatment possible, and therefore the remaining concentrations in the effluent.

The GPS-X™ modelling and simulation package from Hatch (formerly Hydromantis) is a fully dynamic-mechanistic modelling platform that allows users to create treatment plant models by drawing a process flow diagram and entering influent loading information and treatment unit sizes and operation. The system predicts plant performance, effluent quality, and operating costs for both steady-state and dynamic conditions. The MantisIW biological model has been developed for modelling the treatment of refinery wastewaters. This model extends the default municipal wastewater treatment model (Mantis2) by adding and addressing refinery-waste-specific aspects of the treatment process. The MantisIW library revises the categorization of COD to allow tracking of individual groups of hydrocarbons that can be grouped

Table 6.12 Key organic contaminant groups in petroleum refinery wastewater

Model organic group	Typical compounds
Short-chain aliphatic compounds	Hydrocarbons with less than 18 carbon atoms including: alkanes, alkenes, alkynes, fats and oils
Long-chain aliphatic compounds	Hydrocarbons with 18 or more carbon atoms including: alkanes, alkenes, alkynes, fats and oils
Monocyclic aromatic compounds	Aromatic compounds with single benzene ring e.g. benzene, toluene, styrene, xylene
Polycyclic aromatic compounds	Aromatic compounds with multiple benzene rings e.g. naphthalene, phenanthrene, benzo[a]pyrene
Phenolic compounds	Phenolic compounds

together based on their physical properties and processes they undergo in the treatment steps. The model includes biological degradation of additional key organic groups of compounds found in petroleum refinery wastewater as listed in Table 6.12.

In addition to including the fate of individual organic compounds, the reaction processes in the model are expanded to include processes for surface volatilization, air stripping, sorption, sulfur oxidation (including additional biomass types) and inhibition (Schraa et al. 2004). The Fig. 6.19 shows the model modifications including the biological growth processes related to sulfur oxidation, while Fig. 6.20 shows

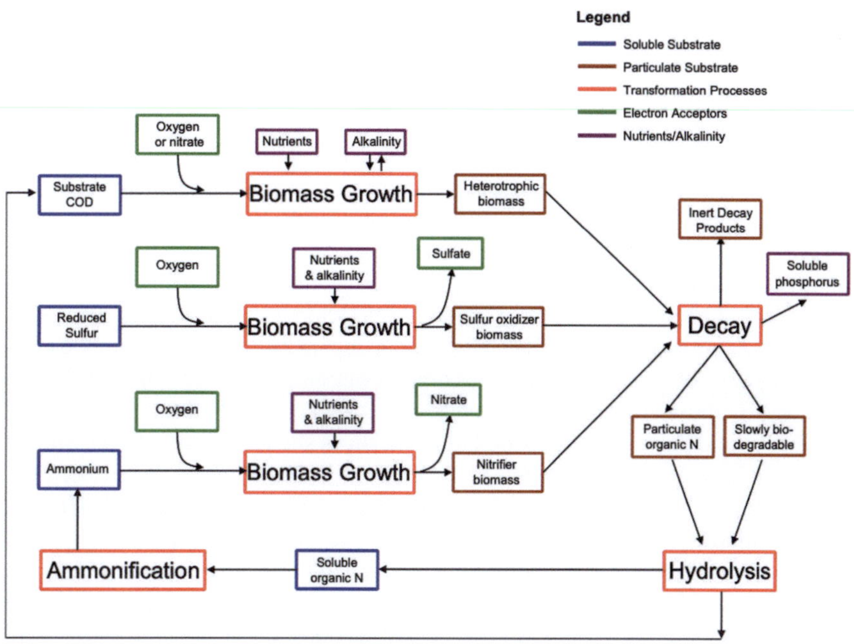

Fig. 6.19 Overview of biological growth processes in MantisIW

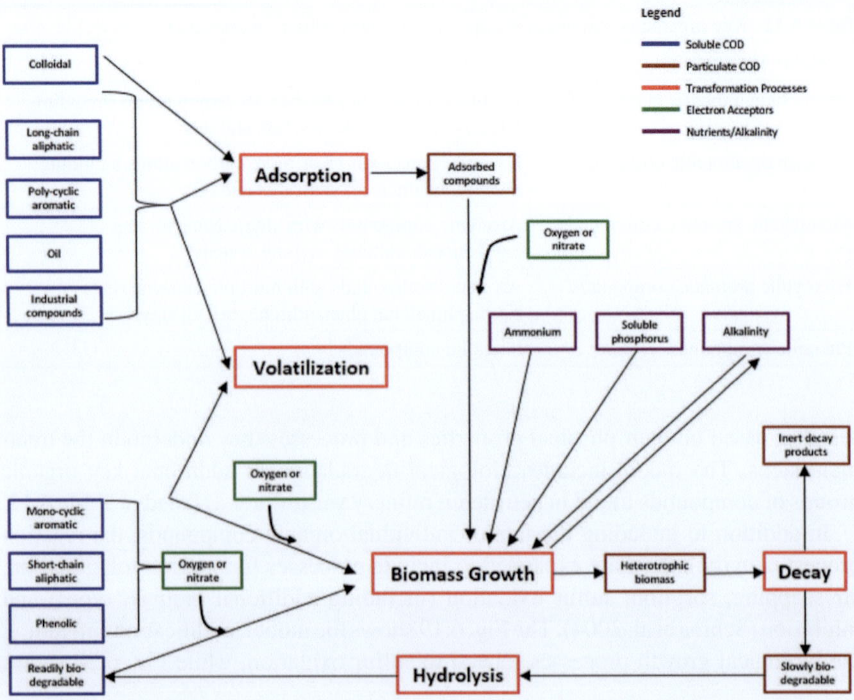

Fig. 6.20 COD transformation processes in MantisIW (adsorption, volatilization, and heterotrophic growth/decay)

the additional processes of adsorption, volatilization included in estimating the fate of additional organic compounds found in petroleum wastewater.

Petwin+ is a mechanistic-dynamic modelling system developed by Envirosim Associates Ltd., which allows for modelling of refinery activated sludge systems. Similar to the MantisIW model, the basic municipal activated sludge model from Envirosim's Biowin package (ADSM) has been extended and adapted for refinery wastewater, with the development of the ADSM model (based on Baker 1994). ADSM contains sulfide-oxidizing and sulfate-reducing biomass populations, as well as additional soluble, volatile substrate components, and the ability to model biomass inhibition on those components.

Other municipal wastewater simulation packages such as WEST (DHI), Simba# (IFAK e.V.) and SUMO (Dynamita) do not have specific refinery wastewater treatment libraries. Other simpler modelling tools such as SimpleTreat (RIVM 2018) are based on simple algebraic (non-dynamic) mass-balance calculations.

The non-mechanistic modelling approaches (such as data-driven models like Artificial Neural Networks) have also been developed and applied (Lira et al. 2021).

These process models can have several applications. By conducting steady-state and/or dynamic simulations, users can understand and investigate the complex relationships between loading, plant size/operation and treatment performance. Through these investigations, facilities can be analyzed and optimized for various metrics such as effluent quality, treatment robustness (against loading dynamics) and operating cost. Mechanistic models allow users to explore the performance of a facility outside of the range of observed behaviors, including different process failure points, such as short solids retention time (SRT) or insufficient secondary settling capacity.

For example, mechanistic models could be used to study the following scenarios:

(1) High BOD_5 in the effluent is caused by nutrient limitation in the bioreactors. High concentrations of readily biodegradable substrate are available to be removed by the biomass, however not enough nutrients are available to allow proper growth. The model can be used to investigate and optimize the dosage of H_3PO_4 as a phosphorus nutrient source, removing the nutrient limitation issue.
(2) Poor performance is observed in the bioreactors, caused by under aeration of the biomass. The model can be used to determine the optimal amount of airflow required to achieve a specific residual dissolved oxygen (DO) concentration (e.g., 2 mg/L) in the bioreactors. Further optimization could be investigated through modelling advanced process control schemes such as ammonia-based aeration control (ABAC) to minimized aeration energy cost while meeting effluent concentration limits.
(3) High solids in the effluent is caused by a combination of high solids loading and poorly settling sludge in the secondary settlers. Adjusting the waste activated sludge (WAS) flow rate could reduce the mixed liquor suspended solids (MLSS) concentration and therefore reduce loading to the settlers. Engineers/operators could find the right balance between enough MLSS to robustly remove influent BOD while still keeping the effluent TSS below the desired limit.
(4) Poor solids removal from primary dissolved air flotation (DAF) causes high solids loading to the bioreactors and subsequently the secondary settlers. Adjustment to polymer dosing could improve DAF solids removal and reduce the solids loading on downstream units.
(5) Increase in the effluent pH above the desired limit due to the presence of significant caustic wash loading in the influent wastewater. H_2SO_4 addition can be optimized to bring down the effluent pH while not overdosing.

As noted above, the process modelling platforms can provide substantial benefits to users during wastewater facility design and operational optimization.

Air-Emission Modelling Platforms

Refinery wastewater treatment plants receive wastewater which contain a mix of volatile organic compounds. As the wastewater flows through the wastewater treatment plant, a fraction of the VOC compounds volatilizes. Studies have shown that VOCs emitted by refineries may pose health and safety risks to exposed populations, such as noncancer health effects, carcinogenic effects, effects on development or reproduction, and flammability (OEHHA and CEPA 2019). Petroleum refining is also a source of greenhouse gas (GHG) release due to combustion-related, equipment leaks and vented emissions (USEPA 2010).

United States Environmental Protection Agency steered the development of Water 7 (now Water 9) software to model the fugitive emissions from wastewater treatment facilities treating wastewater containing VOC compounds.

Toxchem was developed in early 1990s as an alternative to Water7 (now Water9) software to overcome limitations of Water8/9, including improved mass transfer processes, sorption of contaminants to solids, and a compound database containing critically reviewed physical, chemical, and biological properties. Toxchem is based on fundamental mass transfer equations and mass balances including the removal mechanism of stripping and volatilization, biodegradation, and sorption. Thus, it can also be used to determine the fate of any synthetic chemical compounds for which the physical, chemical, and biodegrading properties are known. Toxchem was specifically included in Appendix C of 40 CFR Part 63 as an accepted alternative to Water 8 for estimation of wastewater treatment emissions.

Toxchem is mainly used for estimating VOC air emissions from wastewater collection, preliminary/primary/secondary treatment, and disposal facilities. The site-specific wastewater characteristics, contaminant properties, and the process design and operating information are used to estimate VOC emission rates. In addition to air emission estimates, Toxchem can also be used to estimate the concentrations/loads of contaminants in the treated water effluent, and oil and residual solids streams.

Model Calibration Techniques

Typically, models are calibrated from full-scale historical data, or from special pilot-scale treatability study data. This is typically done by adjusting relevant parameters in order to have the model outputs match the observed performance data. The model can then be applied to confirm, optimize or troubleshoot existing plant operations.

The IWA Guidelines for Using Activated Sludge Models (Rieger et al. (2012)) defines calibration as "an iterative adjustment of the model parameters until simulation results match an observed set of data." The choice of parameters to adjust, and selection of parameters to fit to, is a function of the treatment system, and the purpose of the modelling exercise.

Using historical plant data for model calibration involves selecting a set of typical plant operational and performance data from the set of existing historical data (usually collected for monitoring and/or regulatory purposes). Often these datasets do not cover all the information required for a full steady-state and dynamic calibration of the model but are all that is available in the absence of a special sampling campaign specifically for model calibration.

It is important to select an appropriate and representative set of data to which the model will be fit, avoiding periods of instability or non-process-related functional changes at the facility (e.g., construction, shutdowns, plant upgrades, etc.). The calibration data should be representative of typical performance of the facility for the initial model calibration. Removing data outliers by performing a data cleaning and mass balance check is recommended, such as confirming the mass balance of solids into and out of secondary clarifiers.

In most wastewater modelling applications, the influent characterization is one of the key steps in calibrating the model. The partitioning of the influent COD into soluble and particulate biodegradable, and inert fractions has a significant effect on the downstream concentrations. A discussion on the fractions of wastewater COD has been included in Chap. 1. Setting these influent concentrations is the first step in model calibrations. In municipal wastewater, influent ratios such as VSS/TSS, BOD_5/COD and NH_4-N/TKN are often well known, and do not vary greatly between facilities. In refinery wastewater however, the influent ratios are not as constant or well-specified. The wastewater characteristics can be a function of refinery production choices as well as properties of crude sources and can vary over time. Considering this, it may not be accurate to use typical influent ratios from municipal wastewater. As a good practice these ratios should be measured for specific wastewater.

Refinery wastewaters can have significant soluble inert components, which are not removed by the biological treatment processes. The appropriate amount of inert soluble COD that needs to assign to the soluble inert (S_I) state variable in the influent can be estimated from the effluent COD in the observed calibration data. Refer to the discussion on COD fractionation in Chap 1.

For calibration of the biological and settling processes, historical data on the performance of the system is required. Adjustment of key parameters to fit the model to the data is an iterative process. Pilot treatability studies are often used where historical information is not available, such as in situations where a facility will be receiving a new wastewater stream or where new technology is being considered for a facility upgrade. For example, Marinetti et al. (2009) used a pilot scale membrane bioreactor (MBR) operation to characterize incoming wastewater and test the feasibility of implementing full-scale MBR technology at a refinery in Bahrain to meet future capacity requirements.

Batch respirometry experiments can be carried out on the pilot-scale sludge to determine estimates of model kinetic parameters. Additional batch-scale respirometry techniques were used in the Marinetti et al. (2009) paper above to determine biological model parameters such as heterotrophic yield (Y_H), decay (b_H), anoxic growth factor (η_G), and autotrophic growth rate (μ_{AUT}). Due to the unique wastewater composition, and high temperature of the influent feed, the set of calibrated parameters were considered to be very specific to this facility and wastewater stream.

6.4.2.3 Examples of Practical Application

Dynamic activated sludge models have been applied in the literature across a wide range of situations and technologies. The following section describes applications of models to address specific biological treatment issues.

Maqueda et al. (2006) modelled the activated sludge system of a treatment system in Mexico which was receiving refinery wastewater at temperatures from 33 to 40 °C. The simulations were part of a project to evaluate the upgrade of the existing aeration system from surface to diffused aeration. The model was calibrated using laboratory reactors to determine updated values for kinetic parameters (e.g., maximum specific growth rate, μ_{max}) at higher temperatures.

A simplified dynamic mechanistic model was used by Velasco et al. (2011) to propose a feedback control approach to dynamically adjust sludge wastage to maintain an appropriate solids retention time (SRT) to achieve effluent COD targets.

Ahmadi et al. (2019) used various kinetic modelling approaches (first order, Grau second-order and modified Stover-Kincannon) to model the biological growth process in a membrane bioreactor (MBR) facility treating refinery wastewater. A lab-scale MBR was modelled at different HRT and MLSS concentrations, and the different modelling approaches were compared. An optimal HRT and MLSS concentration set was determined, and the best kinetic model approach was determined.

Huang et al. (2020) used pilot-scale MBR studies to calibrate the kinetic parameters of the Biowin ASDM model, and to subsequently validate the model predictions. MBRs were operated at three different SRTs (10, 20 and 30 days) to observe different operating points. Model calibration showed that the influent characterization (such as unbiodegradable fractions of soluble and particulate components) and kinetic parameters (such as heterotrophic yield, decay rate) had to be adjusted from the default values to match performance. Trojanowicz and Wójcik (2014) developed and calibrated two mechanistic models for a fixed-film process for carbon removal from petroleum refinery wastewaters, using a 1-dimensional model for substrate diffusion and conversion. The models were implemented in VBA-Excel and AQUASIM 2.1f and calibrated from pilot and laboratory studies. The models were validated by comparing them to the results collected from a pilot system installed at an oil refinery in Poland that was using an Aerated Submerged Fixed-Bed Biofilm Reactor (ASFBBR) technology.

Thunnissen et al. (2021) developed a model to simulate the fate of refinery wastewater components during dissolved air flotation (DAF) treatment using a first-principles model in SimpleTreat developed by Dutch National Institute for Public Health and Environment (RIVM). The primary sedimentation model in I-SimpleTreat (an industrial wastewater version of the SimpleTreat package) with a DAF unit model and compared the model against observed removal efficiencies for 14 hydrocarbon constituents.

6.4.2.4 Process Modeling Case Study

Refinery wastewater treatment process models find several applications in practice for wastewater treatment plant design, operational optimization and troubleshooting. The development of treatment plant calibrated models entails an extensive exercise of data collection, flow scheme, reactor volumes, redundancies and various operational controls to make these a good tool for operator training and decision support. In this section, we highlight some of these important practical usage of process models.

Wastewater Treatment Plant Design

Dynamic, mechanistic process models can be used to test the sizing of unit processes under different loads and operating conditions. Typically, engineers will size units such as aeration tanks, DAFs and blowers by traditional design equations based on reaction rates, stoichiometry, and residence time. Note that in this document the terms aeration tanks, aeration basins, biological reactors, and bioreactors have been used interchangeably. Design procedures use the incoming COD load, temperature, and design choices (level of treatment, desired SRT, MLSS) to calculate aeration basin volumes, clarifier surface areas and required airflows. These calculations are done for steady-state conditions, and do not consider short-term changes in loading or operation. Safety factors (e.g., choice of design SRT) are applied to the calculations to account for unknowns associated with loading and process parameter values (e.g., bacterial growth rates).

The design (tank sizes, etc.) determined from the above method can then be configured in a dynamic modelling package to test the treatment performance. The input flow, concentrations and temperature will be entered into the model alongside the flow scheme, unit sizes and operation. The dynamic mass-balance-based model will then predict MLSS, SRT, sludge production, operating cost, and concentrations throughout the plant. Designers can then assess the information contained in the model predictions, to look for opportunities to improve performance, optimize resources and/or reduce cost. Design choices (volumes, areas, etc.) can be adjusted, and the model re-run with new inputs to generate an updated prediction of treatment performance. Iterating in this manner allows designers to take advantage of the sophisticated details present in the activated sludge models and gain a better understanding of the relationship between design choices and expected performance in a complicated system.

In essence, the flow of information in the traditional design equations and process modelling approaches work in opposite directions. Traditional design approach starts with the input of loading, effluent limits and other constraints/choices in the design equations, and solution of those equations produce the size of the units, as shown in Fig. 6.21.

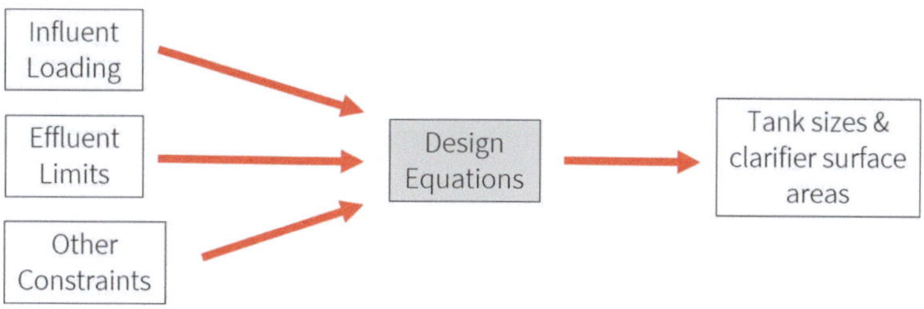

Fig. 6.21 Flow of information in traditional design approach

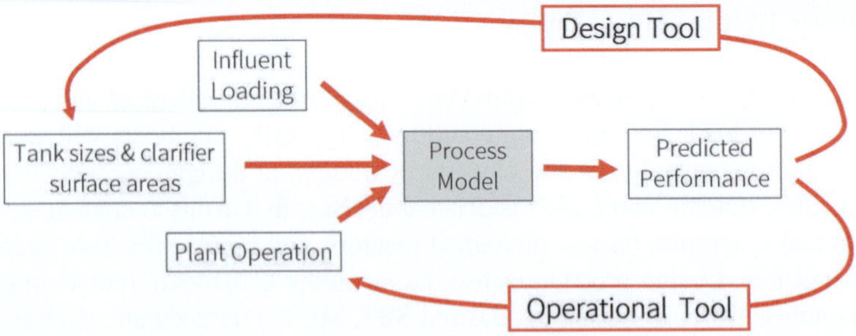

Fig. 6.22 Flow of information in modelling approach for design

With a process model, the starting point is the design loading, and unit sizes, and arrive at a prediction of performance. The sizes and/or operation of the facility can then be adjusted, and iterated through to a final optimized design, as shown in Fig. 6.22.

Some modelling packages are outfitted with advanced simulation tools that allow users to automatically run a large number of simulations and assess the collective results. These can be set up as sensitivity analyses where, for example, predictions of effluent quality can be plotted vs. aeration tank volume and/or airflow rate, aiding the designer in identifying process capacity. Some modelling packages also have search algorithms that can be employed to find the best combination of process design choices to achieve a specified goal (e.g. lowest operating cost) while observing a set of constraints (e.g. meeting effluent limits).

In this section we discuss the use of the GPS-X process simulator in designing an activated sludge treatment plant. A methodology for designing an activated sludge plant using a process simulator is presented in Fig. 6.23.

The proposed design methodology will be used to design an activated sludge treatment plant to treat a petroleum refinery wastewater.

Step 1: Problem Definition

The feed characteristics of the refinery wastewater to the biological treatment is as shown in Table 6.13.

Note that the BOD and COD concentrations used in this case study are much higher than typical (refer to the dataset presented in Chap. 5) and conservative estimates considering the operation under upset condition of refinery process units and/or the primary treatment units. The treated effluent target concentrations from the secondary treatment are shown in Table 6.14.

Based on the desired effluent concentrations in Table 6.14 the facility should be designed for both BOD removal and nitrification. The proposed flow scheme for a completely mixed activated sludge process consisting of an aeration tank and a secondary clarifier is shown in Fig. 6.24.

Design criteria within the typical range of values for the wastewater were selected:

- An SRT of 15 days and an MLSS of 3,500 mg/L in the biological reactor. A high SRT value is typically used in designing petroleum refinery wastewater treatment plant to ensure degradation of organic compounds and complete nitrification.

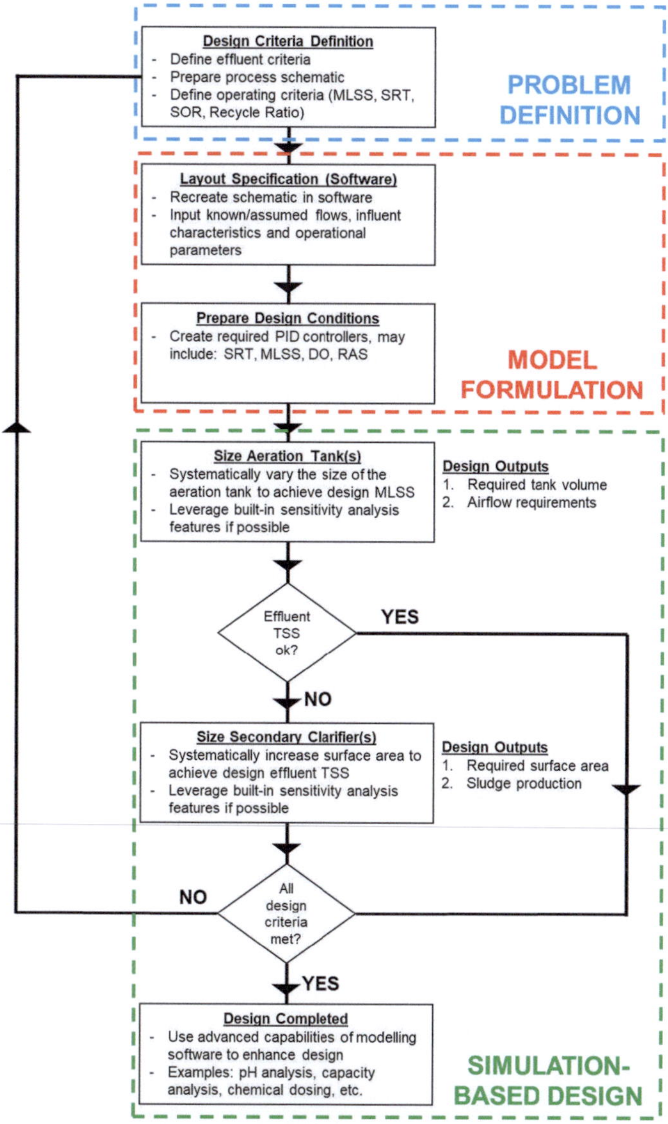

Fig. 6.23 Methodology for process simulator based design

- A surface overflow rate (SOR) of 20 $m^3/(m^2.d)$ for calculation of secondary clarifier surface area.
- A design recycle ratio of 1.0 is used to recycle the return activated sludge (RAS) back to the aeration tank.

The design criteria are summarized in Table 6.15.

Table 6.13 Case study influent characteristics

Influent parameter	Value	Units
Flow	20,000	m^3/d
Temperature	30	°C
pH	7.5	–
BOD_5	785	mgBOD/L
COD	1200	mgCOD/L
NH_4-N	80	mgN/L
TKN	100	mgN/L
Phenolic fraction of COD	0.25	mgCOD/mgCOD
Aliphatic fraction of COD	0.19	mgCOD/mgCOD

Table 6.14 Case study effluent targets

Effluent Species	Value	Units
BOD_5	<15	mgBOD/L
TSS	<15	mg/L
Ammonia	<1	mgN/L

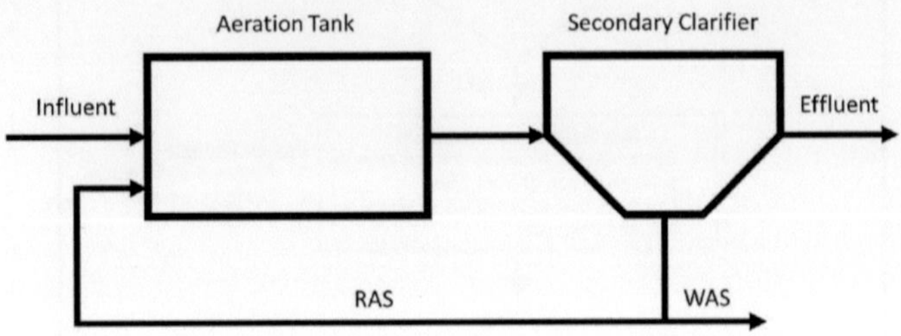

Fig. 6.24 Process schematic for the design case study

Table 6.15 Case study design criteria

Parameter	Value	Units
SRT	15	d
MLSS	3,500	mg/L
DO setpoint	2.0	mg/L
Recycle ratio	1	–
SOR	20	$m^3/(m^2.d)$

The size of the aeration tank, surface area of the clarifier and other design parameters e.g., sludge production rate, aeration requirement, effluent concentrations, and ultimate capacity of the treatment plant will be determined and optimized using the process simulator.

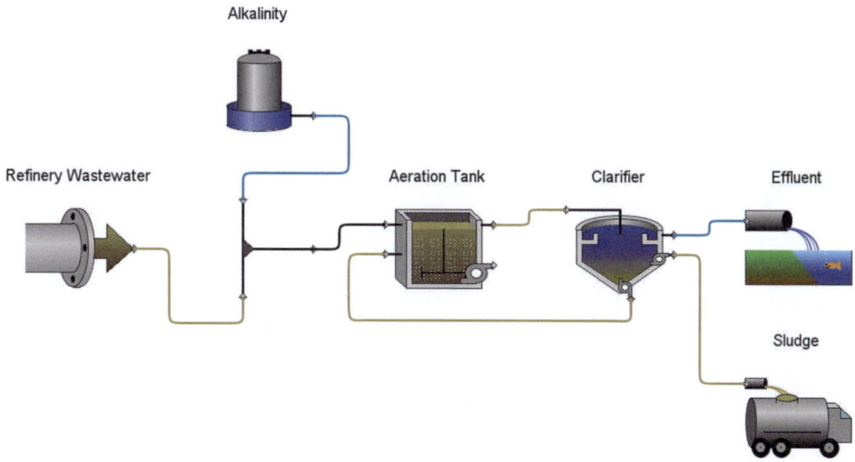

Fig. 6.25 Facility layout for design case study

Step 2: Model Formulation

The process simulator used for the design example was GPS-X. Following the definition of the design criteria, the plant layout must be set up in the process simulator. This consists of selecting the appropriate unit processes, interconnecting the unit processes and setup of selected design criteria. Figure 6.25 shows the facility layout in GPS-X with all the unit processes and inter-connections defined.

The model "MantisIW" (Hydromantis, 2017) representing the biological processes for petroleum refinery wastewater was used in this example. The MantisIW model describes BOD removal, nitrification, denitrification, and considers the volatilization and adsorption of organic compounds found in petroleum wastewater. The model also estimates the pH based on equilibrium chemistry. The typical oxygen transfer efficiency and alpha (α) value assumptions accounted for by design equations (see aeration calculations in Chap. 2) are similarly accounted for in process simulators for refinery wastewater. The secondary clarifier is modelled using a 1-dimensional settler model with a double exponential settling equation.

Petroleum refinery wastewaters typically have high concentrations of ammonia, especially those processing non-traditional or opportunity crudes, thus the layout includes an alkalinity dosage object. In practice, supplemental alkalinity may be required to achieve complete nitrification. For the preliminary design, a neutral pH of 7 will be assumed, pH calculation will be disabled, and no additional alkalinity will be added. The alkalinity object is included so that after the initial design is complete the pH effects can be re-enabled, and the model can be used to investigate pH effects on treatment.

The next step in layout specification is to specify all the known values/assumptions into the appropriate input menus of the model. In this example this includes setting up the influent flow and characterization (Fig. 6.26), entering the initial secondary clarifier surface area and recycle ratio (Fig. 6.27).

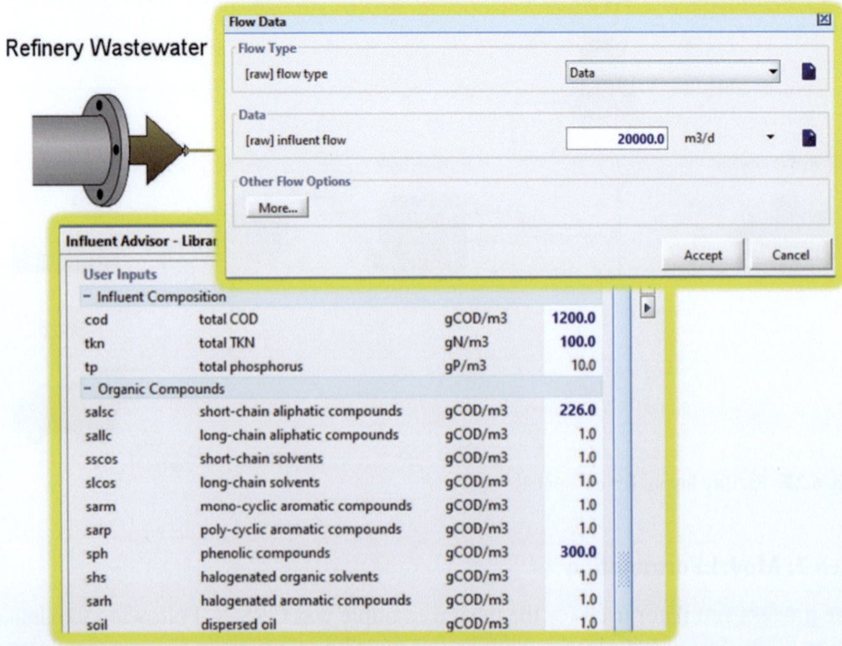

Fig. 6.26 Influent characterization in the process simulator

The influent object in the process simulator allows for a more detailed characterization compared to a typical design approach. A simple design approach typically assumes a uniform substrate with a single set of kinetic parameters. In practice, there are multiple organic/volatile compounds found in petroleum refinery wastewaters, each with its own set of kinetic parameters. The process simulator allows for substrate speciation among many of these compounds.

In this example short-chain aliphatic compounds and phenolic compounds make up a significant portion of the influent BOD. The composition of petroleum refinery wastewater that are used in this case study represents a deviation from the assumptions (municipal wastewater) inherent to the use of traditional design equations.

With the layout specification now complete, the next step of model formulation is preparing the design conditions. This step leverages the advanced features of the process simulator to assist the designer. PID controllers are defined that will automatically control certain operational parameters during the upcoming model simulations in the design process. A DO controller will ensure that the dissolved oxygen in the aeration tank will be 2.0 mg/L throughout the design process. Thus, the maximum airflow requirement can be easily obtained from the model outputs when the design criteria are satisfied. Figure 6.28 displays the operational menu of the aeration tank object and highlights the important settings for setting up a DO controller with a setpoint of 2.0 mg/L.

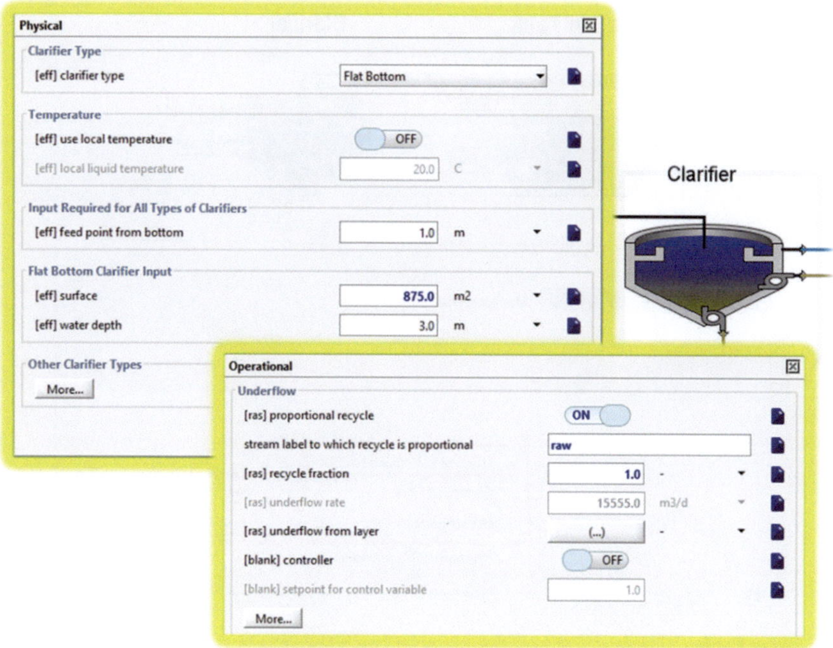

Fig. 6.27 Clarifier set-up in the process simulator

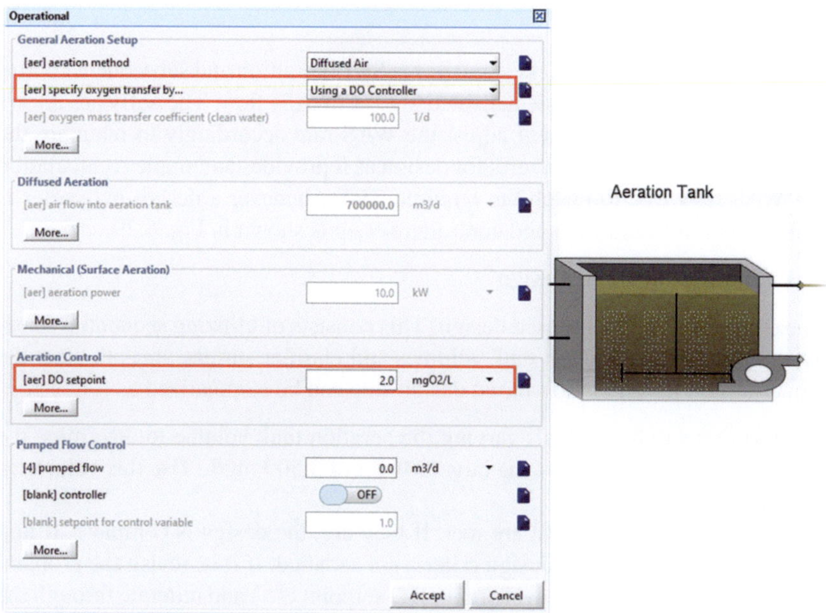

Fig. 6.28 Aeration controller set-up in the process simulator

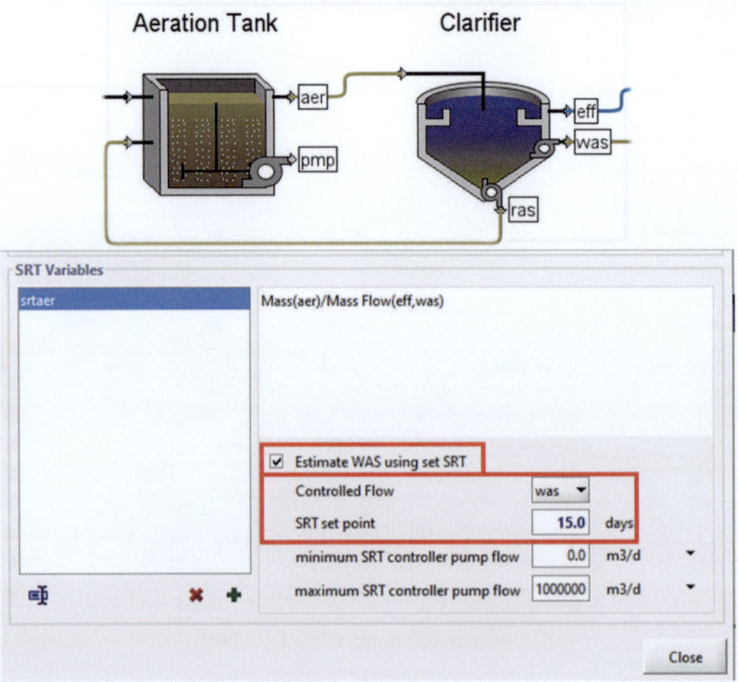

Fig. 6.29 SRT definition and controller set-up in the process simulator

Another feature of the process simulator is SRT definition and subsequent control through manipulating the waste activated sludge (WAS) flow. The software allows the user to set the SRT and will adjust the WAS rate accordingly to maintain the SRT setpoint. This strategy is useful for design as it provides automatic recalculation of the WAS flow rate to match the required SRT whenever a design parameter is modified. The SRT definition and controller set up is shown in Fig. 6.29.

Step 3: Simulation-Based Design

The next step is simulation-based design. This consists of utilizing sequential sensitivity analyses on the aeration tank volumes and clarifier surface areas to meet the design criteria. The simulation-based design step can be summarized as follows:

1. Perform sensitivity analysis varying the aeration tank volume to determine the required volume to achieve the target MLSS of 3,500 mg/L. Use this volume in the subsequent steps.
2. Check if the design criteria are met. If they are, the design is complete. If not, check if the effluent TSS design criteria are satisfied. If it is, revisit the problem definition (change SRT, recycle ratio, DO setpoint etc.) and reiterate through the design methodology.

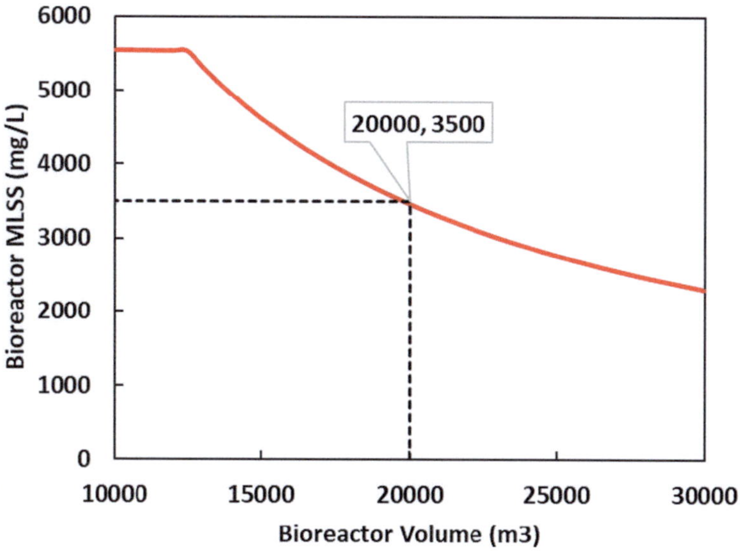

Fig. 6.30 Sensitivity analysis on the bioreactor volume

Table 6.16 Simulation results with "optimized" bioreactor volume

Model output	Value	Units
Effluent BOD_5	5.8	mgBOD/L
Effluent TSS	12.6	mg/L
Effluent ammonia	1.05	mgN/L
Bioreactor MLSS	3,457	mg/L
Bioreactor DO	1.99	mg/L
Total volatile emissions	84.8	kg/d

3. Perform sensitivity analysis varying clarifier surface area to determine the minimum surface area necessary to achieve TSS removal target.
4. Check if the design criteria are met. If they are, the design is complete. If not, revisit the problem definition (change SRT, recycle ratio, etc.) and reiterate through the design methodology.

Figure 6.30 shows the results of the sensitivity analysis on the aeration tank volume as described in Step 1 of the design procedure.

The sensitivity analysis reveals that a bioreactor volume of 20,000 m^3 results in an MLSS of approximately 3,500 mg/L. Model output from a simulation using the optimal volume of the bioreactor are shown in Table 6.16.

Because the TSS requirement is satisfied, but the ammonia is not, the design criteria must be revisited. Our engineering process knowledge tells us that increasing

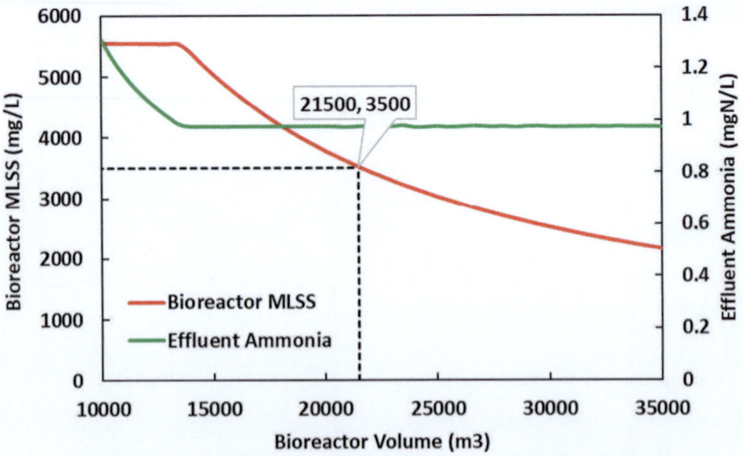

Fig. 6.31 Sensitivity analysis on the bioreactor volume with an SRT of 17 days

Table 6.17 Simulation results with "optimized" bioreactor volume and 17-day SRT

Model output	Value	Units
Effluent BOD_5	5.5	mgBOD/L
Effluent TSS	12.6	mg/L
Effluent ammonia	0.976	mgN/L
Bioreactor MLSS	3,523	mg/L
Bioreactor DO	2.0	mg/L
Total volatile emissions	83.5	kg/d

the SRT can improve nitrification. Thus, the sensitivity analysis on the bioreactor volume was re-run using an SRT of 17 days. The results are shown in Fig. 6.31.

Using an SRT of 17 days and an aeration tank volume of 21,500 m^3 results in the desired MLSS and effluent ammonia concentrations, as shown in Table 6.17.

The effluent TSS and the effluent BOD are both less than 15 mg/L. Thus, the design is within specifications and can be considered complete. Table 6.18 summarizes the physical and operational parameters determined through the model-based design procedure.

The completed design can be explored further using the process simulator. For example, due to the high concentrations of ammonia in the influent supplemental alkalinity may be required to achieve complete nitrification. This is an important design consideration for the facility and thus warrants investigation. The process simulator makes this investigation possible because it includes pH estimation functionality where the pH is estimated based on equilibrium chemistry.

First, the pH functionality is enabled, and the simulation is run using the final design parameters. The model outputs are summarized in Table 6.19.

Table 6.18 Final design parameters

Parameter	Value	Units	Notes
Bioreactor volume	21,500	m^3	Determined via sensitivity analysis
Bioreactor airflow	609,300	m^3/d	Determined from model outputs (DO controller)
SRT	17	d	Increased after revisiting model definition
WAS flow rate	617	m^3/d	Determined from model outputs (SRT controller)
Secondary clarifier surface area	1000	m^2	Design criteria SOR

Table 6.19 Simulation results with pH estimation functionality enabled

Model output	Value	Units
Effluent BOD_5	6.2	mgBOD/L
Effluent TSS	12.7	mg/L
Effluent ammonia	24.2	mgN/L
Bioreactor MLSS	3,582	mg/L
Bioreactor DO	2.0	mg/L
Total volatile emissions	98.3	kg/d
Bioreactor pH	5.8	–

The simulation results show that the influent does not contain sufficient alkalinity for complete nitrification. The pH in the bioreactor is estimated as 5.8, and the effluent ammonia is 24.2 mgN/L, much higher than the design limit of 1 mgN/L. As previously mentioned, an alkalinity dosing object was included in the model layout so the designers can investigate the required quantity of alkalinity to dose should it be needed. A sensitivity analysis was run on the alkalinity chemical flow rate and the results are shown in Fig. 6.32.

The sensitivity analysis reveals that an alkalinity flow rate of 152 m^3/d will bring the pH back up to 7 and lower the effluent ammonia concentration below 1 mgN/L. Model output from a simulation using the optimal flow rate are shown in Table 6.20.

The design is within specifications, thus the alkalinity flow rate provided by the sensitivity analysis can be added to the design parameters and supplemental alkalinity infrastructure should be accounted for in the final design.

Another useful extension that the simulator provides to the design process is capacity analysis. Using process simulator for capacity analysis allows the designer to ascertain what aspects of the treatment are most likely to fail as flows and loads increase (i.e., clarification, nitrification, BOD removal, etc.), and in what order they will likely fail.

For the capacity analysis the final design airflow will be used as the maximum available airflow and the SRT controller will be replaced with an MLSS controller to hold the MLSS at 3,500 mg/L. This allows the designer to observe the effect of

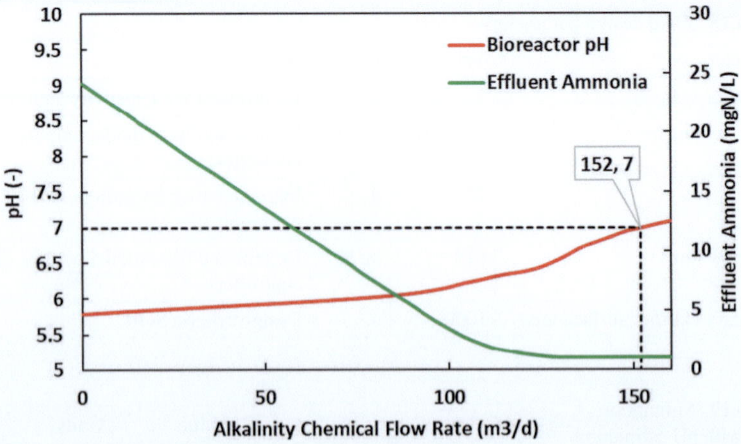

Fig. 6.32 Sensitivity analysis on the supplemental alkalinity flow rate

Table 6.20 Simulation results with pH estimation and alkalinity flow rate of 152 m3/d

Model output	Value	Units
Effluent BOD_5	5.5	mgBOD/L
Effluent TSS	12.7	mg/L
Effluent ammonia	0.98	mgN/L
Bioreactor MLSS	3,522	mg/L
Bioreactor DO	2.05	mg/L
Total volatile emissions	83.6	kg/d
Bioreactor pH	6.99	–

increased flow and load on the facilities SRT. Figure 6.33 shows the results of the capacity analysis.

The capacity analysis reveals the failure mechanisms of the various effluent concentrations of interest. The increasing influent flow reduces SRT, and the effluent ammonia increases gradually until approximately 10 days of SRT is reached (influent flow: ~32,000 m^3/d). At SRT lower than 10 days further decreases in SRT result in sharp increases in effluent ammonia concentration, indicating incomplete nitrification at these conditions. The effluent TSS gradually increases with increasing flow, and the effluent BOD increases proportionally, up until the point that airflow becomes limiting in the aeration tank. At that point, the soluble BOD begins to increase exponential with increasing flow, leading to an analogous increase in total BOD.

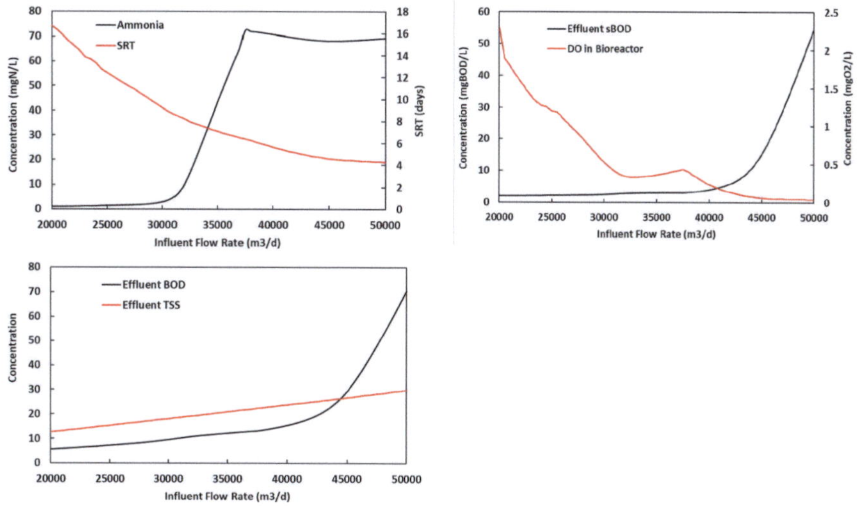

Fig. 6.33 Capacity analysis using the design parameters

One final note: It is important to recognize that the model simulations do not include safety factors. The simulated airflow and pumped flows are calculated as exactly what is required for the treatment. Appropriate safety factors should be applied by the designer when sizing equipment, as is done with a traditional design equation approach.

Troubleshooting and Process Optimization

In addition to testing and optimizing BOD removal and nitrification at the design level, process models can also be useful for troubleshooting operational issues. An example for nitrifying plant operation can be illustrated through modelling the hypothetical refinery wastewater treatment system shown in Fig. 6.34. The system consists of a refinery wastewater influent (plus additional loading from caustic wash), chemical dosage points for acid and nutrients as needed, an equalization tank, API unit (for primary oil/water separation), a DAF unit (for secondary oil/water separation), and biological activated sludge plug-flow bioreactor and clarifier. The sludge from the API, DAF and secondary settlers are disposed along with the oil removed via the API.

In the "Troubleshooting Nitrification" scenario shown below, the influent ammonia is 80 mgN/L and the effluent ammonia is 74 mgN/L (Fig. 6.35), indicates that there is no nitrification happening in the facility. In order to troubleshoot the nitrification process failure, the model can be used to observe the predicted effluent ammonia, nitrite and nitrate, along with the residual DO and SRT of the system.

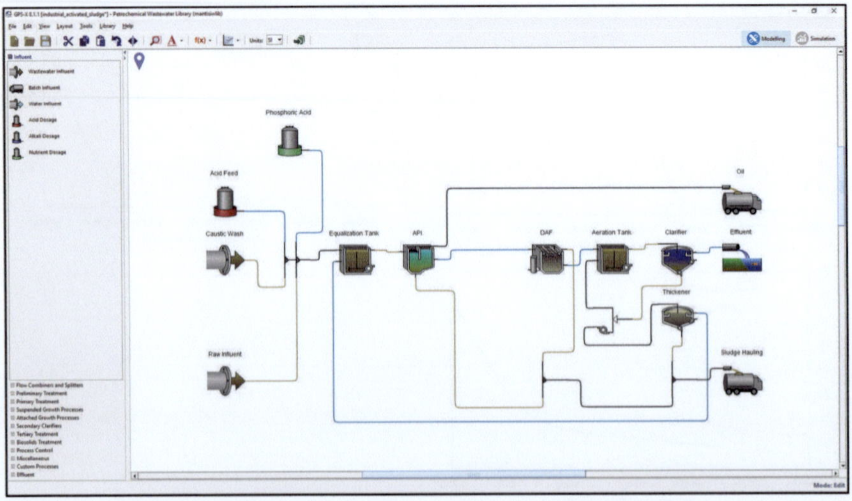

Fig. 6.34 GPS-X version 8.1, showing a model of a refinery wastewater treatment facility

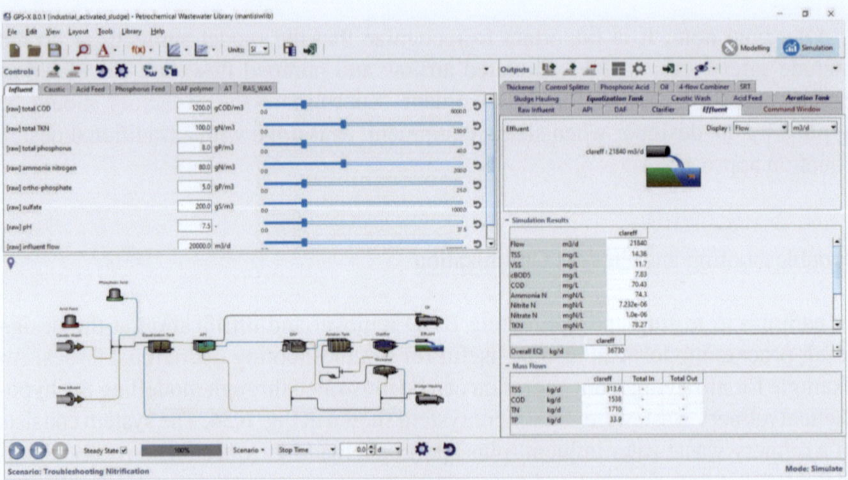

Fig. 6.35 Output from MantisIW model showing high effluent ammonia

Additional process variables such as oxygen transfer rate, biomass composition and concentration of ionized and non-ionized species are also available.

To troubleshoot the nitrification issue, the model can first confirm that there are no autotrophic biomass growing in the system. Inspecting the biomass concentrations (Fig. 6.36) it confirms that there are ordinary heterotrophic organisms (OHOs) growing in the bioreactors, but that there are no ammonia-oxidizing biomass (AOBs) or nitrite-oxidizing biomass (NOBs) present.

Fig. 6.36 Biomass concentrations of OHO, AOB and NOB populations in the bioreactor

Other factors required for nitrification are (i) sufficient dissolved oxygen, (ii) sufficient SRT to allow time for the more slowly growing AOBs and NOBs to grow in the system, and (iii) there is not any significant inhibition of AOBs and NOBs from non-ionized forms of dissociating species (e.g., free ammonia (FA) and free nitrous acid (FNA)). For an in-depth understanding of FA and FNA and their inhibitory effects on autotrophic AOB and NOB populations reader is directed to a paper authored by Anthonisen et al. (1976).

Using the model, the residual DO, SRT and potential inhibitory concentrations can be predicted. In this case study, the DO (2 mg/L) and SRT (13.7 days) are not rate-limiting factors, but the large influent concentration of over 75 mg/L could lead to inhibition. By adjusting the pH of the system, the inhibiting concentrations can be reduced. The model predictions indicate that adding ~2 m^3/day (500 US gpd) of H_2SO_4 to reduce pH from 8.5 to 7.2 was enough to reduce the inhibitory FA to a level that allowed for ammonia oxidation. However, the second nitrification step (nitrite to nitrate) has not happened, showing a high nitrite concentration (25.5 mg/L) and no nitrate (Fig 6.37), a situation commonly referred to as "nitrite lock".

In order to push the nitrification reaction all the way over to the full second step, the model can be used to investigate options. By reducing the WAS rate from 1000 to 380 m^3/day (264,000 to 100,000 US gpd), the SRT of the system was increased from 13.7 to 34.8 days. The MLSS increased from 2430 to 3940 mg/L, but the capacity of the secondary settlers was large enough to accommodate this increase in solids loading without a significant increase in effluent TSS. With the longer SRT, the nitrification reaction completes both steps, and both the ammonia and nitrite concentrations have been reduced to the desired levels (Fig. 6.38).

Table 6.21 Illustrates the predictions at the three stages of simulation, troubleshooting and analysis.

– Simulation Results

		clareff
Flow	m3/d	21840
TSS	mg/L	14.29
VSS	mg/L	11.62
cBOD5	mg/L	7.637
COD	mg/L	70.08
Ammonia N	mgN/L	1.012
Nitrite N	mgN/L	25.53
Nitrate N	mgN/L	1.0e-06
TKN	mgN/L	4.989
TN	mgN/L	30.52
Soluble PO4-P	mgP/L	0.8931
TP	mgP/L	1.711
Total Alkalinity	mgCaCO3/L	401.5
pH	-	7.781

Fig. 6.37 Effluent quality showing "nitrite lock"

Fig. 6.38 Effluent quality, showing full nitrification

– Simulation Results

		clareff
Flow	m3/d	21850
TSS	mg/L	15.8
VSS	mg/L	12.29
cBOD5	mg/L	5.814
COD	mg/L	70.34
Ammonia N	mgN/L	0.6158
Nitrite N	mgN/L	1.77
Nitrate N	mgN/L	28.98
TKN	mgN/L	4.669
TN	mgN/L	35.42
Soluble PO4-P	mgP/L	2.899
TP	mgP/L	3.736
Total Alkalinity	mgCaCO3/L	458.8
pH	-	7.941

Table 6.21 Simulation analysis results

Parameter changes	MLSS (mg/L)	Effluent TSS (mg/L)	Effluent ammonia (mg/L)	Effluent nitrite (mg/L)	Effluent nitrate (mg/L)	Effluent pH (−)	Plant SRT (d)
Base case	2500	14.4	74.3	0	0	8.2	13.7
Increase acid dosage to bring down pH	2430	14.3	1.0	27.0	0	7.8	13.7
Lower WAS to increase SRT	3940	15.8	0.6	0.8	31.4	7.7	34.8

6.4.2.5 Operator Training and Decision Support

The above type of simulation analysis is often used in training and teaching settings, where engineers and operators use a virtual wastewater treatment facility to investigate treatment performance under different operational and loading scenarios. Interactive troubleshooting challenges (such as the "nitrite lock" scenario above) are presented to the trainees, who can change different plant operating setpoints in attempting to meet the required effluent targets.

Often in these situations, a custom "operator oriented" software interface is used in the teaching environment, rather than the interface of the native modelling software. Figure 6.39 shows the SimuWorks™ operator learning interface that is applied to GPS-X models when using them for teaching petroleum refinery activated sludge process concepts.

6.4.2.6 VOC Emission Modelling

Increased permitting, monitoring, and reporting requirements for volatile organic carbon (VOC) emissions from petroleum refinery wastewater facilities has expanded the use of modeling tools such as EPA WATER9 and Toxchem™ software. These tools greatly facilitate the analysis of emissions, enabling the evaluation of different options for plant design, and regulatory reporting purposes. Connery et al. (2013) used Toxchem™ Fate and Transport Modeling software to predict volatile organic compound (VOC) emissions from petroleum industry wastewater, to provide an understanding of the mechanisms that enable effective VOC reduction using High Pressure Oxygen treatment (HPO) systems.

In the study, the activated sludge model was setup using physical and operational parameters as shown in Table 6.22. Tables 6.23 and 6.24 show the influent characteristics and the contaminants used in the model study.

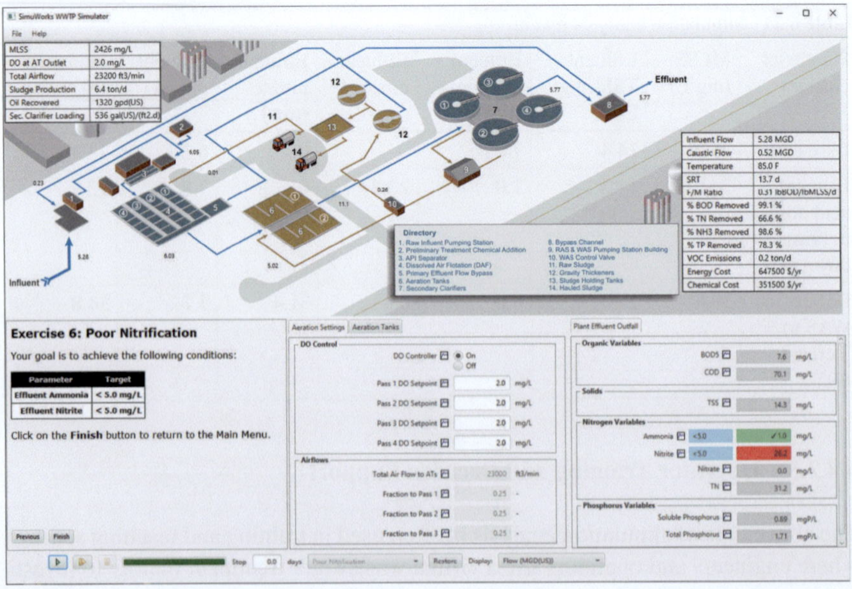

Fig. 6.39 Custom operator-oriented interface for refinery activated sludge learning environment

Table 6.22 Physical and operational parameters

Parameter	Value	Unit
Basin (aeration tank) volume	53,000	m^3
Basin depth	5	m
Basin surface area	10,600	m^2
MLSS	3000	mg/L
DO setpoint	2	mg/L
VSS/TSS	0.85	–
HRT	1.77	d
SRT	23.9	d
O_2 requirement	813	kg/h

The study was done to compare the emissions from mechanical aerated activated sludge system and open basin high purity oxygen (HPO) system. An open basin HPO is used for this example, as closed-basin systems might be at risk for potential explosion due to high concentrations of VOCs in the headspace (as discussed in Table 5.9 of Chap. 5). For the open-basin pure oxygen system, the field oxygen transfer efficiency of the system was set to 85%. For the conventional mechanical aeration system, the clean water aeration efficiency was set to 1.8 kg/kWh which is consistent for low-speed surface aeration system (Metcalf & Eddy/AECOM 2014). The alpha factor for mechanical aeration was set to 0.85 and the total power required was set to 588 kW based on the oxygen requirement. Figure 6.40 is the screen

Table 6.23 Influent characteristics

Parameter	Value	Unit
Flow	30,000	m^3/d
TSS	200	mg/L
COD	940	mg/L
TN	35	mg/L
O&G	43	mg/L
Temperature	20	mg/L

Table 6.24 Contaminant characteristics

Group	Model chemical	Concentration	Unit
Volatile organics	Benzene	48	mg/L
Phenolic/naphthenic	Phenol	48	mg/L
Alkanes	Nonane	68	mg/L
Alcohols/solvents	Ethanol	92	mg/L
Organic acids	Acetic acid	22	mg/L
Amines	Ethanolamine (Mono-)	28	mg/L

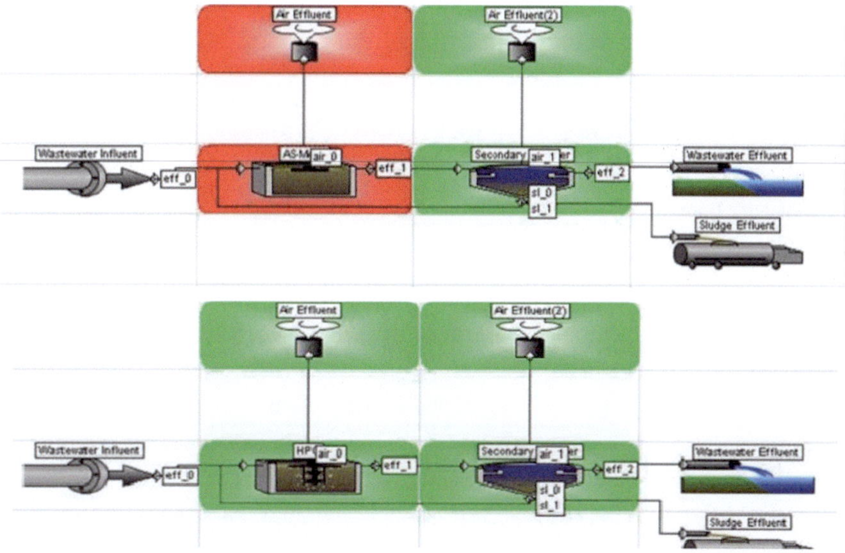

Fig. 6.40 Benzene Air Emission Intensity from (1) Mechanically Aerated System, and (2) Open Basin High Purity Oxygen System

capture from the model that shows the setup of the model (wastewater influent—biological tank—clarifier—wastewater effluent—sludge effluent). The top part of this figure illustrates the result of benzene emissions using mechanical aeration. The red background indicates that emissions from the biological basin are greater than 5%. Bottom part of this Figure shows the result of benzene emissions for the HPO system. A green background indicates emission values are less than 1%.

Table 6.25 shows the results of estimated air-emissions from the Toxchem software model using six model chemicals in the influent stream. The mass flow of the VOCs with air and the mass flow biodegraded are shown along with the percent of total. Values for chemicals leaving with the wastewater effluent and removed by adsorption to sludge were smaller and are not included in the table. Table 6.26 shows the results for the case of open basin HPO treatment system. Although biodegradation is high in both cases, air emissions were much higher for conventional mechanical aeration particularly for benzene (5.3% versus 0.3%) and nonane (4.0% versus 0.5%). Overall, the emissions emitted to the air with pure oxygen aeration were predicted to be reduced by 92.2% compared to conventional mechanical aeration.

Table 6.25 Estimated VOC air emissions from the mechanically aerated bioreactor

Compound	Air emission		Biodegradation	
	(g/d)	% of influent load	(g/d)	% of influent load
Benzene	77,233	5.363	1,360,740	94.50
Phenol	92	0.006	1,438,730	99.91
Nonane	81,666	4.003	1,952,790	95.73
Ethanol	13,483	0.489	2,714,940	98.37
Acetic acid	22	0.003	652,987	98.94
Ethanolamine (mono-)	13	0.002	728,571	86.73
Total	172,509		8,848,758	

Table 6.26 Estimated VOC air emissions from the open basin high purity oxygen system

Compound	Air emission		Biodegradation	
	(g/d)	% of influent load	(g/d)	% of influent load
Benzene	481	0.033	1,437,370	99.82
Phenol	7	0.000	1,434,530	99.62
Nonane	11,064	0.542	2,023,190	99.18
Ethanol	909	0.033	2,727,370	98.82
Acetic acid	8	0.001	653,001	98.94
Ethanolamine (mono-)	7	0.002	728,577	86.74
Total	13,467		9,004,038	

6.4.2.7 Summary and Conclusions of Biological Treatment Process Case Study

Over the past three decades, the domestic wastewater industry has made extensive use of mass-balance-based mechanistic models for plant design, optimization, and analysis. The basic structure of the underlying model equations, parameter calibration and application experience have been based on domestic wastewater treatment, however successful applications to industrial treatment facilities are also possible and are being utilized increasingly.

Over the years, commercial simulators have been developed for specific industries, including refinery wastewater treatment. Developing refinery-specific simulators has required adding additional components and to the underlying mass balances to cover processes not relevant to municipal-only wastewater (such as volatilization, sorption, inhibition, etc.).

Purpose-built models for air emissions have also become popular over the past 20 years (Toxchem, Water9) which allow users to predict air emission rates for a wide range of volatile components. In addition to process analysis and optimization, these models allow users to estimate air emissions for regulatory reporting purposes.

The application of dynamic-mechanistic models of the activated sludge system to refinery wastewater is generally a more challenging task than for domestic wastewater. There is much more uncertainty and variability in the characterization of the influent, as concentrations can be much higher and vary more significantly due to the wastewater treatment being tied to specific facility's product production schedules and crude slate compositions, rather than being evened out over a large watershed of households. This means that the models need to be calibrated with site-specific data, as the variability of influent and potentially acclimatized biomass make each modelling application very specific to the facility.

With careful model calibration and application, process simulators can provide invaluable insight into activated sludge process dynamics.

Only the modeling of activated sludge process is discussed here as this is predominantly the wastewater treatment process operating in existing petroleum refineries. However, other types of treatment processes, both attached and suspended growth, are also in operation at many refineries. In particular, there is a growing trend of adoption of MBR and MBBR in refineries globally. Separate discussion of modeling those processes could not be accommodated in consideration of space and total material to be covered in this book but each of the commercial simulators—GPS-X, Biowin, Sumo, etc. has modules dedicated to most of the commonly used biological wastewater treatment processes and are available for interested users.

Nomenclature

r_v	Rate of volatilization, g/d
$K_L{}^{voc}$	Overall mass transfer coefficient of VOC, m/d
a	Specific area for mass transfer, m^2/m^3

C	Concentration of volatile compound in the water, g/m^3
C*	Equilibrium concentration of volatile compound in water, g/m^3
V	Volume of process vessel, m^3
K_H	Henry's law coefficient, m^3 liquid/m^3 gas
C_g	Contaminant concentration in headspace, g/m^3
D_{voc}	Diffusion coefficient of compound in water, m^2/d
D_{oxy}	Diffusion coefficient of oxygen in water, m^2/d
n	Proportionality parameter between 0.5 and 1.0
$C_{sorb,\ solid}$	Sorbed mass per unit solid mass, g/g solid
K_p	Sorption partition coefficient, m^3 liquid/g solid
X_{VSS}	Solid concentration for sorption, g solid/m^3 liquid
$C_{sorb,\ liquid}$	Sorbed mass per unit liquid volume, g solid/m^3 liquid
K_{OW}	Octanol/water partition coefficient
r	Substrate uptake rate (g/m^3.d)
r_{max}	Maximum substrate uptake rate (g/m^3.d)
S	Substrate concentration (g/m^3)
K_s	Half-saturation coefficient (g/m^3)
K_{IH}	Haldane inhibition coefficient (g/m^3)
K_{IE}	Edwards inhibition coefficient (g/m^3)
K_{IA}	Aiba inhibition coefficient (g/m^3)
S_m	Substrate concentration above which net growth ceases (g/m^3)
n	Luong coefficient
IC50	Half maximal inhibitory concentration, which is a measure of the potency of a substance in inhibiting a specific biological or biochemical function.

Abbreviations and Glossary of Terms

AQ	Aqueous
AOB	Ammonia Oxidizing Bacteria
ASM 1	Activated Sludge Model 1
CFR	Code of Federal Regulations
COD	Chemical Oxygen Demand
DAF	Dissolved Air Floatation
DH	Downhole
E-NRTL	Electrolyte non-random two-liquid
EoS	Equation of State
EPA	Environmental Protection Agency
ES	Excess Solute
HC	Hydrocarbon
HKFT	Helgeson-Kirkham-Flowers-Tanger
IWA	International Water Association
MLSS	Mixed Liquor Suspended Solids

MSE	Mixed Solvent Electrolyte
NESHAP	National Emission Standard for Hazardous Air Pollutants
NIST	National Institute of Standards and Technology
NOB	Nitrite Oxidizing Bacteria
OHO	Ordinary Heterotrophic Organisms
P	Pressure
PW	Produced Water
RAS	Return Activated Sludge
S	Saturation
SI	Scale Index
SR	Scale Ratio
SRT	Solids Residence Time
ST	Scale Tendency
SWS	Sour Water Stripper
T	Temperature
TFA	Total Fatty Acids
TIC	Total Inorganic Carbon
TRC	Thermodynamic Research Center (of NIST)
VOC	Volatile Organic Carbon
WAS	Waste Activated Sludge
WEF	Water Environment Federation.

References

Ahmadi M, Benis K, Faraji M, Shakerkhatibi A (2019) Process performance and multi-kinetic modeling of a membrane bioreactor treating actual oil refinery wastewater. J Water Process Eng 28:115–122

Anderson G (2005) Thermodynamics of natural sciences, 2nd edn. Cambridge University Press

Anthonisen AC, Loehr RC, Prakasam TBS, Srinath EG (1976) Inhibition of nitrification by ammonia and nitrous acid. J WPCF 48(5):835–849

Antonio P, Hamilton J, Koopman B, Jain R, Holloway B, Lyberatos G, Svoronos SA (1990) (1990) Effect of temperature and pH on the effective maximum specific growth rate of nitrifying bacteria. Water Res 24(1):97–101

ASTM International (2019) ASTM D5002-19, standard test method for density, relative density, and API gravity of crude oils by digital density analyzer. https://doi.org/10.1520/D5002-19

Baker AJ (1994) Modelling activated sludge treatment of petroleum and petrochemical wastes. Ph.D. Thesis, McMaster University, Hamilton, Ontario

Blum DJW, Speece RE (1991) A database of chemical toxicity to environmental bacteria and its use in interspecies comparisons and correlations. Res J Water Pollut Control Fed 63(3):198–207

Bury SJ, Groot CK, Huth C, Hardt N (2002) Dynamic simulation of chemical industry wastewater treatment plants. Water Sci Technol 45(4–5):355

Carrera J, Jubany I, Carvallo L, Chamy R, Lafuente J (2004) Kinetic models for nitrification inhibition by ammonium and nitrite in a suspended and an immobilized biomass system. Process Biochem 39(2004):1159–1165

Connery K, Larrea A, DeGracia M, Fabiyi M, Novak R (2013) Practical applications of the integrated HPO solver™ Software Tool. In: Proceedings of the 86th Annual Water Environment Federation Technical Exhibition and Conference, Chicago, Illinois, USA 5–9

Dobbs RA, Wang L, Govind R (1989) Sorption of Toxic compounds on wastewater solids: Correlation with fundamental properties. Environ Sci Technol 23:1092–1097

Edwards VH (1970) The influence of high substrate concentrations on microbial kinetics. Biotechnol Bioeng 12:679–712

Eremektar G, Özgün K, Özlem, Germirli Babuna F, Ovez S, Uner H, Orhon D (2002) Biological treatability of a corn wet mill effluent. Water Sci Technol 45:339–346. https://doi.org/10.2166/wst.2002.0444

Fatone F, Di Fabio S, Bolzonella D, Cecchi F (2011) Fate of aromatic hydrocarbons in Italian municipal wastewater systems: an overview of wastewater treatment using conventional activated-sludge processes (CASP) and membrane bioreactors (MBRs). Water Res 45:93–104

Felmy AR, Rai D, Moore DA (1990) The solubility of $BaSO_4$ precipitates: Thermodynamic equilibrium and reaction path analysis. Geochimica et Cosmochimica Acta 57(18):4345–4363

Goel R, Schraa O (2013) Maximum allowable temperature for nitrification in industrial wastewater treatment plant—model based evaluation. In: WEFTEC proceeding

Goodfellow J, Malyk B, Hahn D, Johnson D (2004) Evaluation of biological phosphorous removal versus chemical phosphorous removal at a potato processing facility. Proceedings of the WEF/A&WMA 10th annual industrial wastes technical and regulatory conference, 22–25 Aug 2004, Philadelphia, Pennsylvania

Han K, Levenspiel O (1988) Extended Monod kinetics for substrate, product, and cell inhibition. Biotechnol Bioeng 32:430–437

Henze M, Grady L, Gujer W, Marais GVR, Matsuo T (1987) Activated Sludge Model, IAW Publishing

Hill GA, Robinson CW (1975) Substrate inhibition kinetics: Phenol degradation by pseudomonas putida. Biotech Bioeng 17:599–615

Hockenbury MR, Grady Jr CPL (1977) Inhibition of nitrification-effects of selected organic compounds. J (Water Pollut Control Fed) 49(5):768–777. https://doi.org/10.1017/CBO9780511840258

Hsieh C-C, Babcock Jr. RW, Stenstrom MK (1993) Estimation emissions from 20 VOCs, Part II: Diffused aeration. J Environ Eng 119(6)

Huang S, Pooi C, Shi X, Varjani S, Ng H (2020) Performance and process simulation of membrane bioreactor (MBR) treating petrochemical wastewater. Sci Total Environ 747:141311

Hydromantis (2017) Water and wastewater treatment model and simulation software. https:\\hydromantis.com

International Organization for Standardization of Natural Gas (1996) Standard reference conditions (ISO 13443) Geneva, Switzerland (as presented in https://en.wikipedia.org/wiki/Standard_temperature_and_pressure#cite_note-ISO13443-7)

Jemal M (1999) An alternative method of introducing fugacity. Chem Educator 4:1–2, Springer Nature. https://doi.org/10.1007/s00897990284a

Jiang C (1996) Solubility and solubility constant of barium sulfate in aqueous sodium sulfate solutions between 0 and 80 °C. J Solut Chem 25(1):105–111

Ky C, Comeau Y, Perrier M, Takács I (2001) Modelling biological phosphorus removal from a cheese factory effluent by an SBR. Water Sci Technol J Int Assoc Water Pollut Res 43:257–264. https://doi.org/10.2166/wst.2001.0145

Lieser KH (1965) Radiochemische messung der loslichkeit von erdalkalisulfaten in wasser und in natriumsulfatiosungen. Zeitschrift fur anorganische und allgemeine Chemie 335(5–6):225–231

Lira DCR, Will IBS, Guardani R, Nascimento CAO (2021) Modeling the kinetics of a petrochemical industrial wastewater treatment using artificial neural networks

Mackay D, Yeun TK (1983) Mass transfer coefficient correlations for volatilization of organic solutes from water. Environ Sci Technol 17:211

Maqueda MAM, Martinez SA, Narvàez D, Rodriguez M, Aguilar R, Herrero V (2006) Dynamical modeling of an activated sludge system of a petrochemical plant operating at high temperatures. Water Sci Technol 53(11):135–142

Marinetti M, Zaffaroni C, Daigger G, Givens S, Ballard R, Grady Jr L, Soydaner S (2009) Pilot testing and modeling of a membrane biological reactor system for refinery wastewater. In: Proceedings of the water environment federation's 82nd annual technical exhibition and conference, Orlando, FL

Melcer H, Bell JP, Thompson DJ, Yendt CM, Kemp J, Steep P (1994) Modeling volatile organic contaminants fate in wastewater treatment plants. J Environ Eng 120(3)

Metcalf & Eddy/AECOM (2014) Wastewater engineering. In: Treatment and reuse, 5th edn. McGraw-Hill, NY

Munz C, Roberts P (1989) Gas and liquid phase mass transfer resistances of organic compounds during mechanical surface aeration, environmental science, chemistry, water research. https://doi.org/10.1016/0043-1354(89)90026-2

Neufeld R, Greenfield J, Rieder B (1986) Temperature, cyanide and phenolic nitrification inhibition. Water Res 20(5):633–642

Nieuwland HF, Collins IR (2004) Groet 1 scale study: analysis, diagnosis, and solution implementation. In: SPE 87472 presented at the 6th international symposium on oilfield scale. Aberdeen, UK

Office of Environmental Health Hazard Assessment (OEHHA) and California Environmental Protection Agency (CEPA) (2019) Analysis of refinery chemical emissions and health effects.

Pincince AB (1995) Toxic air emissions from wastewater treatment facilities. WEF and ASCE Task force on Toxic Air Emissions from Wastewater Treatment Facilities

Rieger L, Gillot S, Langergraber G, Ohtsuki T, Shaw A, Takacs I, Winkler S (2012) Guidelines for using activated sludge models. IWA Publishing

RIVM (2018) Dutch National Institute for Public Health and the Environment website: https://www.rivm.nl/en/soil-and-water/simpletreat (accessed 23 April 2025)

Schraa O, Belia E, Churn CC (2004) The use of process modeling to investigate the performance of a large industrial wastewater treatment plant. Proceedings of the water environment federation's 77th annual technical exhibition and conference, 2–6 Oct 2004, New Orleans, Louisiana

Schroder M, Muller C, Posten C, Deckwer W-D, Hecht V (1997) Inhibition kinetics of phenol degradation from unstable steady-state data. Biotechnol Bioeng 54:567–576

Shammas NK (1986) Interactions of temperature, pH, and biomass on the nitrification process. J (Water Pollut Control Fed) 58(1):52–59

Smith JM, Van Ness HC, Abbott MM, Swihart M (2021) Introduction to chemical engineering thermodynamics, 9th edn. McGraw Hill

Snoeyink VL, Jenkins D (1991) Water chemistry. Wiley & Sons

Stark JM (1996) Modeling the temperature response of nitrification. Biogeochemistry 35:433–445

Stricker AE, Racault Y (2005) Application of activated sludge model no. 1 to biological treatment of pure winery effluents: case studies. Water Sci Technol 51(1):121

Stringfellow WT, Alvarez-Cohen L (1999) Evaluating the relationship between the sorption of PAHs to bacterial biomass and biodegradation. Water Res 33(11):2535–2544

Strubel G (1967) Zur Kenntnis und Genetischen Bedeutung des Systems BaSO4-NaCl-H2O. Neues Jahrbuch fur Mineralogie Monatshefte 15:223–234

Stumm W, Morgan J (1996) Aquatic chemistry, chemical equilibria, and rates in natural waters. J. Wiley & Sons NY (Chapter 4:4)

Takács I, Lockwood S, Caplis JR (1998) Simulation model helps manufacturing facility maintain regulatory compliance, optimize treatment processes, and train operators. Ind Wastewater

Templeton CC (1960) Solubility of Barium Sulfate in Sodium Chloride Solutions from 25 to 95 °C. J Chem Eng Data 5(4):514–516

Thibodeaux LJ (1979) Chemodynamics. John Wiley and Sons, New York, N.Y.

Thunnissen NW, van de Meent D, Struijs J, Hjort M, Redman AD, Smit MGD, Hendriks AJ, van Zelm R (2021) Simulating behavior of petroleum compounds during refinery effluent treatment using the simpletreat model. Chemosphere 263:128081

Tomson MB, Kan AT, Fu G, Wu X, Al-Thubaiti M, Kirk J, Martin R, Prukop G, Esra Inan E, Davis ST (2005) Simultaneous analysis of total alkalinity and organic acid in oilfield brine. In: SPE93266. Society of Petroleum Engineers, Richardson TX

Trojanowicz K, Wojcik W (2014) Dimensioning of aerated submerged fixed bed biofilm reactors based on a mathematical biofilm model applied to petrochemical wastewater—the link between theory and practice. Water SA 40(2):323–330

Uchameyshvili NY, Malinin SD, Khitarov NI (1966) Solubility of barite in concentrated chloride solutions of some metals at elevated temperatures in relation to problems of the genesis of barite deposits. Geochem Int 3(5):951–963

United States Environmental Protection Agency (U.S. EPA) (2010) Greenhouse gas emissions reporting from the petroleum and natural gas industry. U.S. EPA, Washington, DC, USA

van Hulle Stijn WH, Volcke Eveline IP, Lopez TJ, Brecht D, van Loosdrecht Mark CM, Vanrolleghem Peter A (2007) Influence of temperature and pH on the kinetics of the Sharon nitritation process. J Chem Technol Biotechnol 82:471–480

Velasco A, Martinez-Delgadillo S, Puebla H (2011) Regulation of petrochemical at an activated sludge system via a simple feedback control approach. In: AMCA conference

Wang P, Anderko A, Young RD (2002) A speciation based model for mixed solvent electrolyte systems. Fluid Phase Equilib 203:141–176

Weber WJ, Jones BE (1986) Toxic substances removal in activated sludge and PAC treatment systems. EPA-600/S2-86-045

WEF Manal of Practice No. 8 (2009) Design of municipal wastewater treatment plants, 5th edn. McGraw Hill

Weiland RH, Hatcher NA (2012) Sour water strippers exposed. Paper presented at the Laurence reid gas conditioning conference, Norman, OK, 28 Feb 2012

Chapter 7
Environment, Health and Safety, Security and Regulations

Mudumbai Venkatesh, Justin Scott, Tanmoy C. Chakravarty, Thomas E. Rehm, and Somnath Basu

Abstract Health, safety and well-being of employees and neighbors are critical to the success of any industrial operation. This includes operating the facilities in a responsible manner to ensure no damage is done to the environment and ecological balance is not disturbed. The concepts of Environmental Health and Safety; Personal and Process Safety and Security (EHSS) aspects associated with the management of water and wastewater in petroleum industrial facilities are discussed in this chapter. Monitoring, prevention, mitigation of hazards, and reporting impacts of accidental or intentional hazardous releases are also included. The discussions are focused on the overall goal for EHSS in increasing environmental sustainability and preventing harmful events. These relate mainly to the operation of water and wastewater treatment systems for all sectors of petroleum industry including upstream, midstream, downstream and marketing operations.

Keywords Biological hazard · Biomonitoring · Bow tie analysis · CERCLA · Chemicals of emerging concern · Clean air act · Clean water act · Cybersecurity ·

M. Venkatesh (✉)
Process and Energy Consultant, Andover, MA, USA
e-mail: m.venkatesh1@yahoo.com

J. Scott
Cove Environmental LLC, Stillwater, OK, USA
e-mail: justin@covesciences.com

T. C. Chakravarty
ZVI, Inc, Houston, TX, USA
e-mail: tchuckc@yahoo.com

T. E. Rehm
GSOP Climate Collaboration, Inc., Houston, TX, USA
e-mail: tom@tomrehm.com

S. Basu
Jewel Innovations in Energy and Environment, Katy, TX, USA
e-mail: som.basu@jewelenergyenvironment.com

S. Basu et al. (eds.), *Water Management in Petroleum Industries*, Water Resources Development and Management, https://doi.org/10.1007/978-981-19-3159-8_7

Dose response · Ecotoxicology · Environment · Fault Tree Analysis (FTA) · Fire and explosion · Hazard and Operability (HAZOP) study · Health · Layer of Protection Analysis (LOPA) · Lower explosive limit · Maximum achievable control technology · Methyl tert-butyl ether · NESHAP · NPDES permit · Opportunity crudes · OSHA · Personnel Protective Equipment (PPE) · PFAS · Process Hazard Analysis (PHA) · Process Safety Management (PSM) · Produced water · RCRA · Regulations · Risk based process safety · Risk management program · Risk matrix · Safety · Security · Tetraethyl lead · Treatability study · TSCA · WET testing · What if analysis · WHO

7.1 Introduction

The materials in this chapter will cover the concepts of Environmental Health and Safety, Personal, and Process Safety, as well as the Security for the operation water and wastewater handling and treatment of petroleum industrial facilities. Historically, industry and society developed and implemented regulations to address Environment, Health and Safety, and Security (EHSS) related concerns associated with these facilities. EHSS sectors focus on determining and monitoring impacts holistically for air, water, and land, both within and outside petroleum production and processing facilities. Discussions in this chapter relate to water management in both upstream and downstream sectors of the business. While this chapter is limited to only the operation of water and wastewater environments, it is difficult to totally avoid the related references to air, groundwater, and soil environments in the immediate vicinity. Therefore, EHSS discussions have briefly included these matrices, wherever applicable, to account for all sectors of petroleum industry operations pertaining to upstream, midstream, and downstream.

- The Environment sector focuses on monitoring, reporting and preventing impacts associated with accidental or intentional pollutant releases to abiotic and biotic surroundings and local ecosystems.
- The Health sector typically deals with the well-being of the employees in the workplace in relation to the effects of the exposure to chemicals and physical environment on employees.
- The Safety sector focuses on protecting employees and assets. Risks associated with Wastewater Treatment Plant (WWTP) operations are determined and mitigated as much as possible.
- The Security sector in this chapter focuses on risk management in the cybersecurity OT (Operational Technology) area.

In the United States (US) regulatory activities are spearheaded by the Environmental Protection Agency (EPA) and Occupational Safety and Health Administration (OSHA). Similar regulations have been implemented by the European Union (EU) and the World Health Organization (WHO). Additionally, local and regional authorities also set their own regulatory requirements, which oftentimes may be more stringent than the national and international standards. A large part of petroleum exploration and production activities are carried out offshore on platforms. There are many regulations regarding produced water and wastewater discharges from these platforms. The basic intent of these regulations is like those in the downstream operation and are not discussed separately although they have been referred to on a case-by-case basis. This chapter aims to provide an overview and basic understanding of regulatory and permitting requirements with a specific focus on both chemical and biological assays for routine monitoring and reporting, and health, safety, and security sectors for water management in the petroleum industry.

The Health and Safety program for a wastewater treatment plant should be guided by the Health and Safety department of the overall facility. Consult appropriate OSHA, American National Standard Institute (ANSI), ASTM International, and the National Fire Protection Association (NFPA) regulations and guidelines for detailed instructions. For information on biological risks and hazards, consult the Center for Disease Control (CDC) guidelines specific to the wastewater treatment practice.

7.2 Environmental Health and Safety

The ability to monitor and detect harmful impacts of wastewater discharges is not only important for aquatic environments that include native flora and fauna, but subsequently human health as well. Most protected waters are considered fishable, swimmable, and navigable and play large roles in daily human interactions. Ultimately, water pollution from petroleum sources can have major health implications and require stringent monitoring and discharge criteria.

Treated wastewater discharges can have primary impacts in protected water bodies and subsequently may have impacts to other non-protected water sources if accidental spills occur. Importantly, ecotoxicology and human toxicology incidents may occur upon exposure to treated and untreated wastewater based on the discharge variables. These include physicochemical properties of the toxicants of concern such as hydrocarbons relating to bioavailability and the mechanism of toxic action (Sikkema et al. 1995). Additionally, bioaccumulation and biomagnification are concerns when looking at chemicals that tend to degrade at lower rates such as the highly lipophilic compound benzo(a)pyrene (Torres et al. 2012) or metals such as selenium and lead (Besser et al. 2012).

Effluent and receiving stream characteristics can vary in physical and chemical properties such as pH, temperature, hardness, and dissolved oxygen (DO) levels, which can act as complexing factors to sample toxicity. An example would be a decrease in toxicity for some metals with an increase in hardness causing a less bioavailable neutral form to precipitate out (Gensemer et al. 2018). Specifically, complexes can occur with the addition of salts to certain toxic trace metals such as silver (Ag) and chloride (Cl^-). This matrix manipulation changes the more toxic form of Ag^{2+} to the less toxic form AgCl, which is less likely to be taken up by aquatic organisms such as fish (Ibrahim et al. 2020). Another example is an increase in toxicity with an increase in both pH and temperature resulting in the more toxic neutral form as un-ionized ammonia (NH_3) being present (Kleinhenz et al. 2019). Fish eliminate bodily NH_3 waste from their gills and the presence of unionized NH_3 can complicate excretion success and ultimately lead to convulsions and death (Randall and Tsui 2002). Another result of wastewater discharges could be an increase in nutrient load (i.e., nitrogen, phosphorus, sulfur and iron) and may result in harmful cyanobacterial blooms resulting in releases of toxic microcystin (Davis et al. 2009). Moreover, chemicals of emerging concern (CECs) are becoming more relevant as their frequency increases with increased operational use within facilities. Examples include the surfactant benzalkonium chloride (a quaternary ammonium compound; QAC), used in facilities as a biocide for disinfection, and the termed forever chemicals, per-and polyfluoroalkyl substances (PFAS: Sect. 7.3.4), which are used in firefighting foams for industrial application. CECs are not well understood toxicologically compared to legacy chemicals such as heavy metals (i.e., arsenic, cadmium, chromium, lead, and mercury) and organics (polychlorinated biphenyls [PCBs]), polybrominated biphenyl ethers [PBDEs]), and dichloro-diphenyl-trichloroethane [DDT]) and are becoming more stringently regulated.

Depending on the classes of chemicals and mechanism of action, responses may be categorized into chronic and acute endpoints. This relates to the duration and frequency of the dose. Chronic effects can occur over many months to several years with continuous, typically low doses. Acute effects can occur only after a single dose and may occur ranging from almost immediately to a few hours or days given a large enough dose. Toxicological impacts can vary with exposure and result in nonspecific effects such as narcosis (i.e., overall toxicity to the subject), as well as specific effects such as gastrointestinal, reproductive, or neurological toxicity depending on the mechanism of toxicity and entry route. Considering all the variables related to toxicological impacts on environmental and human health, routine monitoring and testing should be established to try and predict adverse outcomes.

7.2.1 Ecotoxicology in Relation to Environmental Degradation

Concerns with environmental degradation and the need for protective measures have historically been linked to anthropogenic effects. Since around the time of World War II, increasing wastewater discharges were fueled by industrial and urban growth (Shapiro 2022). Harmful environmental impacts were becoming more prevalent, resulting in the destruction of aquatic habitats and threatening human health. One well documented example is the Cuyahoga River (Cleveland, Ohio, USA) which at one time was one of the most polluted rivers in history. The Cuyahoga River subsequently caught fire which drew attention numerous times before 1969 due to concentrated chemical and petroleum product discharges, and oil slick accumulation (Stradling

Fig. 7.1 Anthropogenic impacts to aquatic ecosystems as illustrated by wastewater outfall discharges (**a**), fish kills as seen with polluted waters (**b**), and the Cuyahoga River Fire of 1952 (**c**). [Figure adapted and edited from Sciencing, YRT News, and Ohio History Central]

and Stradling 2008). Ultimately, the fire would be documented through publications and articles such as National Geographic's issue "*Our Ecological Crisis*", with an emphasis on the Cuyahoga River fire. Eventually, attention would shift from the direct effects of the river's fires towards the more pressing issues of the presence of pollutants, their origin points, and lesser examined dangers of toxicological impacts.

The Cuyahoga River was symbolic in facilitating the need to hold industrial dischargers accountable for the direct impacts they are causing on aquatic ecosystems. Moreover, there exists a balancing act of economic gains centered on industrial growth and reducing the detrimental impacts plaguing aquatic systems. From this major destructive event, regulations focusing on routine monitoring and reporting were created to reduce wastewater pollutant discharges and maintain the integrity of protected water bodies (Fig. 7.1).

7.2.2 *Ecotoxicology and Regulatory Testing*

Wastewater dischargers fall under proactive monitoring through the National Pollutant Discharge Elimination System (NPDES) to ensure harmful impacts do not occur. This is routinely accomplished through chemical analysis to determine toxicant concentrations are below harmful thresholds. Additionally, as part of NPDES, biomonitoring is one other step to aid in predicting ecotoxicological effects, using total toxicity testing of effluents. While some in situ (meaning place of origin) and field assessments can be made to determine this, more feasible and robust methods are conducted in aquatic toxicity testing facilities using in vivo (meaning within a living organism) methods. Accepted guidelines for test procedures for the analysis of pollutants for both chemical and biological components are found in the Code of Federal Regulations (CFR) at 40 CFR Part 136 (USEPA 2012a, b).

7.2.3 Chemical Analysis

A list of priority pollutants has been established to determine contaminants of concern and how they relate to the Clean Water Act (CWA) and discharge criteria. Petroleum toxicological impacts can consist of toxicants of concern including metals, ammonia, sulfide, cyanide, phenols and hydrocarbons at various concentrations (Wake 2005). A focused approach to monitor and report chemical constituent concentrations specific to the petroleum industry can be found in the *Guide for the Application of Effluent Limitations and Guidelines for the Petroleum Refining Industry* (USEPA 1985).

To have a standardized approach for regulatory testing of wastewater dischargers, *The Standard Methods for the Examination of the Water and Wastewater* is a collaborative effort jointly published by the American Public Health Association (APHA), the American Water Works Association (AWWA), and the Water Environment Federation (WEF) for laboratory use. This includes detailed protocols for common analytical measurements such as gravimetric, electrochemical, and colorimetric methods. Some example constituents include oil and grease, biological and chemical oxygen demand (BOD and COD), nutrients (phosphorus and nitrogen) and solids (total dissolved, suspended, and volatile). More advanced instrumentation may be required to measure concentrations of classes of chemicals such as metals (ion mass spectrometry and chromatography), volatile, semi-volatile and stable organics (liquid and gas chromatography) and surfactants (chromatography and fluorescence spectroscophy). Support analytical methods for biomonitoring assays include common constituents such as ammonia, pH, conductivity, alkalinity, hardness, major cations (sodium, potassium, calcium, and magnesium), and major anions (chloride, sulfate, and bicarbonate). A few examples of ecotoxicity common to petroleum derived chemicals of concern are provided below:

- Ammonia toxicity is ubiquitous with gill toxicity to fish and has shown to be toxic at concentrations of 0.25 to 0.3 and 7.72 to 11.2 μg/ml un-ionized ammonia to adult rainbow trout (*Oncorhynchus mykiss*) (Smart 1976) and sac fry (Brinkman et al. 2009), respectively. Ammonia toxicity of a water contaminated river from petroleum industry discharge to Wels catfish (*Silurus glanis*) was investigated and found that chemical effects can cause peroxidation in muscle and liver tissues (Avci et al. 2005).
- Trace metals, especially iron, manganese, cadmium, and lead, were shown to bioaccumulate in various tissues of the catfish, *Arius arius,* and potentially cause genotoxicity (D'Costa et al. 2017). In another study, at metal concentrations permitting survival but impairing reproduction, *D. magna* showed a decrease in growth (Biesinger and Christensen 1972).

- Phenols have been shown to affect the algal species, *Chlorella vulgaris* with sixteen (16) nitro- and methyl- substituted phenols showing growth inhibition toxicity in the bioassays ranging from 2.88 to 131.39 μg/ml (Tugcu et al. 2017). Phenols have also been found to affect fish metabolism, survival and growth, and reproduction (Saha et al. 1999).
- Dispersant-induced crude oil was evaluated and found increased erthoxyresorufin-*O*-deethylase (EROD) activity in newly hatched mummichog (*Fundulus heteroclitus*) with highest mortality rates seen at 0.5 g/L (water-accommodated fractions of dispersed crude oil) and 479 μg/L (sum of polycyclic aromatic hydrocarbons (PAHs) (Couillard et al. 2005). Hyperbranched polyethylenimine (HPEI) dispersant-like compounds have shown toxicity to *Daphnia magna* at 0.16 μg/ml with a 7-kDa HPEI size (Salehi et al. 2017).
- The group of compounds consisting of benzene, toluene, ethylbenzene, and xylenes (BTEX) have been shown to affect cell growth and chlorophyll inhibition of the alga, *Euglena gracilis* and its mutant SMZ cells down to concentrations ranging between 0.005 and 5 μM (Peng et al. 2015). In another study, water, sediment, and the fish, African arowana (*Heterotis niloticus*) were measured for bioaccumulation instances. Total concentration of BTEX in sediment and water was 32.37 and 49.86 μg/L respectively; while total concentration of BTEX in fish liver and intestine was 33.75 and 40.16 μg/g, respectively (Akinsanya et al. 2020).

7.2.4 Biomonitoring Assays

To further build on the concepts of toxicity and impacts to aquatic ecosystems, one area that is utilized for predictability of harmful impacts to aquatic populations is Whole Effluent Toxicity (WET) testing. WET testing serves to act as a metaphorical "canary in a coal mine" for wastewater and effluents discharged from point sources to protected water bodies. WET testing uses standardized aquatic toxicity methods and guidelines with model organisms to determine effects in the laboratory.

It is often questioned why a biological component is required if the known chemical concentrations for each toxicant of concern fall below toxic thresholds. However, it is observed that due to the complex nature of effluent matrices and seasonal and operational variability, biological assays are important in routine toxicity monitoring to ensure when toxicity occurs. One reason is the potential physicochemical property changes over time and chemical cycling and drifts associated with effluent composition and interactions with its receiving body of water. Examples include chemistry changes such as pH, temperature, and organic loads which can prove difficult for chemicals to remain homeostatic to the aquatic environment. Moreover, with complex effluents containing multiple toxicants, exposure scenarios to aquatic life may lead to additive or synergistic outcomes. Another aspect is that it is challenging and not feasible to measure every potential analyte that is routinely or temporarily present. Therefore, analytical measurement concentrations cannot solely serve to predict toxicological outcomes and potential effects to aquatic populations.

Along with the required support of analytical chemistry, WET testing typically involves an invertebrate and vertebrate component of environmental relevance for the facility. These assays are aimed to be reproducible for both intra- and interlaboratory testing with minimal variability. Specific to discharges containing petroleum products, previous studies have investigated comparability for the preparation of samples for aquatic toxicity of oil, dispersants, and dispersed oil and how results are collected and presented (Singer et al. 2000). Overall, testing guidelines are standardized to reduce variability and consist of controlled testing environments (i.e., temperature, light intensity, and photoperiod), age and number of organisms, test duration, exposure scenarios (i.e., sample volume and static versus flow-through exposures, and renewal frequency). Tables 7.1 and 7.2 list the routinely used model organism methods by the USEPA for WET testing guidelines for both freshwater and saltwater exposure scenarios (USEPA 2012a, b). Additionally, wastewater effluent and chemical guidelines have been established by organizations such as the International Organization for Standardization (ISO), the Organization for Economic Co-Operation and Development (OECD), and ASTM International (formerly the American Society for Testing and Materials; ASTM) that may be incorporated into NPDES permits. Traditionally, model organisms are selected for their ease of aquaculture success, handling size, and well documented toxicological databases.

Based on NPDES requirements, testing frequency, test method, and reporting endpoints are specified for each discharger. Specifically, two major types of tests are conducted depending on discharge details pertaining to the NPDES permit which consist of acute and chronic exposures at various dilution series. WET tests incorporate quality control criteria for test validation consisting of negative control (i.e., the

Table 7.1 USEPA chronic toxicity WET methods for freshwater and marine organisms

Species type	method title	Method number
Freshwater	Fathead minnow, *Pimephales promelas*, larval survival and growth	1000.0
Freshwater	Fathead minnow, *Pimephales promelas*, larval survival and teratogenicity	1001.0
Freshwater	Daphnia, *Ceriodaphnia dubia*, survival and reproduction	1002.0
Freshwater	Green alga, *Raphidocelis subcapitata*, growth	1003.0
Marine	Sheepshead minnow, *Cyprinodon variegatus*, larval survival and growth	1004.0
Marine	Sheepshead minnow, *Cyprinodon variegatus*, embryo-larval survival and teratogenicity	1005.0
Marine	Inland silverside, *Menidia beryllina*, larval survival and growth	1006.0
Marine	Mysid, *Americamysis bahia*, survival, growth and fecundity	1007.0
Marine	Sea urchin, *Arbacia punctulata*, fertilization	1008.0

Adapted from USEPA's Short-term Methods for Estimating the Chronic Toxicity of Effluents and Receiving Waters to Freshwater Organisms and Short-term Methods for Estimating the Chronic Toxicity of Effluents and Receiving Waters to Marine and Estuarine Organisms

Table 7.2 USEPA acute toxicity WET methods for freshwater and marine organisms

Species type	Method title	Method number
Freshwater	Fathead minnow, Pimephales promelas,and bannerfin shiner, *Cyprinella leedsi*	200.0
Freshwater	Daphnia, *Ceriodaphnia dubia*	2002.0
Freshwater	Rainbow trout, *Oncorhynchus mykiss*, and brook trout, Salvelinus fontinalis	2019.0
Freshwater	*Daphnia pulex* and *Daphnia magna*	2021.0
Marine	Sheepshead minnow, *Cyprinodon variegatus*	2004.0
Marine	Silverside minnow, *Menidia beryllina*, *Menidia menidia*, and *Menidia peninsulae*	2006.0
Marine	Mysid, *Americamysis bahia*	2007.0

Adapted from USEPA's Methods for Measuring the Acute Toxicity of Effluents and Receiving Waters to Freshwater and Marine Organisms

receiving body of water and/or a synthetic reagent grade water) and positive controls (i.e., reference toxicants such as NaCl or $CuSO_4$). A dilution series is created by mixing either a sample of the receiving water body (i.e., river, creek, lake, etc.) or synthetic reagent grade water, and concentrations are indicated by permit specifications or method requirements. One example is a 0.5 serial dilution series ranging from 100, 50, 25, 12.5 6.25, and 0% effluent exposure at the designated method volumes. A critical concentration value is assigned representing the pass or fail concentration criteria. An example summary of test parameters for the fathead minnow (*P. promelas*) chronic toxicity test can be found in Table 7.3 (USEPA 2002a, b, c, d).

Acute WET testing typically consists of a 48- or 96-h assay and measures only mortality for both invertebrate and vertebrate species. The mortality endpoint is commonly reported as the lethal concentration of 50% to the population (LC_{50}) of organisms tested and is calculated using the Hill slope equation (Gadagkar and Call 2015). Acute assays are generally permitted to dischargers whose impact on receiving water bodies is considered minor to the overall water quality and would require a large discharge insult to trigger toxicity. Chronic assays are generally permitted to dischargers that would have long term harmful effects due to extended exposure durations even at lower dilution concentrations. An example of the use of a chronic assay may be warranted if the discharger makes up much of the volume of the body of water when threatened or endangered species are present downstream. Chronic endpoints such as growth, and reproduction are typically reported as the lowest observed effect concentration (LOEC) to organism populations and concentrations that are termed no observed effect concentration (NOEC). Reporting endpoint examples can be found in Fig. 7.2.

Selection of acute or chronic assays are based on discharge volume to water body volume interactions, with seasonal variability, flow, and mixing zones, as well as

Table 7.3 Summary of test parameters for the Fathead Minnow (*Pimephales promelas*) USEPA chronic toxicity test 1001.0 (Adapted from USEPA's *Fathead Minnow, Pimephales promelas, Larval Survival and Teratogenicity Test, Chronic Toxicity*)

Test parameter	Parameter summary
Test type	Static renewal
Temperature	$25 \pm 1^{\circ}$C
Light quality	Ambient laboratory illumination
Light intensity	$10 - 20\mu E/m^2/s(50 - 100$ft-c)
Photoperiod	16 h light, 8 h darkness
Test chamber size	500 mL minimum
Test solution volume	250 mL minimum
Renewal of test solutions	Daily
Age of test organisms	Newly hatched larvae less than 24 h old
Larvae per test chamber	10
Replicate chambers/concentration	4 minimum
Food source and feeding	<24 h old hatched Artemia nauplii fed on days 0–6
Cleaning	Siphon daily, immediately before test solution renewal
Aeration	None, unless DO concentration falls below 4.0mg/L
Dilution water	Uncontaminated receiving or other natural water, or synthetic water
Test concentrations	5 (minimum) and a control
Dilution factor	≥ 0.5
Test duration	7 days
Endpoints	Survival and growth (weight)
Test acceptability criteria	$\geq 80\%$ survival in controls and average dry weight per surviving organism ≥ 0.5 mg
Sample volume required	2.5L/day
Sampling requirements	On-site tests: collected daily and used < 24 h of sampling; Off-site tests: a minimum of 3 samples (e.g., days 1, 3 and 5) and holding time of < 36 h before use

native aquatic populations of concern. However, scenarios where a permit consists of both acute and chronic method requirements can occur.

For acute and chronic WET testing, organisms are deemed deceased with lack of response upon prodding in test vessels. For chronic assays, sublethal endpoints are measured daily and depending on the assay at test termination which can range from 96 h to 21 days. A few examples of chronic endpoint analysis consist of the following based on 40 CFR 136:

- Growth success measured as biomass weight associated with the freshwater fish, fathead minnow (*Pimephales promelas*), through feeding of organisms until test take down.

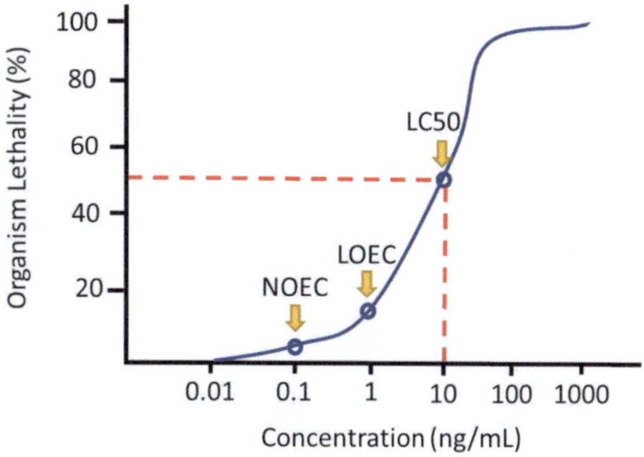

Fig. 7.2 A dose response curve comparing the no observed effect concentration (NOEC), the lowest observed effect concentration (LOEC), and the lethal concentration of 50% organismal population (LC_{50})

- Reproduction success associated with the freshwater water flea, *Ceriodaphnia dubia,* through average number of neonates throughout test duration.
- Chlorophyll content associated with the freshwater microalga, *Rhapidocelis subcapitata,* through absorbance measurement.
- Teratogenicity associated with embryo and larvae development, with the saltwater fish, sheepshead minnow (*Cyprinodon variegatus*), through gross morphological observations.
- Fecundity success associated with the saltwater crustacean, *Mysidopsis bahia,* through egg counts until test take down*; and,*
- Fertilization success associated with the saltwater sea urchin, *Arbacia punctulata*, through microscopic examination of zygote formation.

While the above-mentioned organisms represent native species in certain geographic locations, they do not cover all environmentally relevant scenarios. Thus, these model organisms serve as a best fit approach to cover the basis of their respective phylum with some more specific species such as cold-water species (rainbow trout, *O. mykiss*) and warm water species (fathead minnow, *P. promelas*). However, sensitivities can vary between species and should be noted when comparing toxicological results, even if they are similar phyla. Examples of a few select species for WET testing are presented in Fig. 7.3.

7.2.5 Toxicity Identification/Reduction Evaluations and Ecotoxicological Impacts

Toxicity Identification and Reduction Evaluations (TIE/TRE) are required when an effluent has been deemed toxic based on WET testing criteria. TIE/TRE protocols are broken into three distinct phases (Fig. 7.4). Phase I is the toxicity characterization step which allows the appropriate analytical methods to be chosen to identify toxicants (USEPA 1992). Phase II consists of the toxicant identification process and

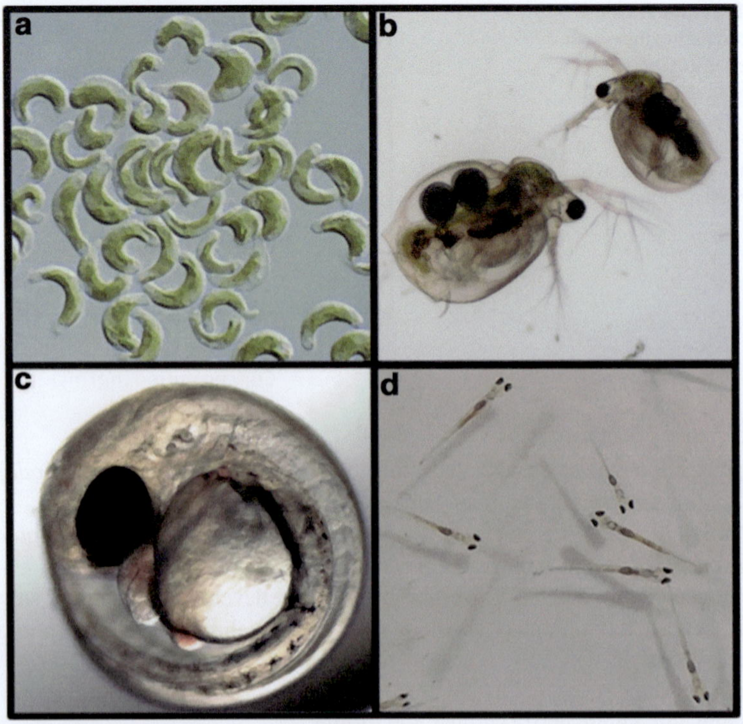

Fig. 7.3 Images of Common WET Aquatic Toxicity Testing Organisms: Green Alga, *Raphidocelis subcapitata* (**a**); Daphnia, *Ceriodaphnia dubia* (**b**) Fathead Minnow, *Pimephales promelas* Embryo (**c**) and larvae (**d**)

is different for each characteristic toxicant with a focus on separating the toxic from the nontoxic constituents (USEPA 1993a). Phase III consists of toxicity confirmation and is designed to provide additional support to the identification of the specific toxicant (USEPA 1993b).

During phase I, aliquots of effluent demonstrating baseline toxicity are subjected to a series of different physical and chemical characterizations, as well as identification and confirmation procedures. A description of the ten specific characterization tests is presented below that are frequently used:

1. **Degradation Test**—Determines the extent of degradation of the toxicant over time due to photolysis, oxidation, and/or evaporation.
2. **pH Adjustment Test and Graduated pH Test**—Determines the effect of pH manipulation on effluent toxicants.
3. **Filtration Test**—Determines the toxicity associated with filterable materials or toxicants;
4. **Aeration Test**—Determines the relative volatility or degradability of the toxicant;

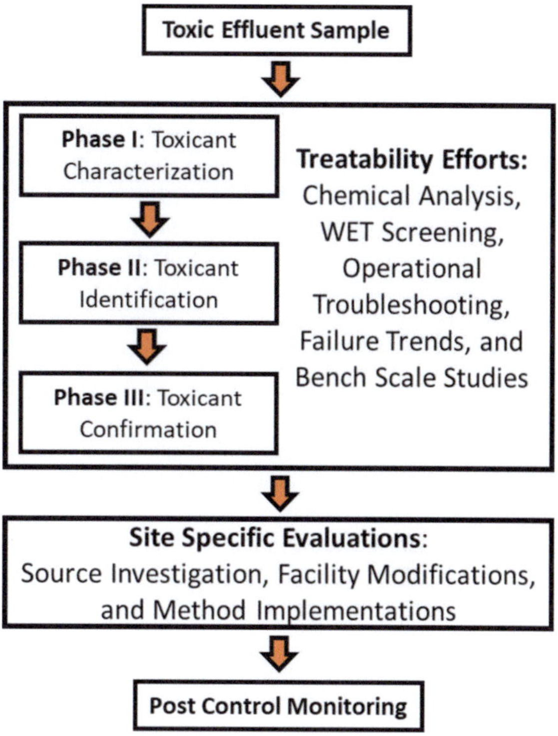

Fig. 7.4 Toxicity identification and reduction evaluation flow diagram

5. **C18-Solid Phase Extraction (SPE)**—Determines the toxicity attributable to semi-polar or non-polar organic constituents.
6. **Oxidation/Reduction Test**—Determines the extent to which the toxicants can be oxidized or reduced to non-toxic forms;
7. **Ethylenediaminetetraacetic acid (EDTA) Chelation Test**—Determines how much toxicity is attributable to certain cationic toxicants such as heavy metals;
8. **Ammonia Removal Test**—Determines if ammonia is responsible for WET, and to what extent;
9. **Mixed Bed Ion Exchange Test**—Determines if the toxicant is associated with an ionic constituent, such as Total Dissolved Solids (TDS) and heavy metals; and,
10. **Powdered Activated Carbon (PAC)**—Determines the toxicity attributable to constituents that adsorb to activated carbon.

7.2.6 Bench Scale and Treatability Studies

Once toxicants of concern have been identified, phase III allows the development of engineering and operational troubleshooting or upgrades to be considered. If the toxicant of concern is not associated with a minor change in facility operations (i.e., chemical dosage errors, operator negligence, misused equipment, etc.) then site-specific engineering modifications and upgrades may be warranted to alleviate and reduce toxic instances. Operational modifications can be incorporated into bench-scale treatability tests to further improve upon or remediate issues and deficiencies of current treatment operations. This includes mock studies of effluent manipulation through model replication to extrapolate actual facility modifications based on TIE treatment success, and physical and chemical changes required to reduce toxicity occurrences.

A few examples of this include adjusting feed rates, adjusting biological reactor specifications, increased aeration, chemical and flocculant addition, pre-acidification, and temperature adjustments. Figure 7.5 shows a bench scale treatability used to monitor influent characteristics to determine impacts on effluent composition for WET testing. Bench scale approaches can aid laboratories in allowing real-time measurements of chemical parameters and WET species sensitivity at a feasible rate. Importantly, this proactive approach allows treatment facilities to determine appropriate steps for facility modification without actual changes or potentially harmful impacts occurring with discharges.

It should be noted here that biological treatment using anaerobic reactors e.g., UASB reactors, are relatively uncommon for refinery wastewater treatment. It is presented here as an example of a biological treatment process. However, SBR reactors that use aerobic treatment are quite extensively used in refinery wastewater treatment. Documented case studies specific to petroleum wastewater treatment have highlighted TIE/TRE and treatability management and their success at toxicant reduction. A few specific examples are highlighted below:

- TIE/TRE strategies of a petroleum refinery wastewater facility indicated metals such as copper (Cu), zinc (Zn) and nickel (Ni) toxicity to be main drivers of chronic toxicity to *C. dubia* as well as the polycyclic aromatic hydrocarbon, benzo(a)pyrene (Daflon et al. 2017). In the study, C18 columns were an effective means of toxicity remediation with EDTA, and sodium thiosulfate was most beneficial for metal toxicity reduction.
- Study included the use of biodegradation and solid-phase micro-extraction-biomimetic extraction (SPME-BE) methods to address potentially bioaccumulative substances (PBS) in oil refining (Comber et al. 2015). SPME-BE methods determine the relationship between the log K_{ow} of the specific chemical and its bioaccumulation potential and subsequent toxicity. Specifically, desorption of the chemicals from fiber material is performed and quantification using chromatography is utilized with an external standard (2,3-dimehtylnaphtalene). In this way, it is then possible to address the persistence and toxicity of these PBS constituents of an effluent over time.

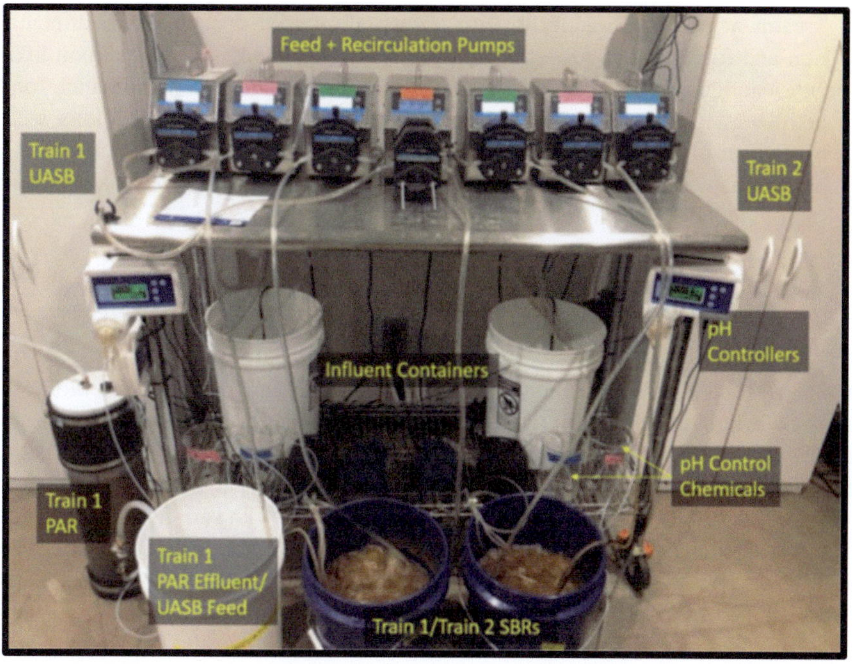

Fig. 7.5 Image of a Bench scale treatability setup used to determine WET Toxicity and biological loading of reactors for engineering and operational purposes. Setup Indicates influent feed, pre-acidification reactor (PAR) Up flow anaerobic sludge blanket (UASB) reactors, and sequencing batch reactors (SBR) for effluent decanting and sampling

- The additions of aerobic granular sludge have previously been shown as a promising biological method for wastewater treatment (Chen et al. 2019). Improvements were found with granulated sludge additions and resulted in reduced COD by 95% and petroleum compound contents by 90%. Ammonia and total nitrogen removal were also more efficient with the addition of glucose to sludge.
- Mock studies have focused on the treatment of selenium in a simulated refinery effluent by constructed wetland treatment systems (CWTSs) for treatment with organic carbon addition (Spacil et al. 2011). The study found that using CWTSs they were able to reduce the Se concentrations from approximately 50 μg/L to less than 10 μg/L and highlight the novelty of their use for stringent discharge limits.
- Investigations looked at the use of iron-tetra-amido macrocyclic ligand (TAML) activators to catalyze toxic and recalcitrant naphthenic acids (NAs) present in effluents from the oil industry with TAML catalyzation of oxidation of NAs through hydrogen peroxide activation (Pinzón-Espinosa et al. 2021)

- Treatment processes using flotation, coagulation with flocculation, sedimentation and filtration of sediments were successful at removing hydrocarbon toxicity (Steliga et al. 2015). During biological treatment, the contaminants: total petroleum hydrocarbons, BTEX, and polycyclic aromatic hydrocarbons were treated on a flow-line installation. Reduction in toxicity was finalized with a biological module consisting of an active biopreparation of a microbial consortium (non-pathogenic indigenous bacteria and fungi).

7.2.7 Toxicity Modeling and Novel Alternative Advancements

Modeling tools such as the Biotic Ligand Model (BLM) which examines the bioavailability of metals and their affinity to accumulate on gill surfaces of aquatic organisms can be used to predict potential toxicity (Niyogi and Wood 2004). Overall, the BLM can help to better understand potential impacts that may occur due to chemical availability in solution. However, as mentioned earlier, issues with variability of effluents to environmental conditions may make the BLM not that reliable for predictability in actual exposure scenarios. Previous studies used the marine copepod, *Acartia tonsa*, to determine BLM predictability of copper toxicity and found salinity and food were needed in the model to better facilitate its novelty (Leães Pinho and Bianchini 2010). This could be the case with not only metals but organics as well due to chemical competition at biological surface receptor sites, as well as chemical complexation of the external environmental conditions.

Additional predictive tools such as Advanced Outcome Pathways have gained momentum in being useful to try and predict toxicological outcomes to fish populations from the chemical of concern (Ankley et al. 2010). These pathways are models that categorize the sequence of events required to produce toxicological effects from a substance's molecular and cellular interaction up through the biological organizations to organismal and population effects. The classical example of this is looking at the chemical fadrozole, which is an aromatase inhibitor and responsible for decreasing the conversion of testosterone to 17β estradiol in fish (Conolly et al. 2018). Due to adverse impacts to the endocrine and reproductive systems and subsequently decreased egg clutch size and viability, a decrease in fish populations can be predicted using the framework. Mechanistic modeling has also been suggested based on oil refinery wastewater and measured bioavailability of hydrocarbons in toxicity testing (Whale et al. 2022). The previous study evaluated the biomimetic extraction method to determine baseline narcosis toxicity due to the total sum of hydrocarbons in lipids.

Moreover, animal alternatives such as in vitro (Latin meaning in glass) are becoming an increasingly popular topic among both private and public sectors. In vitro assays are typically conducted in an artificial environment (i.e., petri dish, test tube, well plates, etc.) using microorganisms, molecules, and cells. Recently, cytotoxicity assays using the rainbow trout (*O. mykiss*) gill cell line, RTgill-W1, and Fish acute Embryo Toxicity (FET) tests have been established as ISO and OECD guidelines and approved as replacements to the acute fish test (OECD 2013, 2021; ISO 2019). RTgill-W1 cells have shown good correlations with common WET chemicals of interest and identification of toxicants of concern based on subcellular effects (Scott et al. 2021). The use of cytotoxicity assays highlights the novelty of toxicological assays that can specifically identify which toxicants might be inducing failures in WET discharges to help with remediation and insult frequency. Figure 7.6 shows the effects of ammonia toxicity on the lysosomal integrity of RTgill-W1 cells. Previous studies have investigated the use of more sensitive species such as invertebrates (i.e., algae and daphnids) to predict toxicity to higher trophic levels such as fish in petroleum wastewater examples (Hughes et al. 2021). Additionally, the use of fathead minnow FETs has shown the ability to predict toxicity to WET common chemicals of interest and is useful in highlighting toxicity at a more sensitive developmental life stage to fish (Scott et al. 2022).

Overall, WET testing provides a limited amount of mechanistic information regarding toxicological insults to the organisms of interest. Improving on current methods with more advanced and detailed mechanistic assays could facilitate a more effective step for TIE and TRE strategies. The use of modeling and in vitro techniques offers strategies worth incorporating as novel forensic tools aimed at increasing the success of toxicant identification in WET testing.

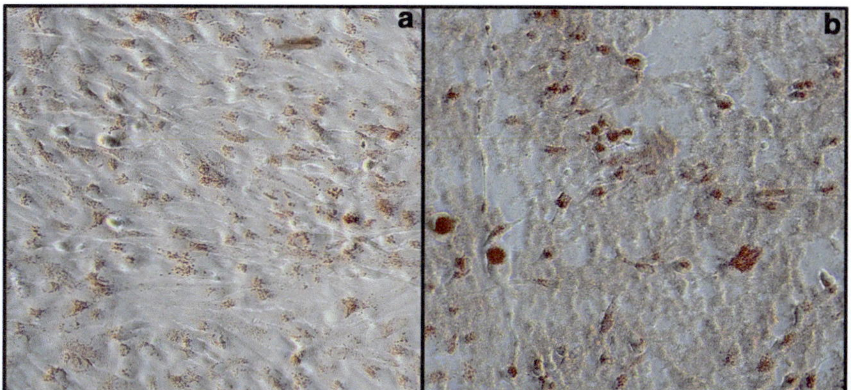

Fig. 7.6 RTgill-W1 Cells Used for Toxicity Identification Evaluations to Detect Ammonia Toxicity. Cells have been Stained with a Fluorescent Dye used to Detect the Morphological and Enzymatic Effects of Chemicals on the Lysosome Control (**a**) and Exposed (**b**) Cells. Exposed Cells Indicate Increases in Vacuolization and Uptake of the Dye

7.3 Environmental Regulations

7.3.1 Permitting and Discharge Guidelines Overview

Petroleum exploration, production and processing activities have a significant impact on the environment and the surrounding community. Produced water from upstream operations and wastewater from oil refineries contain numerous pollutants such as sulfides, ammonia, phenols, cyanides, and other organic compounds plus an array of heavy metals that need to be addressed and/or treated before discharge. Additionally, typical air pollutants released from oil refineries include volatile organic compounds (VOCs), sulfur oxides (SO_x), nitrogen oxides (NO_x), carbon oxides (CO and CO_2), hydrogen sulfide (H_2S), and particulate matter (PM). Many of the pollutants emitted by oil refineries are hazardous air pollutants such as BTEX and some greenhouse gases e.g., CO_2, methane and nitrous oxide to a lesser extent.

In the United States, petroleum industrial operations are governed by the provisions of both the Clean Water Act (CWA) and the Clean Air Act (CAA). When petroleum facilities are newly built, or changes made to existing facilities, they are subjected to a New Source Review (NSR) under the CAA. Before the facility is constructed an NSR permit needs to be obtained. The details of this permitting program are quite involved and beyond the scope of this book.

7.3.1.1 Regulations for Produced Water from Upstream Operations

The discussions in this section are primarily geared towards the regulations for waters produced from offshore exploration and from shale oil reserves.

For offshore oil and gas production activities in the United States, requirements are covered under the U.S. EPA offshore effluent guidelines for the Oil and Gas Subcategory (USEPA, 1993c). The Primary Drinking Water Regulations for radionuclides (40 C.F.R. Part 141.15, 1976) (Federal Government 2021) established the limits of radioactivity in water.

The offshore production locations are distributed globally, and regulations are specific to those locations. One of the most active areas in North America is the Gulf of Mexico. Marine discharges from offshore platforms covering the coast of Louisiana and Texas are regulated by the EPA Region 6 office which issues the NPDES Permits. It became effective from May 2023 and will expire in May 2028. The regulations specify the discharge limits for each kind of fluid including cutting fluid, drilling fluid, produced water and sanitary wastes. Produced water effluents require monitoring and compliance with the discharge limits for Flow Rate, Oil & Grease, Visual Appearance of Oil Sheen, Temperature, Chronic and Acute Toxicity by WET Testing Protocol and MEG (if it is used as an antifreeze). Oil and Grease limits are daily maximum 42 mg/L and monthly average 29 mg/L. Details of each limit, monitoring and sampling and analytical methods are provided in the EPA document GMG290000.

Another active region in the continental shelf of USA is the territorial waters off the coast southern California covered by the EPA Region 9. The permit requirements are—no visible oil sheen, Oil and Grease limits of daily maximum 42 mg/L and monthly average 29 mg/L, limits on eight metals and two non-metals (arsenic and selenium), cyanide, phenols, monitoring of 26 chemicals and WET testing. Details of NPDES permit document are provided in the EPA document CMG280000.

International efforts for the protection of marine environment from oil and gas exploration activities is largely conducted through regional treaties and agreements between multiple countries covering the waters in those regions. The OSPAR Convention for the Protection of the Marine Environment of the North-East Atlantic was originally convened in Oslo in 1972 and ratified in Paris between the member states Belgium, Denmark, Finland, France, Germany, Iceland, Ireland, Luxembourg, Spain, The Netherlands, Norway, Portugal, Sweden, Switzerland and the United Kingdom in 1994. One of the main goals set in 2001 was to reduce oil concentration in produced water to 30 mg/L by 2007. This resulted in 20% reduction of oil discharges between 2005 and 2009 and 19% reduction between 2009 and 2018, and 50% reduction of hazardous chemicals discharges between 2005 and 2009 and 28% reduction between 2009 and 2018. A risk-based assessment (RBA) was initiated in 2012 with the goal of completing the assessment of all the offshore installations within the countries by 2020. Acute Whole Effluent Toxicity (WET) procedure was followed to determine the critical dilution for the RBA. Review of the results of the RBA program was scheduled for 2024. The key conclusions (Smit 2024) so far are:

- The methods followed for the risk-based analysis (RBA) are overly conservative.
- The ability for risk assessment of offshore chemicals discharge is limited by availability and access to data which are critical to the success of OSPAR program.
- Acute toxicity results from the tests may not adequately represent controlled risks.
- The RBA method needs further refinement rather than major modification to achieve meaningful results.

Another important international treaty is the Barcelona Convention which was enacted in 1978 under the United Nations Environment Program (UNEP). It was renamed in 1995 as the Convention for the Protection of the Marine Environment and the Coastal Region of the Mediterranean. The signatories are 21 countries in Asia, Europe and Africa sharing the waters of the Mediterranean Sea, and the European Union. Details of their program and publications are available on their website unep.org/unepmap/.

The other important resource for oil and gas production is the unconventional shale reserves. The health, safety, and environmental concerns associated with production by hydrofracturing operation are as follows:

- Potential for groundwater aquifer contamination by hydrofracturing chemicals.
- Potential for groundwater contamination by NORM.
- Health and safety concerns for workers exposed to radiation hazard.
- Environmental concerns associated with disposal of dissolved solids (TDS) and radioactive materials.

Petroleum exploration and production activities are continuously increasing due to the rising demand of energy globally. Associated with it is the continuously increasing need for reuse and disposal of produce water. Environmental and social impact is an important consideration because of the heavy traffic caused by trucking in fresh water for hydrofracturing and trucking out produced water for treatment and disposal has led to several fatal accidents and injuries to residents. Even transportation by pipelines also may encounter social conflicts as some of the shale plays are in native reservation areas and infringe upon the land rights of those communities. These will need careful consideration.

Producers are adopting new technologies with the aim of increasing reuse and reducing disposal. As the characteristics of produced water vary widely depending upon the production method and the geochemistry of the formation the regulations are becoming increasingly stringent to ensure no harm to the environment is caused by the production activities. Some of the most productive shales around the globe are located within the US which include Permian Basin covering West Texas and Southeast New Mexico; Marcellus Shale in Appalachian Mountains region covering parts of Kentucky, Maryland, New York, Ohio, Pennsylvania, Virginia and West Virginia; and Bakken Shale between North Dakota and Montana. The hydrocarbon production activities simultaneously produce large quantities of water which varies significantly from shale to shale and also from time to time. For example, the produced water generated from the Appalachian formation is on an average ~0.1 barrel per barrel of oil equivalent, at Bakken the factor is 0.9, and at Permian basin the rate is ~3.9 barrels per barrel of crude (Nowak 2024). The producers largely depend on underground injection as the means of disposal as this water is not readily usable due to the excessively high salinity. However, disposal by underground injection is strongly being discouraged due to the recent events of seismicity, especially in Texas and New Mexico, caused by the injection pressure on the rock structure. The magnitude and frequency of seismicity in the Permian basin dramatically increased from 2012 to 2022. In Texas, the Railroad Commission is actively working on reducing injection by revising the Rule 8 (16 Texas Administrative Code 3.8) and rolling it over to Code 4 of the 16 Texas Administrative Code of the Railroad Commission for the Permit requirements for underground injection. Solicitation for public comments on the of the revised draft permit started in October 2023 and final rule making is expected to be in late 2024. To reduce injection the State of Texas is promoting increase of recycle, beneficial reuse, and disposal to surface water bodies by implementation of new and innovative desalination technologies (Wooten 2024). Beneficial reuse by land application for agriculture requires permit from Texas Railroad Commission (TRRC) and disposal to bodies of surface water requires NPDES Permit from the Texas Commission for Environmental Quality (TCEQ).

At Marcelus shale individual states take the lead in rulemaking and implementing those for the protection of human health, safety and environment. Interested readers are referred to the websites of the states concerned for detailed information.

The third major means of oil production is from unconventional resources by application of thermal methods. These are primarily located in the Canadian Rockies. Emissions and effluents from thermal facilities in Canada are regulated through several acts and directives issued by the provincial and federal regulatory agencies and are not separately discussed here.

7.3.1.2 Regulations for Wastewater Discharges from Downstream Operations

The Clean Water Act (CWA) has established regulations aimed at restoring and maintaining the chemical, physical, and biological integrity of the Nation's waters. Specifically, the CWA's laws and regulations focus on fishable, swimmable, and navigable waters and are administered and enforced through collaborated efforts of the US Environmental Protection Agency (USEPA) and state governments through their respective Departments of Environmental Quality (DEQ) and Departments of Natural Resources (DNRs). Much improvement has been made since the implementation of the CWA for aquatic life and water quality through stringent guidelines and regulatory efforts.

The year 2023 marked the 50th anniversary of CWA. National Pollutant Discharge Elimination System (NPDES) was created in 1972 by the CWA, and under it, dischargers are issued permits and required to maintain compliance following criteria subject to their discharge type and frequency (USEPA 2022). The USEPA has regulated a list of 126 priority pollutants (USEPA 1991) that may be used for initial permit applications or renewals, or in some cases a specific set of chemicals may be monitored given the type of permit issued. This includes initially developing a list of chemicals of concern about the specific facility operations and applicable wastewater matrices.

The NPDES permit program provides two types of levels of control, Technology Based Effluent Limitations (TBEL) and Water Quality-Based Effluent Limitations (WQBEL) (USEPA 2010). TBELs utilize either a national effluent limitation criterion or a case-by-case basis of professional judgment for point source discharge using Best Available Control Technologies (BACT). When these technologies are inadequate to meet their treatment criteria, agencies may develop a Total Maximum Daily Load (TDML) which requires discharges not to exceed a safety margin of load to the receiving body of water. It must be recognized that TMDL limits are based on the total daily load of a pollutant in question, which in turn is based on both the discharge flow and concentration, as opposed to only the concentration.

Overall, two classes of permits are issued, and depending on the applicability of the facility, they could fall under a general permit (i.e., broad requirements based on facility type) or a specific permit (i.e., more complicated due to less common conditions). Specifically, petroleum wastewater guidelines such as the Petroleum Refining Guidelines and Standards (40 CFR Part 419; USEPA 2012a, b) are incorporated into NPDES permits for wastewater direct dischargers as well as other control mechanisms for indirect dischargers.

7.3.1.3 Regulations for Air Emissions from Petroleum Production and Refining Activities

The Clean Air Act (CAA) has established regulations, through the USEPA, that establishes National Ambient Air Quality Standards (NAAQS) to regulate emissions of hazardous air pollutants in order to protect public health and public welfare (42 US Code Chapter 85 1970). Under the CAA, NAAQS were established in every state by 1975 along with directing each state to develop their own State Implementation Plans (SIPs). These were developed to be appropriate for industrial sources in the state, in order to be applicable and achievable to these standards. The Act was later amended to set timeframes for completing NAAQS since many states had failed to meet their original deadlines.

Specifically, section 112 of the CAA discusses emissions of hazardous air pollutants. In 1990, amendments to the CAA revised Section 112 to primarily require implementation of technology-based standards for major sources of emissions and certain area sources. For major sources, Section 112 requires USEPA to establish emission standards which require the maximum degree of reduction in emissions of hazardous air pollutants also known as, Maximum Achievable Control Technology (MACT) standards.

Additionally, in the US, petroleum refineries are subject to the National Emission Standards for Hazardous Air Pollutants (NESHAP) regulations which were promulgated by USEPA in 1993. Although these are air pollution regulations, they apply exclusively to wastewater streams in oil refineries and are intended to reduce benzene emissions from refineries. These are also commonly referred to as the Benzene Waste Operations NESHAP (BWON) rules. BWON rules are complex and can be somewhat confusing. However, some definitions under BWON are provided below:

Total Annual Benzene (TAB)—The sum of the annual benzene quantity for each waste stream at the facility that has a flow-weighted and volume-based annual average water content greater than 10% mixed with water (or other wastes), at any time. The mixture must also have an annual average water content greater than 10%.

Waste Stream—The waste generated by a particular process unit, product tank, or waste management unit. The characteristics of the waste stream (e.g., flow rate, benzene concentration, water content) are determined at the point of waste generation.

Examples of a waste stream include process wastewater, product tank drawdown, sludge and slop oil removed from waste management units, and landfill leachate.

Waste—Any material resulting from industrial, commercial, mining, or agricultural operations, or from community activities that is discarded or is being accumulated, stored, or physically, chemically, thermally, or biologically treated before being discarded, recycled, or discharged.

Examples of waste include slop oil routed to slop tanks, spent caustics, sludge, Dissolved Aeration Flotation (DAF) float sent offsite, pump drainages, lab wastes, and unit wash-down material.

Waste Management Unit (WMU)—A piece of equipment, structure, or transport mechanism used in handling, storage, treatment, or disposal of waste. Examples of a waste management unit include a tank, surface impoundment, container, oil–water separator, individual drain system, steam stripping unit, thin-film evaporation unit, waste incinerator, and landfill.

Point of Waste Generation (POG)—The location where the waste stream exits the process unit component or storage tank before handling or treatment in an operation that is not an integral part of the production process. Additionally, for waste management units that generate new waste after treatment, the location where the waste stream exits the waste management unit component.

Process Wastewater—Water that comes in contact with benzene during manufacturing or processing operations conducted within a process unit. Process wastewater is not classified as organic wastes, process fluids, product tank drawdown, cooling tower blowdown, stream trap condensate or landfill leachate.

The compliance requirements of the BWON rule are determined by calculating the TAB quantity for the facility. The annual benzene quantity for a waste stream is calculated by multiplying the annual waste quantity of the waste stream times the flow-weighted, annual average benzene concentration (weight %) of the waste stream. The annual benzene quantity and TAB are calculated in units of megagrams per year (Mg/yr.) of benzene where one megagram is equal to one metric ton. The benzene quantity for each waste stream is determined by estimating the stream's annual waste quantity and annual benzene concentration. Periodically the quantity of waste may not be known. In this situation, engineering judgment and/or operational experience must be used to determine the quantity. When determining a waste quantity, the oil and water fraction of the waste must be determined.

The BWON rule allows two ways for determining the benzene concentration for a waste stream, process knowledge or direct measurement. As noted in the definition of TAB, the water content of the waste stream is a key threshold. Wastes with <10% water are considered organic waste and are not included in the calculation of the TAB. However, if organic waste is mixed or co-mingled with an aqueous waste (a waste with >10% water content) then the mixed/co-mingled waste is treated as an aqueous waste and the contribution of benzene in the organic phase must be included in the TAB.

In summary, facilities with a TAB quantity of greater than 10 Mg/yr. are required to control and treat selected facility waste streams per one of the available BWON compliance options. Generally, either the 2 Mg/yr. segregate-and-treat option or the 6 Mg/yr. treat-to-target option, also referred to as the '2BQ' and '6BQ' compliance options, accordingly.

In upstream operations produced water is subject to new and upcoming regulations. These include the New Mexico Title 20, Part 2, Chapter 50 regulation which took effect from August 2022 for controlling VOC and NO_x emissions. This stipulates beginning monitoring emissions within 90 days by all existing facilities emitting over 2 tons/year of VOC, reduction of VOC emissions by 20% within 2 years and by 35% within 3 years for all existing facilities. New facilities are required to submit VOC emissions control plan.

A global initiative was undertaken at the COP 28 summit in 2023 in Dubai aiming methane emissions reduction. Emission Factors were established for produced water as a source of methane emissions. An Oil and Gas Decarbonization Charter was launched during this summit where 52 oil and gas companies around the world pledged to stop routine flaring and control upstream methane emission to near zero level by 2030.

7.3.2 Risk Assessment and Impacts of Environmental Regulations

Most of the previous examples have discussed controlled discharges with a focus on regulations and preventing harmful impacts based on the CWA. Additionally, the CWA covers accidental releases of petroleum and hazardous substances. This includes spill prevention, reporting and planning, and required responses regarding environmental and community impacts, and if applicable civil and criminal penalties associated with violations (Clayton et al. 2015).

Petroleum wastewater facilities can help reduce enforcement actions by development and implementation of compliance programs that include stringent internal quality assurance and quality control programs. This can include safety and facility internal audits and checks to ensure compliance requirements are being properly implemented, discrepancies noted, and corrective actions taken. Additionally, a risk assessment and risk management program should be included to help prevent and address issues with accidental discharges or events of noncompliance such as WET failures. Moreover, proactive steps are essential to ensure that minimal impacts occur to the environment, remediation and responses to regulatory agencies are appropriate, and penalties and violations are averted.

7.3.3 Technologies for Regulatory Compliance and Remediation

The handling and discharge of water and wastewater streams associated with petroleum production and refining activities are regulated by the USEPA due to their adverse impacts on environment and human health. As discussed above, most of these regulations are driven by the CWA that regulates the quality and characteristics of discharges into receiving water bodies through the implementation of the NPDES permit. However, several other regulations also oversee the water quality protection programs at public utilities and industries. This includes the following:

- Clean Air Act (CAA)
- Resource Conservation and Recovery Act (RCRA)
- Toxic Substances Control Act (TSCA)
- Comprehensive Environmental Response Compensation and Liability Act (CERCLA)
- Federal Insecticide, Fungicide and Rodenticide Act (FIFRA)

As discussed above, the BWON Rule was primarily driven by the CAA to protect the air quality from emission of hazardous air pollutants from industrial water and wastewater streams. To address these regulatory restrictions, while maintaining safe and uninterrupted operation, the concept of covering tanks and vessels was developed and implemented. Details of these measures are discussed below under the section on Process the section on Safety. Environmental regulations have spurred the application and innovation of technology to address regulations such that the petroleum production and processing facilities continue to contribute to the economic wellbeing of ~~the~~ society while operating safely and causing no harm to the environment and human health. A few examples are presented below.

As discussed in Chaps. 3 and 5, upstream produced water and refinery wastewater streams contain naphthenic acids (NA) which are recalcitrant to biological treatment. This is especially more pronounced when such waters are associated with heavy oils. As a result, treated effluent from refineries that process high NA crudes will have low BOD but high COD. Thus, it will be easy for a refinery to comply with the regulatory requirement if the refinery is required to fulfill only the effluent BOD limit. However, most refineries are required to comply with multiple discharge quality criteria including COD, and/or TOC heavy metals concentrations, and effluent toxicity as determined by WET testing procedures, described earlier in this chapter. In such cases, for effluent streams containing high NA concentrations, it may be difficult to comply with the regulatory requirements. To overcome such situations, it will require removal of NA from secondary effluent before final disposal.

Three advanced treatment processes for NA are—(i) GAC Adsorption, ii) Advanced Oxidation Processes (AOP), and (iii) Reverse Osmosis (RO). Each of these processes has been described in Chaps. 2 and 3. GAC is most commonly used and can remove organic compounds including NA that escape biological treatment. AOPs involving ultraviolet (UV)-Peroxide and UV-Ozone are effective in removing organic compounds, however metal ions pass through untreated. RO produces the highest quality treated effluent, which may surpass the discharge limits. Additionally, RO requires a pretreatment step by an ultrafiltration (UF) membrane system. Both GAC and RO treatment processes are based on phase separation and a concentrated reject stream will require disposal. On the contrary, AOPs are destructive processes and do not produce a reject stream that require disposal. Each of these processes needs to be pilot tested before field scale application to determine the technical and economic feasibility, and optimum operating conditions of the selected option.

Another example of a CAA regulation that influenced wastewater and groundwater treatment can be attributed to the quest for high octane in gasoline. The octane rating of a motor fuel directly influences its smooth combustion in Internal Combustion (IC) engines. Gasoline is typically marketed by retailers in three octane ratings—87 as Regular Gasoline, 89 as Mid-Grade and 93 as Premium. With a higher octane rating, the anti-knock performance is more efficient, which minimizes wear and tear of the engine. Refiners traditionally blended tetraethyl lead (TEL) to enhance antiknock property of gasoline. However, this caused widespread air emission of lead as fine particulate matter from automobile exhaust lines. Lead is a pollutant and toxic to humans, and lead poisoning by inhalation became a major health issue arising from use of TEL as an additive to motor fuels. Starting in 1973 EPA started mandating control of TEL in gasoline ultimately enforcing its total ban in 1996 through the provisions of the CAA.

In the 1990's the regulations under the CAA amendments required increased use of Reformulated Gasoline (RFG) that contained oxygenated organics to reduce the air emission of exhaust gases resulting from incomplete combustion in the internal combustion engines. Oxygenates help to ensure complete combustion of gasoline thereby reducing emission of ozone precursors in the environment. Methyl tert-butyl ether (MTBE) was introduced as the chosen product due to its ease of production in the refineries, blending and transportation. Over time it was observed that MTBE in gasoline stored in the Underground Storage Tanks (UST) caused severe health effects due to the contamination of groundwater and soil from leakage of corroded tanks. Some of the major oil companies were cited by the USEPA as polluters resulting in imposition of fines (US Department of Justice 2005).

In the year 2000, the USEPA announced the gradual phasing out of MTBE and its replacement by ethanol (EtOH). A major issue with MTBE is its difficulty in being removed from water sources. As an oxygenated compound it is highly soluble in water with low volatility and is difficult to remove economically by physicochemical processes (e.g. air stripping, GAC, etc.). Aerobic biotreatment by an enriched culture has proven to be effective, but is a slow process (Cowan and Park 1996). Therefore, remediation of MTBE contaminated groundwater cannot be conducted by traditional activated sludge-based wastewater treatment systems and requires a dedicated MTBE treatment system by specialized inoculum that need to be developed and cultivated. On the contrary, ethanol is prone to biodegradation by both aerobic and anaerobic microorganisms. In the case of underground storage tanks (UST) leakage incidents, ethanol can be degraded by indigenous microbes in soil and groundwater. Because of the faster rate of biodegradation of ethanol than BTEX in gasoline, ethanol will consume a large amount of air in aerobic treatment with the possibility of leaving untreated BTEX in the groundwater plume. From this viewpoint, anaerobic biodegradation of contaminated groundwater and soil is advantageous as it does not create a competing environment between ethanol and BTEX compounds (Powers et al. 2001). A historical review of progressive octane enhancement in gasoline by chemical additives, their environmental impacts, and mitigation by regulatory and technological initiatives has been presented in a comprehensive report by Stolark (2016).

Increasing use of opportunity crudes significantly impacts both the water and air environments. Many of such crudes contain high concentrations of H_2S and organic sulfur compounds that cause the emission of high concentrations of SO_x upon combustion. Over the past decade, regulators have been imposing increasingly stringent air emissions standards. In the US, from 2017, the USEPA has started phasing in the Tier 3 Standard for maximum sulfur content of 10 μg/mL in gasoline, which is much lower than the Tier 2 standard of 30 μg/ml. To address such requirements refiners around the world have been upgrading and retrofitting the facilities to produce clean burning transportation fuels. A large part of such upgrades includes the incorporation of new hydrogenation and hydrocracking units in refineries to convert organic sulfur compounds to H_2S, followed by the removal of H_2S by amine treatment of gaseous streams.

In addition to the reduction of SO_x, clean fuels significantly reduce the emission of air pollutants CO, hydrocarbons (HC), NO_x, and PM from passenger vehicles that meet the Euro 4, Euro 5, and Euro 6 standards for gasoline and diesel respectively (Nichols 2016). The use of high sulfur opportunity crudes, coupled with the low sulfur content in refined products result in the removal of large quantities of sulfur in the form of H_2S, a part of which is also associated with the wastewater streams. This imposes several challenges to the biological treatment process which include: (i) odor and safety concerns; (ii) corrosion of metallic parts; (iii) high aeration demand; (iv) incomplete oxidation of H_2S to sulfate resulting in escape of partially oxidized products with the treated effluent; and v) growth of filamentous organisms leading to bulking and foaming on activated sludge basins as discussed in Chap. 5.

This situation can be overcome by conducting a pretreatment step of refinery wastewater containing high H_2S concentrations by chemical oxidation prior to biotreatment. Oxidation of H_2S yields multiple products containing sulfur in several oxidation states between −2 (for H_2S) and SO_4^{2-} (+6). These include elemental sulfur (S, oxidation number 0), thiosulfate ($S_2O_3^{2-}$, oxidation number + 2), sulfite (SO_3^{2-}, oxidation number + 4), and sulfate (SO_4^{2-}, oxidation number + 6). Complete oxidation to sulfate is a slow process. The results of a study conducted by Santo et al. (2016) on primary treated wastewater demonstrated that the reaction rate is significantly increased by oxidation of H_2S using Fe^{3+} as the catalyst and oxygen as the oxidant. It achieved a maximum 90% removal of H_2S in about 30 min in the temperature range of 25–40 °C. These findings can be applied on a full scale by providing a catalytic oxidation reactor between the primary and secondary treatment of refinery wastewater containing large concentrations of H_2S. This process has been discussed in detail in Chap. 5.

7.3.4 *Impacts and Treatment of Chemicals of Emerging Concern*

Large refinery complexes include their own fire departments consisting of fire trucks and firefighters to respond to on-site fire-related emergencies. They conduct routine exercises of simulated fire events to keep the equipment and personnel in readiness. Each time such an exercise is conducted a fire-retardant foam is sprayed to put out the simulated fire event. As mentioned earlier, commercial fire-retardant products use substances such as Per or Poly fluorinated alkyl substances (PFAS) as their active ingredients. Two PFAS chemicals are primarily used for this purpose, the perfluorooctane carboxylate acid known as perfluorooctanoic acid (PFOA) and perfluorooctanesulfonate (PFOS). Their molecular structures are presented in Fig. 7.7. These chemicals drain into the surface drainage system, eventually percolate through the

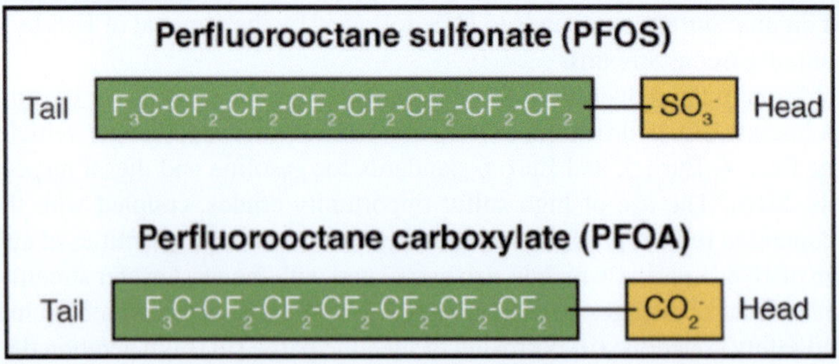

Fig. 7.7 Chemical structure of PFOA and PFOS compounds that are active ingredients of fire-retardant foams (IRTC 2020)

ground and contaminate the groundwater. They are extremely resistant to degradation and remain a long time in a water matrix without breakdown, warranting the term "forever chemicals".

PFAS group of chemicals are known toxicants and exposure to those by drinking and contacting contaminated water causes adverse health effects including cancer, birth defects, and reproductive disorders. In April 2024 the USEPA announced the enforceable limits for PFOA and PFOS of 4 parts per trillion individually as parts of the Drinking Water Standard (also known as Safe Drinking Water Act, or SDWA). In addition, EPA has also introduced enforceable Maximum Contaminant Level (MCL) of 10 parts per trillion individually for other chemicals belonging to the PFAS group - Perfluorononanoic acid (PFNA), Perfluorohexane sulfonate (PFHxS), and hexafluoropropylene oxide (HFPO) dimer acid and its ammonium salt (referred to as "**GenX Chemicals**"). Refiners use treatment by GAC, Ion Exchange and RO to remove PFAS from surface water and groundwater contaminated by routine firefighting training activities and actual fire incidents in the refineries. However, these are not destructive technologies and create small, but concentrated streams of PFAS that need ultimate disposal. A large number of destructive technologies are currently under various stages of development, some of which are close to full scale deployment A comprehensive discussion of various emerging technologies for PFAS control, their working principles and efficacy has been presented in a publication on this subject by the Water Environment Federation (WEF) (Basu et al 2022).

The examples presented above are a few where water, wastewater and groundwater in refinery complexes are indirectly contaminated to support regulatory requirements for air pollution control and firefighting activities. Technologies to mitigate those effects are also applicable to other similar situations and responses to those are considered on a case-by-case basis.

7.4 Plant Safety Overview

Situations are inherent in wastewater treatment that can result in serious personnel injuries and fatalities, loss of physical plant assets, exposure to toxic releases in nearby communities, business interruption loss, etc. A good safety program focuses on identifying loss scenarios and reducing risks from those scenarios.

There are primarily two aspects of safety in a manufacturing process: Occupational health and safety, and Process safety. Occupational health and safety are concerned with hazards that can cause harm to the health and safety of the personnel. Process safety is concerned with risks (to personnel, the business assets, and the environment) from highly hazardous chemicals. Companies often have Environmental, Health and Safety (EHS) programs to manage waste or air emissions to minimize impact to nearby residents and ecosystem.

7.4.1 Occupational Health and Safety Overview

The goal of Occupational Health and Safety is to control the occupational hazards to the health and safety of the personnel. Occupational health is generally addressed by the specialized field of industrial hygiene which is concerned with various types of chronic injuries and illnesses, such as nonfatal illnesses, respiratory conditions, skin diseases, poisoning, hearing loss, and repetitive stress injuries among others. The field of industrial hygiene is beyond the scope of this section. However, it is important to be aware of toxic industrial chemicals that OSHA has identified (https://www.osha.gov/emergency-preparedness/guides/toxic-industrial-chemicals), including data in OSHA Table Z-1 *Limits for Air Contaminants*, and OSHA Table Z-2, *Ceiling Concentrations*. Should any of these hundreds of chemicals be present at a wastewater treatment site, decisions should be made to address exposure to these chemicals. Toxic chemicals that lead to many of these chronic conditions may or may not be subject to OSHA Rule 29 CFR 1910.119, Process Safety Management of Highly Hazardous Chemicals, which has a different focus (see Sect. 7.6).

Occupational safety addresses all other risks to personnel that are unrelated to hazardous chemicals. There are numerous hazards unrelated to chemicals, such as those associated with surfaces that could cause falls, slips, and trips; rotating equipment; all electrified equipment; confined spaces; working at heights; poor ergonomics; extreme heat/cold; ionizing radiation; and vibration. It is beyond the scope of this section to discuss each of these hazards. However, safety personnel must be provided training in these hazards, as many of them are likely common to most wastewater treatment plants.

Sewage Treatment Plant (STP) and Wastewater Treatment Plant (WWTP) are part of infrastructure-related processing in a facility, which includes petroleum WWTPs. Under the Occupational Safety & Health Act (OSHA) the WWTP needs a written program to cover the Process Safety Management (PSM) for wastewater treatment operations (29 CFR 1910.119 and 40 CFR 68). In addition, each state may have its own rules to be covered. Although some WWTPs may not have employees stationed at the site and may be exempted from PSM standard coverage under paragraph 29 CFR 1910.119, a refinery's WWTP facilities are almost always manned and will require OSHA PSM coverage.

Maintaining the Health and Safety program for any organization is the responsibility of every member starting from the highest management level to individual employees. The program consists of four core components: (i) reduction of risk and liability; (ii) compliance with regulations; (iii) protection against financial loss of the organization; and (iv) protection against personal loss to the employee and family members. A healthy and accident-free workplace is in the best interest of the organization and its employees. For a WWTP, the program must be established collectively by the utility management based on local regulations, operating experience, and common health and safety hazards experienced in the wastewater treatment industry.

All staff members working routinely with, or around, the WWTP system should read and familiarize themselves with the information in health and safety documents and standards established by the utility. This applies to everyone associated with the operation, maintenance, sampling, and analytical activities at and around the WWTP. This encompasses both employees and contract staff members. Additionally, it must be noted that Health and Safety guidelines are not limited only to any specific unit operation or process and the associated equipment. Instead, these should be considered in totality for the wastewater treatment system.

7.4.2 Occupational Safety Program

Several risk assessment methods are employed for mitigating risks, each requiring specific training. A detailed discussion is beyond the scope of this section. At a fundamental level, virtually all activities within wastewater treatment plants benefit from two powerful techniques:

1. Job Hazard Analysis (Job Hazard Analysis, U.S. Department of Labor, Occupational Safety and Health Administration, OSHA 3071, 2002 (Revised))—There are four basic stages in conducting a Job Hazard Analysis:
   a. Select the job to be analyzed.
   b. Break the job down into a sequence of steps.
   c. Identify potential hazards for each step.
   d. Determine preventive measures to overcome these hazards.
2. Work permitting—Operations must not allow work to be done until the work plan has been thoroughly reviewed, and preparations at the work site have been inspected. Several types of work permits may be employed:
   a. Hot work permits are used when heat or sparks are generated by work such as welding, burning, cutting, drilling, and where work involves the use

of pneumatic hammers and chippers, non-explosion proof electrical equipment (lights, tools, heaters), and internal combustion engines. Three types of hazardous situations need to be considered when performing hot work:
   i. The presence of flammable materials.
   ii. The presence of combustible materials that burn or give off flammable vapors when heated.
   iii. The presence of flammable gas in the atmosphere, or gas entering from an adjacent area, such as sewers that have not been properly protected.
b. Cold work permits are used in hazardous maintenance work that does not involve 'hot work'. Cold work permits are issued when there is no reasonable source of ignition, and when all contact with harmful substances has been eliminated or appropriate precautions taken.
c. Confined space entry permits are used when entering any confined space such as a tank, vessel, tower, pit or sewer.
d. Special permits:
   i. Radioactive materials employed, e.g., nondestructive testing, work on industrial gauges employing radioisotopes.
   ii. Critical crane lifts
   iii. Breaking containment that might result in a release.
   iv. Excavations.

Both job hazard analyses and work permitting benefit greatly from a detailed field review checklist. An example of a checklist used at a Saudi Arabian refinery in the 2000s is provided in a checklist inserted in Appendix D Example D.1. The checklist was used for inspection of the work site before Operations department permitting the work and at any time during the work. 'Required' indicates the importance of a line item, which if not met, is a major failure identified during the field review. Some line items in this checklist that are not identified as 'required' may need to be required by Operations at some WWTP. Any major failure from a field review in connection with the work permitting process must be resolved before the work permit is issued. Any major failure identified during the work must result in a shutdown of the work until resolved.

Operators, maintenance staff members, laboratory and service technicians are reminded to read and review all information contained in the vendor manual(s) before beginning any work on any piece of equipment. Everyone working in and around the wastewater treatment system is required to follow strict health and safety guidelines. Personal Protective Equipment (PPE) must always be worn following appropriate and required guidelines based on Health and Safety specifications as well as OSHA guidelines. The three most common types of occupational hazards at industrial workplaces, including wastewater treatment plants, are fall, slip, and trip hazards. Exposure to other personal safety hazards such as mechanical maintenance, confined space, electrical maintenance, as well as occupational noise, and emergency stations, are described below. While process hazards such as gaseous chemical, biological, and fire and explosion can apply to this section, they have been separated and discussed below in Sect. 7.5 with remedial measures summarized.

7.4.2.1 Personal Protective Equipment (PPE)

Eye protection, hard hats, steel-toed shoes, gloves, protective clothing, respirators, ear plugs are part of any workplace safety accessories and must always be used by workers where needed as part of procedures written in the Owner's Safety Manual.

7.4.2.2 Fall, Slip, and Trip Hazard

Employees must be cognizant about potential for the fall, slip and trip hazards during the activities at a WWTP. Although it may appear as a matter of common sense, accidents arising from these can cause major injury, and in certain cases may lead to dismemberment of body parts, and even death. Personnel working at and around WWTPs need to be extremely cautious and careful to avoid such incidents. Leakage of wastewater carrying pipes, equipment, spillage of polymer solutions that are typically utilized for coagulation and flocculation to aid settling in clarifiers, sludge thickening, and dewatering operations can cause a slip and fall. Similarly, falls and trips can occur under many situations, for example while working at an elevated platform, climbing a vertical ladder in a workplace can result in serious injury or fatality. Additionally, aerated water in activated sludge reactors have a reduced density which greatly increases the hazard due to loss of buoyancy in the event someone falls into the reactor.

7.4.2.3 Mechanical Maintenance Hazard

Direct contact with running rotating equipment such as pumps, blowers, mixers, etc. is to be avoided. The rotating equipment under maintenance should be carried out under the supervision of a qualified professional. Before any maintenance work is undertaken on any rotating machinery the equipment should be shut down, de-energized, locked out and tagged out.

7.4.2.4 Confined Space Hazard

Confined space is defined as: (i) having limited or restricted entry or exit; (ii) is large enough for a worker to perform assigned work; and (iii) is not designed for continuous occupancy. Entry into confined space should only be performed after following all the specified procedures, permits issued, and precautions. Importantly, checks must be conducted to ensure that no toxic gases are present and oxygen concentrations are present at ambient air levels that support life (29 CFR 1910.146).

7.4.2.5 Electrical Maintenance Hazard

All rotating equipment, instrumentation and control systems, and any other equipment driven by electrical power must be properly grounded and checked routinely, and electrical maintenance should be conducted under the supervision of a qualified professional. Before any maintenance work is undertaken on any electrical equipment, the equipment should be isolated electrically and locked out from the power supply from the Master Command and Control (MCC) center.

7.4.2.6 Occupational Noise and Emergency Stations

Occupational noise exposure needs administrative controls at >85 dBA (29 CFR 1910.95). In areas where rotating equipment, pumps, compressors, and fluid flow machinery are operated, monitoring of noise level and use of ear plugs should be enforced for the long-term health benefit of the employees. Periodic testing of the hearing of the employees shall be used as a measure of the success of the occupational noise management program.

Provision for sprinkler eyewash facilities and showers must be made available close to the wastewater treatment system and staff members need to be made aware of these before working with corrosive and toxic materials. All operators should be familiar with the location and correct operation of emergency wash facilities.

In summary, occupational safety falls under the overall EHSS division of a company and, more importantly, is not limited to following OSHA guidelines to avoid penalties. Many leading companies keep a dashboard of the metrics such as lost time rate, total accident rate, working days since last incident, etc. Regarding personal safety, it is necessary to perform a regular management review and publish newsletters to create awareness for employees. This will help employees to be mindful while at work and allow the execution of responsibilities to be safely completed on time. The EHSS management must understand and agree to the intent of the recommended practices. Frequent workshops to improve the knowledge of personal safety procedures, workplace hazards, and process knowledge will help reduce workplace injury, illness, and death. Moreover, a safe workplace is a sound business.

7.5 Process Safety Program

The goal of a Process Safety program is to recognize, evaluate, and control hazards and associated risks in a process. The integrity of the assets in any industry is based on (i) design integrity; (ii) technical integrity; and (iii) operating integrity. Design integrity is where equipment design keeps the risks at an acceptable level. Technical integrity consists of maintenance, inspection, repair, and assurance activities to contain hazards. Operating integrity involves operating within the approved operating envelope, compliance with procedures and standards, management of change, and adherence to the performance requirements of several other process safety elements.

7.5.1 Process Hazards

Hazards from chemical substances such as liquids and gases, which are used in process under various operating conditions (i.e., high temperature, high pressure, etc.) are inherent with the technology. Technology determines the chemicals and the process conditions such as temperature, pressure, concentrations, etc. which constitute some of the hazards in a process. In this section, we will focus on the specific hazards that can arise from the process and human error.

The hazards related to the process chemicals and the specific operating conditions are gaseous, chemical, biological, fire, reaction, and explosion hazards. The management of these hazards is known as PSM (Process Safety Management) as outlined by the 1972 OSHA standard (29 CFR 1910.119). These standards have been made into a law on May 26, 1992.

Various standard techniques are available as tools for PSM, such as What-if analysis, Hazard and Operability Study (HAZOP), Process Hazard Analysis (PHA; 29 CFR 1910.119 [e]), Risk Matrix, Layer of Protection Analysis (LOPA), and Fault Tree Analysis (FTA). These serve to evaluate the process hazards, identify, and quantify the risk, and control or mitigate the overall risk. Companies may have their own preferred methodologies that may include one or more of the above techniques.

Below are a few common terms for personal and process safety practices:

Hazard: A physical situation with a potential for human injury, damage to the environment, or some combination of these.

Risk: A combination of the severity of the consequences of a scenario and the likelihood of the scenario occurring. This includes individual and societal risks.

Safety: Control of recognized hazards to achieve an acceptable level of risk. Also termed, As Low As Reasonably Practicable (ALARP).

Managing hazards comprise of:

- Identifying hazards
- Assessing risk
- Implementing Control measures by
  - Eliminating hazards
  - Implementing engineering controls
  - Implementing administrative controls such as establishment of Standard Operating Procedures (SOP), employee training, and follow Management of Change (MOC)
- Personal Protective Equipment (PPE) and evacuation serves as the last line of defense

The hazard analysis should cover the consequences of failure of engineering and administrative controls (29 CFR 1910.119 [e]). The SOP shall address at least the following (i) initial startup, (ii) normal operations, (iii) temporary operations, (iv) emergency shutdown, (v) normal shutdown, (vi) startup following a turnaround, or after an emergency shutdown.

7.5.1.1 Gaseous Hazard

PSM covered gases in WWTP operation consist of various VOCs such as sulfur dioxide (SO_2), methane (CH_4), carbon dioxide (CO_2), NH_3, and H_2S. These gases are dissolved either in upstream produced water or downstream refinery wastewater streams, or both in many cases. Chlorine gas (Cl_2) is also covered by PSM but is very unlikely in water and wastewater streams associated with petroleum industrial operations. The only places Cl_2 is relevant are in the disinfection of sanitary water for use by employees and in its use as a biocide in injection water in upstream production by hydrofracturing. Chlorine is dosed into water in the form of chlorine gas (Cl_2) or sodium hypochlorite solution (NaOCl). There are serious health and environmental effects of these gases in water and warrant caution.

As explained in earlier chapters, H_2S is very commonly associated with produced water. In refinery wastewaters, it is common, especially at the refineries that process sour crude. The gas has a typical "rotten egg smell", but at harmful concentrations, it paralyzes olfactory nerves and disables the sense of smell. Excessive exposure to H_2S will impair the nervous system and can lead to death. 700 to 1000 ppm H_2S can cause collapse and unconsciousness within just 1 or 2 breaths.

SO_x are products of the combustion of high sulfur fuel oils in fired heaters and boiler furnaces in refineries. SO_x are very toxic and corrosive to the respiratory tract. Short-term severe exposure can cause long-term respiratory effects and can be fatal if inhaled. Acute exposure to high concentrations of Cl_2 can lead to a build-up of fluid in the lungs (i.e., pulmonary edema) and severe shortness of breath that could lead to death if untreated.

Ammonia is typically associated with refinery wastewater streams. It is primarily produced as a co-product of H_2S from hydrogenation and hydrocracking operations. Ammonia in the gas phase is toxic to humans as it causes damage to the eyes, throat, and lungs. Discharge of excessive ammonia in refinery wastewater to the water bodies is toxic to aquatic life and causes DO deficiency. Internationally, regulators have been imposing strict discharge limits of ammonia into receiving waters.

OSHA and the American Conference of Governmental Industrial Hygienists (ACGIH) have established workplace limits for Cl_2, SO_2, and H_2S. An eight-hour Permissible Exposure Limit-Time Weighted Average (PEL-TWA) for Cl_2, SO_2, and H_2S are, respectively, 0.1, 2 and 10 μg/ml respectively. A 15-min Permissible Exposure Short-Term Exposure Limit (PEL-STEL) for the above gases are, respectively, 1, 5, and 15 μg/ml. However, the National Institute for Occupational Safety and Health (NIOSH) recommends a PEL-STEL of 0.5 μg/ml. Threshold Quantities (TQ) for

Cl_2, SO_2 (liquid), and H_2S respectively are 1500, 1000, and 1500 lbs (680, 454 and 680 kg respectively). (OSHA 1910a, b.119 subpart H).

7.5.1.2 Chemical Hazard

In WWTPs, employees may be exposed to hazardous materials and chemicals, particularly to dosing chemicals. The flocculant and the defoamer chemicals can be slippery and their leakage and/or spillage on the floors and walkways can add to the slip and trip hazards and accidents, as discussed above (Sect. 7.4.2). Cyanide and acids are chemicals covered in the PSM for WWTP operations. DOT 49 CFR 172.101 is also concerned with hazardous materials.

Heavy metals and metalloids (e.g., mercury, arsenic, cadmium, chromium, lead, etc.) are covered in 29 CFR 1910.1026–1027. Other hazards are sulfuric acid (H_2SO_4) and sodium hydroxide (NaOH) which are commonly dosed into wastewater for pH adjustment. Additionally, these two chemicals are used for the regeneration of cation and anion exchange resins in ion exchangers for boiler feed water treatment applications.

A review of the Safety Data Sheet (SDS) provided by the manufacturer or supplier of each chemical must be conducted, individuals need to become familiar with the risks in handling these, and the safety precautions required before working with any chemical. Additionally, health and safety risk information and mitigation measures of commonly used chemicals are available from online resources.

7.5.1.3 Biological Hazard

WWTPs and biological reactors can be a source of numerous bacteria and viruses. Infectious diseases are transmitted by these micro-organisms. It is important to note that bacteria do not need a living host, and wastewater is an ideal environment for bacteria to grow and multiply. On the contrary, viruses can multiply only in the living cells of animals and humans. In general, refinery wastewater, as opposed to sewage, does not contain and transmit bacteria and viruses of enteric origin. Typically, this is one of the reasons why sewage treatment systems in petroleum production and processing facilities are kept separate from treatment of sanitary and cafeteria wastes in the refinery complex. This is even more important since the large-scale outbreak of the COVID-19 pandemic, as sewage is a major carrier of SARS-CoV-2 virus strains (Randazzo et al, 2020).

Staff members in wastewater treatment operations and maintenance activities may be vulnerable to infectious diseases due to exposure to wastewater microorganisms. The most common infection routes are ingestion and inhalation of splashes, through food handled by unprotected hands, and direct contact with cuts on the body. As treated refinery wastewater is not disinfected, strict adherence to personal hygiene, and use of PPE (e.g., gloves, clothing, face covers, respirators, and goggles, etc.)

are essential to ensure the health, safety, and well-being of staff and close contacts outside of the facility.

7.5.1.4 Fire and Explosion Hazard

Fire and explosions can occur from gases that accumulate in the headspace of closed tanks, vessels, and process equipment. In a refinery WWTP, most of the equipment and tanks conducting primary treatment are covered to control air emission of hazardous gases as required by BWON and other similar regulations. This results in accumulation of flammable gases in the headspace of tanks and equipment. Electrical sparks and flames from other sources (i.e., welding during maintenance activities) can cause fire and explosions. Storage of combustible materials, (e.g., wood, paper, etc.) and flammable liquids and gases in the enclosed spaces should be avoided. To mitigate this situation, several safety measures are taken in the design and installation of covered tanks and vessels in wastewater treatment service. These include the following:

- Installation of Lower Explosive Limit (LEL) detection probes in the headspace of tanks that sound an alarm when the concentration of the most explosive component in the headspace gas reaches 25% of the LEL limit. The actual alarm setting is adjustable and 25% is a typical number. This prompts the operator to take the necessary steps to stop the situation from escalating further, leading to an actual explosion.
- Continuous nitrogen purging of the headspace creating an inert environment under a split range control action such that, from time to time, as the headspace pressure rises due to the generation of gases from the wastewater, a vent valve automatically opens at a set pressure to release that gas, and another valve on the nitrogen purge line also opens to introduce nitrogen that takes the place of the released explosive gas. This helps maintain an inert atmosphere in the headspace continuously.
- Installation of a lightning arrestor system on the top of the covered tanks to prevent fire and explosion resulting from lightning.

In addition to the above, fire-prone areas should have provisions for fire extinguishers and fire hose lines to prevent the possibility of fire and explosion hazards. Each large refinery complex should have its own fully equipped fire department to respond to emergencies.

7.6 Process Safety Management (PSM)

On February 24, 1992, OSHA promulgated the Final Rule for Process Safety Management of Highly Hazardous Chemicals, 29 CFR 1910.119. This standard went into effect on May 26, 1992. Before acceptance of the final PSM Rule, stakeholder meetings and public hearings were conducted, which resulted in a rule considered by some process safety pioneers to be inadequate. As such, the Center for Chemical Process Safety (CCPS), which was formed following the 1984 Bhopal, India disaster, worked to improve the PSM Rule, resulting in numerous guidance documents and the first CCPS guidelines book in 2007, Guidelines for Risk Based Process Safety. The PSM Rule has 14 elements. Risk Based Process Safety (RBPS) has 20 elements.

The Process Safety documents prepared following the 14 elements of PSM are completed initially at the start-up of the plant. It is a living document. To be certain that PSM is effective, owners must certify that they have evaluated compliance with the provisions of PSM at least every three years. This will verify that the procedures and practices developed under the standard are adequate and are being followed. OSHA requires that Process Hazard Analysis (PHA) is revalidated every five years (CCPS 2018, API 521 2020).

The high-level focus of PSM is the safety and health of all employees, following the permitting standards and rules, and providing resources for more information to all employees. The WWTP operations should cover all the elements of the PSM in cases where it may apply. Routine health and safety training is also a very important component.

The PSM Rule, 29 CFR 1910.119, Process Safety Management of Highly Hazardous Chemicals, is focused on controlling highly hazardous chemicals that could result in devastating losses to personnel, physical plant assets, surrounding communities, and the continued commercial viability of the operating company itself.

The PSM Rule is performance-based. It is not prescriptive. A company structures its Process Safety Management Program to meet the performance expectations of the PSM Rule.

The PSM Rule itself has been demonstrated to be inadequate resulting in the development of Risk-Based Process Safety. OSHA has issued more than two hundred interpretation letters and 15 compliance directives to add clarification and guidance to the PSM Rule.

The PSM Rule requires compliance if the on-site inventory of a highly hazardous chemical is met or exceeded. 222 chemicals are now specified by OSHA as requiring compliance with the PSM Rule or RMP Rule. A hyperlink providing the table of these chemicals is available in Appendix D, Example D.6. If a chemical is not on this list or its inventory is below its threshold, then compliance in regard to of that chemical is not required.

Information about the hazard of a highly hazardous chemical includes:

- Flammability data
- Reactivity data
- Corrosivity data
- Thermal and chemical stability data
- Hazardous effects of inadvertent mixing of dissimilar materials that could foreseeably occur
- Toxicity information
- Permissible exposure limits.

An entire operating facility may or may not be covered by the PSM Rule. Only the processing areas of the facility are covered depending on the inventories of highly hazardous chemicals. The PSM Rule and guidance from OSHA should be reviewed to determine if a facility, or a portion of a facility, is covered by the Rule.

Any process that is covered by the PSM Rule must undergo periodic risk assessment to determine if controls are adequate. The frequency of risk assessment depends on the inherent level of risk. Scheduled risk assessments may need to be moved up if incidents occur in the process.

7.6.1 Risk Assessment

A risk assessment focuses on hazards that could result in loss of containment. A fundamental principle of process safety is that no loss will occur if the highly hazardous chemical remains contained in piping and equipment. The loss of containment could occur from mechanical failure of piping and equipment, or from rapid overpressure within piping and equipment that exceeds its maximum allowable working pressure limit. A risk assessment typically considers hundreds of possible hazards. For example, a temperature exceeding a specified safe level is a hazard. The question is posed as to what would need to occur for that event to occur. Several possible scenarios are evaluated by the risk assessment team. Each scenario is assessed, with the result that the team may determine that improvements are needed in the plant to reduce risks associated with that scenario.

In process operations, Process Safety is an important and integral part of the safety analysis and risk assessment. There are many ways the risk of a process can be assessed.

Several risk assessment techniques are commonly used for qualitative brainstorming:

- Risk Matrix Analysis
- Process Hazard Analysis (PHA)
- What-if Analysis
- Hazard and Operability Study (HAZOP)
- Bow-tie Analysis
- Layer of Protection Analysis (LOPA)
- Fault Tree Analysis (FTA).

Examples of some of these techniques are included in Appendix D for the benefit of the readers.

The WWTP owner's PSM program should have a written manual covering the 14 elements of the PSM that are relevant to their operations. The regulated chemicals in WWTP are mentioned above in Sects. 7.5.1.1 and 7.5.1.2 of 29 CFR 1910.119 lists over 130 hazardous chemicals with threshold quantities ranging from 50 lbs. (~23 kg) to 15,000 lbs. (~680 kg). Exemptions are also mentioned, but all facilities that have chemical processes should assess their operations for coverage under the PSM standard.

7.6.1.1 Risk Matrix Analysis

For each of the above techniques, the objective is to identify the risk of current operations and determine if changes to those operations are needed to reduce that risk. For each scenario studied by the assessment team, the risk is estimated from expected scenario impacts (consequences) and scenario frequency (likelihood). The likelihood and severity of an incident determine the risk rank of an incident in the matrix. The risk rank then allows the business to decide on how to manage it. A matrix such as Fig. 7.8 is used to determine risk. In this figure, five levels of risk are noted: Negligible, Minor, Moderate, Serious, and Critical. A company may set any level as being unacceptable. For example, a company may choose to accept no risk level exceeding Minor. If the initial risk assessment of a scenario results in a risk level of Moderate, the assessment team must figure out what needs to be done to reduce either the consequence (scenario impact) or the likelihood (scenario frequency). For example, if the likelihood is very low (not expected to occur during the lifetime of the facility) but the consequence is very severe (unacceptable category), the business may consider escalating it to the senior management to bring down the risk ranking to at the least 'Minor' level. On the other hand, if the likelihood is common (once or more than once in a year) and the consequence is very low or negligible (tolerable with routine procedures), the business may fully mitigate it using routine procedures. This is a simple mechanism to increase the visibility of risks and assist management decision-making (Fig. 7.8).

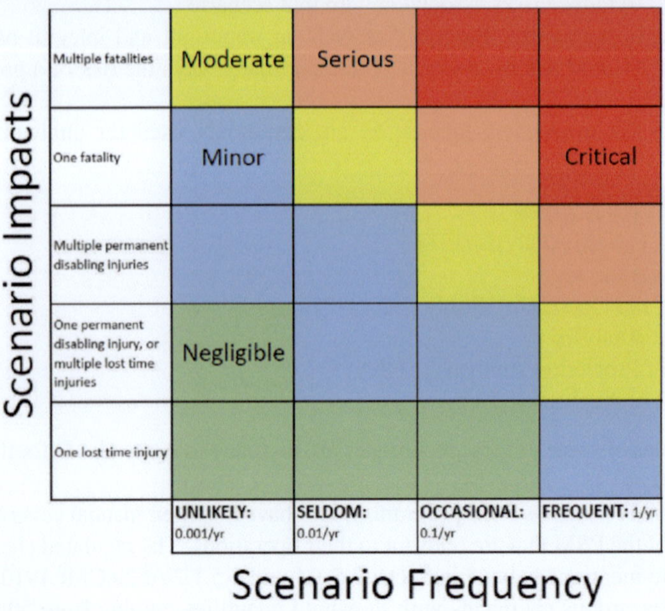

Fig. 7.8 A standard business risk matrix (Guidelines for Hazard Evaluation Procedure, 2nd ed.)

Table 7.4 Example frequency categories for risk matrix

	Scenario frequency categories
Category	Description
Unlikely	Not expected to occur during the facility's lifetime
Seldom	Expected to occur no more than once during the facility's lifetime
Occasional	Expected to occur several times during the facility's lifetime
Frequent	Expected to occur more than once in a year

Table 7.5 Example consequence categories/scenario impacts for risk matrix

Scenario consequence categories	
Category	Description
Multiple fatalities	Should be mitigated with engineering and/or administrative controls to a risk ranking of III or less within a specified time such as six months
One fatality	Should be mitigated with engineering and/or administrative controls to a lesser risk ranking within a specified time such as 12 months
Multiple permanent disabling Injuries	Should be mitigated with engineering and/or administrative controls to a lesser risk ranking within a specified time such as 15 months
Single permanent disabling or multiple lost time Injuries	Should be verified that procedures or controls are in place and operations understand how to follow them
One lost time injury	Provide additional training. No additional mitigation required

The work of the risk assessment team continues until all hazards have been addressed.

Examples of scenario frequency and consequence categories are shown in Tables 7.4 and 7.5.

7.6.1.2 Process Hazard Analysis (PHA)

Process Hazards Analysis (PHA) is a study of process hazards that includes several methods such as HAZOP, What-If, and Checklist. Hazard and Operability Analysis (HAZOP) is a structured and systematic technique for identifying hazards in a work process. A Process Hazard Analysis (PHA) is performed to determine hazards and potential accident scenarios. The terms HIRA, Hazard Identification, and Risk Analysis are more desirable than PHA. See Appendix D Example D.4 for an example of a table on HIRA. When HIRA is applied at the conceptual design phase of a process, it is often termed Preliminary Hazard Analysis.

Brief examples of What-If, HAZOP, Bowtie, and other techniques are described below to conceptually understand these techniques and their applications. At the conceptual design stage of a process, a preliminary Hazard Analysis is customarily performed to decide if the major hazards can be eliminated, minimized, or controlled using principles of inherent safety rules such as elimination, replacement, and reduction.

7.6.1.3 What-If Analysis

This analysis usually reviews the process, beginning with the introduction of feed and following till the end of the process. The results of a What-If analysis usually address potential scenarios by asking What-If questions, such as What if the pump fails, or what if the cooling fails. Examples of What-If questions for a WWTP are:

"What if the wastewater has the wrong composition the plant is not designed for?"

or

"What if the SCADA (Supervisory center and data acquisition system) control malfunctions or is compromised?"

The team might then recommend (1) installing an emergency shutdown system or taking special precautions when feeding the raw material to the process, and (2) having a backup system, putting a firewall, or anomaly detection system with an alarm. The questions and the answers, including hazards, consequences, engineered safeguards, and viable solutions for important issues are all recorded. A typical format for a What-If Analysis worksheet is found in Table 7.6.

Table 7.6 Typical format for a what-If analysis worksheet

Area: ____________ Meeting ____________

Drawing Number: ____________ Team Members: ____________

What-If	Consequence/Hazard	Safeguards (Note)	Recommendation

7.6.1.4 Hazard and Operability Study (HAZOP)

The HAZOP Analysis technique was developed by Imperial Chemical Industries (ICI) in the UK in 1968. A team, usually termed a HAZOP team, is formed by drawing experts from several disciplines. The analysis is based on brainstorming by several experts with different backgrounds in a creative, systematic fashion to identify problems, if any, with an existing construction-ready Piping and Instrumentation Diagram (P&ID) or P&IDs from an existing operating process. The team works through plausible deviations from the design intention, brainstorming on causes that could result in the deviations, and their effects. The HAZOP study focuses on specific points of the process or operation called study nodes, process sections, or operating steps. The development of HAZOP and its application in hazard analysis in chemical processes has been discussed by Lawley (1974). A process safety specialist should be on the HAZOP team. An expert in emergency relief systems (especially reactive relief) should be either readily available for consultation or on the team. This way, the number of action items can be optimized, their scope, and usefulness can be enhanced.

Some common process parameters in HAZOP analysis are listed below:

- Flow
- Pressure
- Temperature
- Level
- Time
- Composition
- pH
- Speed
- Frequency
- Viscosity
- Voltage
- Information
- Mixing
- Addition
- Separation
- Reaction

Guide words are applied to both the more general parameters (e.g., mixing, reaction) and the more specific parameters (e.g., temperature and/or pressure). Table 7.7 shows HAZOP analysis guide words and process parameters.

A typical example of a HAZOP study on a storage tank containing a liquid polymer is shown in Appendix D.2.

Table 7.7 Common guide words in HAZOP study

Guide words	Meaning
No	Negation of the design intent
Less	Quantitative decrease
More	Quantitative increase
Part of	Qualitative decrease
As well as	Qualitative increase
Reverse	Logical opposite of the intent
Other than	Complete substitution

7.6.1.5 Bow-Tie Analysis

All processes contain hazards. If a hazard is not controlled, a loss event can happen. Common loss events are fire, explosion, or toxic exposure. Based on the above risk matrix, a Bow-Tie model is a graphic representation (Fig. 7.9) of how the ultimate consequence (loss of life, assets, business) from the Top Event is managed. The undesirable consequence is an event that should not happen. Common threats are loss of containment, loss of control, or human errors that lead to the loss event.

Bow Tie analysis, as mentioned earlier, uses a graphic format as shown in Fig. 7.9. This analysis is set up to flow from a left-to-right format. At the center is the Hazardous Event, also called the Top Event. On the diagram's left is a list of potential hazards and threats that can cause harm to human health and safety, as well as the process and equipment integrity. When one of more of those are triggered, a set of preventive barriers tries to mitigate those to prevent the occurrence of the Hazardous Event. These barriers are the measures in place in the form of safety provisions in the engineering and design of the system, and proactive operational and maintenance measures. However, if those measures are unable to control the event, the Bow Tie model lists a set of reactive barriers on the right to minimize the extent of damage because of the occurrence of the Hazardous Event. Finally, on the diagram's right is a list of consequences if both the proactive and reactive sets of barriers fail. The graphic format of the model facilitates its ease and simplicity to visualize and understand the details of the Bow Tie analysis.

The same information is rather time consuming and mentally exhausting for the HAZOP team (see Sect. 7.6.1.4 above) as it is prepared in text format. Among other limitations of a HAZOP is the fact that in its existing format, it is difficult to document hazardous incidents caused by multiple independent events occurring simultaneously. As a result, a new version of the HAZOP study has evolved which structures

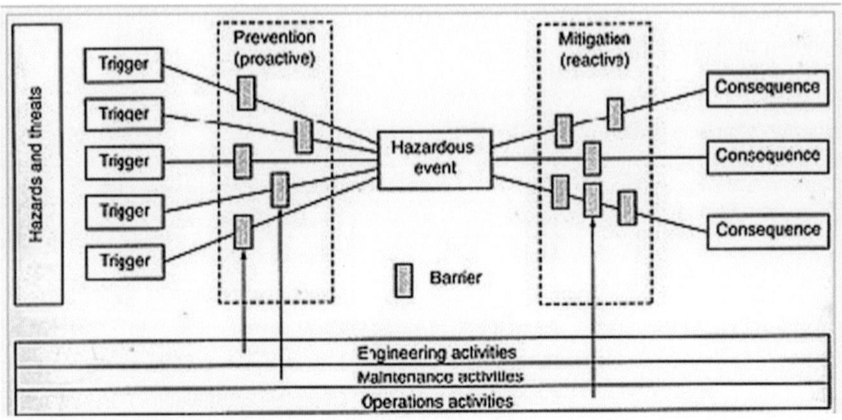

Fig. 7.9 A basic structure of a typical bow tie analysis studying a potential hazardous (Top) event, triggers (Threats), barriers and consequences (adapted from Hatch et al. 2019)

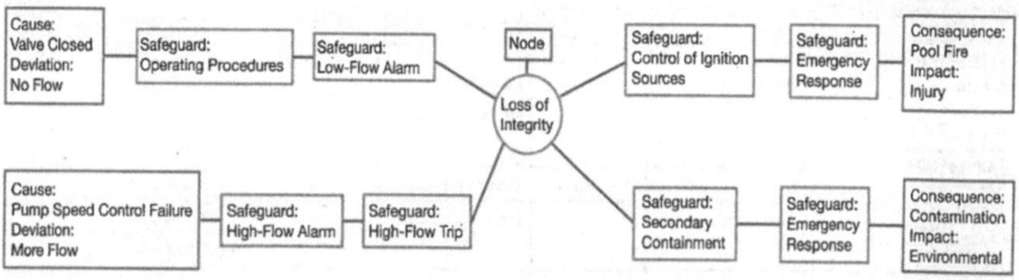

Fig. 7.10 The bow-tie model as a visual HAZOP showing the relationships between the components (Adapted from Hatch et al. 2019)

it in the Bow Tie format while using the terminologies for HAZOP. However, it is still difficult to overcome the weaknesses just by changing only the terminologies and using the Bow Tie structure for HAZOP studies. It is necessary to use specialized software(s) and an experienced team to get the best results from a HAZOP study. A detailed and in-depth discussion on this subject is available by Hatch et al. 2019 and CCPS 2018. Figure 7.10 demonstrates Bow Tie analysis that can be used as a visual HAZOP combining the guideword/deviation approach of a HAZOP.

7.6.1.6 Layer of Protection Analysis (LOPA)

The layer of Protection Analysis (LOPA) is often applied to scenarios that have been identified as elevated risk, e.g., serious, or critical as shown in the matrix in Fig. 7.8. LOPA is a quantitative analysis, as opposed to the risk assessment techniques noted previously, which are typically conducted as qualitative brainstorming. LOPA requires a team that understands which safeguards are independent protection layers (IPLs) and the probability of failure on demand (PFD) for IPLs. Not all safeguards are IPLs. The LOPA team also needs to understand the concept of enablers, which are events or conditions that must be present or active for the scenario to proceed, e.g., an alarm that is bypassed. Enablers do not by themselves initiate hazard scenarios, but they make them possible. Some special types of enablers need to be understood by the team, e.g., at-risk factors, and conditional modifiers. As with the other risk assessment techniques, LOPA requires training and an experienced facilitator.

In LOPA, the risk is quantified by using consequence severity (e.g., multiple fatalities, or negligible), order of magnitude values for Initiating Event (IE) frequency (e.g., once/yr. is 10^{-1}), likelihood of human errors (e.g., once/yr. is 10^{-1}), and likelihood of failure of independent protection layers (IPL) along with the Corporate risk tolerance criteria to determine a risk level (quantified by a target factor or TF value) for any given scenario. This TF value is matched against the frequency or likelihood factors of the initiating event, human errors along with the probability of failure on demand (PFD) of the IPLs (e.g., Process Safety Valves have a PFD of 1/100 or 10^{-2} i.e., a SIL (Safety Integrity level) of 2). If a gap still exists in TF by using all the IPLs, the risk still needs to be reduced by adding more IPLs, or modifying administrative procedures, or addressing by escalating to a higher level. Refer to the Example D.5 in Appendix D showing the Swiss Cheese concept.

Note that LOPA calls for the independence of IPLs. It does not allow credit for protection layers that share components with Initiating Event, or multiple protection layers that have common components. As a result, LOPA can result in recommendations for unnecessary and costly added instrumentation safeguards to achieve the tolerable Target Factor. Therefore, LOPA needs to be followed by FTA (Fault Tree Analysis) to get the optimum cost-effective result in such cases.

7.6.1.7 Fault-Tree Analysis (FTA)

In an FTA, a list of failure modes is identified that can cause the Top Event to occur. For example, toxic gas exposure may occur if a vent valve is not opened within 15 min of the high-high level alarm. The likelihood of the failure of the alarm and the failure of the operator to complete his task combine (frequencies will be multiplied) to produce the result of the frequency of the toxic gas exposure event. This is an example of multiple failures needing to occur. On the other hand, if the likelihood of a fire in a vessel can occur due to a pipe leak fire, a pool fire, of a nearby pool, the frequencies will be added. This example is about any failure leading to the ensuing consequence. If the frequency is less than once in a million years, the risk is tolerable (as a comparison, the average person has about a 1/1000 probability of dying). Otherwise, FTA is to be reviewed, the TF and the probability factors revisited, or the risk accepted with a variance. Moreover, the management needs to be involved in the decision-making.

A recent publication (Marszal and Applegate, 2024) comparing LOPA with FTA shows a specific example of overfilling a Butane Tank that the FTA could save on the number of safeguards that were needed with LOPA to achieve a lower frequency of Top Event. This is accomplished by considering the contribution of a single component shared as part of an initiating event and one or more protection layers. FTA can accurately determine the contribution of the shared component to the overall frequency.

7.7 Risk-Based Process Safety (RBPS)

Risk-Based Process Safety is the gold standard for process safety practitioners. Inherent deficiencies in the PSM Rule led to the development of the 20 RBPS elements gathered under four RBPS pillars.

A catastrophic industrial accident occurred at the BP refinery in Texas City, Texas on March 23, 2005, in which 15 persons died and more than 170 others were injured. Investigations, including one led by former U.S. Secretary of State James Baker, confirmed the need for major improvements to the principles of process safety.

Although the PSM Rule has not been updated in the Code of Federal Regulations (29 CFR 1910.119), Risk-Based Process Safety (RBPS) has become the standard by which diligent companies practice process safety. Thirteen of the fourteen PSM Rule elements are contained within RBPS (excluding PSM Rule 14, Trade Secrets), one PSM Rule element is covered by two RBPS elements (PSM Rule 2, Process Safety Information), and six new elements have been added (RBPS elements 1, 3, 5, 15, 18, 20). The applicable PSM counterparts are indicated in the list below. Significant improvements have been made to each of the PSM counterparts.

RBPS Element Pillar 1—Commit to Process Safety

1. Process Safety Culture
2. Compliance with Standards (PSM Rule element 2)
3. Process Safety Competency
4. Workforce Involvement (PSM 1)
5. Stakeholder Outreach.

RBPS Element Pillar 2—Understand Hazards and Risks.

6. Process Knowledge Management (PSM 2)
7. Hazard Identification and Risk Analysis (PSM 3).

RBPS Element Pillar 3—Manage Risks.

8. Operating Procedures (PSM 4)
9. Safe Work Practices (PSM 4 and PSM 9)
10. Asset Integrity and Reliability (PSM 8)
11. Contractor Management (PSM 6)
12. Training and Performance Assurance (PSM 5)
13. Management of Change (PSM 10)
14. Operational Readiness (PSM 7)
15. Conduct of Operations
16. Emergency Management (PSM 12).

Pillar 4—Learn from Experience.

17. Incident Investigation (PSM 11)
18. Measurement and Metrics
19. Auditing (PSM 13)
20. Management Review and Continuous Improvement.

RBPS addresses PSM Rule deficiencies with improvements to all PSM Rule elements, plus the addition of six new elements:

1. Process safety culture (RBPS Element 1). The Center for Chemical Process Safety (CCPS) defines process safety culture as the combination of group values and behaviors that determine the manner in which process safety is managed". More succinct definitions include, "How we do things around here," "What we expect here," and "How we behave when no one is watching." Catastrophic accidents, such as the 2005 Texas City BP unconfined vapor cloud explosion, reinforced the need for this element.

2. Process safety competency (RBPS Element 3). Each Process Safety Management Program, whether structured on the PSM Rule or the RBPS elements, must have competent employees assigned to lead each element. In the years following the promulgation of 29 CFR 1910.119, concern grew that those charged with leading elements were not always competent. Simply attending process safety training is not enough.
3. Stakeholder outreach (RBPS Element 5). A stakeholder is any individual or organization who could be, or believes they could be, affected by company operations. Stakeholders include employees, contractors, community groups, neighboring industrial facilities, other facilities in the company, other companies within the industry, and regulatory authorities. RBPS Element 5 describes stakeholder outreach as a process for seeking out individuals or organizations and engaging them in a dialogue about process safety, establishing a relationship with community organizations, other companies and professional groups, local, state, and federal authorities, and providing accurate information about the company and facility's products, processes, plans, hazards, and risks.
4. Conduct of operations (RBPS Element 15). Under PSM Rule 4, Operating Procedures, a facility must have both operating procedures and maintenance procedures in place for any work involving highly hazardous chemicals. However, it became apparent that having procedures in place does not guarantee that they will be followed in the plant. RBPS Element 15 seeks to establish means to ensure safe operations, a key factor being awareness of acceptance of unsafe operations known as normalization of deviance.
5. Measurement and Metrics (RBPS Element 18). Management guru Peter Drucker said, "If you can't measure it, you can't manage it." This element establishes performance and efficiency indicators to monitor the near-real-time effectiveness of the RBPS management systems and its constituent elements and work activities.
6. Management Review and Continuous Improvement (RBPS Element 20). This element requires routine evaluation of whether management systems are performing as intended and producing the desired results as efficiently as possible. It fills the gap between day-to-day work activities and periodic formal audits. Management reviews are typically conducted by focusing on one RBPS element. They should be conducted based on the following factors: The current life cycle stage of the facility, maturity of the RBPS management system, level of management performing the review, experience (incident history, previous reviews, audit results), and risk posed by the activities.

One important PSM Rule element, Process safety information (PSM Rule 2), is handled by two RBPS elements, Process Knowledge Management (RBPS Element 6) and Compliance with Standards (RBPS Element 2). In addition to this change, improvements in all other PSM Rule elements, and the six new elements, the Center for Chemical Process Safety has provided several guideline books that are valuable in the proper management of RBPS programs.

In the CCPS book Guidelines for Risk Based Process Safety, for each RBPS element, guidance is provided on the level of resources to commit to an element, based on the risk level.

The Health and Safety program for a wastewater treatment plant should be guided by the Health and Safety department of the overall facility. Consult appropriate OSHA, American National Standard Institute (ANSI), ASTM International, the National Fire Protection Association (NFPA) and local regulations and guidelines for detailed instructions. For information on biological risks and hazards of wastewater, consult the Center for Disease Control (CDC) guidelines specific to the wastewater treatment practice.

7.8 OSHA Rules on Personnel Safety

OSHA Rules cover the safety to personnel at operating facilities.

In Sect. 7.6, we discussed 29 CFR 1910.119, Process Safety Management of highly Hazardous Chemicals. Note that Risk Based Process Safety (RBPS) is not an OSHA Rule, but it is the process safety standard promoted by the Center for Chemical Process Safety. We discussed RBPS in detail in Sect. 7.7 above.

Other pertinent sections of 29 CFR 1910 are shown below.

Table 7.8 shows certain OSHA plans covered by specific CFR codes.

In Sect. 7.9, we discuss 40 CFR 68 Subpart G, Risk Management Plan, which is an EPA Rule not an OSHA Rule.

Table 7.8 OSHA plans covered by specific CFR codes

CFR	Description
1910.38	Emergency action and evacuation plan
1910.119	Lists hazardous materials and their threshold quantities in Appendix D
1910.120	Hazardous waste operations and emergency response
1910.147	Control of hazardous energy, confined spaces such as storage Tanks, Manhole, Sump, Sewer/Drain (lockout/tagout)
1910.176	Materials handling and storage

Note that this is not a comprehensive list of 29 CFR 1910 sections and the regulations are updated routinely. The latest versions can be obtained from respective websites

7.9 Risk Management Program (RMP)

The United States Environmental Protection Agency (USEPA) established RMP in 1990 as a part of the Clean Air Act amendments. This program emphasizes the environmental risks and how to mitigate them and is covered by 40 CFR 68. Certain aspects of the PSM program elements, specifically PHA and Emergency Response Planning (ERP), share many common requirements with the projected RMP requirements. Therefore, pre-planning and execution of the PSM elements with an awareness of the RMP requirements can result in cost savings from reduced compliance efforts and improved consistency.

The core questions guiding risk management are: What-If; What-Then; and So-What? HAZOP and PHA techniques primarily are driven by What-Ifs and are usually qualitative. LOPA, FTA, Quantitative Risk Analyses (QRA), and Consequence Analysis are risk assessment techniques and have certain quantitative aspects to them. These techniques address the question What-Then? Finally, So-What is the area of judgment, informed but not constrained by the earlier inputs. This is a decision process considering the economic, political, and human dimensions. The risk assessment techniques are discussed in Sect. 7.6.1. Process Safety Management also applies here.

Safety and Security of the WWTP operations are part of the critical infrastructure at the petroleum industrial facility. It is also recommended that the facility's Health and Safety department has in its possession a copy of the latest edition of the WEF manual *Safety Health and Security in Wastewater Systems, Sixth Edition, MOP 1* (WEF Manual of Practice) as a reference material for personal safety. QRA is a formal, systematic method to estimate the likelihood and consequences of hazardous events. The result is expressed quantitatively as a risk to people, the environment, and business. QRA uses the consequence analysis, risk contours, F-N curves (see Sect. 7.9.2), or other measures to compare the corporate or regulatory criteria for acceptability.

7.9.1 Risk Contours

Risk contours are maps of various risk rankings superimposed on a specific facility. It allows one to determine the risk of the employees working in various areas of the plant.

7.9.2 F-N Curves

An F-N curve is a plot of cumulative frequency versus consequence (often measured as having N + fatalities per year) in an event. An example of an F-N plot shown below shows that the probability of 10 + fatalities is one in 1000 years (Fig. 7.11).

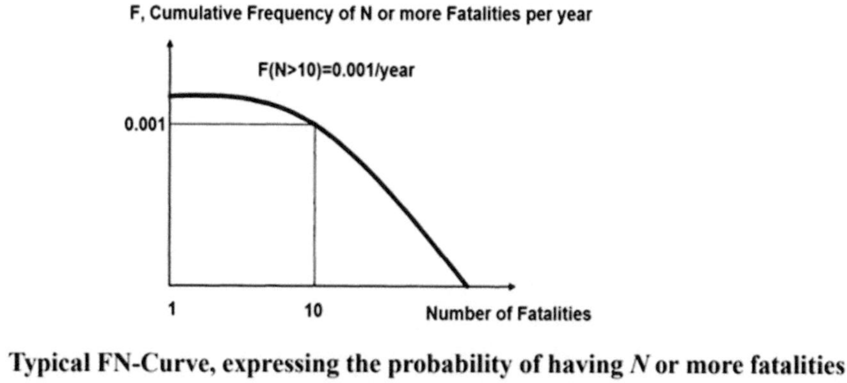

Fig. 7.11 Cumulative frequency F(N) versus the number of fatalities. [*Source* CCPS—Guidelines for Developing Quantitative Safety Risk Criteria, John Wiley & Sons, Inc., New Jersey, USA (2009)]

7.9.3 Consequence Analysis

Consequence analysis is a technique where the routine and episodic emissions are modeled to determine the effect inside and outside the fence. A typical cloud dispersion side view of ethylene glycol from a PSV, at a petrochemical plant developed using a commercially available model, is shown in Appendix D, Example D.3. A similar approach can be utilized for consequence analysis in the dispersion of hazardous chemicals from petroleum processing industries.

7.10 Security of Water and Wastewater Treatment Plant Operational Technology (OT)

The WWTP in a petroleum processing facility is part of the infrastructure for the overall facility. Safety and Security of these infrastructures, therefore, are part of its priority. Cyber (network) security is of prime importance due to the use of Information Technology (IT), Internet-of-Things (IoT), and Industrial Internet of Things (IIOT)—smart devices in all or most of the processes. Operational Technology (OT), unlike IT, is the network of connected smart devices used to run the WWTP operations. It includes Programmable Logic Controllers (PLCs), IoT devices, and mobile devices that are used to communicate with Supervisory Control and Data Acquisition (SCADA) system to control and run the processes. In Industrial Cyber Security (ICS), OT space is the prime target for hackers to either simply damage or harm the routine operation of the plant, or to take control of the operation for other malicious purposes. The subject is vast and of prime importance and only a brief outline is provided in this section.

7.10.1 Cybersecurity Risk Management

A cybersecurity contingency plan is a written risk management document that provides instructions, recommendations, and considerations for a company on how to recover its IT services and data in the event of a security breach, disaster, or system disruption. SCADA systems are the commonly used control systems in WWTP operations. The National Institute of Standards and Technology (NIST) has developed a cybersecurity framework for protection against cyber threats. NIST has several publications like API to cover different topics in Cybersecurity and is routinely used in the IT sector in each industry.

Many incidents (Colonial Pipeline, GA), Oldsmar, FL, San Francisco, CA, Queensland, Australia) have occurred in the last few years. Colonial Pipeline Company is an example of a total collapse of oil distribution pipeline operation on the east coast of the US in February 2021. It collapsed the entire distribution and delivery of oil products for a week which could be recovered only after paying ransom to the hackers resulting in the loss of millions of dollars to get back to normal operation. Caesars Entertainment paid a $30 million ransom that hackers demanded after a cyberattack in late summer of 2023. Tools are available to detect anomalies, protect (anomaly avoidance), and respond to an anomaly (real-time remediation) in the WWTP OT operations. This ensures the security and safety of Global Critical Infrastructure (i.e., the physical plant and distribution system) by protecting the OT network and preventing data loss or leaks, damage to processes, service disruption, and loss of operational control.

The tools used to manage cybersecurity risks are based on data-centric Machine Learning (ML) and design-centric (i.e., simplified process model developed based upon process design) models. The tools can be applied to the actual operating WWTPs. In case this is not practical, a digital twin (i.e., an online electronic facsimile of the real plant) can also be deployed to show the efficacy of the protection. This approach has been shown to result in an ultra-high rate of attack detection and ultra-low rate of false alarms (Gauthama Raman et al. 2017).

7.10.2 Business Impact Analysis (BIA)

A Business Impact Analysis (BIA) is a process that helps to identify critical business functions that are running within an organization. Using the BIA process, one can minimize the potential loss due to an unexpected event. After risks have been determined, one can calculate the effect of these risks. One can then prioritize the functions that need more attention and protection. At that point, one can calculate the Recovery Point Objective (RPO) and Recovery Time Objective (RTO) to prioritize the functions.

RPO defines the tolerance level for an organization to which they can revert to an older version of data. Just as an ALARP criterion (see Sect. 7.5.1) is used in risk analysis in Process Hazard, similarly in the digital world some loss of data cannot be avoided in certain cases. RPO is the maximum amount of data you can afford to lose in case of a disaster.

RTO is the time during which the disrupted service or server must be restored. It should be the maximum time to restore service to its operations.

BIA has multiple steps that must be performed to complete the process:

- Risk Assessment
- Business Impact Analysis (BIA)
- Business Continuity Plan Development
- Strategy and Plan Development
- Plan Testing and Maintenance.

The BIA considers various scenarios. For example, it would consider the following (not exhaustive, not only technical, and can also be intentional by outsiders or insiders):

- **Natural disasters:** Flood, hurricane, and earthquake;
- **Accidental disasters:** Fire and power outage;
- **Human-made errors:** Data deletion;
- **System errors:** System failure and hard drive failure; and,
- **Intentional errors:** System hacks and data thefts.

In summary, the cybersecurity threats in a typical WWTP are: Intentional errors, natural and accidental disasters. The Risk is a product according to the formula:

$$Risk = Threat * vulnerability$$

$$Risk = Threat * (exploitation\ likelihood + asset\ exposure + impact)\ and\ vulnerability$$

Given the vulnerabilities are similar in almost all WWTPs, the risk depends on the level of threat along with the probability of exploitation. Unfortunately, threat levels have recently increased significantly and therefore so has the risk.

The BIA identifies mission-essential functions and critical systems that are essential to the organization's success. It also identifies maximum downtime limits for these systems and components, various scenarios that can impact these systems and components, and the potential losses from an incident. BIA is not designed specifically for a particular scenario. For example, you will not design the BIA to only work in a data breach scenario. It is not important what disrupted a web application. The more important factor is identifying the critical processes, such as the web application, and ensuring that it continues to run no matter what type of disaster takes place.

Just like life, water is precious to all mankind. To sustain life on this planet, we must take steps to develop ways to reuse and recycle water. Use and reuse of water through the application of AI, data analytics, and machine learning interface can be termed as digital water. Digital water is the key enabler for a sustainable water industry. Data analytics can help in long range planning and investment strategies by identifying opportunities, managing risks, and improving customer service. Digital technology reduces human error, enhances safety – both personal and process, increases productivity, resolves issues and machine downtimes quickly thus saving valuable time and money.

7.11 Conclusion

In closing, EHSS regulations are in place to reduce the harmful impacts of wastewater discharges and the negative impact that can occur if the proper steps are not in place to prevent them. Routine regulatory monitoring and reporting assist in the ability to predict adverse outcomes to aquatic ecosystems, flora, fauna, and human health.

WET testing serves as a worst-case scenario, but ultimately is necessary to correctly identify toxicity when analytical analysis fails to account for toxicological effects holistically. TIE/TRE strategies are vital to determine how a WWTP can safely and efficiently remove toxicity and prevent further impacts downstream. Risk assessments, proper engineering efforts, and technologies should be applied to reduce risks associated with environmental releases, and safety, both personal and assets.

Personal and process safety fall under the overall EHSS responsibility in a company. This is not about only following OSHA guidelines and avoiding penalties but the overall health and wellness of employees, and the integrity of the facility and its equipment. Safety and security of the WWTP operations are part of the infrastructures of industrial facilities, and a priority for industrial facilities.

The cybersecurity threats in a typical WWTP primarily are intentional errors and natural and accidental disasters. *The Risk is a product of Risk = Threat x (exploitation likelihood + asset exposure + impact) and vulnerability.* Given the vulnerabilities are similar in almost all WWTPs, the risk depends on the probability of the nation-state cyber threat actors exploiting vulnerabilities with the most risk and that's the threat. Nation-state attackers usually target critical infrastructures because they have the greatest negative impact on a nation when compromised. Overall, more work is in progress and needs to be done to advance this field to keep the petroleum production and processing facilities, including WWTPs and WTPs, safe.

Leading companies keep a dashboard that all employees can view. This dashboard should include metrics, both leading and lagging (for example, lost time rate, total accident rate, preventative maintenance, MOC, PHA), results of management reviews, and newsletters, all to create awareness of all employees and resident contractors that the company is taking all steps necessary to maintain a safe workplace management.

Conduct frequent workshops to improve the knowledge of the safety procedures, and workplace hazards, and process knowledge to help reduce workplace injuries, illnesses, and deaths. Ultimately, a safe workplace is a sound business. EHSS sectors are each a crucial component that encompasses and ensures industrial and societal progress is achieved safely and responsibly. By taking the time to develop and enforce the concepts of this chapter, petroleum companies, and their WWTPs can offer more awareness, compliance, and increased sustainability towards the future. The overall goal for EHSS is to reduce impacts and prevent harmful instances that may be irreversible, both on our planet and those that inhabit it.

Abbreviations and Glossary of Terms

ACGIH	American Conference of Industrial Hygienists
ALARP	As Low As Reasonably Practicable
ANSI	American National Standards Institute
AOP	Advanced Oxidation Processes
APHA	American Public Health Association
ASTM	American Society for Testing and Materials
AWWA	American Water Works Association
BACT	Best Available Control Technologies
BIA	Business Impact Analysis
BLM	Biotic Ligand Model
BPCS	Basic Process Control System
BOD	Biological Demand
BTEX	Benzene, Toluene, Ethylbenzene and Xylene
BWON	Benzene Waste Operations NESHAP
CAA	Clean Air Act
CCPS	Center for Chemical Process Safety
CDC	Center of Disease Control
CEC	Contaminants of Emerging Concern
CERCLA	Comprehensive Environmental Response Compensation and Liability Act
CFR	Code of Federal Regulations
CH_4	Methane
Cl	Chlorine
Cl_2	Chlorine Gas
CO	Carbon Monoxide
COD	Chemical Oxygen Demand
CWA	Clean Water Act
CWTS	Constructed Wetland Treatment Systems
DAF	Dissolved Aeration Flotation
DAP	Di Ammonium Phosphate
DDT	Dichloro-diphenyl-trichloroethane
DEQ	Departments of Environmental Quality
DNR	Department of Natural Resources
DO	Dissolved Oxygen
EtOH	Ethanol

EDTA	Ethylenediaminetetraacetic acid
EHSS	Environment, Health and Safety, and Security
EPA	Environmental Protection Agency
ER	Emergency Relief
EROD	Erthoxyresorufin-*O*-deethylase
ERP	Emergency Response Planning
EU	European Union
FET	Fish Embryo Acute Toxicity
FIFRA	Federal Insecticide, Fungicide and Rodenticide Act
FMEA	Failure Modes and Effects Analysis
FTA	Fault Tree Analysis
H_2S	Hydrogen Sulfide
H_2SO_4	Sulfuric Acid
HA	Health Advisory
HAZOP	Hazard and Operability Study
HIPS	High Integrity Protection System
HC	Hydrocarbons
HPEI	Hyperbranched polyethylenimine
IC	Internal Combustion
ICS	Industrial Cyber Security
IE	Initiating Event
IPL	Independent Protection Layer
IRTC	Ingredients of Fire Retardant Foam
ISO	International Organization for Standardization
IT	Information Technology
IOT	Internet of Things
IIOT	Industrial Internet of Things
KPI	Key Performance Indicator
LC_{50}	Lethal Concentration to kill of 50% of the population of organisms tested
LEL	Lower Explosive Limit
LOC	Loss of Containment
LOEC	Lowest Observed Effect Concentration
LOPA	Layer of Protection Analysis
MACT	Maximum Achievable Control Technology
MAWP	Maximum Allowable Working Pressure
MCC	Master Command and Control
ML	Machine Learning
MTBE	Methyl tert-butyl ether
MOC	Management of Change
NAAQS	National Ambient Air Quality Standard
NaClO/NaOCl	Sodium Hypochlorite
NaOH	Sodium Hydroxide
NESHAPS	National Emission Standards for Hazardous Air Pollutants
NFPA	National Fire Protection Association

NH_3	Ammonia
NIOSH	National Institute for Occupational Safety and Health
NIST	National Institute of Standards and Technology
NSR	New Source Review
NOEC	No Observed Effect Concentration
NO_x	Nitric Oxides
NPDES	National Pollutant Discharge Elimination System
OECD	Organization for Economic Co-Operation and Development
OSHA	Occupational Health and Safety Administration
OT	Operational Technology
PAC	Powdered Activated Carbon
PAHs	Polycyclic Aromatic Hydrocarbons
PAR	Pre-acidification Reactor
PBDEs	Polybrominated Biphenyl
PBS	Potentially Bioaccumulative Substances
PCBs	Polychlorinated Biphenyls
PEL-STEL	Permissible Exposure Limit, and Short-Term Exposure Limit, on a Time Weighted Average
PFAS	Per-and Polyfluoroalkyl Substance
PFD	Probability of Failure on Demand
PFOA	Perflurooctanoic Acid
PFOS	Perfluorooctanesulfonate
PHA	Process Hazard Analysis
PHAST	Process Hazard Analysis Software Tool
PLC	Programmable Logic Controller
PM	Particulate Matter
POG	Point of Waste Generation
PPE	Personal Protective Equipment
PSM	Process Safety Management
PSV	Process Safety Valve
RCRA	Resource Conservation and Recovery Act
RFG	Reformulated Gasoline
RMP	Risk Management Program
RO	Reverse Osmosis
RPO	Recovery Point Objective
RTO	Recovery Time Objective
QAC	Quaternary Ammonium Compound
QRA	Quantitative Risk Analyses
SBR	Sequencing Batch Reactors
SCADA	Supervisory Control and Data Acquisition
SDS	Safety Data Sheet
SIL	Safety Integrity Level
SIS	Safety Integrity System
SO_2	Sulfur Dioxide
SO_x	Sulfur Oxides

SOPs	Standard Operating Procedures
SPE	Solid Phase Extraction
SPME-BE	Solid Phase Micro Extraction Biomimetic Extraction
SIP	State Implementation Plan
STP	Sewage Treatment Plant
TAB	Total Annual Benzene
TDS	Total Dissolved Solids
TAML	Tetra-Amido Macrocyclic Ligand
TBEL	Technology Based Effluent Limitation
TDML	Total Maximum Daily Load
TEL	Tetraethyl Lead
TF	Target Factor
TIE/TRE	Toxicity Identification and Reduction Evaluations
TQ	Threshold Quantities
UASB	Up Flow Anaerobic Sludge Blanket
UF	Ultrafiltration
UV	Ultraviolet
US	United States
UST	Underground Storage Tank
VOC	Volatile Organic Compounds
WHO	World Health Organization
WQBEL	Water Quality-Based Effluent Limitations
WEF	Water Environment Federation
WET	Whole Effluent Toxicity
WMU	Waste Management Unit
WTP	Water Treatment Plant
WWTP	Wastewater Treatment Plant

References

Akinsanya B, Ayanda IO, Onwuka B, Saliu JK (2020) Bioaccumulation of BTEX and PAHs in Heterotis niloticus (Actinopterygii) from the Epe Lagoon, Lagos. Nigeria Heliyon 6(1):e03272. https://doi.org/10.1016/j.heliyon.2020.e03272.doi:10.1016/j.heliyon.2020.e03272

Ankley GT, Bennett RS, Erickson RJ, Hoff DJ, Hornung MW, Johnson RD, Mount DR, Nichols JW, Russom CL, Schmieder PK, et al (2010) Adverse outcome pathways: a conceptual framework to support ecotoxicology research and risk assessment. Environ Toxicol Chem 29(3):730–741. https://doi.org/10.1002/etc.34.API; Standard 521, 7th edition, June 2020.

Avci A, Kaçmaz M, Durak I (2005) Peroxidation in muscle and liver tissues from fish in a contaminated river due to a petroleum refinery industry. Ecotoxicol Environ Saf 60(1):101–105. https://doi.org/10.1016/j.ecoenv.2003.10.003

Basu S, Thomas T, Ali P (2022) Emerging technologies for PFAS control (Book Chapter 19), understanding PFAS: developing management practices and treatment options for industry, water environment federation, Alexandria, VA

Besser JM, Brumbaugh WG, Papoulias DM, Ivey CD, Kunz JL, Annis M, Ingersoll CG (2012) Bioaccumulation and toxicity of selenium during a life-cycle exposure with desert pupfish (Cyprinodon macularius). Sci Investig Rep 36. http://pubs.er.usgs.gov/publication/sir20125033

Biesinger KE, Christensen GM (1972) Effects of various metals on survival, growth, reproduction and metabolism of Daphnia magna. J Fish Res Board Canada 29(12):1691–1700

Brinkman SF, Woodling JD, Vajda AM, Norris DO (2009) Chronic toxicity of ammonia to early life stage rainbow trout. Trans Am Fish Soc 138(2):433–440. https://doi.org/10.1577/t07-224.1

Chen C, Ming J, Yoza BA, Liang J, Li QX, Guo H, Liu Z, Deng J, Wang Q (2019) Characterization of aerobic granular sludge used for the treatment of petroleum wastewater. BioresourceF Technol 271:353–359. https://doi.org/10.1016/j.biortech.2018.09.132

Clayton C, Gold HD, Gurney B, Kalpin M, White A (2015) Minimizing risk under the clean water act 69–94. http://0-eds.a.ebscohost.com.wncln.wncln.org/eds/pdfviewer/pdfviewer?sid=48cfe33f-d9fb-4b79-93a3-b7ae7793569c%40sessionmgr4002&vid=2&hid=4213

Comber MHI, Girling A, den Haan KH, Whale G (2015) Oil refinery experience with the assessment of refinery effluents and receiving waters using biologically based methods. Integr Environ Assess Manag 11(4):653–665. https://doi.org/10.1002/ieam.1639

Congress US (2002) Federal water pollution control act. 33 USC 1251 seq.:234.

Conolly RB, Ankley GT, Cheng W, Mayo ML, Miller DH, Perkins EJ, Villeneuve DL, Watanabe KH (2018) Quantitative adverse outcome pathways and their application to predictive toxicology 1–24. https://doi.org/10.1021/acs.est.6b06230.Quantitative

Couillard CM, Lee K, Légaré B, King TL (2005) Effect of dispersant on the composition of the water-accommodated fraction of crude oil and its toxicity to larval marine fish. Environ Toxicol Chem 24(6):1496–1504. https://doi.org/10.1897/04-267R.1

Cowan RM, Park K (1996) Biodegradation of the gasoline oxygenates MTBE, ETBE, TAME, TBA and TAA by aerobic mixed cultures. In: Proceedings of the 28th mid-Atlantic industrial and hazardous waste conference; July; Buffalo, NY. Lancaster, PA: Technomic Publishing Co., Inc.; pp 523–530

D'Costa A, Shyama SK, Praveen Kumar MK (2017) Bioaccumulation of trace metals and total petroleum and genotoxicity responses in an edible fish population as indicators of marine pollution. Ecotoxicol Environ Saf 142(April):22–28. https://doi.org/10.1016/j.ecoenv.2017.03.049. doi:10.1016/j.ecoenv.2017.03.049

Daflon SDA, Guerra IL, Reynier MV, Cerqueira AC, Botta CR, Campos JC (2017) Toxicity identification and evaluation (TIE) of a petroleum refinery wastewater. J Environ Sci Heal - Part A Toxic/hazardous Subst Environ Eng. 52(9):842–848. https://doi.org/10.1080/10934529.2017.1312186

Davis TW, Berry DL, Boyer GL, Gobler CJ (2009) The effects of temperature and nutrients on the growth and dynamics of toxic and non-toxic strains of Microcystis during cyanobacteria blooms. Harmful Algae 8(5):715–725. https://doi.org/10.1016/j.hal.2009.02.004

DOT 49 CFR 172.101, Purpose and Use of Hazardous Tables for Transportation

Gadagkar SR, Call GB (2015) Computational tools for fitting the Hill equation to dose-response curves. J Pharamacological Toxicolical Methods 71:68–76

Gauthama Raman MR, Dong W, Mathur A (2017) Deep autoencoders as anomaly detectors: Method and case study in a distributed water treatment plant, iTrust. Singapore University of Technology and Design, Singapore, Centre for Cyber Security Research

Gensemer RW, Gondek JC, Rodriquez PH, Arbildua JJ, Stubblefield WA, Cardwell AS, Santore RC, Ryan AC, Adams WJ, Nordheim E (2018) Evaluating the effects of pH, hardness, and dissolved organic carbon on the toxicity of aluminum to freshwater aquatic organisms under circumneutral conditions. Environ Toxicol Chem 37(1):49–60. https://doi.org/10.1002/etc.3920

Guidelines for Risk Based Process Safety, Center for Chemical Process Safety (2007). ISBN-978-0-470-16569-0

Hatch D, McCulloch P, Travers I (2019) Enhancing PHAs: the power of bowties, chemical engineering progress

Hughes SA, Maloney EM, Bejarano AC (2021) Are vertebrates still needed in routine whole effluent toxicity testing for oil and gas discharges? Environ Toxicol Chem 40(5):1255–1265. https://doi.org/10.1002/etc.4963

Ibrahim M, Oldham D, Minghetti M (2020) Role of metal speciation in the exposure medium on the toxicity, bioavailability and bio-reactivity of copper, silver, cadmium and zinc in the rainbow trout gut cell line (RTgutGC). Comp Biochem Physiol Part - C Toxicol Pharmacol 236:108816. https://doi.org/10.1016/j.cbpc.2020.108816; International Process Safety Management Conference and Workshop, September 1993, San Francisco, CA, CCPS.

IRTC (2020). PFAS technical and regulatory guidance document and fact sheets PFAS 1, interstate technology & regulatory council, PFAS Team, Washington DC. https://pfas-1.itrcweb.org/

ISO (2019) Water quality—determination of acute toxicity of water samples and chemicals to a fish gill cell line (RTgill-W1) ISO 21115;2019

Kleinhenz LS, Humphrey CL, Mooney TJ, Trenfield MA, van Dam RA, Nugegoda D, Harford AJ (2019) Chronic ammonia toxicity to juveniles of 2 tropical Australian freshwater mussels (Velesunio spp.): toxicity test optimization and implications for water quality guideline values. Environ Toxicol Chem 38(4):841–851. https://doi.org/10.1002/etc.4370

Lawley HG (1974) Operability Study and Hazard Analysis. Chem Eng Prog 70(4):45–56

Leães Pinho GL, Bianchini A (2010) Acute copper toxicity in the euryhaline copepod Acartia tonsa: Implications for the development of an estuarine and marine biotic ligand model. Environ Toxicol Chem 29(8):1834–1840. https://doi.org/10.1002/etc.212

Marszal EM, Applegate J (2024) Supporting LOPA with fault tree analysis, chemical engineering progress. aiche.org/cep

Nichols L (2016) Business trends: clean fuels – a global shift to low sulfur world, hydrocarbon processing

Niyogi S, Wood CM (2004) Biotic ligand model, a flexible tool for developing site-specific water quality guidelines for metals. Environ Sci Technol 38(23):6177–6192. https://doi.org/10.1021/es0496524

Nowak NC (2024) Produced water and flowback water recycling, storage and evaporation, presented at the produced water society annual conference, February 8, 2024, Houston, TX

OECD (2013) Oecd guidelines for the testing of chemicals nr 236: Fish Embryo Acute Toxicity (FET) Test. OECD Guidel Test Chem Sect 7.2, OECD Publ 1–22. https://doi.org/10.1787/9789264203709-en

OECD (2021) Test no. 249: fish cell line acute toxicity—the RTgill-W1 cell line assay. Guidel Test Chem Sect 2.(249). https://www.oecd-ilibrary.org/environment/test-no-249-fish-cell-line-acute-toxicity-the-rtgill-w1-cell-line-assay_c66d5190-en

OSHA Chromium and Cadmium Hazards (1910a) 29 CFR 1910.1026–1027

OSHA Control of hazardous energy (1910b) 29 CFR 1910.147

OSHA (1990) permit required confined spaces, 29 CFR 1910.146

OSHA (2002) hazardous waste operations and emergency response, 29 CFR 1910.120

OSHA (2011) Emergency action plan, 29 CFR 1910.38

OSHA (2013) Process safety management standard, 29 CFR 1910 a, b.119

Peng C, Lee JW, Sichani HT, Ng JC (2015) Toxic effects of individual and combined effects of BTEX on Euglena gracilis. J Hazard Mater 284:10–18. https://doi.org/10.1016/j.jhazmat.2014.10.024.doi:10.1016/j.jhazmat.2014.10.024

Pinzón-Espinosa A, Collins TJ, Kanda R (2021) Detoxification of oil refining effluents by oxidation of naphthenic acids using TAML catalysts. Sci Total Environ 784. https://doi.org/10.1016/j.scitotenv.2021.147148

Powers SE, Rice D, Dooher B, Alvarez PJJ (2001) Will ethanol-blended gasoline affect groundwater quality? Environ Sci Technol 35(1)

Randall DJ, Tsui TK (2002) Ammonia toxicity in fish. Mar Pollut Bull 45:17–23. https://doi.org/10.1016/s1095-6433(99)90247-7.www.elsevier.com/locate/marpolbul

Randazzo W, Truchado P, Cuevas-Ferrando E, Simon P, Allende A, Sanchez G (2020) Water Res 181:115942

Safety, Health, and Security in Wastewater Systems, MOP 1 (WEF Manual of Practice)
Saha NC, Bhunia F, Kaviraj A (1999) Toxicity of phenol to fish and aquatic ecosystems. Bull Environ Contam Toxicol 63(2):195–202. https://doi.org/10.1007/s001289900966
Salehi M, Rodriguez R, Boettcher A, Powers S, Geitner N, Ladner DA, Rikard S, Whelton AJ (2017) Impact of dispersant on early life stages of the water flea Daphnia magna and the eastern oyster Crassostrea virginica. J Appl Toxicol 37(12):1464–1470. https://doi.org/10.1002/jat.3494
Santo CE, Vilar VJP, Bhatnagar A, Kumar E, Botelho CMS, Boaventura RAR (2012) Sulphide removal from petroleum refinery wastewaters by catalytic oxidation. Desalin Water Treat 46(2012):256–263
Scott J, Belden JB, Minghetti M (2021) Applications of the RTgill-W1 cell line for acute whole-effluent toxicity testing: in vitro–in vivo correlation and optimization of exposure conditions. Environ Toxicol Chem 40(4):1050–1061. https://doi.org/10.1002/etc.4947
Scott J, Grewe R, Minghetti M (2022) Fish embryo acute toxicity (FET) Testing and the RTgill-W1 cell line as in-vitro models for whole effluent toxicity (WET) testing: an in vitro/in vivo comparison of chemicals relevant for WET testing. Environ Toxicol Chem. https://doi.org/10.1002/etc.5455
Shapiro JS (2022) Pollution trends and US environmental policy: lessons from the past half century. Rev Environ Econ Policy 16(1). https://doi.org/10.1086/718054
Sikkema J, De Bont JAM, Poolman B (1995) Mechanisms of membrane toxicity of hydrocarbons. Microbiol Rev 59(2):201–222. https://doi.org/10.1128/mmbr.59.2.201-222.1995
Singer MM, Aurand D, Bragin GE, Clark JR, Coelho GM, Sowby ML, Tjeerdema RS (2000) Standardization of the preparation and quantitation of water-accommodated fractions of petroleum for toxicity testing. Mar Pollut Bull 40(11):1007–1016. https://doi.org/10.1016/S0025-326X(00)00045-X
Smart G (1976) The effect of ammonia exposure on gill structure of the rainbow trout (Salmo gairdneri). J Fish Biol 8:471–475
Smit MGD (2024) OSPAR's regulatory framework for produced water assessment—still fit for purpose? Presentation at the produced water society annual conference, February 28, 2024, Houston, TX
Spacil MM, Rodgers JH, Castle JW, Murray Gulde CL, Myers JE (2011) Treatment of selenium in simulated refinery effluent using a pilot-scale constructed wetland treatment system. Water Air Soil Pollut 221(1–4):301–312. https://doi.org/10.1007/s11270-011-0791-z
Steliga T, Jakubowicz P, Kapusta P (2015) Changes in toxicity during treatment of wastewater from oil plant contaminated with petroleum hydrocarbons. J Chem Technol Biotechnol 90(8):1408–1418. https://doi.org/10.1002/jctb.4442
Stradling D, Stradling R (2008) Perceptions of the burning river: deindustrialization and Cleveland's Cuyahoga River. Environ Hist Durh N C 13(3):515–535. https://doi.org/10.1016/B0-12-369400-0/01076-0
Stolark J (2016) A brief history of octane in gasoline: from lead to ethanol, fact sheet, environmental and energy study institute, Washington, DC
Torres RJ, Cesar A, Pereira CDS, Choueri RB, Abessa DMS, do Nascimento MRL, Fadini PS, Mozeto AA (2012) Bioaccumulation of polycyclic aromatic hydrocarbons and mercury in oysters (Crassostrea rhizophorae) from two Brazilian Estuarine Zones. Int J Oceanogr 1–8. https://doi.org/10.1155/2012/838320
Tugcu G, Ertürk MD, Saçan MT (2017) On the aquatic toxicity of substituted phenols to Chlorella vulgaris: QSTR with an extended novel data set and interspecies models. J Hazard Mater 339:122–130. https://doi.org/10.1016/j.jhazmat.2017.06.027
USEPA (1985) USEPA guide for the application of effluent limitations guidelines for the petroleum refining industry
USEPA (1991) US EPA, national primary drinking water regulations, federal register; 40 CFR Chapter I, Part 141:663–664
USEPA (1992) Characterization of chronically toxic effluents, phase I. (May)

USEPA (1993a) Methods for aquatic toxicity identification evaluations: phase II toxicity identification procedures for samples exhibiting acute and chronic toxicity 63
USEPA (1993b) Methods for aquatic toxicity identification evaluations: phase III toxicity identification procedures for samples exhibiting acute and chronic toxicity 63
USEPA (1993c) Offshore effluent guidelines for oil and gas Subcategory
USEPA (2002a) Method 1001.0: Fathead Minnow, Pimephales promelas, larval survival and teratogenicity test; chronic toxicity
USEPA (2002b) Short-term methods for estimating the chronic toxicity of effluents and receiving waters to freshwater organisms, 4th edn
USEPA (2002c) Short-term methods for estimating the chronic toxicity of effluents and receiving waters to marine and estuarine organisms, 3rd edn
USEPA (2002d) Methods for measuring the acute toxicity of effluents and receiving waters to freshwater and marine organisms, 5th edn
USEPA (2003) 40 CFR 136 guidelines establishing test procedures for the analysis of pollutants. Water Pollution Control
USEPA (2010) NPDES permit writer's manual
USEPA (2012a) 40 CFR 419 Petroleum refining point source category. (c):363–398.
USEPA (2012b) 40 CFR Parts 136,260, et al. guidelines establishing test procedures for the analysis of pollutants under the clean water act; analysis and sampling procedures. Final Rule 77(97):1–90. https://doi.org/10.2753/RSH1061-1983310140.
USEPA (2022) Subpart D—certification by the subpart E—consultations PART 122—EPA administered permit programs: the national pollutant discharge elimination system (iv):167–295
US Department of Justice (2005) Oil companies pay EPA to Settle Santa Monica MTBE Clean-up Costs, Press release on February 16, 2005. www.usdoj.com
Wake H (2005) Oil refineries: a review of their ecological impacts on the aquatic environment. Estuar Coast Shelf Sci 62(1–2):131–140. https://doi.org/10.1016/j.ecss.2004.08.013
Whale GF, Hjort M, Di Paolo C, Redman AD, Postma JF, Legradi J, Leonards PEG (2022) Assessment of oil refinery wastewater and effluent integrating bioassays, mechanistic modelling, and bioavailability evaluation. Chemosphere. 287(P3):132146. https://doi.org/10.1016/j.chemosphere.2021.132146; 42 U.S. Code Chapter 85. 1970. Air Pollution Prevention and Control
Wooten T (2024) Railroad commission of texas, oil and gas division technical permitting section, 2024 Permian basin water in energy conference, Midland, TX, March 6, 2024

Important Websites for Information on OSHA Guidance

https://www.osha.gov/emergency-preparedness/guides/toxic-industrial-chemicals
https://www.osha.gov/sites/default/files/publications/osha 3071.pdf
https://sitemate.com/resources/articles/safety/safe-work-permits/
https://www.osha.gov/laws-regs/regulations/standardnumber/1910/1910.119
https://www.osha.gov/sites/default/files/OSHA_FlowChart.pdf
https://www.aiche.org/resources/publications/books/guidelines-risk-based-process-safety
OSHA interpretation letters: https://www.osha.gov/laws-regs/standardinterpretations/standardnumber/1910/1910.119%20-%20Index/result.OSHA compliance directives: https://www.osha.gov/enforcement/directives/search?title_body=PSM&field_date_range_radio=single&field_fr_date_single_value=
https://www.osha.gov/chemical-executive-order/psm-terminology

Important Websites for Information on Cybersecurity Framework and Federal Government Guidance

1. www.nsc.org
2. www.nfpa.org
3. www.wef.org
4. www.epa.gov
5. www.dnr.wi.gov

Chapter 8
Sustainability—Water Recycle and Reuse

Mudumbai Venkatesh, Fernando Almada Calvo, Richard T. Grubbs, and James Cunningham

Abstract Water is a precious resource both for the sustenance of life on the planet and also for the economic growth of society. However, unrestrained use of water to support the needs of a growing population with ever increasing demand for higher standards of life globally creates a stress on the rapidly diminishing supply of fresh water. Sustainable management of water resources ensures its responsible use by reducing consumption and maximizing recycle and reuse of treated wastewater. Many parts of the world, including the western states of the US, are actively pursuing programs to treat municipal wastewater to the primary drinking water standards for direct potable reuse (DPR) or indirect potable reuse (IPR). Similarly, industries also have been implementing reuse measures to minimize withdrawal of fresh water and reuse of treated wastewater. Petroleum industries have set their own sustainability goals by following high standards of treatment to move well beyond meeting their discharge compliance requirements and producing a valuable resource for beneficial reuse. This chapter presents the steps taken by petroleum industries for sustainable management of water with the help of several case studies provided by Chevron Corporation. Similar sustainability initiatives have also been undertaken by other major energy industries.

Keywords Advanced oxidation · Blowdown · Brackish water · Brine · Crude · GAC · Permeate · Permian Basin · Phenols · Recycle and reuse · Refining · SAR · Stripped sour water · Sustainability · Terrestrial ecosystems · Total Dissolved

---

M. Venkatesh (✉)
Process and Energy Consultant, Andover, MA, USA
e-mail: m.venkatesh1@yahoo.com

F. A. Calvo
Chevron Corporation, San Ramon, CA, USA
e-mail: FAlmadaCalvo@chevron.com

R. T. Grubbs
Chevron Corporation, Midland, TX, USA
e-mail: RTgrubbs@chevron.com

J. Cunningham
Chevron Corporation, Houston, TX, USA
e-mail: jhcunningham@chevron.com

S. Basu et al. (eds.), *Water Management in Petroleum Industries*, Water Resources Development and Management, https://doi.org/10.1007/978-981-19-3159-8_8

Solids (TDS) · Total Suspended Solids (TSS) · Upstream operations · Water resources · Well completions

8.1 Introduction

The oil and gas industry has historically played a pivotal role for economic growth. Increased availability of affordable energy for transportation and heating has accelerated economic development and the rise of standards of living around the globe. At the same time, the industry has the potential to impact the environment in various ways.

Few topics are as impactful to communities and the environment as the topic of water resources management. Water is needed to sustain healthy aquatic and terrestrial ecosystems and to meet basic human needs like food production and sanitation. In addition, surface water bodies (lakes, rivers, oceans) also have an important recreational, religious, and cultural value for the local communities. Water is also used in a myriad of industrial processes, such as electric power generation, food and beverage industry, textile, and semiconductor manufacturing, just to name a few, playing an important role in economic development. Likewise, the oil and gas industries require water not only throughout the production and refining processes, but also manage water recovered from both upstream and downstream operations. Appropriate management of water resources by all actors in the community is a prerequisite for ensuring strong ecosystems, healthy communities, and a thriving economy.

Water use in the general population has been increasing by 1% each year since the 1980s, mostly propelled by population growth, economic development, and changing consumption patterns (United Nations 2018). Moving forward, it is expected that between 2000 and 2050, water use will increase by about 60%, due to rising demand in the industrial and domestic sectors. While industry accounts for roughly 20% of the world's water withdrawals, it is expected to quadruple by 2050 (Willet et al. 2019). Presently, approximately 3.6 billion people (about half the global population) live in regions that are water scarce at least one month per year, and this number is expected to increase to 4.8–5.7 billion by 2050 (United Nations 2018).

In the U.S, the refining industry uses about 2 billion m^3/year (528 billion gal/year) of water, much of which is used for heating and cooling demand (EPA, 2022). According to the United States Geological Service, industrial self-supplied withdrawals, excluding cooling water for electric power generation, were estimated at 20 billion m^3/year in 2015. Therefore, refining is estimated to make up about 10% of the industrial water self-supplied withdrawals in the U.S. For upstream operations, water consumption as well as produced water production can be significant. For example, in the Permian Basin, in the United States between 2009 and 2017, the oil industry used about 137 billion gallons (~518 million m^3) for well completions and generated about 177 billion gallons (670 million m^3) of produced water. Estimated projections of water use and produced water management needs in the Permian basin are staggering. During the next 20 years the industry will need about 18 billion m^3 (4,800 billion gallons) of water for well completions, while 50 billion m^3 (13 trillion gallons) of produced water will need to be managed (Scanlon et al. 2020).

The potential negative impacts of using large amounts of water by industries are twofold: (1) the withdrawal itself can result in competition among users (human use and ecosystem requirements) of a finite resource, particularly in water stressed areas, and (2) after use, wastewater disposal can have negative effects on water quality if not properly treated, affecting the aquatic ecosystem and the communities that rely on the resources provided by the water body (drinking water, irrigation, fishing, recreation).

While there are many ways to assess the sustainability of water management practices by industrial operations, a sound water stewardship program should aim to: (1) increase the efficiency of water use across the operations, by reducing use and increasing reuse and recycling, (2) avoid ecotoxicological and human health risks associated with the return of the used water to the environment, (3) optimize energy use to reduce the carbon footprint of water treatment/management processes, (4) engage other water users, local regulators, and policymakers at a watershed level to develop shared solutions to water resource constraints and increase water circularity within the community, and finally, (5) seek a continuous improvement approach, where lessons learned are used to improve the water management approach in a cyclical way.

This chapter discusses the issues related to recycle and reuse of water in the upstream and downstream sectors of the petroleum supply chain. For both upstream and downstream sector various options available for recycling water are discussed in detail. This is supported by descriptions of water management case studies that Chevron Corporation has pursued in the last 25 years to improve aspects of water management across different business units in the corporation.

8.2 Upstream Sector

As discussed in Chap. 3, produced water quality varies widely depending on the oil production method—conventional versus unconventional, location—on-shore versus off-shore, and type of unconventional production method—hydrofracturing versus thermally enhanced operations. Streams from conventional production wells that contain lower TDS, hydrocarbon and metal ions can be reused for various beneficial purposes after primary treatment to remove suspended solids and hydrocarbons. However, produced waters from unconventional production such as hydrofracturing operations contain high TDS and other contaminants which require extensive and expensive treatment before making them suitable for reuse. There are two different avenues available for reuse:

- Reuse of treated produced water in hydrofracturing operations
- Use treated water for land applications such as irrigation, landscaping, and agriculture.

Use of treated produced water for re-fracturing has become more common in recent years. This helps in reducing the demand for fresh water, disposal of produced water, and trucking water/wastewater through communities to centralized treatment plants. The knowledge of the chemistry for hydrofracturing formulations has advanced significantly so that extensive levels of treatment, particularly TDS removal, is no longer necessary to the extent that was in the past. This has made reuse of treated produced water much more feasible and economical than ever before.

In some cases, the produced water is of sufficient quality that it can be reused directly without any treatment for hydrofracturing. Water from mining tailing ponds, where chlorides are limited, can be reused for thermal operations replacing brackish water sources. It must be noted that although technologies exist for treatment, a significant portion of produced water is still disposed off in underground wells as this method is considered the most economical.

As the unconventional oil production industry has grown, water management companies have also flourished that provide comprehensive water solutions encompassing both water supply and produced water disposal for production sites. It is likely that the shift of control in the water market from individual operators to larger companies that handle the water logistics for many sites simultaneously will also result in the growth of produced water reuse and reduction of saltwater disposal. An example of one such large operating companies is Oklahoma based Devon Energy which owns and operates eight produced water treatment plants across the Delaware Basin, west of the Permian Basin in New Mexico (Carlsbad Current Angus, 2019).

In the province of Alberta, Canada, a produced water recycle rate of at least 90 percent is mandated for thermally enhanced oil recovery (EOR) production methods like steam assisted gravity drain (SAGD), cyclic stream stimulation (CSS) and Steam Flooding.

For certain industrial reuse applications such as power plants that generate power through combined heat and power cycles, or those using steam turbines, extremely high-quality water is required. For these applications, enhanced treatments consisting of one or a combination of multiple unit operations such as ion exchange, reverse osmosis, and electrodialysis, may be necessary.

Land application of produced water is also being considered, especially in drought-prone regions of the United States. The Sodium Adsorption Ratio (SAR) is an important consideration for land use applications. SAR is defined by the following equation:

$$SAR = \frac{(Na^{+})meq/l}{\left\{\sqrt{\left[\frac{(Ca^{2+}+Mg^{2+})meq/l}{2}\right]}\right\}} \tag{8.1}$$

An SAR value of 13 is the upper limit to protect soil from severe deterioration of vegetation (USDA 2012). If the reuse stream has a higher SAR value than this range, it may need to be diluted with fresh water to adjust the SAR value. Other potential opportunities for reuse include livestock watering, surface water augmentation and groundwater recharge. However, for each application careful consideration of the treatment technology required for enhancement of the quality of the water and the capital and operating costs to achieve the quality to fulfill the reuse goal needs to be taken into account. As water scarcity continues to shape the economy of water, it is likely that further investigations and advancements will allow for an even better understanding of and the ability to reuse produced water in applications outside of oil and gas production.

Recycle and reuse of treated produced water is finding increasing urgency among producers in the Permian Basin of West Texas and New Mexico in the U.S. due to the potential negative impact of disposal by injection into the ground on the seismicity of the region. Earthquakes resulting from hydrofracturing for production activities and well injection for disposal of produced water have been discussed in Chap. 7. The Texas Railroad Commission (TRRC), the regulatory authority for proper handling and control of produced water in Texas, promulgated a regulation in 2021 restricting the underground disposal of produced water to prevent the seismic activity in a 650 square kilometer (250 square mile) region surrounding the cities of Midland and Odesa Texas, limiting the disposal volume of 1,600 cubic meters/day (10,000 bbl/day) to each existing permitted well (GWI 2021). Although there are disposal wells that are not utilized at their full capacity, this restriction encourages the owners and operators to seek new avenues to handle large volumes of produced water, and the obvious choice is to treat it for beneficial reuse. Currently TRRC is working on revising the permit requirements for produced water injection into the ground. The new permit is expected to be in force late 2024.

Due to the excessively large amount of water produced at ~3.9 barrel of water per barrel of crude, proper management of water is of critical importance at Permian basin. There are four ways of water management—(a) treatment and recycle of produced water for well completion, (b) beneficial reuse in agriculture and industrial applications, (c) disposal into natural water bodies and (d) disposal by underground injection. The current focus of the industry and policymakers is to develop a sustainable water management strategy by minimizing the underground injection and enhancing the other three options with the help of innovative technologies.

Referring to Fig. 3.19 in Chap. 3, for recycling produced water for well completions, only the suspended solids and Oil & Grease removal by primary separation steps are sufficient. However, if the water contains hardness and other divalent cations then secondary separation steps will be required. If the water contains reduced iron and manganese ions then an additional oxidation step needs to be added to remove those ions. It is essential to disinfect the produced water before it is reinjected for refracturing operation to protect the equipment from corrosion and biofouling, and more importantly, to prevent fouling of microporous structure of the rocks bearing oil and gas. Other chemicals listed in Table 3.6 of Chap. 3 also need to be added to perform their respective functions. If an oxidation step is included in the treatment process for oxidizing soluble iron and manganese ions, the oxidant (sodium hypochlorite, chlorine dioxide, etc.) may also perform the role of a disinfectant and separate biocide addition may not be necessary. The recommended specification (Kidder 2023) for treated produced water recycled for refracturing are listed in Table 8.1.

New Mexico has implemented the new "Clean Brine Standard" to promote recycle by imposing the requirement of 80% produced water, 15% saline water and only 5% fresh water for hydrofracturing. However, there is a large gap between the water production and recycle in well completion activities. Even if 100% of produced water could be utilized for refracturing there would still remain about 40 to 60% excess water that needed to be properly disposed or reused (Crain 2023). This gap has forced the policymakers to consider alternate options including beneficial reuse. Given the geography of the Permian basin in the semi-arid regions of Texas and New Mexico, where groundwater is the only source of freshwater, this water is currently being viewed as a valuable resource rather than a liability. Research and development activities have been undertaken by industry, academia, and various state agencies to upgrade the quality of produced water. New Mexico Produced Water Research Consortium (NMPWRC), managed by the New Mexico State University (NMSU), has been conducting pilot scale research to develop treatment technologies for making produced water fit for agricultural reuse, discharge to surface water bodies, and use by industry (Hightower 2023). A Joint Industry Partnership (JIP) pilot R&D program between the energy companies is currently in progress.

For beneficial reuse or disposal to a surface water body, produced water has to go through a tertiary treatment step by evaporation/crystallization or NF/RO as discussed in Chap. 3. Pilot tests were conducted by Transmissive Water Services (TWS) using RO process for desalination after primary treatment steps of Oil & Grease and TSS removal by coagulation/flocculation and gravity separation. Post RO steps consisted of granular activated carbon (GAC) adsorption for removing any residual organics and UV disinfection to ensure that the chemical and microbiological characteristics of the recovered water was safe for land application. Results of the test program are presented in Table 8.2. Water produced from this pilot set up is currently applied to vegetations grown in three experimental glass cases to examine the impact on their growth at different levels of TDS. Water at TDS 1,500 mg/L is applied to soil from Reeves County with native vegetation, at TDS 1,000 mg/L it is applied to soil

Table 8.1 Recommended specifications of Produced Water for recycling for hydrofracturing

Parameter	Unit	Recommended specification
Total Oil and Grease (TOG)	mg/L	< 10–100
TSS	mg/L	< 50–1,000
Total iron (Fe)	mg/L	< 5–10
H_2S	mg/L	Non-detect
ORP	mV	> 0–300
pH	Su	6–8
Bacteria cell count as ATP	ng/L	< 50–500

Table 8.2 Produced water treatment pilot test results of TWS

Parameter	Unit	TWS pilot test results		
		Influent	Effluent	RRC target values
TDS	mg/L	111,000	824	1,000
TSS	mg/L	42	< 2.50	–
Hardness	mg/L as $CaCO_3$	12,800	76	150
Alkalinity	mg/L as $CaCO_3$	279	< 20	100
pH	su	6.62	7.01	6.5–8.4
O&G by HEM	mg/L	7.4	< 5.7	35
Chloride	mg/L	101,000	528	100
Nitrate	mg/L	< 1	0.43	45
Sulfate	mg/L	701	5.0	–
Sulfide	mg/L	0.16	< 0.05	–
Ammonia-N	mg/L	66.1	4.65	30
Arsenic	mg/L	0.1	< 0.01	0.1
Barium	mg/L	3.58	0.02	–
Boron	mg/L	42.3	1.18	0.75
Cadmium	mg/L	0.005	< 0.0002	0.01
Chromium	mg/L	< 0.1	< 0.01	0.1
Iron	mg/L	33	0.05	5
Lead	mg/L	< 0.06	< 0.006	5 ara>
Manganese	mg/L	0.97	0.009	0.2
Mercury	mg/L	< 0.0002	< 0.0002	–
Potassium	mg/L	802	5.4	–
Phosphorus	mg/L	6.59	0.056	5
Selenium	mg/L	< 0.1	< 0.01	0.02
Silver	mg/L	< 0.05	< 0.005	–
Sodium	mg/L	55,600	309	100
Benzene	mg/L	1.1	< 0.001	–
Radium 226	pCi	96.1	0.43	30
Radium 228	pCi	315	0.56	30
Gross α	pCi	2,580	11.6	15
Gross β	pCi	816	8.4	15

from Loving County with native vegetation, and at TDS 500 mg/L it is applied to a mixture of 50% soil from Reeves County and 50% Loving County with Alfalfa plants. The next step is for TWS to embark on a larger scale pilot with produced water feed rate of 15,000 to 20,000 Barrels/day (~2,400 to 3,200 cu.m./day) using the same treatment train and apply the product water to a land owned by the Texas Railroad Commission (RRC) and study the impact on the plant growth.

New Mexico Produced Water Research Consortium (NMPWRC) conducted their pilot test program consisting of primary/secondary treatment, followed by tertiary treatment (Hightower 2023). Two different primary treatment processes—(a) membrane separation, and (b) ozone oxidation followed by filtration were tested and the results are presented in Table 8.3.

The primary treated water was then desalinated using evaporation. The distillate fulfilled the TDS target but contained some ammonia and volatile organic compounds, primarily as total BTEX (combined benzene, toluene, xylene and ethyl benzene) ~2 mg/L, total petroleum hydrocarbons (TPH combined C6–C35) ~4 mg/L and some other heavier organic molecules in insignificant quantities. The distillate was then treated by GAC which removed all organic compounds to trace level in the final product.

The final product from this pilot test program was intended for discharge to the public water system. Two options are—(a) discharge to surface water into Pecos River which passes through New Mexico and Texas, and (b) discharge into the groundwater table. Since both of these water bodies are used for beneficial human consumption, the treated water from this pilot test program was assessed for Whole Effluent Toxicity (WET) and Human Cell Line Toxicity. For WET testing the methods used were—(a) EPA 1003.0 for chronic toxicity utilizing the microalgae *S. capricornutum*, (b) EPA 1002.0/2002.0 for acute and chronic toxicity utilizing water flea species *C. Dubia*, (c) OECD 236 for acute toxicity utilizing freshwater fish species *D. Rerio*, and (d) ISO 11348 for acute toxicity utilizing marine bacterium species *V. Fischeri*. For human cell line toxicity assessment methods used were MCF-7 for breast cancer cell line, HEK 293 for embryonic kidney cells and Caco-2 for colorectal adenocarcinoma cells.

Table 8.3 Produced water primary treatment pilot test results of NMPWRC

Parameter	Unit	Influent	Effluent
TDS	mg/L	101,000	101,000
TSS	mg/L	790	N/R
Turbidity	NTU	200	< 25
Particle size	Mm	N/R	< 25
Alkalinity	mg/L as $CaCO_3$	870	N/R
pH	Su	6.6 – 8.1	6.0 – 8.0
COD	mg/L	3,100	N/R
TOC	mg/L	250	N/R
Ammonia	mg/L	750	N/R
H_2S	mg/L	N/R	N/D
Total iron	mg/L	N/R	< 10
Ferrous iron	mg/L	N/R	< 5
ATP	ng/L	N/R	< 200

Notes N/R = Not Reported, N/D = Not Detected

All toxicity studies were performed at NMSU and were reported by Deanka-Pedige (2024).

Regardless of the method used for desalination – NF/RO, or evaporation/crystallization, recovery is typically limited to only 50 to 60%. Therefore, the remaining part, the concentrated brine, still needs disposal by underground injection. As this stream is a byproduct of the final treatment step the constituents that are removed by the desalination process accumulate in this concentrated brine. This includes the metals, naturally occurring radioactive materials (NORM) and some of the heavy organic compounds in the produced water. Some of the residual materials are critical minerals that have high commercial potential as valuable resources. One such resource is lithium, the demand for which is very rapidly increasing as a raw material for production of battery for energy storage. Majority of lithium is currently supplied by mining from a small region in South America within the countries of Argentina, Bolivia, and Chile. Other sources are located in China and Australia. Given the sharp rise in the demand for lithium in electric vehicles, renewable energy generation, storage batteries, and numerous other applications, this results in a pressure on the price of lithium. Produced waters and brines from unconventional oil fields are alternate resources for extraction of lithium that is expected to close the large gap between demand and supply. In the US, lithium concentrations in these streams have been reported to be between 50 and 500 mg/L (Kumar, et al. 2019) in the states of Texas, Arkansas, and North Dakota. Various hydrometallurgical technologies for lithium recovery from these liquid streams are currently under development in pilot and demonstration scales.

Another element of importance is iodine. Globally, iodine and its derivatives are used in X-Ray contrast media (28%), biocides and antiseptics (12%), pharmaceuticals (12%), optical polarizing films (12%), catalysts (8%) and various other applications. Due to its high demand, there is a sharp rise in its price over the last few years—from $20/kg in 2016 to $60/kg in 2022. Iofina, the second largest iodine production company in North America uses produced water as a valuable resource to extract iodine by chemical processing in its production plants located in Oklahoma and Kentucky (Dom 2024).

8.3 Downstream Sector

It is recognized that there is a scarcity of fresh water in many parts of the world and oil refineries are large consumers of water. In fact, at some locations there are severe restriction on the amount of water that can be withdrawn from aquifers, rivers and other fresh water sources, and in some regions any water that is needed by the refinery is required to be withdrawn from the sea and desalinated which is an expensive process. The desalination is carried out using either multistage evaporation or reverse osmosis. Refer to Chap. 2 for a detailed description of desalination processes.

In order to explore opportunities for recycle and reuse of wastewater in a refinery it is important to first understand the sources of wastewater so that candidates for recycle and destinations for recycled water can be identified. Figure 8.1 shows the process units within a refinery where wastewater is generated. It followed by a brief description.

The first step is raw water treatment where solids and sediments are removed and discharged. Then the water is sent to the cooling towers as makeup and to the demineralizer system for dissolved solids removal. Cooling water that is sent to the various units in the refinery is returned to the cooling tower. The cooling water return stream is cooled in the cooling towers by evaporation. In order to prevent a buildup of dissolved solids in the recirculating cooling water system a blowdown is taken out and sent to the refinery wastewater treatment plant. Water lost with blowdown is compensated as make-up water in the recirculating cooling water system. This make up water is sourced from raw water after preliminary and primary treatment. Raw water after primary treatment is demineralized by following advanced treatment steps and sent to the boiler system. Steam that is generated is sent to the various units as required. An overview of the preliminary, primary and advanced treatment processes is presented as Fig 2.1 in Chap. 2.

Wash water is used in desalters to remove the salts contained in the crude oil and the brine is sent to wastewater treatment. Water drained from the crude tanks, which contains dissolved and free hydrocarbons as well as drilling muds and sludge, is also sent to wastewater treatment.

Steam is used in many units as stripping steam and also as motive steam for vacuum ejectors. This steam is condensed and comes into contact with hydrocarbons ammonia and hydrogen sulfide in the process units and becomes sour water. This water is sent to the sour water stripper for removing ammonia and H_2S. As shown in Fig. 8.1, sour water is generated in the following units:

- Crude Distillation Unit (CDU)
- Vacuum Distillation Unit (VDU)
- Fluid Catalytic Cracker (FCC)
- Coker (Delayed Coking)
- Hydrotreater (Hydrodesulfurization)
- Other Units (such as sulfur plant etc.)

Stripped sour water is sent to wastewater treatment.

The first option for recycle and reuse is to recycle within the battery limits of the refinery. An example of this is illustrated and described as shown in Fig. 8.2.

Phenolic sour water that is produced primarily in the FCC and Coker units is sent to a dedicated sour water stripper. Phenols are not stripped in the tower; they remain in the stripped water and are also difficult to biologically treat in the wastewater treatment plant. This stripped water from sour water stripper can be recycled to the desalter as wash water where up to 90 percent of the phenols are recovered and returned to the crude oil to get distributed into the distillate products in the crude distillation unit. This can be fairly easily implemented since most refineries have multiple sour water strippers and the feed system can be re-piped for the new configuration. The challenge is to match up the capacities available in the existing

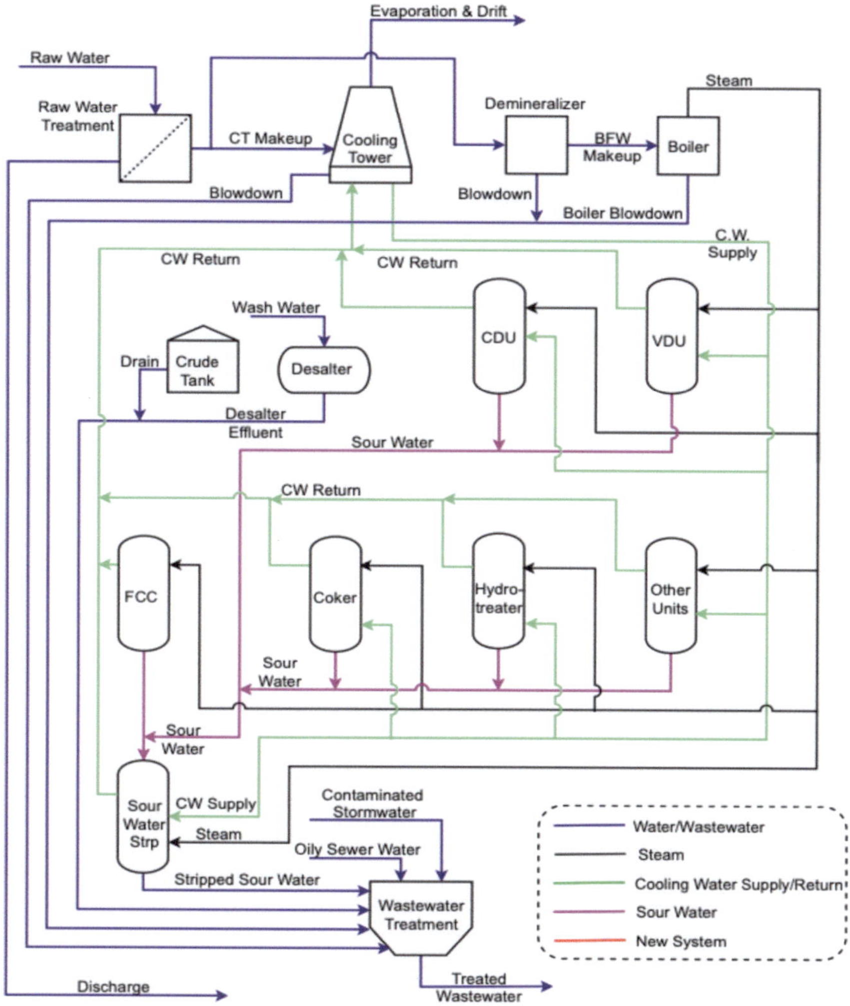

Fig. 8.1 Refinery wastewater sources

strippers with the flow of the phenolic and non-phenolic sour water produced in the refinery.

The second option for recycling is to further treat the treated effluent from the refinery wastewater treatment plant so that it can be reused as makeup water to the cooling towers and as makeup to the boiler feedwater system. This Option is illustrated in Fig. 8.3.

In this option the treated wastewater from the refinery WWTP is sent for further treatment which is shown and described in the Fig. 8.4.

The treated refinery effluent is sent first to a micro filtration system where any residual suspended solids are removed. The water is then sent to a Granular Activated

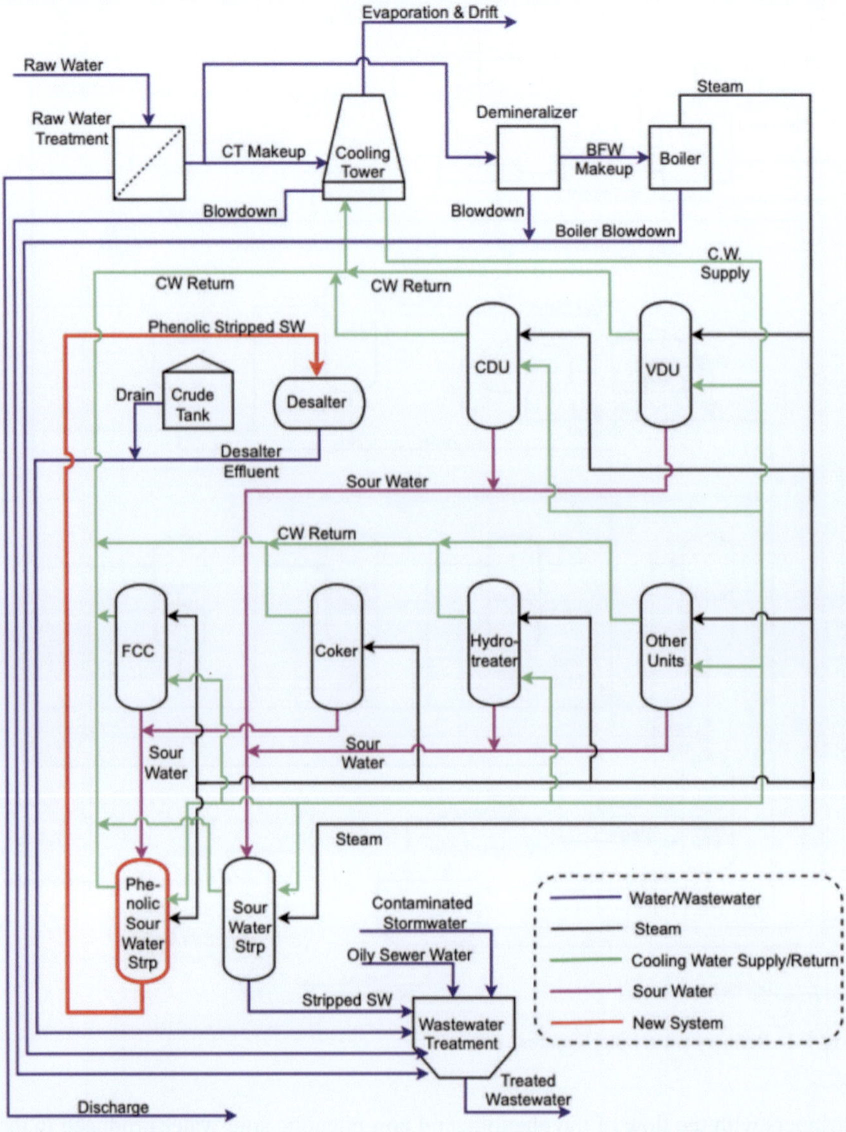

Fig. 8.2 Recycle of phenolic sour water

Carbon (GAC) bed where the COD is lowered further. The water then passes through RO units where dissolved solids are removed after passing through Cartridge Filter (to protect the RO beds). The permeate from the 1st stage RO can be used as cooling tower makeup whereas the permeate from the second stage can be used as the refinery boiler feed make up water.

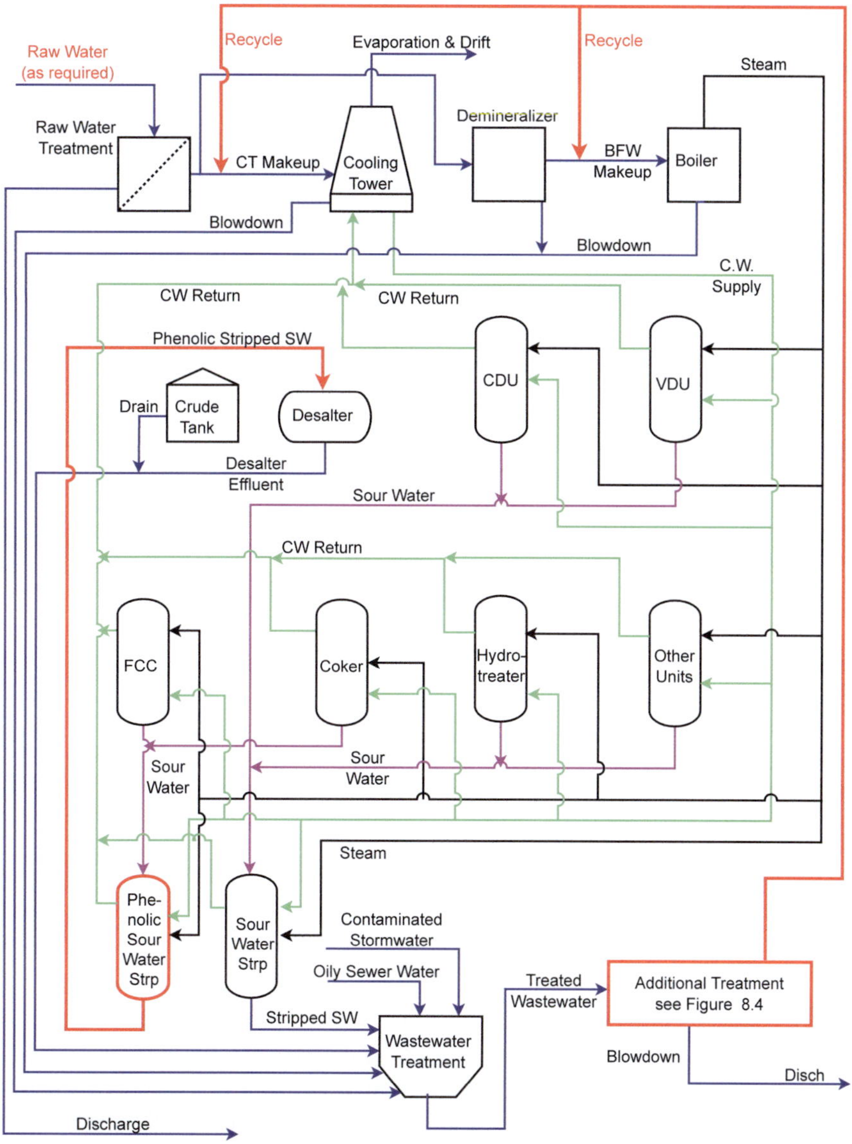

Fig. 8.3 Recycle of treated refinery wastewater

Table 8.4 shows the concentrations of contaminants in the effluent from a typical refinery wastewater treatment plant.

Table 8.5 shows typical chemical characteristics of the permeate (recycle stream to boilers) from the Post Treatment system.

A case study illustrating the use of the above option is shown in Sect. 8.4.3.

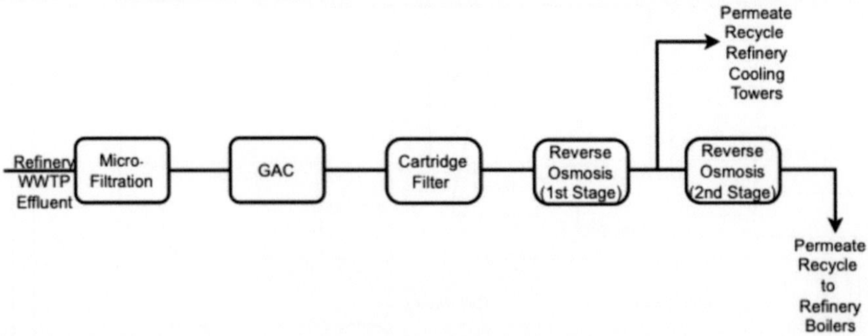

Fig. 8.4 Refinery effluent post treatment

Table 8.4 Typical refinery effluent concentrations

Contaminant	Units	Value
Chemical Oxygen Demand (COD)	mg/L	100–150
Biochemical Oxygen Demand (BOD)	mg/L	< 50
Oil & Grease (O&G)	mg/L	< 10
Sulfides	mg/L	< 5
Ammonia	mg/L	< 10
Phenol	mg/L	< 2
Total Suspended Solids (TSS)	mg/L	< 40
Total Dissolved Solids (TDS)	mg/L	500–1,000

Table 8.5 Permeate contaminant concentrations

Contaminant	Units	Value
Chemical oxygen demand	mg/L	< 50
Total suspended solids	mg/L	< 5
Total dissolved solids	mg/L	< 150

The third option for recycle and reuse is importing treated municipal wastewater from offsite and reusing it in the refinery as makeup water to the cooling towers. This option is illustrated in Fig. 8.5. Table 8.6 shows the typical characteristics of the effluent from municipal treatment plants.

The imported water is treated further to remove residual contaminants to make it suitable for reuse. Figure 8.6 illustrates the additional treatment that might be required.

The imported water is first sent to a micro filtration system for removal of suspended solids and then to an advanced oxidation system where UV and ozone are used for removal of residual dissolved organic matter. A part of effluent can be sent to the refinery for use as cooling tower makeup water. If ammonia concentration in the treatment plant effluent is excessive, a nitrification step might be required to

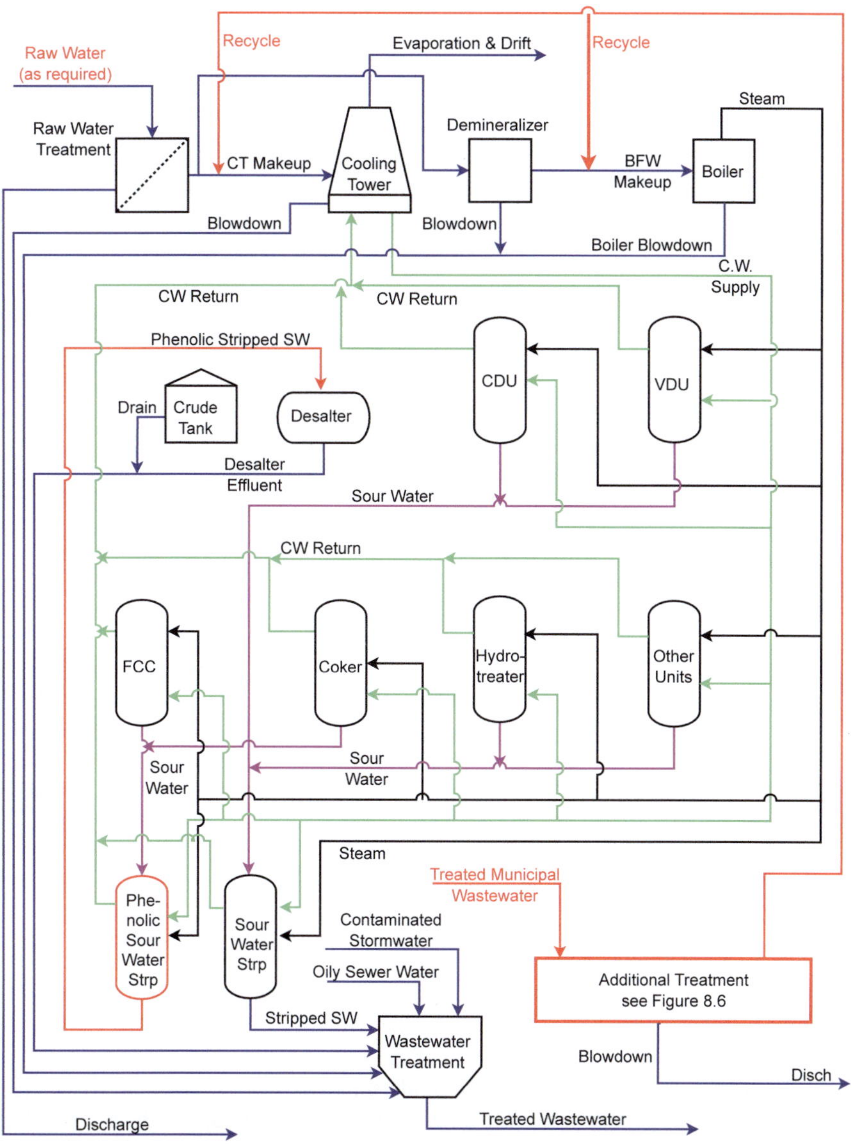

Fig. 8.5 Import of treated municipal wastewater

protect against corrosion of copper in heat exchanger and condenser tubes in the refinery cooling water system. In some instances, additional chemical dosing might be required to supplement alkalinity for supporting nitrification for the the remaining part of the effluent which then can be sent to RO units for dissolved solids removal. The permeate from the first stage can be used in low pressure boilers in the refinery,

Table 8.6 Typical Municipal Treatment Plant Effluent

Contaminant	Units	Value
Biochemical oxygen demand	mg/L	< 30
Total suspended solids	mg/L	< 30
Ammonia Nitrogen (Note 1)	mg/L	< 4
Total nitrogen (Note 2)	mg/L	< 5
Total phosphorous	mg/L	< 1

Note 1 For treatment plants that conduct both BOD oxidation and nitrification
Note 2 For treatment plants that conduct both BOD oxidation and total nitrogen (TN) removal by nitrification and denitrification

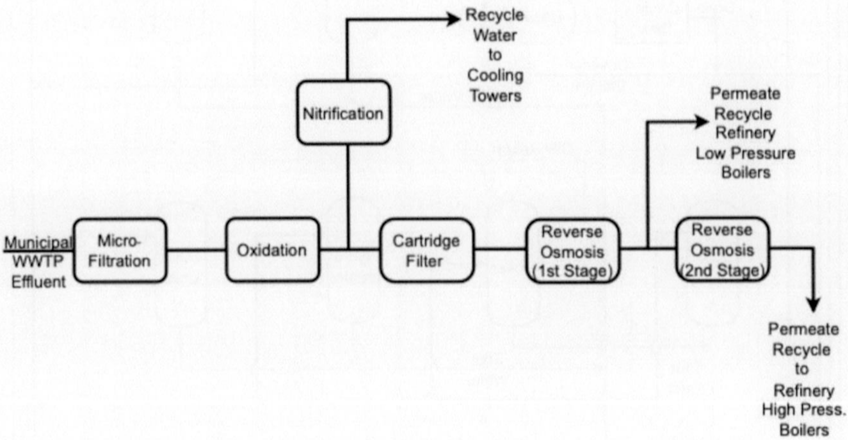

Fig. 8.6 Water treatment train

whereas the permeate from the second stage can be sent to the high pressure boilers in the refinery. It should be noted that the required treatment will need to be evaluated on a case-by-case basis depending on the quality of the imported municipal wastewater as well as the intended reuse of the water.

In the above option, the dissolved solids removal step is expected to be less technically challenging (than reusing treated refinery effluent) due to the fact that the TDS content of treated municipal wastewater is likely to be significantly lower than that in treated refinery wastewater. The following issues need to be considered and deliberated when considering this option for implementation:

- The capital cost for the pipeline required from the municipal treatment plant to the refinery.
- The environmental and social issues (right-of-way etc.) associated with the routing of this pipeline also needs to be taken into account
- If the treatment systems associated with this option is located at the refinery, then the refinery could become responsible for contaminants that are present in the blowdown from these systems even though they were not generated in

the refinery. This might conflict with the contaminants included in the existing refinery discharge permit.

- One way to alleviate this issue is to locate the treatment systems within the battery limits of the municipal treatment plant with the refinery taking the responsibility for the capital costs involved in implementing this system at the municipal treatment plant.

Case studies illustrating the use of the above option is shown in Sects. 8.4.1 and 8.4.2.

A fourth recycle option may be considered to grass-roots plants or refineries that may have multiple treatment trains in the wastewater treatment plant. This concept is illustrated in Fig. (8.7).

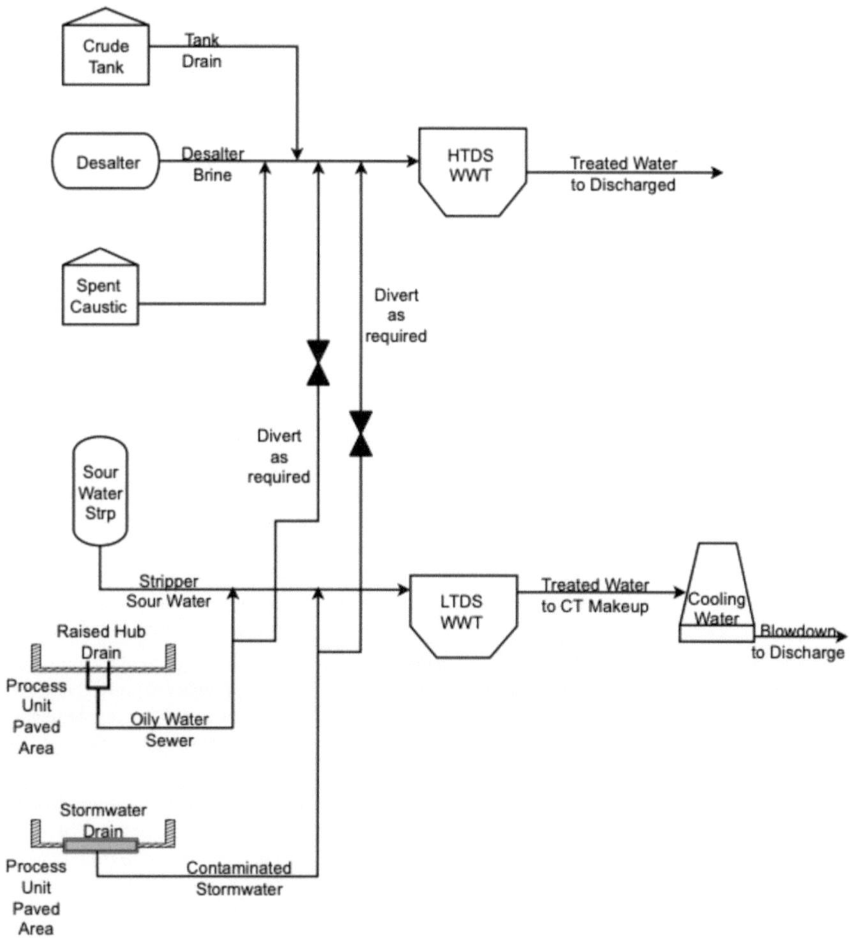

Fig. 8.7 Segregated treatment plants

In this type of system, the High Total Dissolved Solids (HTDS) streams, namely, the crude tank drain, the desalter brine and spent caustic are segregated and sent to the high TDS treatment train. The treated water from this train is discharged. The low TDS streams like stripped sour water, other oily water generated in the refinery and contaminated stormwater are sent to a second treatment train. The treated water from this train is entirely recycled as makeup water to the cooling towers.

In certain situations when the refinery is operating with opportunity crudes, which results in the discharge of increased contaminants via the desalter brine, it could be challenging to effectively treat the HTDS wastewater. In this case the HTDS wastewater could be diluted with water from oily water sewer, or the contaminated stormwater treatment as required to make the treatment more feasible.

The issues related to recycle and reuse of water/wastewater can be very refinery and location specific and can be quite complex. A number of factors need to be taken into account and deliberated before reaching a final decision. Table 8.7 shows the advantages and disadvantages of each recycle option described.

Table 8.7 Recycle options

Option	Advantages	Disadvantages
Segregated sour water and recycling stripped phenolic sour water	Results in a significant reduction in use of fresh water in the refinery	
	Should be relatively easy to implement since most refineries have multiple sour water strippers	It might be challenging to match the capacities of existing strippers with the quantities of phenolic and non-phenolic sour water produced in the refinery
	Minimizes the quantity of phenol discharged to wastewater treatment thus making it easier to comply with tight phenol discharge limits imposed at most refineries	Sour water stripper performance is subject to failures in absence of careful and routine operation and maintenance (O&M) oversight resulting in higher than expected ammonia and H_2S concentrations in the effluent
Recycle and reuse of treated refinery wastewater	Results in substantial decrease in use of fresh water in the refinery	Requires significant additional capital investment for the installation of the supplemental treatment system (TDS removal)
	Treated refinery wastewater can be blended with stormwater and stored on site for use as a source of fire water	Feasibility and permit issues related to the discharge of concentrated brine from the TDS removal system needs to be evaluated

(continued)

Table 8.7 (continued)

Option	Advantages	Disadvantages
Import treated municipal wastewater for reuse in the refinery	Results in substantial decrease or entirely eliminates the use of fresh water in the refinery (a relatively small quantity of potable water for personnel use will still be required)	Requires additional capital investment for the installation of residual contaminant and TDS removal system
		Requires additional capital for the cost of the pipeline for importing WW
		Will need to evaluate the environmental and social impacts of the routing of the pipeline from the municipal treatment plant to the refinery
		Refinery may become responsible for discharge of contaminants that come from outside the refinery
Segregated wastewater treatment trains and reuse of treated LTDS wastewater	Results in substantial decrease in use of fresh water in the refinery	Increased capital and operating costs due to multiple treatment trains
		Additional capital costs due to the routing of segregated sewers in the refinery
		Treatment of concentrated HTDS wastewater could be challenging

8.4 Case Studies—Chevron Facilities

8.4.1 Richmond Refinery—Richmond, California

Background

California's weather is conspicuously unreliable for providing a steady supply of rainfall to recharge rivers, lakes, reservoirs, and groundwater. Rainfall can vary greatly throughout the state and from year to year. After the statewide drought that lasted from 1987 through 1992, there has been a concerted effort at a state level to build drought resilient infrastructure (Jones 2000). During this time, Chevron Richmond Refinery, situated in the northeast of the bay area near San Francisco, partnered with West County Wastewater District (WCWD) and East Bay Municipal Utility District (EBMUD), to do a pilot study to explore the idea of using reclaimed municipal wastewater from WCWD for use in industrial cooling.

Technical Challenges

The initial pilot study results provided treatment recommendations to meet water quality parameters (scaling, corrosion, fouling) and address human health concerns (Oil & Gas Journal, 1992). In 1994, the project was approved and funded by East Bay Municipal Utilities District (EBMUD) and the State of California revolving fund. The North Richmond Water Reclamation Plant began operations in 1996 and has been in use ever since. The plant consists of a chemical softening unit (high pH precipitation) followed by sand filtration. This water is used as make up water for 4 cooling towers in the refinery. The plant has a design capacity of ~19,000 m^3/day (5 MGD) and is one of the largest industrial cooling water reuse projects in the U.S. (East Bay Municipal Utility District 2022).

In the early 2000's, with a continued threat of droughts on the horizon, to increase the use of reclaimed wastewater, EBMUD and Richmond Refinery sought to partner to secure a reliable source of ~ 13,300 m^3/day (3.5 MGD) high quality, low total dissolved solids (TDS) water for meeting the needs of boiler feed water. This was a task more difficult to accomplish than the previous cooling water project.

Solutions

Chevron and EBMUD built the Richmond Advanced Recycled Expansion plant (RARE) to produce the needed amount of boiler feed water, using the secondary effluent as feed. The new plant consisted of a microfiltration unit to remove total suspended solids (TSS), followed by reverse osmosis to remove TDS (MF-RO). The agreement involved Chevron allowing the plant to be built inside refinery grounds, while EBMUD operates the plant. This project involved the construction of new conveyance pipelines to and from the refinery, new pump stations, and improvements to the EBMUD wastewater treatment plant to meet desired water quality prior for MF-RO treatment.

Results

After both projects started, around 46% of the total amount of water use in the refinery comes from reclaimed water, that would otherwise be discharged to the bay without a second use. A total of about 6.4 million m^3/year of reclaimed water was used in 2021 at the refinery in lieu of potable freshwater.

8.4.2 El Segundo Refinery—El Segundo, CA

Background

Chevron's El Segundo Refinery is located south of the Los Angeles International Airport, California. It is the largest producing oil refinery on the U.S. West Coast with daily demands for water to meet refinery process needs including boilers and cooling towers. With the water supply, water demand and drought conditions in the late-1980s becoming an increasing concern, the refinery started discussions with the local water and wastewater agencies—West Basin Municipal Water District (a water agency providing water to the cities in Los Angeles county) and the City of Los Angeles. This resulted in an agreement in 1991 for the City's Hyperion wastewater treatment plant to deliver treated municipal wastewater to a new West Basin Water Recycling Facility in El Segundo (West Basin Municipal Water District 2022a) with the refinery making long-term capital investment funding commitments to West Basin to enable their reclaimed water delivery capacity. By 1995, West Basin started delivering reclaimed water to Chevron following advanced wastewater treatment processes.

Technical Challenges

Use of treated sanitary wastewater to replace or offset potable water use for industrial demand was a new and growing opportunity pushed forward in southern California by the consumption/demand and drought issues being faced. The wastewater treatment industry was developing new treatment technologies to meet a wide range of treatment needs, while considering its risks in receiving alternative water sources. Could treated sanitary wastewater be delivered to consistently meet the quality and quantity of demand needed to keep industry operating and replace the historically reliable and consistent potable water as its source?

Solutions

The West Basin Water District El Segundo Water Reclamation Plant, built in 1995, specializes as a "boutique water provider" processing secondary-treated effluent received from the City of Los Angeles' Hyperion Wastewater Treatment Plant. The West Basin Water District then feeds three satellite plants including the Chevron Nitrification Plant, each of which polishes the water for a specific industrial use. The five types of designs produced as a result of these studies include:

- Tertiary Water (Title 22 of the California Code of Regulations) for a wide variety of industrial and irrigation uses
- Nitrified Water for industrial cooling towers
- Softened Reverse Osmosis Water: Secondary treated wastewater purified by micro-filtration (MF), followed by reverse osmosis (RO), and disinfection for groundwater recharge
- Reverse Osmosis to produce water for refinery low-pressure boiler feed water
- Ultra-Pure Reverse Osmosis Water for refinery high-pressure boiler feed water.

The industrial users, including Chevron, pay specific rates for the recycled water, but this cost is offset and justified by the consistency of water quality and quantity. When compared to existing potable water supplies, whose reliability can be affected by droughts, having a reliable and drought resistant water supply source is a very economical option (West Basin Municipal Water District 2022b).

Results

Chevron El Segundo is the largest single user of recycled wastewater in West Basin's service area (West Basin Municipal Water District 2022c) and one of the largest in the state, and the use of recycled water has provided a considerable offset from reliance on potable water. Both West Basin of California and Chevron have learned through the process, as West Basin replaces and modernizes as technology allows.

8.4.3 Singapore Refining Company (SRC)—Singapore

Background

Singapore's National Water Agency PUB (Public Utilities Board) reports that water consumption for Singapore is currently 1.6 million m^3/day (~430 MGD), with the domestic sector consumption at 45%, and non-domestic sector use at 55%. Singapore's water consumption is projected to double by 2060, with the non-domestic sector use at 70% of total water demand (PUB, Singapore's National Water Agency 2018a). Water conservation and reuse are key strategies for non-domestic users such as the Singapore Refining Company (SRC), a Chevron's joint venture refinery. Refining, petrochemicals, and chemical sector is one of the largest water-consuming industries in Singapore, with the majority of that consumption for use in boilers and cooling towers. SRC is working in conjunction with the national efforts to both conserve and reuse water to offset non-domestic demand (PUB Singapore's National Water Agency 2018b).

Technical Challenges

In 2013, SRC's refinery process wastewater underwent a basic oil removal process, activated sludge process and clarification treatment before being discharged to the sea. A water audit conducted in 2013 identified the potential for implementing tertiary treatment of this wastewater to replace or offset water consumption. With the implementation of advanced tertiary treatment process, effluent from SRC's existing discharge would be further treated to remove suspended solids, dissolved solids, oil and grease and other contaminants to effectively reclaim its own water for process reuse, thereby reducing SRC's uptake from PUB's piped supply (PUB Singapore's National Water Agency 2018b).

Solutions

Detailed Engineering for the system revealed a design capacity of up to 2,500 m^3/day of water for reuse, and this facility allowed the refinery to reduce PUBs piped potable water supply by 30%. The key enablers for the success of this project included a strong commitment of SRC towards long term water sustainability, technical support from PUB during the implementation and operation of the tertiary treatment process, and technology advancements in the field of membrane materials that are robust enough to treat refinery effluent water. With the success of this project, the technology of using ceramic MF-RO to treat and reclaim process water from refinery effluent has been tested and proven to be viable to meet the quality requirements for SRC's local process reuse. This process has a good scalability and viability for future full-scale implementation in a similar industry.

Results (Reduced Fresh Water Use and Increased Recycle)

Over the past five years, Chevron and joint-venture SRC have increased the recycling and reuse of its process water. In 2018, Singapore's national water agency presented SRC with the agency's inaugural Water Efficiency Award and Watermark Award in the Refining category for efforts in water recycling. SRC's journey to reusing treated effluent in refining operations was the first of its kind in Singapore. Using flat-sheet ceramic membranes and a two-stage reverse osmosis process has increased SRC's capability to recycle its own water. Government incentives and technological advances have driven SRC's recycling initiatives, and greater advances in water recycling are forthcoming (Chevron 2022).

8.4.4 Mid-Continent Business Unit (MCBU)—United States

Background

The activity of extracting oil and gas from shale formations through horizontal wells requires hydraulic fracturing. Historically, fresh water had been used for this purpose. In the arid high plains of West Texas and Southeastern New Mexico, the geographical location of the Permian Basin, surface and freshwater resources are limited, and the area is prone to periodic drought. By 2014, it was apparent that alternative water sources were needed to produce energy without competing with others for the natural resource of fresh water.

Technical Challenges

Selection and evaluation of alternative water sources included considerations of quality, availability, cost, surface management and transfer, and any detrimental impacts of the use of the alternative sourced water in the reservoir. Alternative water sources exceeding the freshwater threshold of a TDS higher than 2,000 ppm, were considered as an option to reduce competition for fresh water. Optimal availability was near the well development area. Any large volume storage on the surface needed to meet criteria such as no sheen, while surface storage containment and transfer needed to address environmental impacts associated with the potential leaks and spills to the surface.

Solutions

Brackish ground water with higher than 2,000 ppm Total Dissolved Solids (TDS); and produced water that was stored in ponds or tanks on the surface prior to downhole disposal were both evaluated. Brackish water was available from groundwater aquifers but came as a purchased commodity. Redirecting and reusing produced water before underground disposal offered advantages of no fees but required treatment to store it safely on the surface in large volume recycled water containment structures. On-demand needs for large volumes of water required optimizing treatment and transfer costs and shaped customized solutions for various development areas. Two levels of produced water treatment were developed that included treatment to remove free oils and other key constituents to aid in surface storage in large volumes; and minimalistic treatment for produced water that was sent directly to the location on demand. The blending of the three water sources (brackish ground water, treated produced water, and direct reuse produced water) quickly became the optimal combination to address the high demand required.

Results

Use of alternative water resources conserves water that could be used for human consumption, agriculture or for habitat conservation. By 2018, more than 99 percent of the water used in the Chevron Permian Basin well completions was either non-fresh (brackish) or recycled produced water from operations, compared to about 25% in 2014 (Chevron 2019). Chevron also utilized centralized infrastructure and a pumping and piping network to transfer fluids to further reduce surface footprint by allowing pipelines, infrastructure, and hydraulic fracturing ponds to serve multiple well pads.

8.5 Summary

Chevron's water recycling projects discussed in this chapter offer a few lessons: (1) The refinery projects required the partnership of the local governments and utilities. State and local governments were key to provide the right incentives to work with the local utility companies and public–private joint investments were essential to get the projects off the ground. (2) A real need and a sense of urgency was felt from all actors to move the projects forward. Dire drought conditions and future water demand projections drove action. (3) Economic needs, and water savings and sustainability goals were aligned in all projects. It made sense to make the investments from both, a business and sustainability perspective.

Chevron continues exploring new ways to sustainably manage water resources. For example, Richmond and El Segundo continuously revisit the idea to expand recycling, either through reclaimed municipal/sanitary wastewater or internal recycling, to decrease the use of potable freshwater. MCBU's Permian basin operations also continues to expand the use of produced water.

Moving forward, as the amount of water recycling and reuse increases in its operations, Chevron aims to do it with the lowest carbon footprint possible, which makes it more challenging yet. Water conservation objectives must not only be weighed on the business case, but also on the project's impact on carbon footprint. Another recycled water challenge is that as water is recycled more, the effluent streams become more concentrated with salts and other constituents, which poses some problems when performing toxicity tests and meeting concentration based effluent limits. Working closely with the regulators is key to understand the limits set by the permits and making sure that any recycling project does not interfere with current effluent limits.

Examples of recycle and reuse of water in both upstream and downstream operations of Chevron are used as case studies. Similar efforts are ongoing at other major oil and gas companies globally.

Abbreviations and Glossary of Terms

BTEX	Benzene, Toluene, Ethyl Benzene, Xylene
CSS	Cyclic Steam Simulation
EBMUD	East Bay Municipal Utility District
EOR	Enhanced Oil Recovery
GAC	Granular Activated Carbon
GWI	Global Water Intelligence
ISO	International Organization for Standardization
MCBU	Chevron Mid Continent Business Unit
MF	Microfiltration
NF	Nanofiltration
NMPWRC	New Mexico Produced Water Research Consortium
NORM	Naturally Occurring Radioactive Materials
OECD	Organization for Economic Cooperation and Development
PUB	Singapore Public Utility Board
RO	Reverse Osmosis
SAGD	Steam Assisted Gravity Drain
SAR	Sodium Absorption Ratio
SRC	Singapore Refining Company
GAS	Granular Activated Sludge
TDS	Total Dissolved Solids
TPH	Total Petroleum Hydrocarbon
TRRC	Texas Railroad Commission
TSS	Total Suspended Solids
USDA	United States Department of Agriculture
UV	Ultraviolet Radiation
WCWD	West County Wastewater District
WET	Whole Effluent Toxicity

References

Carlsbad Current Angus (2019) Published on September 13, 2019, Issue of the Newspaper
Chevron (2019) 2018 corporate responsibility report highlights. (chevron.com)

Chevron (2022) 2021 Corporate sustainability report. chevron-sustainability-report-2021.pdf
Crain R (2023) Permian basin water landscape and outlook. In: Presentation at the annual conference of produced water society, August 2023, Houston, TX
Deanka-Pedidge HMK (2024) Produced water treatment, Analysis and reuse effort by New Mexico produced water research consortium. In: Presented at the Permian Basin water in energy conference, March 5, 2024, Midland, TX
Dom F (2024) Iodofina: Technology leader in Iodine, presented at the Permian Basin Water in Energy Conference, March 5–7, 2024, Midland, TX
East Bay Municipal Utility District (2022) Recycled water projects. East Bay Municipal Utility District (ebmud.com)
EPA (2022) Water & energy efficiency by sectors. Oil Refineries| Region 9: Sustainable Water Infrastructure, US EPA
GWI (2021) Global Water Intelligence Report 2021
Hightower M (2023) Permian Basin produced water quality drivers for treatment and reuse, Produced Water Society—Permian Basin Summit, August 14–16, Midland, TX
Jones J (2000) Preparing for California's next drought, Changes Since 1987–92. State of California, The resource Agency, Department of Water Resources. Preparing for California's next drought
Kidder M (2023) Groundwater Protection Council (GWPC) produced water report—Update 2023 (Permian Focus). In: Presented at annual conference of produced water society, August 15, 2023, Midland, TX
Kumar A, Fukuda H, Alan Hatton T, Lienhard JH (2019) Lithium recovery from oil and gas produced water: A need for a growing energy industry. ACS Energy Lett 4(6):1471–1474
Oil &Gas Journal (1992) Chevron plans revamp of richmond refinery. Published in 1992. Website accessed in May 2022. Oil & Gas Journal (ogj.com)
PUB Singapore's National Water Agency (2018a) Our water our future. PUBOurWaterOur-Future.pdf
PUB Singapore's National Water Agency (2018b) Best practice guide in water efficiency refineries, petrochemicals and chemicals sector. Water_Efficiency_Refineries_Petrochemicals_Chemical_Sector.pdf (pub.gov.sg)
Scanlon BR, Ikonnikova S, Yang Q, Reedy RC (2020) Will water issues constrain oil and gas production in the United States? Environ Sci Technol 54:3510–3519 (acs.org)
United Nations (2018) The United Nations World Water Development Report 2018, Nature based solutions for water. nature-based solutions for Water|United Nations Development Programme (undp.org)
US Department of Agriculture (2012). Natural soil survey handbook, title 430-VI
West Basin Municipal Water District (2022a) History. History—West Basin
West Basin Municipal Water District (2022b) Facilities. Facilities—West Basin
West Basin Municipal Water District (2022c) Recycled water master plan, West-Basin-Master-Plan_final_rev4.pdf (westbasin.org)
Willet J, Wetser K, Vreeburg J, Rijnaarts HHM (2019) Review of methods to assess sustainability of industrial water use. Water Resour Ind 21:100110. Review of methods to assess sustainability of industrial water use—ScienceDirect

Chapter 9
Conclusions and Future Direction of the Industry

Somnath Basu

Abstract Petroleum and natural gas as the main sources of energy have played a critical role in the development of modern society. However, it also contributes to global warming and the associated climate change. Increasingly, renewable energy resources are being deployed to mitigate the impacts of climate change, with the goal of achieving carbon neutrality by 2050. Green and Blue Hydrogen are projected to replace natural gas as the main energy resource. As with the production of energy from traditional resources, water will continue to play a pivotal role in energy transition. In the face of an increasing shortage of freshwater alternative wastewater streams like treated wastewater, produced water, and seawater will become significant resources for the production of hydrogen.

Keywords Alternative sources of water · Blue hydrogen · Carbon capture · Crude oil · Energy transition · Fossil fuels · Gray hydrogen · Green hydrogen · Greenhouse gas · Renewable energy

Since the discovery of crude oil in the late 1800's petroleum has fueled the growth of industry and economy the world over as the raw material for transportation fuels, heat and electric power generation. Today, petroleum derived products have become a part of our daily lives in modern society. This rapid growth has been possible due to the technological innovations leading to the development of new products and processes. Starting in January 2022 the magazine Hydrocarbon Processing celebrated its 100 years anniversary by publishing a series of ten articles (Nichols 2022) that chronicled the decade-by-decade technological advancements driving the growth of the oil and gas industry. However, all these benefits to society came at a cost—climate change.

Petroleum is a fossil fuel and its combustion to produce energy also produces CO_2, which is a Greenhouse Gas. Also, all major industrial production processes, e.g. steel making, cement manufacturing, etc. require energy that is supplied by

S. Basu (✉)
Jewel Innovations in Energy and Environment, Katy, TX, USA
e-mail: som.basu@jewelenergyenvironment.com

S. Basu et al. (eds.), *Water Management in Petroleum Industries*, Water Resources Development and Management, https://doi.org/10.1007/978-981-19-3159-8_9

burning fossil fuels and emit CO_2 to the environment contributing to the overall Greenhouse Gas emissions. To mitigate this situation various types of renewable energy are increasingly being adopted in the energy mix with the target of achieving a carbon neutral planet by 2050 as set by the Paris Climate Accord of 2015.

In 2023 about 4.18 trillion kWh of electrical energy was generated in the US. About 60% of this was produced by fossil fuels, 19% by nuclear energy and 21% by renewable energy including solar, wind, hydropower, biomass and geothermal resources (https://www.eia.gov/tools/faqs/faq.php?id=427&t=3). But this fraction of renewable energy is on a path to a growing trend.

Hydrogen is poised to play a key role in becoming an alternate fuel in the near future. This is because it can be relatively easily transported by pipelines and barges, and also hydrogen can be stored so that it can be used for power generation as needed. Water is intricately tied with production of hydrogen. Hydrogen is classified in different shades based on the greenhouse gas emission potential of the production process as presented in Table 9.1.

Table 9.1 Hydrogen by color

Shade of hydrogen	Production technology	Feed source	GHG footprint
Grey	Steam Methane Reforming (SMR)	Natural gas	Medium
Blue	SMR + Carbon Capture and Sequestration (CCS)	Natural gas, heavy oil	Low
Purple/Pink	Electrolysis	Nuclear energy	Minimal
Green	Electrolysis	Renewable energy e.g., Hydropower, wind, solar, marine power	Minimal

Traditionally hydrogen is produced from fossil fuels by steam methane reformation, followed by water gas shift reaction, as follows:

Reformation:

$$CH_4 + H_2O \rightarrow CO + 3\,H_2 \tag{9.1}$$

Water–Gas Shift Reaction:

$$CO + H_2O \rightarrow CO_2 + H_2 \tag{9.2}$$

Overall:

$$CH_4 + 2\,H_2O \rightarrow CO_2 + 4\,H_2 \tag{9.3}$$

CO_2 is a co-product of hydrogen. It is removed from the gas mixture by separation processes and is released in the atmosphere which adds to the greenhouse gas emissions from the production facility. Hydrogen produced by this process without any control of CO_2 emissions to the environment is "Grey Hydrogen". When the CO_2 is removed and immobilized in an underground geologic formation using the process of carbon capture and sequestration (CCS) hydrogen produced by this process is "Blue Hydrogen". An alternative route for hydrogen production is via electrolysis of water. At a basic concept level, it is a process to produce hydrogen by splitting water molecules by the application of electrical energy in electrolyzers.

$$H_2O + \text{Electrical Energy} \rightarrow H_2 + 1/2\,O_2 \tag{9.4}$$

Oxygen is a valuable co-product which can be utilized for various purposes including industrial and medical uses and can be sold in the market. Hydrogen produced by electrolysis of water where the electrolyzers are powered by renewable energy e.g. wind, solar, hydropower, etc., is "Green Hydrogen". This hydrogen does not co-produce CO_2 contributing to greenhouse gas to the environment. To qualify for being a green energy the feed energy should be supplied from a grid powered by a renewable resource. Table 9.1 lists hydrogen of different shades based on their production process and the potential to emit CO_2.

To meet the goal of total decarbonization by the year 2050 a large part of energy will be derived from a combination of Green Hydrogen and Blue Hydrogen. While the emphasis will be to utilize Green Hydrogen as much as possible, for practical reasons certain heavy manufacturing industries may need to use fossil fuels to fulfill their energy needs and capture and immobilize CO_2 by CCS.

For long distance transportation and storage by barges and ships hydrogen in its gaseous form may not be the best suited option. An alternative is to convert it to Green ammonia and methanol derived from Green hydrogen and store and transport to long distances in those forms.

A large amount of water is necessary to feed the electrolyzers for the production of Green Hydrogen. As freshwater resources are dwindling and groundwater table receding in many parts of the world there will be a competition between water demand for agriculture, sanitation and hygiene and that for hydrogen production. Therefore, it is necessary to explore alternative sources of water that are available in abundance but not utilized for beneficial purposes. Three such sources are:

- Effluent from domestic wastewater treatment plants
- Seawater
- Produced water from oil and gas exploration and production sites.

With advanced level treatments it is possible to improve the quality of water from the above sources to make them suitable for feeding into the electrolyzers. The treatment processes have been discussed in the earlier chapters of this book.

As the share of renewable energy keeps increasing over the coming years it is imperative that the demand for petroleum as a raw material for refined products, which are primarily used as fuel, will diminish with time. However, based on the current trend, crude production has been on the rise for the last several years. As petroleum serves as the raw material for refined products and petrochemicals this increased production is driven by the growth of petrochemicals demand globally.

Over the next 25 years, global petrochemicals demand is forecast to increase exponentially, rising from approximately 450 MMtpy in 2025 to > 500 MMtpy in 2035 and upwards to nearly 600 MMtpy by 2050 (Hydrocarbon Processing, March 2024). As a result, many refiners are modifying their equipment design and operation to integrate refining with production of olefins which are the building blocks of various polymers and synthetic chemicals. However, as far as water management is concerned there is expected to be minimal change resulting from these changes in the refinery product slate. There will be no change in upstream and downstream water treatment processes. There may be some changes to the downstream wastewater treatment due to the shift in product mix from fuels to chemicals.

Energy transition will significantly impact the use of petroleum as one of the main sources of energy. However, that is not expected to impact crude production as the focus is shifting towards petroleum as a feedstock for chemicals, as opposed to fuels. On the contrary, there are signs that novel processes, technologies and products will continue to drive the surge in petrochemical production. Synthetic materials are used for manufacturing medical devices and are even used for making human body implants. It will be difficult to replace raw materials for manufacturing products in these applications.

Water will play an increasingly important role in energy transition as it is the raw material for production of Green Hydrogen. Given the gradually dimishing fresh water resources water will be utilized from non-trsditional resources. Innovations in technology to achieve recycle and reuse of wastewater economically will be the key to ensure the path to a carbon neutral future with Green hydrogen becoming the fuel gas of choice. In conclusion, the principles of water management as discussed in this book will remain relevant and useful for years to come unaffected by the ongoing energy transition.

Abbreviations and Glossary of Terms

CCS Carbon Capture and Sequestration

References

Nichols L (2022) History of hydrocarbon processing industries in a series of 10 articles starting from January 2022 for the next 10 months chronicling the growth of the industry over time starting from 1920's through 2000's. These were published in commemoration of 100 years of publication of the magazine Hydrocarbon Processing

Nichols L (2024) Editorial column of the March 2024 issue of Hydrocarbon Processing

Appendix A
Basic Concepts of Chemistry and Definitions

A.1 Structure of Atoms and Molecules, Periodic Table

The basic concepts of physical and biological sciences, especially chemistry, biochemistry and microbiology are essential to gain a good understanding of water and wastewater characteristics which form the basis of planning and designing appropriate treatment systems. It starts with an overview of the periodic table. It is a table of elements organized in increasing atomic numbers, from the lowest atomic number, Hydrogen (H), to the highest atomic number, Oganesson (Og) (Refer to Fig. A.1). Russian scientist Dimitri Mendeleev originally developed this table in 1869 based on his observation of periodicity of physical and chemical properties of elements in relation to atomic number. The original version created by Mendeleev has expanded over time to include new elements discovered from time to time. Note that the following discussions provide the reader a basic level understanding necessary in the context of water chemistry. It is not a detailed and in-depth treatise of the subject.

An atom is the smallest unit of an element that can participate in a chemical change. The diameter of an atom is in the order of 10^{-10} m. An element consists of only one type of atom, which has a mass that is characteristic of the element and is the same for all atoms of that element. Although many elements consist of single atoms, some exist as molecules made up of two or more atoms of the element chemically bonded together. The atoms in turn consist of the building blocks—protons, neutrons, and electrons, which are extremely small. Atomic number of an element is the number of protons in the nucleus of an atom of that element. Atoms are electrically neutral as they contain the same number of positively charged protons and negatively charged electrons. When the numbers of these subatomic particles are not equal, the atom is electrically charged and is called an ion. When one or multiple electrons are removed from an atom it becomes positively charged and is called a positive ion and similarly, and when one or multiple electrons are added to an atom it becomes negatively charged and is called a negative ion. Hydrogen has only one

S. Basu et al. (eds.), *Water Management in Petroleum Industries*, Water Resources Development and Management, https://doi.org/10.1007/978-981-19-3159-8

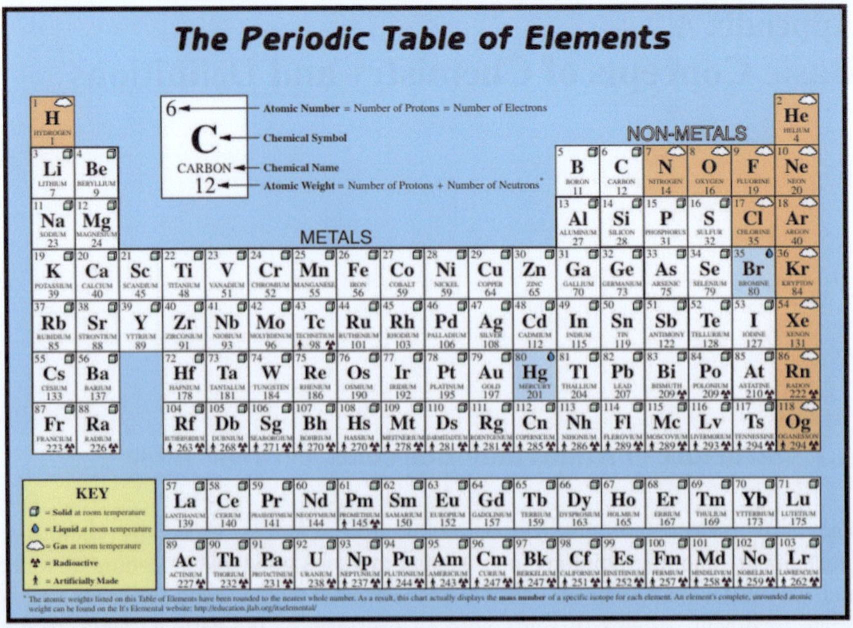

Fig. A.1 Periodic table (*Source* The periodic table of elements (nasa.gov))

proton and one electron but no neutron. When a hydrogen atom loses its electron, it becomes a positive hydrogen ion (H^+) which is same as a proton.

In the Periodic Table elements are arranged in seven (7) horizontal rows, called Periods, and eighteen (18) vertical columns, called Groups. Due to the recurring pattern in their properties, called the "periodic law", elements in the same group have similar properties.

Chemical symbols in the periodic table are used to represent atoms of the elements. Atoms of the same element or different elements combine by chemical bonding to form molecules. Four main types of bonding mechanisms are ionic, covalent, co-ordinate covalent and hydrogen bonding. Positive and negative ions combine by **ionic bonding** to form a neutral molecule. An example is hydrochloric acid (hydrogen chloride) which is formed by ionic bonding of a positive ion of hydrogen with a negative ion of chlorine (chloride ion). By accepting an electron from a hydrogen atom chlorine attains the stable electronic structure of a Group 8 element. When an atom shares its electrons with those of another atom of the same element or a different element to form a bond and attain stable electronic structure the type of bonding is **covalent bonding**. Examples are hydrogen molecules that are formed by covalent bonding between two atoms of hydrogen and oxygen molecules formed by a combination of two atoms of oxygen. A carbon dioxide molecule is an example of covalent bonding between an atom of carbon with two atoms of oxygen. Water molecules are formed by **hydrogen bonding** between hydrogen and oxygen atoms as shown in Fig. 1.1 a and b in Chap. 1.

Atomic weight of an element is its relative weight with respect to hydrogen atom. In the scale of atomic weights hydrogen is the lightest with atomic weight of 1. For molecules comprising of a single atom the molecular weight is identical to its atomic weight. However, for molecules comprised of multiple atoms, as in the case of multi-atomic molecules of the same element, or atoms of different elements, as in the case of a chemical compound, molecular weight is the sum of the weights of all the atoms making up that molecule. Molecular weight is represented in short form as MW. For example, molecular weights of oxygen (O_2) = 2 (*) atomic weight of oxygen = 2 * 16 = 32. Molecular weight of hydrogen (H_2) = 2 * atomic weight of hydrogen = 2 * 1 = 2. Molecular weight of water (H_2O) = 2 * atomic weight of hydrogen + 1 * atomic weight of oxygen = (2 * 1) + (1 * 16) = 18.

A.2 Units and Dimensions

All physical and chemical parameters used in this book are expressed in terms of International Standard (SI) units. However, in the petroleum industry US standards are also widely used. These are listed in Table A.1.

A.3 Relation Between Atomic Weight, Molecular Weights, and Equivalent Weights

Equivalent weight (EW) of a compound is its molecular weight (MW) divided by the number of hydrogen atoms (n) or its equivalent contained in it. It is expressed as:

$$\mathrm{EW} = \mathrm{MW}/\mathrm{n}$$

For compounds that do not contain hydrogen, the number ‘n’ is the oxidation number of the element that participates in an oxidation-reduction reaction. For example, considering that a hydroxide ion is equivalent to a hydrogen ion the following example of calcium carbonate ($CaCO_3$) shows the relation between atomic weight, molecular weight, and equivalent weight.

Example 1

Atomic Weight of calcium, carbon and oxygen are 40, 12 and 16, respectively.

Molecular Weight (MW) of $CaCO_3$ = 40 + 12 + 3*16, or 100 gm/mol.

Molecular Weight (MW) of carbonate (CO_3^{2-}) ions = 12 + 3*16, or 60 gm/mol.

Molecular Weight (MW) of bicarbonate (HCO_3^-) ions = 1 + 12 + 3*16, or 61 gm/mol.

As calcium carbonate comprises of one Ca^{2+} and CO_3^{2-} ion, each oxidation number ‘n’ of both ions = 2.

Table A.1 Common physical and chemical parameter in SI and English units and their conversion factors

	SI unit	English unit	Conversion factor
Linear dimension	Meter (m) Micron (μm) Nanometer (nm) Angstrom (Å)	Foot (ft) N/A N/A N/A	1 m = 3.28 ft 1 μm = 10^{-6} m 1 nm = 10^{-9} m 1 Å = 10^{-10} m
Area	Square Meters (m^2)	Square Foot (ft^2)	1 m^2 = 10.76 ft^2
Volume	Cubic Meter (m^3), Liter (L)	Cubic Feet (ft^3), Gallon, US Barrel	1 m^3 = 1000 L = 35.31 ft^3 = 264.17 Gallons 1 US Barrel of Oil = 42 Gallons
Mass	Gram (gm), Kilogram (kg)	Pound (lb)	1 kg = 2.2 lbs
Mass in Mols	gmol, kilo mol	lbmols	1 gmol = gram/MW 1 kilo mol = 1,000 gmol = 2.2 lbmol
Temperature	°C, °K	°F, °R	F = 1.8 C + 32 K = C + 273.15 R = F + 460
Pressure	Bar, Kilopascal (KPa)	Pounds/sq. in. (psi)	1 Bar = 100 kPa 1 Bar = 14.5 psi
Force	Newton (N)	Pound Force (lb_f)	1 N = 0.225 lb_f
Density	g/mL, kg/L	lb_mcu.ft	1 kg/L = 62.43 lb/ft^3
Dynamic viscosity	N.s/m^2	Lb_f.s/ft^2	1 N.s/m^2 = 0.02 lb_f/ft^2
Energy	Joule (J)	ft.lb_f	1 J = 0.736 ft.lb_f
Power	Kilowatt (kW)	Horsepower (HP)	1 kW = 1.34 HP
Concentration (in mass)	mg/L, g/m^3	lb/gal	1 g/m^3 = 1 mg/L 1 lb/gal = 119.826 g/L
Concentration (in molarity) (Note)	mM (mg/L/MW)	N/A	1 lb mol = 453.6 gmol 1 M = 1,000 mM
Concentration (in normality) (Note)	N (g/L/EW)	N/A	Discussed in paragraph 3 below and in Sect. 2.3.5

Note MW = Molecular Weight, EW = Equivalent Weight

Therefore, Equivalent Weight (EW) of calcium (Ca^{2+}) ions = 40/2, or 20 gm/equivalent.

Equivalent Weight (EW) of carbonate (CO_3^{2-}) ions = 60/2, or 30 gm/equivalent.

Equivalent Weight (EW) of $CaCO_3$ = 100/2, or 50 gm/equivalent.

As bicarbonate ions possess a single charge, n = 1 for bicarbonate ions.

Therefore, Equivalent Weight of bicarbonate ions (HCO_3^-) = 61/1, or 61 gm/equivalent.

Equivalent weight is very important in water chemistry as two important parameters—hardness and alkalinity are expressed in terms of calcium carbonate equivalents. These are discussed in further details in Chaps. 1 and 2.

Notes:

1. Concentration of contaminants in aqueous solutions are very commonly referred to as parts per million (ppm). It is essentially same as mg/L as the density of water is 1 kg/L, which is unaffected by the presence of insignificant amount of contaminant, such that 1 mg/L is equal to 1 mg/kg, or one part per million (ppm) on weight per weight (w/w) basis
2. Concentration of constituents (solutes) in a solution is also expressed in terms of mole fraction. In a solution, if there are N_A, N_B and N_C number of moles of solute are dissolved in N_S number of moles of solvent, then the mole fraction of each solute in the solution is expressed as:

$$x_m = N_i/N_T$$

where, N_i = moles of individual solutes A, B, C, etc.

and N_T = Total number of moles in the solution = $N_S + \Sigma N_i$

For a gaseous solution, the mole fractions are expressed by the symbol y_i such that:

$$y_i = N_i/N_T$$

Water is a liquid solution and air is a gaseous solution. In both cases the amounts of contaminants are insignificant relative to the amount of solvent such that:

$$N_T \sim N_S$$

and

x_i for liquid solution, and y_i for gaseous solution ~N_i/N_S, and

$$\Sigma x_i = 1$$

and

$$\Sigma y_i = 1$$

3. Another means of expressing concentration of constituents, especially acids and alkali in dilute aqueous solutions, is normality, expressed as meq/L (miliequivalents/Liter), where meq = mg/L/Equivalent Weight (EW).
4. For gaseous solutions, for example, in the case of oxygen, or carbon dioxide, or hydrogen sulfide, the gas phase concentration is expressed as parts per million (ppm) in terms of volume per volume (v/v), not on the basis of weight per weight

(w/w). This is identical to mole/mole, following the ideal gas law:

$$PV = nRT$$

Under constant temperature and pressure, in a gaseous solution:

$$V_i/V_T = N_i/N_T$$

5. The total pressure P_T of a gas is the cumulative partial pressure (P_i) of its constituents:

$$P_T = \sum P_i$$

Mole fraction of each component $y_i = P_i/P_T$, or $P_i = y_i * P_T$

For example, concentration of oxygen in air, $y_i = 21\%$ (v/v) (rounded up number). Therefore, partial pressure P_i of oxygen in air = 0.21 * atmospheric pressure = 0.21 bar.

A.4 Rate of Chemical Reaction and Chemical Equilibrium

Chemical formulas depict molecules as well as the composition of compounds. Chemical equations provide information about the changes associated with chemical reactions. In a chemical reaction the "reactants" are converted into "products" over a period, the reactants are used up and products are formed.

Chemical reactions take place as a result of collision of similar or dissimilar atoms. However, the reactants are not always completely converted to products because a reverse reaction takes place in which products atoms collide to form the reactants back again- making most chemical reactions "reversible". A chemical reaction is said to have reached equilibrium when its rate of forward reaction is equal to the rate of the reverse reaction. At chemical equilibrium no further changes occur in the concentrations of reactants and products, even though the forward and backward reactions continue at equal rates but opposite in directions. This can be further visualized by considering a chemical reaction between reactants A and B to produce products C and D:

$$aA + bB \leftrightarrow cC + dD \tag{A.1}$$

The rate of forward reaction is expressed as:

$$r_f = k_f * [A]^a * [B]^b \tag{A.2}$$

The rate of reverse reaction is expressed as:

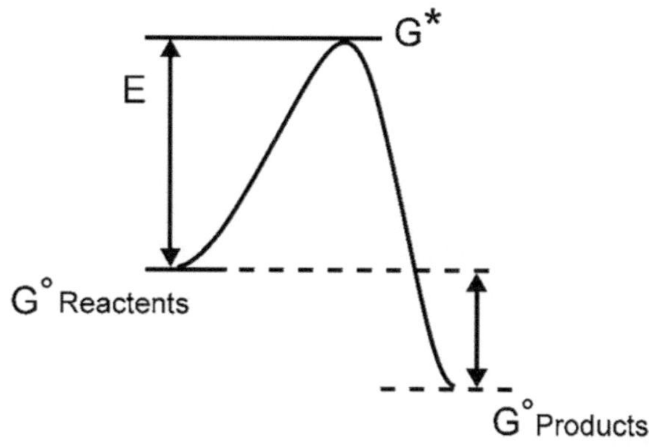

Fig. A.2 A conceptual representation of activation energy, and standard free energies of reactants and products

$$r_r = k_r * [C]^c * [D]^d \tag{A.3}$$

where k_f and k_r are the forward and reverse rate constants, and [A], [B], [C] and [D] are the concentrations of A, B, C and D respectively in terms of molarity (gmol/L).

When the reaction reaches equilibrium

$$r_r = r_f, \text{ such that } k_f * [A]^a * [B]^b = k_r * [C]^c * [D]^d \tag{A.4}$$

The ratio $[C]^c*[D]^d/[A]^a*[B]^b$ $r_f = k_r/ k_f$ termed the Equilibrium Constant, K_{eq}.

Two important questions may arise while following the course of chemical reactions—how far, and how fast? The science of chemical thermodynamics helps answer the first question by establishing the equilibrium conditions that determine the extent of a chemical reaction before it reaches equilibrium. Chemical kinetics, or the rate of reaction, is the other important consideration as that dictates the time necessary to complete a reaction. The rate of a reaction is a function of the activation energy to initiate the reaction. This can be visualized from Fig. A.2 considering a simple reaction between reactants A and B producing the products C and D.

$$A + B \rightarrow C + D$$

A chemical reaction can proceed when the total Standard Free Energy of reactants (ΣG^o Reactants) is higher than the total Standard Free Energy of products (ΣG^o Products) such that:

$$\Delta G^\circ = \Sigma G^\circ \text{ Products } - \Sigma G^\circ \text{ Reactants} = -\text{ve} \tag{A.5}$$

However, high activation energy, E creates a barrier to a reaction from moving forward, which can be overcome by supplying energy externally. G* is the Standard Free Energy of a conceptual transition state, such that:

$$\text{Activation Energy, E} = G^{*} - \Sigma G^{0}\ \text{Reactants} \quad (A.6)$$

The rate of reaction is expressed as:

$$r = k * [A] * [B] \quad (A.7)$$

The reaction rate constant, k is a function of temperature, T and Activation Energy, E

$$k = k_o * \exp(-E/RT) \quad (A.8)$$

where,

k_o Frequency Factor, a constant

E Activation Energy

R Universal Gas Constant

T Absolute Temperature

Therefore,

$$r = k_o * [A] * [B] * \exp(-E/RT) \quad (A.9)$$

This equation for temperature dependence of reaction rate is called Arrhenius equation. It can be observed from this relationship that, at the given concentrations of the reactants, the rate of reaction decreases with increase in activation energy, E and decrease of temperature, T. On the contrary the rate of reaction increases with the decrease in activation energy and increase in temperature. An example is the combustion of fuels, which does not proceed automatically without the addition of activation energy externally in the form of a flame or another ignition source. Catalysts are agents that reduce the activation energy and the barrier to a chemical reaction. Catalysts alter the reaction pathway, and thereby enhance the rate of reaction, but they are not consumed themselves, nor they do become a part of the products. Therefore, catalysts are used for long periods of time in chemical and biochemical reactions until their effectiveness is compromised due to loss of chemical and/or mechanical integrity.

There are three types of catalysis—homogeneous, heterogeneous, and enzymatic. Homogeneous catalysts participate in chemical reactions in the same phase, typically liquid phase. A common application is sulfuric acid or hydrofluoric acid catalysis of alkylation reactions in petroleum refining processes. Heterogeneous catalysis is most widely used in industrial chemical, petrochemical and refining processes which include fluid catalytic cracking (FCC), hydrogenation, reformation etc. Enzymatic catalysis is the primary mechanism of all biochemical processes for pharmaceutical production, and for water and wastewater treatment applications.

A.5 Effects of Temperature and Pressure on Chemical Reaction Rates and Equilibrium

Temperature is one of the most important parameters as chemical reaction rates and equilibrium constants vary with temperature. These include the important water quality parameters e.g., solubility and solubility product as discussed in Chap. 1. Several other important water quality parameters such as density, viscosity, solubility, surface tension, vapor pressure etc., are also functions of temperature. These are major inputs in engineering designs.

Chemical and biochemical reactions in water and wastewater treatment take place under atmospheric pressure. Therefore, pressure is not a major parameter influencing water and wastewater treatment processes in aqueous phase. However, in vapor-liquid equilibria involving water and gases e.g., oxygen, CO_2, H_2S, etc. pressure plays a very important role as solubility of these gases increases with pressure.

Standard state in thermodynamics literature refers to 1 bar (10^5 Pa) pressure and 0 °C (273.15 K) temperature. This standard is primarily followed for gases. In case of dilute solutions, like water and wastewater, the standard temperature is 25 °C. (298.15 K). References made in this book to all parameters e.g., concentration, equilibrium constants are under standard conditions of 1 bar pressure and 25 °C.

The materials presented in this Appendix to Chap. 1 serves as the foundation for the subsequent chapters.

Appendix B
Chemical Composition and Characteristics of Petroleum

B.1 Introduction and Historical Perspective

Modern civilization owes its progress to a large extent to petroleum. Starting from transportation fuels to multitude of consumer goods there is hardly any product that the society beneficially uses that is not derived from this invaluable natural resource. In the US, petroleum was first commercially drilled at the Drake Discovery well in Northern Pennsylvania in 1859. Since then, this site has been recognized as a National Historic Landmark and earned its name listed as a Historic Mechanical Engineering Landmark.

Petroleum has been known to humans throughout the World from prehistoric times for its value as a fuel and for its other beneficial properties. Drake's discovery spurred the phenomenal growth of an industry devoted to producing valuable products targeted for use in multitude of applications. The first commercial scale refinery started operation in Titusville, PA in the year 1860.

Petroleum formed from decomposition and transformation of organic matter contained in plants and animals underground over a span of millions of years. High temperature and pressure conditions promoted decomposition of the complex organic matters into crude petroleum, containing simpler compounds of primarily carbon and hydrogen atoms. It has been hypothesized that bacterial activity is a significant contributor to the complex decomposition reactions underground. Crude petroleum exists in the pores of underground rock formations. A collection of several crude bearing rock structures adjacent to each other in a large area constitutes a petroleum reservoir.

S. Basu et al. (eds.), *Water Management in Petroleum Industries*, Water Resources Development and Management, https://doi.org/10.1007/978-981-19-3159-8

B.2 Chemical Composition of Petroleum

Petroleum is available throughout the world in sedimentary rocks, but its composition is not uniform and varies over a wide range. It is primarily a liquid mixture of different groups of hydrocarbon molecules consisting of hydrogen and carbon atoms. Apart from hydrocarbons, petroleum also contains small amounts of molecules comprising of oxygen, sulfur, and nitrogen atoms. Trace amounts of organometallic compounds of vanadium, nickel, iron, arsenic, and mercury are also present in petroleum. The term crude is used for petroleum in its raw form before any processing. Typical elemental composition ranges of crude oils are presented in Table B.1.

Table B.1 Typical elemental composition of petroleum crudes (Adapted from API 2011)

Element	Composition range (% w/w)
Carbon (C)	84–87
Hydrogen (H)	11.0–14.0
Sulfur (S)	0.06–8.0
Nitrogen (N)	0.02–1.5
Oxygen (O)	0.08–1.82
Metals	0.00–0.14

Hydrocarbons in crude can be classified into a few major classes—paraffin, olefin, naphthene, diolefin, aromatic and acetylene. Paraffins are saturated hydrocarbons of the general formula C_nH_{2n+2}. The simplest of these molecules is methane (CH_4) corresponding to the value of n = 1. It is the lightest of the paraffinic hydrocarbons. It is a gas under ambient temperature and pressure conditions. Ethane (C_2H_6), propane (C_3H_8) and butane (C_4H_{10}) are paraffins corresponding to the values of n = 2, 3 and 4. Boiling point, density and viscosity of hydrocarbons increase with increasing molecular weight. Compounds with the same molecular formula but different molecular structures, depending upon how the carbon atoms are joined to each other, exist as isomers. For paraffins, butane is the compound of lowest molecular weight (n = 4) having isomers - normal butane (n-butane), a straight chain paraffin, and iso-butane, or 2 methyl propane, a branched chain paraffin. The number of isomers increases disproportionately in large numbers with the increase of carbon atoms.

Low molecular weight, or lighter paraffins are commonly present in most crudes and high molecular weight paraffins, both straight and branched chain, are associated with wax. Lighter paraffins, primarily C1 through C4 molecules, exist dissolved in crude under high pressure conditions in the reservoirs. When the crude is pumped out, lighter hydrocarbons are released at lower pressure conditions and separated from the liquid.

Olefins are unsaturated hydrocarbons with a double bond. They have the general formula of C_nH_{2n}. The members of this series are ethylene (C_2H_4), propylene (C_3H_6), butylene (C_4H_8), etc. corresponding the values of n = 1, 2, 3, etc.

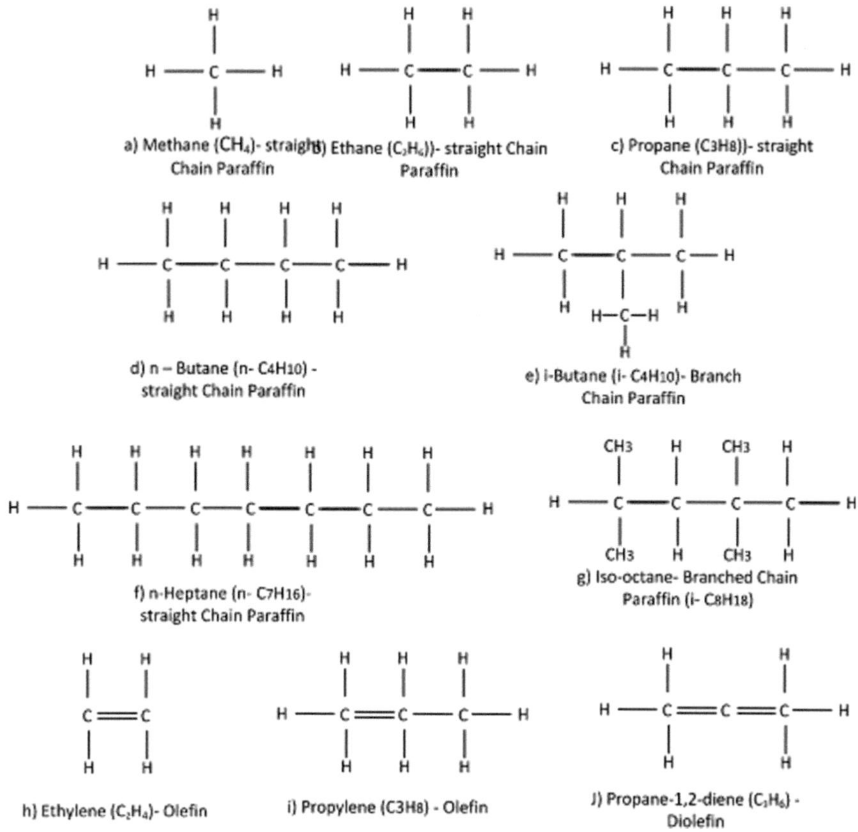

Fig. B.1 Examples of paraffins, olefin and diolefin hydrocarbons

Diolefins are unsaturated hydrocarbons with two double bonds. Their general chemical formula is C_nH_{2n-2}. Because of the presence of double bond, olefins and diolefins can react by direct addition, in addition to substitution of hydrogen atoms. Molecular structures of a few representative paraffin, olefin and diolefinic hydrocarbons are presented in Fig. B.1.

Cyclic hydrocarbons consist of two groups—naphthenes and aromatics. Although the general formula of naphthenic compounds (C_nH_{2n}) is identical to that of olefins they are cyclic and saturated (where the atoms of carbon are tied with hydrogen atoms by single bonds). The lowest member of this group is cyclopropane (C_3H_6) corresponding to n = 3. Higher members of this group are cyclobutene (C_4H_8), cyclopentane (C_5H_{10}), cyclohexane (C_6H_{12}), etc. corresponding to n = 4, 5 and 6, respectively. Naphthenic hydrocarbons are less reactive than their isomeric olefins because they are saturated compounds. These are present in higher boiling components of crude petroleum.

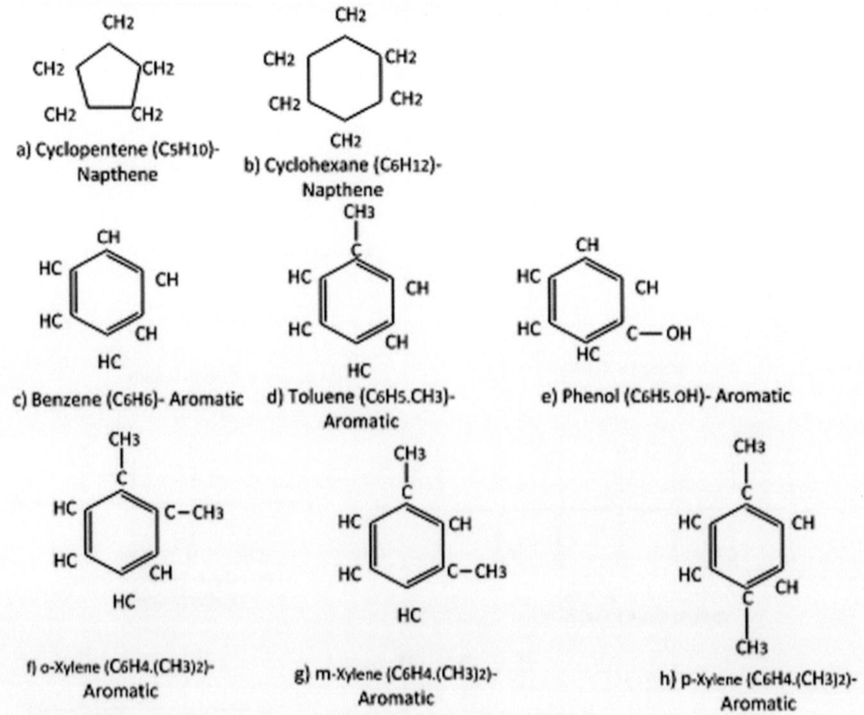

Fig. B.2 Examples of ring structure hydrocarbons

Aromatic compounds are the other group of cyclic compounds. These are unsaturated compounds with the general chemical formula C_nH_{2n-6}. Benzene (C_6H_6) is the lowest molecular weight aromatic compound with the value of $n = 6$. The presence of three double bonds renders them highly reactive. All aromatic compounds are derived from benzene and have similar ring structures. Crudes at only a few locations in the world contain lighter aromatics, e.g., benzene, toluene, xylene, and ethyl benzene (BTEX) in their free forms. Polyaromatic hydrocarbons (PAH), contain multiple benzene rings fused together. These are present in the high molecular weight fractions of crude. Figure B.2 presents the molecular structures of a few representative naphthenic and aromatic hydrocarbons.

Paraffinic, naphthenic, and aromatic hydrocarbons are present in naturally occurring crudes. Olefins and diolefins are generally formed in the products of cracking. Apart from hydrocarbons, petroleum also contains heteroatomic organic compounds consisting of oxygen, sulfur, and nitrogen, which influence its physical and chemical properties. Oxygenated organic compounds in crude petroleum are present as naphthenic acids (NA). These are alkyl substituted acyclic or cyclic carboxylic acids with a general chemical formula of $C_nH_{2n-Z}O_2$ where $n =$ number of carbon atoms and $Z = 0$ or a negative even number. The range of carbon atoms ('n' value) is 5 to 30 and Z is 0 to -12 (Misiti, 2012). All NAs with $Z = 0$ are acyclic compounds. The simplest compound in this group is butane-1-carboxilic acid (also called valeric acid)

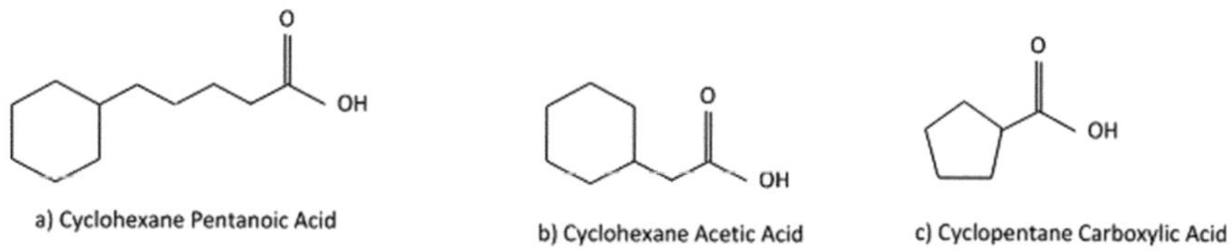

Fig. B.3 Example of common naphthenic acids in crude

C_4H_9.COOH corresponding to n = 5 and Z = 0. For higher values of Z these acids contain single or multiple five and six member rings. Structural formulas for some simple acyclic and cyclic naphthenic acids typically present in crude petroleum are shown in Fig. B.3. Numerous acids exist in crude petroleum with molecular weights ranging from 100 to 450. The presence of NA in a crude makes it corrosive to the metallurgy of processing equipment. Moreover, these are not easily biodegradable and are toxic to aquatic life. Therefore, disposal of treated effluent from refineries processing NA containing crudes may cause ecological toxicity to the aquatic environment. Concentration of NA in a crude is estimated by titration with standard potassium hydroxide (KOH) solution and expressed as its total acid number (TAN) which is the milligrams of KOH to neutralize each gram of the crude. Optimum TAN of a crude of 0.3 to 0.5. Crudes containing NA of higher than this threshold are blended with other types of crude to achieve TAN value within the desired range. However, TAN value may underestimate the NA content in a crude if there is a presence of metal ions that can react and consume the acids and indicate lower than actual TAN concentration.

Several types of sulfur compounds are present in petroleum crude. These are hydrogen sulfide and organic sulfur compounds, which can be further classified into several groups as mercaptans, straight chain organic sulfides and disulfides, cyclic sulfides, thiophenes, sulfonic acids and alkyl sulfonates, sulfoxides, sulfones and thiophene. Their molecular structures are presented in Fig. B.4. Presence of sulfur compounds adversely affect the properties of crudes in multiple ways. These include corrosion of metals that come in direct contact with products derived from processing of high sulfur crude, bad odor in the areas handling and processing sulfur containing crudes and creating serious health and safety risk to the operating and maintenance personnel. Moreover, processing of high sulfur containing crudes also produces wastewater streams with high sulfur concentrations that fosters growth of undesirable microbiological population in biological treatment processes. This has been discussed in detail in Chap. 5. Sulfur content of crudes ranges from 0 to 8% (Nelson, 1969). High sulfur concentrations are typically associated with heavy crudes.

Petroleum crudes contain various organic nitrogen compounds. Globally, these crudes can be classified based on their total nitrogen content. Those lean in nitrogen contain < 0.25% N versus those with higher nitrogen content e.g., crudes from California, South America and Alberta that have concentrations ranging from 0.25 to 1.5% N (Prado et al., 2016). Nitrogen compounds in crudes can be basic and non-basic types. The basic type includes pyridine and its derivatives, e.g., Azapyrene.

Non-basic type includes pyrrole and its derivatives, e.g., quinoline, indole, etc. Their chemical structures are presented in Fig. B.5. Amines also become associated with crudes as these are added during petroleum exploration and production processes.

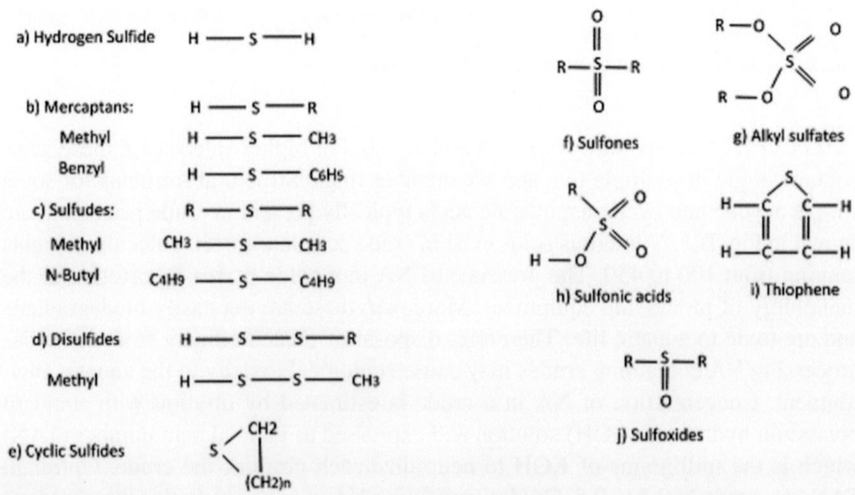

Fig. B.4 Examples of sulphur compounds in crude (Adapted from Nelson, 1969)

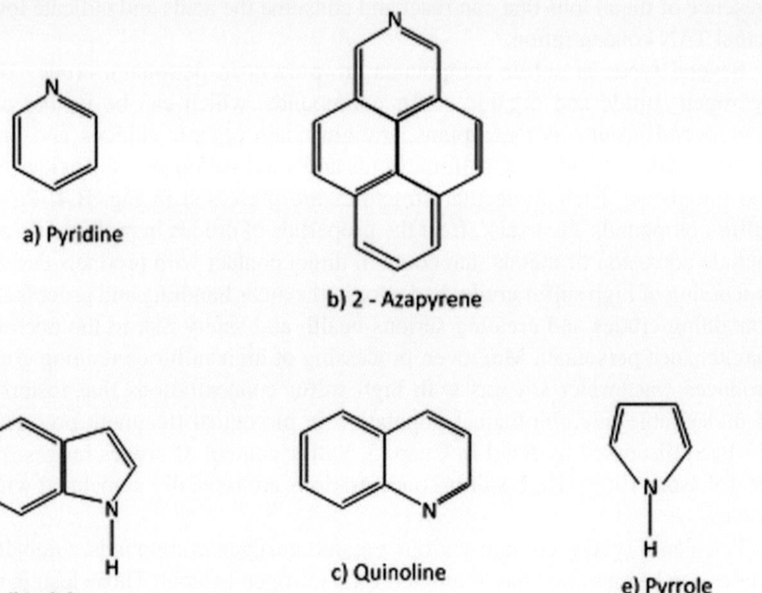

Fig. B.5 Example of nitrogenous compounds in crude

Nitrogen in crudes have several deleterious effects on processing steps. The basic nitrogenous compounds can engage in acid-base reactions with acidic sites of cracking and hydrocracking catalysts and slow down the reaction rates or poison the catalysts altogether. In catalytic reactions the basic nitrogenous compounds can also react with naphthenic acids in the crude and precipitate high molecular weight salts that can potentially reduce the active surface areas of catalysts. All nitrogenous compounds, in general, compete with the reactants for adsorption on to active catalyst sites and reduce available surface area thereby affecting the rates of reaction.

Various metals are present in crude in trace amounts in the form of organo-metallic compounds. Most common among those are compounds of nickel and vanadium. Nitrogen participates in the formation of these complexes. In the presence of salt (sodium chloride), vanadium also combines with sodium as sodium vanadate. Apart from fouling catalyst surfaces, vanadium also causes fusion of refractory materials in furnaces that use heavy oils as fuel. Other metals and metalloids detected in crude oils are calcium, magnesium, iron, barium, copper, mercury, boron and arsenic in organo-metallic complex forms.

B.3 Crude Constituents

Apart from hydrocarbons and heteroatomic compounds crudes also contain several other species that impact their production and processing. Depending upon their source they can be broadly classified as follows.

- Naturally occurring
- Additives to facilitate exploration and production
- Additives to facilitate transportation of crudes.

Water is a naturally occurring constituent in crude petroleum that exists both in soluble and dispersed forms. Water in finely dispersed form, called formation water or connate water, is trapped in the porous rock structure of the reservoirs. This water is present in crudes as very fine drops of 1 to 10 μm size and in 0.1 to 1% (v/v) concentrations. Its content increases with increase in the specific gravity and viscosity of crudes. A part of this water is freed up and forms an aqueous layer inside storage tanks prior to their processing. However, some amount of water forms emulsion with oil and remains with the crude as it moves through the downstream processing steps.

Water is also associated in dissolved state in crude petroleum. Solubility of water in crude is low ($<0.05\%$ v/v) under ambient conditions and increases with increasing temperature. Therefore, as crudes are fed to the desalter operating at 140 to 150 °C they carry a large amount of dissolved water for processing in the atmospheric column, which is undesirable considering the operating energy necessary to pump large amounts of water. Desalting and downstream processing are discussed in detail in Chaps. 4 and 5.

Crude contains solid sediments as one of its constituents. Sand and fine clay particles are silica based and originate from two sources—naturally occurring reservoir rock structures, and solids introduced from upstream exploration and production activities. Solids containing iron sulfide and ferric oxide are the products of corrosion of pipes and metallic containers in crude shipping vessels as a part of midstream transportation activities.

Sediment concentration in crudes is tested by membrane filtration following ASTM Method D 4807 (2020). Collectively, water and sediments together are called Basic Sediment and Water, or BS&W. It refers to the non-hydrocarbon component of crude oil in cargo. It is measured following ASTM Method D 4007 (2022). Acceptable level of BS&W in crude is $<1\%$.

Metal salts are another naturally occurring constituent of crudes. These salts are dissolved in water associated with crude. Primarily, these are chloride, sulfate and carbonate salts of sodium, calcium and magnesium dissolved in formation water with the crude. The types and concentrations of salts depend upon the geology of formation. Some crudes contain high concentrations of sodium chloride. Depending upon the rock structure and mineral content of the reservoir, other types of salts are also observed in crudes at various locations. Salts can also be present in crystalline form in crudes when they precipitate from water at high concentrations.

The presence of high concentrations of salts can create various operational difficulties in downstream processing of crudes and their derivatives. Chloride salts of magnesium and calcium can hydrolyze under the operating conditions of the atmospheric column to form HCl (hydrochloric acid), thus creating corrosive environment to the column internals. On the other hand, presence of carbonate and sulfate ions can produce scaling salts calcium carbonate and magnesium sulfate, which can foul heat exchanger tubes impeding heat transfer rates. Certain crudes contain high concentrations of calcium ions. These can react with naphthenic acids to form calcium naphthenates, which are strong emulsifying agents. These emulsions are difficult to break and escape desalters untreated, thereby causing fouling of heat exchanger tubes and also obstruction to flow.

Scaling salts calcium carbonate, magnesium sulfate and barium sulfate in formation water originating from natural sources can precipitate under supersaturated conditions. Regardless of their source solids create significant operational difficulties by accumulating sludge that is deposited at the bottom of process equipment, thereby reducing the effective processing time. Solids also create fouling and scaling of equipment and pipes. Additionally, presence of fine solids promotes emulsion formation that traps water and the solids.

Transporting oil and gas by multiphase pipelines from offshore wells can be challenging due to clogging by solid natural gas hydrate. As a midstream treatment process, methanol, and ethylene glycol (EG) are added as hydrate inhibitors to facilitate free flow through long pipelines. These become associated with the liquid phase consisting of crude and water after gas-liquid separation onshore. In the storage tanks and downstream processing units, methanol and EG become distributed between the aqueous and organic phases. They contribute to the chemical oxygen demand (COD) of the aqueous phase which is eventually treated in the wastewater treatment plant for final disposal. This is discussed in detail in Chap. 5. Handling and transportation of crudes containing high hydrogen sulfide concentration pose health and safety risk to operating personnel and communities due to air emissions potential. This is mitigated by adding ethanolamine solutions.

Mercury is a Priority Toxic Substance according to the Canadian Environmental Protection Act of 1999. In 2007 Environment Canada conducted a detailed survey of the concentration of mercury in the crudes processed in all the refineries throughout Canada. The results indicated average mercury concentration in all crudes processed in Canada 2.6 ± 0.5 ng of Hg/g of all oil processed in Canada in 2002 (Hollebone and Yang 2007).

Several other trace elements are present in crudes in varying quantities. Concentrations vary widely depending upon the geology of the rock structure of reservoirs. One such constituent is arsenic which poses adverse environmental effects if it partitions into water during processing and escapes with discharge from refinery effluent treatment plants. Among crudes of US origin Louisiana crude contains 46 μg/L, Wyoming crude contains 111 μg/L and California crude 63 to 1112 μg/L concentrations of arsenic (Tang et al. 1994).

B.4 Characterization of Petroleum and Products

Characteristics of petroleum crudes vary widely around the globe. Even at the same location the characteristics vary from well to well and also from time to time. Various methods have been developed over time to classify crudes broadly based upon their constituents and properties, as follows.

B.4.1 Characterization Based on Chemical Composition

One way of traditional classification is the base of a crude depending upon its wax content. If paraffin wax separates upon cooling of a crude it is termed *paraffin* base crude. However, if no such separation of wax occurs on cooling it is an *asphaltic* crude, which has high naphthene content. On a quantitative basis, a crude is defined as *paraffin* base if upon distillation it leaves a residue containing > 5% wax versus *bitumen* base if the residue contains <2% wax. Pure aromatic molecules are not common in crudes, instead they are generally attached with paraffin side chains (Speight 2014). Many crudes do not fall strictly in *paraffinic* or *naphthenic* types and belong to intermediate categories. Therefore, it is difficult to characterize crude oils based strictly on chemical composition alone and different classifications based upon physical properties have been developed by the industry over time.

B.4.2 Characterization Based on Physical Properties

1. Density, Specific Gravity and API Gravity

Density of crude petroleum and refined products is its weight per unit volume and is expressed as kilogram per cubic meter (kg/m^3) at 15.6 °C (60 °F). Specific Gravity of a crude is the ratio of its density at 15.6 °C (60 °F) to that of water at 15.6 °C (60 °F). Therefore, Specific Gravity is unitless. API Gravity is a characteristic of a crude. It is calculated from its specific gravity and expressed as:

$$^{\circ}\text{API Gravity} = (141.5/\text{Sp. Gr. at } 15.6\,^{\circ}\text{C } (60\,^{\circ}\text{F})) - 131.5$$

API gravity of crudes is determined by hydrometers following ASTM Method D 287. In case of highly viscous liquids that do not flow easily a pycnometer can be used for specific gravity determination following ASTM Methods D 941 and D 1217 and API gravity can be calculated.

API gravity is an important property of crudes that is widely used globally as a standard to quantify the gravity (heaviness or lightness) of crudes. Its value varies on a scale of 0 to 100 in a direction opposite to that of the specific gravity. Costs of trading, shipping, and pumping are assessed based upon this property.

Although crudes are generally classified based on their API gravity, there is no clear demarcation between the types of crudes on this basis. Broadly, on the API scale crudes with values >35 are considered light. Those with API gravities between 25 and 35 fall into medium grade category. The values for heavy oils range from 10 to 25 and those below 10 are thick, semisolid materials falling under the category of bitumen.

Apart from the usefulness of API gravity as a means of assigning the costs of transportation, it is also an important indicator of the quality of crude and ease of processing to produce high value products. Typically, heavy crudes of low API gravity values contain high concentrations of heteroatomic organic compounds with sulfur, nitrogen, oxygen, and metals that require additional processing.

2. Viscosity

Viscosity is the internal resistance of a fluid to flow. It is an important property as it determines the ease or difficulty of flow of crude and liquid petroleum products. Higher viscosity liquids require more pumping energy for production and transportation of crudes. In SI units it is expressed as (Pascal) * (Second), or Pa * s. In CGS system the unit of viscosity is Poise, which is equal to 0.1 Pa * s. Viscosity of liquids increases with decreasing temperature. At 16 °C the viscosity of water is 1.002 centipoise (cP). This is defined as *absolute viscosity* (μ). Another way of expressing viscosity is *kinematic viscosity* (ν), which is measured in terms of stokes (m^2/Sec). It is absolute viscosity divided by the density (ρ) of a liquid at the same temperature. The kinematic viscosity of water at 20 °C is 1 cS (cS is centistokes).

The viscosity of petroleum liquids increases with increase in density and decrease in API gravity. The demarcation between heavy crude and bitumen is 10,000 cP.

Bituminous materials of API gravity below 10 have viscosity above 10,000 cP. The viscosity of petroleum liquids is determined by using Saybolt viscometer following ASTM Method D 88.

3. Vapor Pressure

Vapor pressure of a liquid is the pressure developed by the liquid when it is stored in a closed container. For petroleum liquids it is termed Reid Vapor Pressure (RVP). It is measured in a volume of air that is four times the volume of the liquid following ASTM Method D 323. Vapor pressure is a function of temperature. In SI system its unit is Newton/sq. meter, or Pascal (Pa) and in US system it is expressed as lb_f/sq. inch (psi). It is particularly important in devising the safety systems for storage and transportation of particularly light end petroleum liquids.

4. Flash and Fire Points

When the temperature of a petroleum liquid is elevated to the Flash Point the vapor flashes momentarily upon exposure to a flame. If the temperature is further increased, when it reaches the Fire Point, it generates sufficient flammable vapor that sustains its combustion continuously. Like vapor pressure, these are important properties in the design of safety systems for storage and transport of flammable liquids. Flash points for heavy and light oils are determined by ASTM Method D 92 and D 93 respectively.

5. Lower and Upper Explosive Limits

When petroleum crude or product(s) is stored or transported in an enclosed and partly filled container the volatile hydrocarbons occupy the headspace in a mixture with air. The lowest and highest concentrations of the vapor in air (v/v) to support combustion are the Lower and Upper Explosive Limits (LEL and UEL). The compositions of the vapor in air within these two limits are within the range of explosion. The concentrations of the vapor below LEL are too lean and above UEL are too rich to support and sustain combustion.

6. Cloud and Pour Point

Cloud point of a liquid refers to a temperature below which it undergoes a transition from clear liquid to a suspension of fine solid and colloidal particles in liquid. Pour point is the minimum temperature below which a liquid ceases to flow. The presence of wax in oil causes solidification and high pour point of a crude indicates high concentration of paraffins. Knowledge of these temperatures for a crude is important to determine the suitability of its fractions as a motor fuel or a lubricant. Pour point is determined by ASTM Method D 97.

7. Knocking of Engines and Octane Number of Motor Fuels

Knocking involves irregular detonation and vibration of internal combustion engines resulting from sub optimal ignition of gasoline. Knocking is harmful to engines as it creates mechanical damage and deposits carbon due to incomplete combustion. The knocking characteristics of a fuel is quantified by matching the intensity of knocking it creates in a standard test engine with that created by a mixture of iso-octane and n-heptane. In the knocking scale iso-octane is assigned a number 100 and n-heptane as 0. The percentage of iso-octane in the mixture is defined as an octane number. As an example, a fuel of octane number 85 implies that its knocking characteristics match that of a mixture of 85% i-octane and 15% n-heptane. Details of the test methods and test engine specifications are provided in literature (Nelson, 1969).

Improvement of the antiknock property of gasoline was traditionally done by addition of tetraethyl lead (TEL). However, that practice has been discontinued since the late 1990s due to health hazards and replaced by other octane enhancing chemicals. The regulations relating to the discontinuation of TEL and the use of substitute chemicals are discussed further in Chap. 7.

8. Color

The color of a petroleum crude indicates its type and origin, and that of the refined products indicates the product fraction after processing. Lighter crudes are light colored translucent liquids. Heavier crudes are dark in appearance ranging from gray to deep brown in color. Semisolid bituminous materials (API < 10) are even darker and approaching black in color.

For refined products color varies from lighter to darker in order of increasing boiling point ranges of distillation fractions. Chemical conversion of certain refined products, e.g., cracking of residue appears dark due to the presence of carbon particles. Color for petroleum products is quantified by following ASTM Methods D 1500 and D 156. Method 156-12 uses a Saybolt Chromometer to determine the color of gasoline, naphtha, diesel, aviation fuels and waxes. For darker products that exceed Saybolt rating 16 on the color scale, e.g., lubricating oils, heating oils, etc., Method D 156 is used. Details and instruments for these methods are available in ASTM literature.

B.5 Petroleum Resources

Petroleum is explored and produced from various resources in nature. Physical properties of crudes also vary considerably based upon its source. Crudes produced from conventional resources belong to the light and medium grade. These are light to gray color and free-flowing liquid with physical properties associated with API Gravity 25 and higher. When a recovery well is drilled, typically the pressure in the reservoir is sufficient to ensure the flow of crudes from conventional resources for production. In certain cases, it may be necessary to apply pressure to the reservoir by nitrogen or CO_2 externally to maintain flow.

Crudes of API Gravity 10 to 25 are heavy crudes. These are not free flowing and have high viscosity. Heavy crudes are characterized by a high proportion of high molecular weight and high boiling point components. Due to high viscosity, typically in the range of 100 to 10,000 cP under reservoir conditions, they are not free flowing and require non-traditional and expensive methods to recover from reservoirs. In

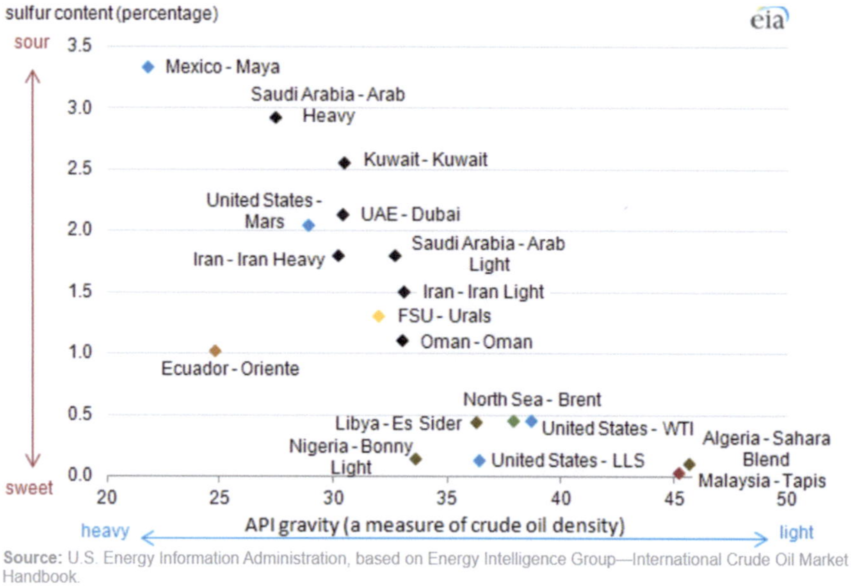

Fig. B.6 Density and sulfur content of crude oils

such cases reservoir stimulations are provided by thermal and/or chemical methods for Enhanced Oil Recovery (EOR) as discussed in detail in Chap. 3. Heavy oils are generally associated with high sulfur contents as shown in Fig. B.6. Crudes are called sweet versus sour based on low versus high sulfur content. Sweet crudes are more expensive than sour crudes as extensive pretreatments are necessary to process sour crudes.

Steady economic growth in the last several decades, especially in the developing world, spurred the rising living standards globally, resulting in rapid increase in the demand of petroleum products for transportation fuels, power generation, and production of consumer goods. This, coupled with dwindling conventional petroleum resources, led the industry to explore oil from unconventional resources. These include opportunity crudes, tar sands, and tight oil and gas.

Opportunity crudes are attractive to refiners because of their significant price advantage over the conventional crudes. However, they contain many undesirable components which render them difficult to process by distillation and associated equipment. These include high concentrations of sulfur and nitrogen compounds, naphthenic acids, metals, and sediment. Pretreatment steps may be necessary to remove certain contaminants that seriously impede storage and processing of crude in a refinery. While pretreatment is effective in removing sediments, other contaminants can pass through to the downstream processing steps and produce fractions that may be corrosive, scale forming or otherwise deleterious for end use. Examples are distillation products containing high concentrations of sulfur compounds that can be corrosive to metals, create health and safety hazard to humans. Similarly, presence of excessive dissolved metal salts can cause corrosion or scaling on metal surfaces. Naphthenic acids are corrosive to metals and can cause ecotoxicity in the receiving water if discharged with treated wastewater from the refinery. These effects are discussed in detail in Chap. 7.

Tar sands are rock structures that contain extra heavy semisolid bituminous materials filled into the pores. They are dark brown to black in appearance, highly viscous under reservoir conditions (>10,000 cP) and do not flow freely. Typically, they need thermal stimulation to recover crude in liquid form. Due to the presence of high concentration of impurities e.g., sediments, sulfur compounds and naphthenic acids these crudes need significant pretreatment before processing for the production of refined products.

Tight oils have properties like the conventional oil, but they are immobile and difficult to recover as these are trapped in very fine pore structures (pore size <1 milli Darcy, or mD) within the reservoirs. The tight oil resources are shale plays of sedimentary rock formation at depths in the range of 1.5 to 5.5 km (5,000 to 18,000 ft) (see Fig. 3.7, Chap. 3). The combination of low permeability and large depths of these formations make these resources rather inaccessible. However, exploration in these difficult terrains has been made possible by the technology of Hydraulic Fracturing of rocks, loosely called Hydrofracking, or simply, Fracking. In principle, this involves application of high-pressurized water to fracture the rocks to release the oil and gas trapped in its fine pores. Hydrofracking, in combination with Horizontal Drilling, has enabled one single well to reach out horizontally to large distances even greater than 4.5 km (15,000 ft.) around the production well (see Fig. 3.14 in Chap. 3) and has made drilling for fracking a cost-effective operation. In the US, the Bakken Shale in North Dakota, Permian Basin, Eagle Ford and Barnett Shales in Texas and New Mexico; and Marcellus Shale in the Appalachian region are examples of a few shale plays producing light tight oil (LTO) crudes of high API gravity (~40).

Further description of production of oil from unconventional resources is provided in Chap. 3.

References

ASTM (2019) D 287 standard test method for API gravity of crude petroleum and petroleum products (hydrometer method), West Conshohocken, PA

ASTM (1988) D 941 standard test method for density and relative density (specific gravity) of liquids by Lipkin Bicapillary Pycnometer, West Conshohocken, PA

ASTM (2020) D 1217 standard test method for density and relative density (specific gravity) of liquids by Bingham Pycnometer, West Conshohocken, PA

ASTM (2020) D 88 standard test method for Saybolt viscosity, West Conshohocken, PA

ASTM (2020) D 323 standard test method for vapor pressure of petroleum products (Reid Method)

ASTM (2013) D 92 Standard test method for flash and fire points by Cleveland Open Cup Tester

ASTM (2020) D 93 standard test methods for flash point by Pensky-Martens Closed Cup Tester

ASTM (2018) D 97 standard test method for pour point of petroleum products

ASTM (2012) D 156 and D 1500 standard test method for Saybolt color of Petroleum products (Saybolt Chromometer Method)

ASTM (2022) D 4007 standard test method for water and sediment in crude oil by centrifuge method
ASTM (2020) D 4807 standard test method for sediment in crude oil by membrane filtration
API (2011) Robust summary of information on crude oil, published January 14, 2011, Washington, DC
Hollebone BP, Yang CX (2007) Mercury in crude oil refined in Canada. Environment Canada, Ottawa, ON
Misiti TM (2012) Fate and effect of naphthenic acid in biological systems. PhD Dissertation, Georgia Institute of Technology, Atlanta, GA
Nelson WL (1969) Petroleum refinery engineering, 4th edn. McGraw Hill, New York
Prado GHC, Rao Y, de Klerk A (2017) Nitrogen removal from oil: a review. Energy Fuels 31:14–36
Speight JG (2014) Chemistry and technology of petroleum, 5th edn. CRC Press
Tang WT, Gumulka J, Sun PY (1994) Source control and treatment of contaminants found in petroleum tank bottoms, API Publication No. 4606, American Petroleum Institute. Washington, D.C.

Appendix C
Mechanism of Biofilm Development in Attached Growth Processes: Role of Exocellular Polymeric Substances (EPS)

C.1 Introduction

Attached growth wastewater treatment processes are carried out by microbial population attached on solid surfaces as biofilms in reactors discussed in Chap. 5. Biofilms consist of high cell densities (10^8 to 10^{11} cells/g) and are complex structures comprising of a variety of species. A biofilm is a collection of microbial cells which are kept together by a self-produced matrix of extracellular polymeric substances (EPS). These groups of cells are immobile (sessile), tightly attached to the surface and are not removed by gentle rinsing. The matrix formation and its stability depends on shear stress, secretion of EPS, nutrient availability, synthesis, and social competition. The biofilm matrix has a predominantly anionic charge which helps it in creating an efficient system for trapping and concentrating essential materials and nutrients for growth. These essential materials include metal ions and other positively charged ions into the film structures by virtue of attractive forces between oppositely charged ions.

The biofilm formation cycle includes the following four steps:

- Development of adhesion surface for initial attachment on to the support,
- Attachment of bacterial cells, their multiplication and accumulation on surface,
- EPS production, dynamics of biofilm development and sloughing, and
- Biofilm maturation.

The initial development consists of formation of films of seed phytoplanktonic (suspended) population at the start-up of a reaction system. Within a few days the films grow, and the support surface becomes rough, which fosters attachment of new cells.

After the first stage, the biofilms grow further as a result of the interaction between the sum of attractive or repulsive forces generated between the bacteria and the supporting surface. The forces that affect the growth are hydrodynamic hinderances, steric hinderances, hydrophobic and electrostatic interactions. In the second stage,

S. Basu et al. (eds.), *Water Management in Petroleum Industries*, Water Resources Development and Management, https://doi.org/10.1007/978-981-19-3159-8

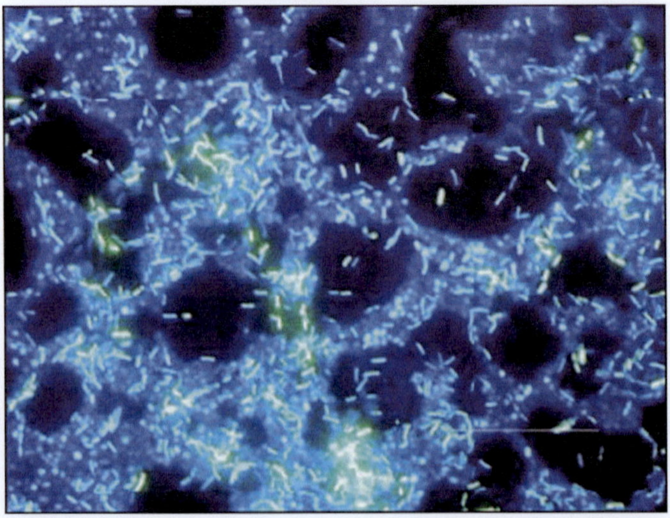

Fig. C.1 Biofilm grown on the stainless-steel surface (Adapted from: Rodney M. Donlan (2002): Biofilm: Microbial Life on Surfaces, Emerging Infectious Diseases, v. 8, No. 9)

microbial cells are brought together by a matrix of self- secreted polymeric substance. It takes about 10 days to achieve structural stability of the biofilm.

This matrix is responsible for the structural and functional stability of the biofilm and helps in interactions between the microbial cells. The matrix which is created outside the cells for its attachment and protection consists mainly of water (97%); soluble components like gel-forming polysaccharides (PS), protein (PN) and eDNA (extracellular DNA); and insoluble components like cellulose, fimbriae (long filamentous polymeric protein structures located at the surface of bacterial cells), pili (hair-like structures on cell surfaces) and flagellae. Extracellular polymeric substances (EPS) are considered to be the primary material of the matrix. EPS may account for 50–90% of TOC of biofilm comprising of polysaccharides (PS), proteins (PN), glycolipids, humic acid, extracellular DNA (e-DNA) and other extracellular components. The e-DNA is generated from remains of lysed cells. It gives the biofilm antibiotic resistance. It is observed that the PS/PN ratio is an important factor in determining the biofilm adhesion and fouling on membranes. According to some investigators it is suggested that biofilm cohesion on membrane aerated surfaces may increase with high concentrations of PS.

Figure C.1 shows a polymicrobial biofilm which has been developed on stainless steel surfaces and examined by epifluorescence microscopy in a laboratory. The microbial colonies are depicted in the matrix structure confining the film. In this figure the blank spaces demonstrate water channels within a biofilm. These water channels allow diffusion of nutrients, oxygen, and other essential materials into, and carbon dioxide and other products out of the biofilm

The ability of biofilm to attach on surfaces depends on the ionic strength of the liquid medium, substratum surface charge and osmolarity of the surface. Biofilms will decay on surfaces with medium or high osmolarity. Biofilms often trap metals, organic and inorganic substances within the matrix. Metal ions such as calcium, iron and manganese contribute towards stabilization of biofilm matrix by bridging carboxyl groups across EPS molecules. EPS are usually isolated using cationic ion exchange resins. The detachment of biofilms takes place resulting from shearing, abrasion or sloughing. Due to the presence of EPS and the matrix the biofilms can co-operate and transfer valuable information among the cells within the attachment. This matrix also allows them to adapt to the environment. For example, oxygen is taken up quickly by upper layer microbial cells in the biofilm, then it diffuses through the biofilm. This creates aerobic upper layer and anoxic/anaerobic lower layer. The presence of matrix also allows microbial cells to be resistant to toxic environments and shock loads in their environment. All these advantages are due to the presence of the matrix which is absent in the case of planktonic cells (suspended bacteria).

It has been noted through experiments that the extent of biofilm increases with increase in surface roughness. It is also found that the biofilms are formed more easily on hydrophobic, non-polar surfaces like Teflon and other polymeric materials compared to hydrophilic surfaces like metals. Fimbriae, other proteins, and components of specific gram-positive bacteria (mycolic acids) are dominant factors during attachments on hydrophobic surfaces whereas EPS and lipopolysaccharides (LPS) play an important part during attachments to hydrophilic surfaces.

Many different techniques have been used to understand the effect of environmental conditions on the physical and biochemical structures of biofilm and its development. Microscopy, Magnetic resonance imaging, Raman microscopy, scanning electron microscopy (SEM), optical coherence tomography and confocal laser microscopy are some of the commonly used analytical techniques to study biofilms in situ. Confocal laser microscopy is the most common technique used as it provides a better in-depth understanding of the biofilm structure.

C.2 Applications of Biofilms in Wastewater Treatment

Biofilms are produced in fixed film systems like Moving Bed Bioreactor (MBBR), Integrated Fixed Film Activated sludge (IFAS), Rotating Bioreactor (RBC), and Trickling Filters.

In MBBR biofilm is developed on a protected media surface. These biofilms are protected from sloughing, erosion and other toxic environments. It has been observed that the biofilm growth at a younger stage during MBBR startup is facilitated by low aeration rate. It has also been suggested that high EPS presence in the biofilm increased resistance to sloughing caused by aeration. The biofilm development in an MBBR starts near the corners of the carriers and then moves towards the center. The carriers with small compartment size provide faster development of biofilm.

EPS in either bound or soluble/colloidal form has been considered as the major cause of membrane fouling in MBR or RO based treatment systems. According to some reports, it was found that the filter cakes from conventional membrane systems have presence of filamentous bacteria in suspended growth reactors upstream of membrane separation system. The filamentous bacteria were said to produce soluble/colloidal EPS. It was also noted that filamentous bacteria produced higher concentration of EPS than floc forming bacteria (planktonic cells). There was no clear relationship found between the bound EPS and membrane fouling according to this study.

There have been many contradictory reports that have been published about the effect of EPS on membrane fouling. It is mentioned in some reviews that fouling tends to occur if there is a higher concentration of polysaccharides in the EPS (Drews et al. 2006). Some even have reported that it depends on the SRT of the reactor and spatial distribution of EPS in the biofilm.

Reference

Donlan RM (2002) Biofilms: microbial life on surfaces. Emerg Infect Dis 8(9)

Drews A, Lee C-H, Kraume M (2006) Membrane fouling—A review on the role of EPS, Desalination, 200:186–188

Appendix D
Examples of Occupational and Process Safety Measures at Refinery Wastewater Treatment Facilities

Presented below are examples that have been reffered to in different parts of Chap. 7 to provide more detailed information on the indiviual statements made there.

Example D.1: Worksite Field Review Checklist (Refer to Sect. 7.4.2, Chap. 7)

1. Administrative and general Personal Protective Equipment (PPE) requirements:
   - Correct type of permit issued? (Required)
     - Joint site inspection conducted before work began? (Required)
     - Issuer's certificate current?
     - Receiver's certificate current?
     - Receiver's copy at the job site? (Required)
     - Management countersignatures obtained when required by Operations? (Required)
     - 'Work location' adequately described?
     - 'Equipment to be worked on' adequately described?
     - 'Work to be done' adequately described?
     - Permit closure/extension adequately noted?
   - Receiver or delegate at the job site? (Required)
   - If a safety device was removed from service and/or defeated, was the appropriate supervisor and affected parties notified, the device tagged, and the action properly documented? (Required)
   - Gas tests
     - Proper tests were taken? (Required)
     - Gas testing equipment in good condition? (Required)

S. Basu et al. (eds.), *Water Management in Petroleum Industries*, Water Resources Development and Management, https://doi.org/10.1007/978-981-19-3159-8

- Entry into work areas shall not be allowed in atmospheres of 100 ppm or above hydrogen sulfide (H_2S), even with air-supplied respirators. Is this being followed? (Required)
- Fall protection considered? (Required)
- Body protection noted (e.g., flame retardant clothing, aprons)?
- Respiratory protection requirements noted?
- Is PPE worn by workers adequate and worn? (Required)
  - Head protection?
  - Eye/face protection?
  - Foot protection?
  - Hearing protection?
  - Hand protection?
- Is housekeeping okay?

2. Cold work:
   - Warning signs and barricades?
   - Standby man?
   - Chemical Hazard Bulletins, or Safety Data Sheets at the work site, when chemical exposure is possible? (Required)
   - When chemicals are being used at the work site, are workers using the appropriate PPE/respiratory equipment as set out in the relevant Safety Data Sheets/Chemical Hazard Bulletins? (Required)
   - Eye wash and safety showers accessible, when chemical exposure is possible?
   - Gas cylinders
     - Secured upright?
     - Integrity of gas cylinders/valves?
     - Properly labeled?
     - Capped if not in use?
     - Cylinders protected from sunlight and heat sources?
     - Is a lifting cage used to lift gas cylinders?
   - Are chemical or hazardous material containers properly labeled? (Required)
   - Are hand tools adequate?
   - Are air and hydraulic hoses okay?
   - Are safety pins in air hose connections?
   - Power tools
     - No frayed/damaged electric cords, or tool damage? (Required)
     - Portable tools plugged into ground fault circuit interrupters? (Required)
     - Deadman's switch? (Required)
     - Only 3-wire grounded extension cords used, with UL/similar label? (Required)
     - Protective guards in place?

  - Portable tools rated <110/120V?
  - Grounded or double insulated?

3. Hot work:

   - Equipment freed of hydrocarbons? (Required)
   - Other fuel sources removed?
   - Sewers covered?
   - Equipment physically isolated and method checked (e.g., chaining)? (Required)
   - Circuit breaker disconnected and operations locks/tags installed? Is each lock uniquely identified? (Required)
   - Permit receiver's lock installed on the breaker? Is each lock uniquely identified? (Required)
   - Electrical field start of equipment attempted, to verify isolation? (Required)
   - Static electricity grounding? (Required)
   - Fire protection

     - Fire extinguisher?
     - Additional fire protection equipment?
     - Fire watch?

   - Standby man?
   - Warning signs and barricades?
   - Vehicle backup alarm?
   - Welding

     - Welder certificate current? (Required)
     - Welding equipment undamaged, including cables? (Required)
     - Welding screens adequate?
     - For arc welding, ground leads are close to the work?
     - Layout of welding hoses/leads adequate?
     - Are electric welding machine outlets 110 V, 3-pronged GFCI type?
     - Ventilation adequate?
     - For gas welding, are flashback flame arrestors installed?
     - Generators >50 feet from buildings or combustible materials?
     - A portable fire extinguisher within 10 feet of the welding location?

4. Potential release of hazardous liquids or gases:

   - Planned regarding wind/drainage directions? (Required)
   - No 'open flame' hot work? (Required)
   - System tested after isolation and lockout? (Required)
   - Workers on the upwind side?

5. Confined space entry:

   - Rescue plan, with communications, how to get the workers out, rescue phone numbers? (Required)

- Is the standby person equipped and does s/he know what s/he's supposed to do? (Required)
- Engine exhaust excluded from confined space? (Required)
- Is there a confined space log sheet for entering the confined space?
- Is there a safe means of access/egress?
- Ventilation adequate? For example, are movers required?
- Is all equipment, including air movers, properly grounded/bonded?
- Process equipment properly prepared for work?
- Process equipment isolated?
- Electrical equipment locked and tagged out? Is each lock uniquely identified?
- Lighting intensity adequate? Lights explosion-proof?
- Ground fault circuit interrupters for 110V use?
- Excavations
  - Hand dig within 10 ft of underground items, e.g., pipes, cables? (Required)
  - Scaffolds kept 1.5X depth away? (Required)
  - Cranes kept 1X depth away? (Required)
  - Other heavy equipment kept 6 feet away? (Required)
  - Workers not in excavation while excavating machines operating? (Required)
  - Slope/bench/shore if >4 feet deep? (Required)
  - Traffic flagman, near excavation? (Required)
  - Ladders okay and at most every 25 feet travel distance along excavation? Do they extend at least 3 feet above the excavation?
  - Underground utilities/hazards marked?
  - Spoil material at least two feet away from excavation?

6. Elevated work:

- Personnel using a personal fall arrest system (PFAS) where required? (Required)
- Lanyards secured adequately? Are lanyards shock-absorbing? Are lanyards secured at locations to prevent a person from impacting the surface below? (Required)
- Personal fall arrest system equipment integrity okay?
- Scaffolds
  - Debris net, etc., in place?
  - Scaffold erecting personnel only on red-tag scaffold?
  - Green/yellow tags at each access, to exclude access by unauthorized personnel?
  - Are Green/yellow tags still valid?
  - Guardrails okay?
  - Scaffold field inspection checklist on file with the permit issuer?
  - Toe boards okay?
  - Are Access points acceptable?
  - Planking okay on work platforms?

- Ladders
    - Nonslip feet and supported on firm ground?
    - Nonconductive ladders near energized electrical equipment?
    - No bent, broken, missing parts?
    - Sturdy? Joints, rivets, hardware tight?
    - Extend >3 feet above elevated surface?
    - Secured at the top?
    - Workers not standing on the top step of a stepladder?
    - Ladders not supported by temporary platforms, e.g., scaffolds?

7. Heavy equipment:
    - Equipment clearance at excavation >1.5X depth? (Required)
    - Engine exhaust kept from entering a confined space? (Required)
    - Protective guards?
    - On flat, solid, sound support?
    - Operator has an appropriate license?
    - Flagmen?
    - Spotter to guide drivers towing trailer-mounted equipment? Are wheels locked when parked?
    - Adequate room to maneuver?
    - Cranes:
        - Rigger assisting on all crane lifts? Proper rigger certification level?
        - Crane inspection sticker?
        - Crane hook safety latch? Anti 2-block prevention device?
        - Load chart?
        - Operation at wind speed okay?
        - Is signalman used for all lifts?
        - Operation near power line >20 feet up to 250 kV, >25 feet over 250 kV? (Required)
        - Outriggers fully extended, tires off ground, crane level, floats solid, and locking pins/locks set?
        - Workers not under suspended loads?
        - Load not lifted over processing equipment or pipelines, unless done following a critical lift plan? (Required)
        - Crane operator clearly visible, and never leaves crane during a lift? Is cab glass clean without obstructions or damage?
        - Approved lift plan for critical lifts? (Required)
        - Mobile cranes: presence of safe load indicator, load moment indicator, and spark arrestor?
        - Rigging equipment:

            Integrity of slings, etc.?
            Safe Working Limit (SWL) and serial number on each piece of rigging equipment?

Hardware not distorted, gouged, or heat damaged?
Lifting hooks have a safety latch?
Wire rope slings not used with knots to form eyes?
Original pin in shackles?
Tag lines used?

- Manbaskets and manlifts:
  - Current inspection sticker? (Required)
  - Proof load test lift done?
  - Manlift lift plan?
  - Manlift permit?
  - Workers stay in manbasket/manlift while it is elevated?
  - Are men wearing full-body safety harnesses that are secured to the manbasket/manlift?
  - Manbasket/manlift used only for lifting workers? (i.e., no tools or materials allowed)
  - Signalman used during lifts?
  - Operation at wind speed okay?
- Forklifts:
  - Are forklift operators wearing seat belts?
  - When a forklift is loading or unloading a trailer or truck, are the trailer/truck brakes set and are the wheels blocked (chocked)?
- Are boom trucks loaded or unloaded by a certified rigger?
- Does heavy equipment have an audible backup alarm?

8. Hydrotesting:
   - Number of persons at site limited?
   - Temporary connections undamaged?
   - Relief device properly rated? (Required)
   - Is All equipment properly lined up and prepared for hydrotesting? (Required)
   - No work done on the equipment during hydrotesting?
9. Ionizing radiation:
   - Radiographer's certificates current? (Required)
   - Unauthorized persons kept away? (Required)
   - Each piece of equipment has valid calibration certificates?
   - Area barricaded, radiation warning signs, and exposure limit posted?
10. Asbestos work:
    - Workers appropriately trained and competent?
    - Workers use disposable coveralls, boots, head coverings, and gloves?
    - Cartridge-type respirator worn that is designed for use with asbestos?

- Self-contained breathing apparatus (SCBA) or cascade system used if heavy dust expected?
- Dust concentrations monitored?
- Asbestos stored and transported in approved containers with warning labels attached?
- Asbestos material kept wet?
- Area roped off? Asbestos hazard warning signs posted?
- Only essential personnel allowed in area?
- All asbestos cleaned up at end of each work shift?

11. Spray painting

- No ignition sources within 75 feet?
- Paint storage, mixing, and equipment cleaning areas away from ignition sources?
- Lower Explosive Limit (LEL) reading, from gas tester, of 0.0 before painting?
- Adequate ventilation?
- Air-supplied respirator inside confined space; cartridge respirators in all other areas?
- No painting on surfaces warmer than 80 °C (176 °F)?

12. Abrasive blast cleaning

- Operating wearing air-supplied hood?
- Assistant wearing SCBA while exposed to dust?
- Other workers exposed to abrasive materials wearing cartridge style respirators for dust?
- Have wall thickness been measured on equipment that will undergo abrasive blasting?
- Proper safety pin and connecting lines shall be in place at all potential disconnect points.
- Dead man's switch installed on abrasive blasting nozzle?
- Abrasive blast equipment electrically bonded?
- Delicate equipment (e.g., process instruments) protected from dust?
- Is blasting media approved?
- Were air compressors used to supply breathing air third-party certified?

13. Blinding equipment for equipment isolation

- A Blind List shall be prepared where blinds are used to ensure proper installation and removal. The Blind List shall include the blind number, location, date installed, date removed, and appropriate signatures. (Required)
- Blinds shall be installed as close as possible to the vessel or equipment to be isolated and using the required number of bolts.
- Selection of blinds shall correspond to requirements of ANSI B31.3.

D.2 Example D.2: HAZOP Study on a Storage Tank (Refer to Sect. 7.6.1.4, Chap. 7)

Modified Process Vessel Process Schematic after HAZOP Analysis.

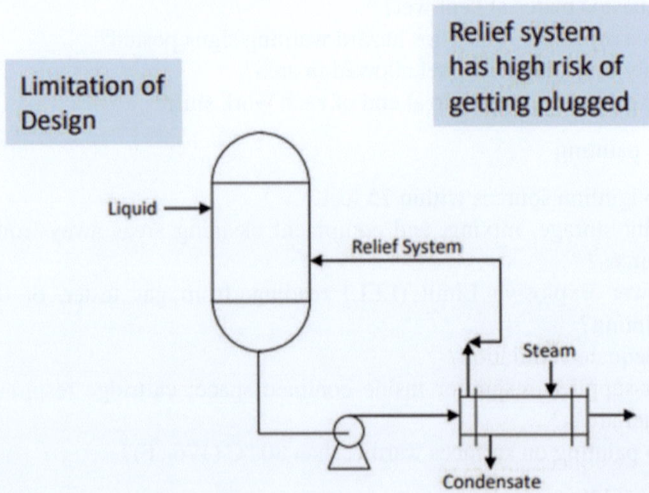

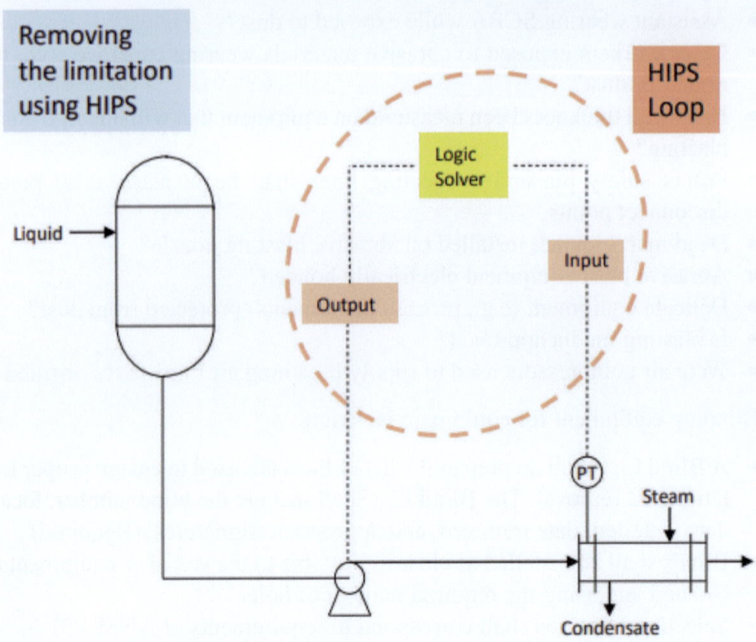

Sample Page from the HAZOP Analysis Example for the Process Vessel Above.

Team: HAZOP Team — Drwg No.: 20-EOM-10001
Meeting Date: 9/18/20 — Revision No.: 3

Item No.	Deviation	Causes	Consequences	Safeguards	Actions
1.0 Vessel - Monomer Solution Process Tank. Contains liquid oxide monomer at 30 psig and 10C					
1.1	Less Cooling	Cooling fails liquid polymerizes	Hazardous hot polymer causes personal injury	Temperature control on cooling with an alarm	Review Vessel P&ID to add independent TI to the control
			the pump seizes the relief valve gets plugged	2oo2 voting	Allow 2oo2 voting to ensure higher reliability
					Replace PRV with a HIPS loop See schematic below

Example D.3: Ethylene Glycol Dispersion Cloud Using Consequence Analysis (Refer to Sect. 7.9.3, Chap. 7)

A typical cloud dispersion side view of ethylene glycol from a PSV developed using a commercially available model as shown in the figure below. The results demonstrate that the offsite concentration is acceptable. However, the onsite ER-3 concentration is exceeded at the tall structure at 40 ft distance from the relief point.

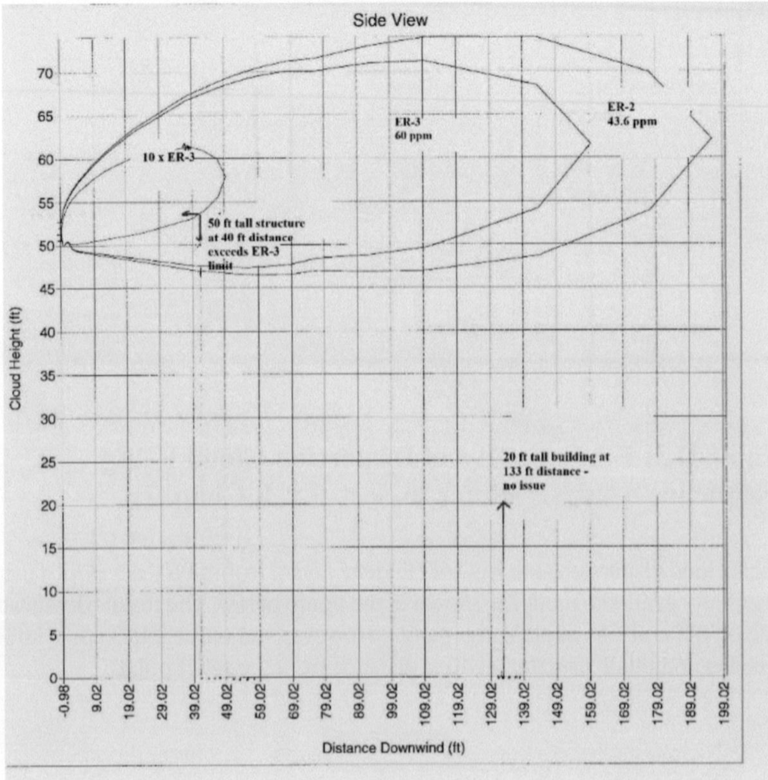

Glycol Time-weighted Average Dispersion Cloud Developed Using PHAST (Reproduced from Chakravarty, T., Reactive Relief Report, 2015).

Example D.4: HIRA Method for Process Hazard Analysis (Refer to Sect. 7.6.1.2, Chap. 7)

Note that Process Hazard Analysis (PHA) can be used at any phase of a process, and therefore, not separately included. For example, if PHA is used at the conceptual phase of a process, it is termed Preliminary Hazard Analysis, also abbreviated often as PHA in literature.

Phase	Checklist Analysis	What-If Analysis	What-if/ Checklist Analysis	Hazard and Operability Studies
Conception	**X**			
Process development	**X**	**X**	**X**	x
Detailed design	x	x	x	**X**
Operation	x	x	x	**X**
Decommissioning	x		**X**	

Example D.5: Swiss Cheese Model for Analysis of Process Safety (Refer to Sect. 7.6.1.6, Chap. 7)

Theory of Process Safety analysis with Swiss Cheese model is that the holes in Swiss cheese ordinarily do not stack up to go through from one end to the other. Similarly in Process safety, if all the safeguards are properly designed and in place, the risk of the final event occurring is reduced. Different layers of safeguards can be engineering controls, administrative controls, behavior controls, and HIPS acting as IPLs.

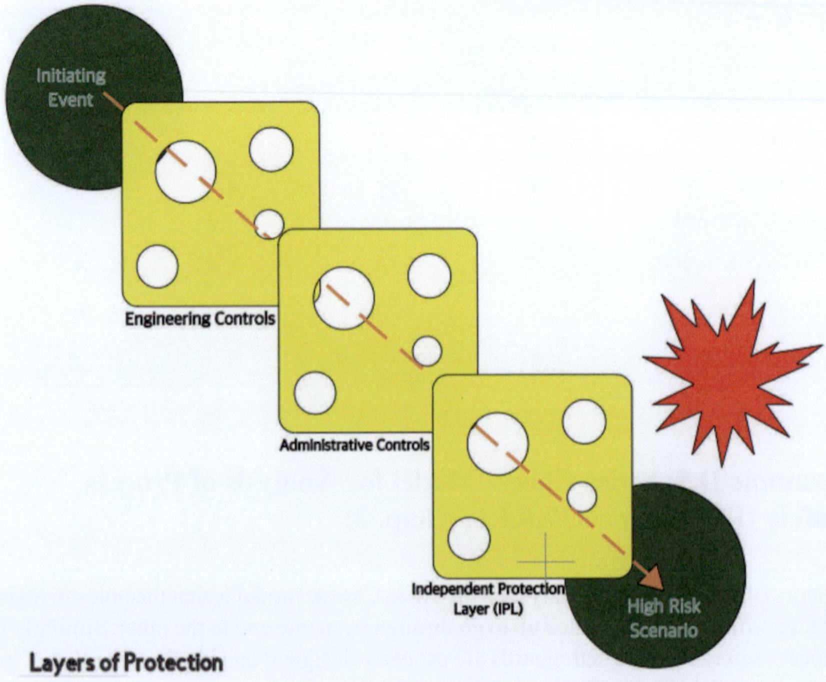

D.6 Example D.6: PSM and RMP Covered Chemicals are Available in the Following Hyperlink Referenced in Chap. 7, Sect. 7.6

https://www.osha.gov/chemical-executive-order/psm-terminology.

Index

S. Basu et al. (eds.), *Water Management in Petroleum Industries*, Water Resources Development and Management, https://doi.org/10.1007/978-981-19-3159-8

W

Z